Pelvic Floor Dysfunction

G. Willy Davila, Gamal M. Ghoniem and
Steven D. Wexner (Eds.)

Pelvic Floor Dysfunction

A Multidisciplinary Approach

With 178 Figures

Springer

G. Willy Davila, MD
Chairman, Department
of Gynecology; Head, Section
of Urogynecology and
Reconstructive Pelvic Surgery
Cleveland Clinic Florida
Weston, FL, USA
Affiliate Associate Professor, University
of South Florida, USA

Gamal M. Ghoniem, MD, FACS
Head, Section of Voiding Dysfunction,
Female Urology and Reconstruction
Clinical Professor of Surgery/Urology,
NOVA Southeastern University
Ohio State University and University of
South Florida
Cleveland Clinic Florida
Weston, FL, USA

Steven D. Wexner, MD, FACS, FRCS, FRCSED
Chief of Staff, Cleveland Clinic Hospital
21st Century Oncology Chair in Colorectal
Surgery
Chairman, Department of Colorectal Surgery
Cleveland Clinic Florida
Professor of Surgery, Ohio State University
Health Sciences Center at Cleveland
Clinic Foundation
Clinical Professor of Surgery, University of
South Florida College of Medicine
Research Professor, Charles E. Schmidt
College at Florida Atlantic University

A Catalogue record for this book is available from the British Library.

Library of Congress Control Number: 2005924370

ISBN 978-1-84800-347-7 ISBN 978-1-84800-348-4 (eBook)
DOI 10.1007/978-1-84800-348-4

First published 2006 in hardcover as ISBN 978-1-85233-730-8

Printed on acid-free paper

9 8 7 6 5 4 3 2 1

Springer Science+Business Media
springer.com

I dedicate this work to those who have been instrumental in the development of my urogynecologic career: Dr. Donald Ostergard for teaching me the basics and beyond, Dr. Oscar Contreras Ortiz for encouraging me to become involved, and Dr. Harold Drutz for showing me how to achieve "the balance." To these mentors and friends I am forever in debt, and no words can express my appreciation for their warmth and interest in my professional and personal growth. I attempt to emulate their efforts on a daily basis.

G. Willy Davila, MD

My work on this book is dedicated to the memory of the great men who influenced my life's work, and who passed in December 2003. To the memory of my father, Mostafa Ghoniem and my mentor, Mohamed Al-Ghorab, to whom I will be eternally grateful for their love and support.

Gamal Ghoniem, MD

This book is dedicated to the most important people in my life – my family. To my parents, for having taught me by their example to succeed by hard work, honesty, and perseverance. Their love and guidance made it possible for me to have achieved my career. To my wife Nicolette, whose devotion to me and to our children and whose intellectual stimulation and love have made it possible for me to continue to be productive and innovative. To my children Wesley, Trevor, Marisa, and Gabriella, as it was precious time away from them that allowed me to produce these pages.

Steven D. Wexner, MD

The editors and authors express their gratitude to Ms. Kristin Dunn and Ms. Stacy Kopka for their exhaustive efforts in preparation of the manuscripts for publication; and Ross Papalardo and Beth Halasz for their superb medical illustration skills.

Foreword

All the characteristics and driving force of The Cleveland Clinic are to be found in this book on pelvic floor function. The Cleveland Clinic is a group practice founded in 1921 on the principles of cooperation, collaboration, and collegiality. Its founders believed that many physicians working together will discover better solutions to medical problems than physicians working in isolation. They believed that the combination of disciplines, with their inherent differences in philosophy and skills, will produce a better outcome than might have evolved singularly.

The power of the collaborative approach is on full display in this book. The pelvic floor unites three separate organ systems. Before this time, each has been approached individually. Urologists, gynecologists, and colorectal surgeons are each trained in their own disciplines, and the pelvic floor is subsumed in these larger fields of study. When they combine their focus on the pelvic floor, they bring their unique perspectives and different approaches to a common goal: the relief of pelvic floor syndromes such as incontinence and pelvic organ prolapse.

The notion that three disciplines can combine for a better outcome than they would achieve singly, is a modern-day version of an old fable. The American poet John Godfrey Saxe (1816–1887) adapted that fable in his poem, "The Blind Men and the Elephant." It tells the story of six blind wise men who encounter an elephant for the first time, and how they try to make sense of it by touch alone. The first blind man falls against the side of the elephant, and declares that the elephant is like a wall. The second man feels the tusk and concludes that anything so sharp is like a spear. The third man grabs the trunk and announces that the elephant is like a snake. The fourth man feels the knee and states that the elephant is like a tree. The fifth blind man touches the ear and decides that it is like a fan. The six man grabs the tail and decides it is a like a rope. Each wise man feels the elephant from his own small perspective and visualizes something different. As individuals, they think they have the answer. But if they had combined their impressions, they would have reached a far different conclusion.

The practice of joining with colleagues to visualize a problem from your own perspective and the group perspective, can lead to magnificent conclusions and interesting solutions. That is precisely what happens in this book. By combining the views of three different disciplines, which have furthered our knowledge of pelvic floor function, it offers a new and better approach to dysfunction. Medicine has advanced tremendously in the past half century, more specifically in the past two decades. The dominance of traditional specialties such as urology, gynecology, and general surgery has given way to an emphasis on subspecialties such as urogynecology, female urology, and colorectal surgery. This has allowed experts, such as the authors, to focus on a narrower field, and enjoy greater success than they might have if their concentration had been diluted by many broad interests. In this book, they are the personification of teamwork. Their characteristics of focus, innovation, collaboration, and dedication are the very qualities that will someday advance medicine far beyond anything we can imagine today.

By combining their expertise in this book, the authors have made a tremendous advance in the understanding of the pelvic floor and its dysfunction. They have raised medicine to yet a higher level, and performed a service that will benefit patients with pelvic floor dysfunction for years to come.

Robert Kay, MD
Vice Chairman, Board of Governors
Chief of Staff
The Cleveland Clinic Foundation
Cleveland OH, USA

Introduction

The aging of our population and arrival of the baby-boomers to advanced adulthood, with expectations for high quality of life, have led to a greater public awareness and help-seeking behavior regarding symptomatic dysfunction of the organs found within the pelvic cavity in women, the female pelvic floor. Women are increasingly less willing to simply accept incontinence or prolapse as a normal part of the aging process. New technologies and a greater understanding of the pathophysiology underlying these dysfunctions have provided us with a number of effective tools for treatment of these patients. Unfortunately, the traditional fragmentation of health care duties among specialists has led to significant gaps in the treatment of pelvic floor dysfunction. The concept of the female pelvic floor as a single functional unit has not yet gained wide acceptance. Importantly, postgraduate training programs have not adapted their curricula to meet the growing need to produce clinicians with an expertise in the management of the various aspects of pelvic floor dysfunction. Thus, there is not a sole clinician who can care for all of the problems that can develop within the pelvic floor. Under our current medical care environment, a multidisciplinary team approach will best serve the needs of symptomatic women. The need for such a team is gaining increasing degrees of acceptance at referral centers. Unfortunately, multiple barriers are in place to prevent such teams from forming and functioning efficiently.

There are currently efforts around the world to delineate training requirements for certification in Urogynecology/Female Pelvic Medicine/Reconstructive Pelvic Surgery (after training in Obstetrics and Gynecology) and Female Urology (after training in Urology). Colorectal surgeons and gastroenterologists have long had an interest in anatomic and functional problems of the lower intestinal tract. Nevertheless, clinicians caring for adjacent pelvic organ systems have yet to achieve a consensus regarding the importance and value of the evaluation and management of pelvic floor problems existent on either side of an anatomic system. As a result, patients may undergo sequential operative procedures and/or achieve only limited quality-of-life improvement with therapy.

Realizing the above shortcomings and the frequent coexistence of pelvic floor dysfunction symptoms among many patients referred for care, the Cleveland Clinic, at its various campuses, has developed a team approach to the care of such patients. Either within one Pelvic Floor Center, such as at our Fort Lauderdale/Weston, Florida campus or in very close proximity, such as in our Naples, Florida or Cleveland, Ohio campuses, patients are evaluated and treated by a team of clinicians with expertise in the various aspects of symptomatic dysfunction. After coordination of evaluation procedures, a treatment plan is designed. Whether care involves a combined surgical procedure or medical intervention, a patient's medical care is streamlined and patients benefit from a global quality-of-life improvement. This text represents a compilation of the clinical approaches of the staff at Cleveland Clinic in the management of disorders involving the lower urinary, genital, and intestinal tracts. It should be a valuable reference for all clinicians involved in the care of women with symptomatic dysfunction of these systems. It will be apparent to clinicians from various fields that there are remarkable similarities and analogies in terms of presenting symptoms, evaluation modalities, and treatment approaches. It is the hope of the authors that clinicians will begin to see the various organ systems as part of a combined unit. As such, recognizing the presence of symptoms involving adjacent organ systems will encourage clinicians to recruit and involve other clinicians with expertise in addressing such symptoms in order to optimize the medical care being provided to a symptomatic patient. Our experience has demonstrated that improvements in quality of life and patient satisfaction can be greater when a comprehensive, horizontally integrated approach is utilized, and we look forward to other centers adopting our model and philosophy of patient care.

G. Willy Davila, MD

Contents

SECTION IV **Anatomic Correlates**

SECTION V **Sexual Function**

SECTION VI **Incontinence Therapy**

SECTION VII **Urgency/Frequency Syndromes Therapy**

SECTION VIII **Prolapse Syndromes Therapy**

SECTION IX **Pain and Irritative Syndromes Therapy**

SECTION X **Evacuation Disorders**

SECTION XI **Hormonal Influences on the Pelvic Floor**

SECTION XII **Physiotherapeutic Approaches**

Contributors

George Abate, DO
Department of Obstetrics and Gynecology, The Cleveland Clinic Foundation, Cleveland, OH, USA

Joseph Abdelmalak, MD
Glickman Urological Institute, The Cleveland Clinic Foundation, Cleveland, OH, USA

Matthew D. Barber, MD, MHS
Section of Urogynecology and Reconstructive Pelvic Surgery, Department of Obstetrics and Gynecology, The Cleveland Clinic Foundation, Cleveland, OH, USA

Daniel H. Biller MD
Section of Urogynecology and Reconstructive Pelvic Surgery, Department of Gynecology, Cleveland Clinic Florida, Weston, FL, USA

Gregory F. Bonner, MD
Department of Gastroenterology, Cleveland Clinic Florida, Weston, FL, USA

Susan M. Cera, MD
Department of Colorectal Surgery, The Cleveland Clinic Foundation, Weston, FL, USA

Firouz Daneshgari, MD
Center for Female Pelvic Medicine and Reconstructive Surgery, Glickman Urological Institute, The Cleveland Clinic Foundation, Cleveland, OH, USA

Giovanna M. DaSilva, MD
Department of Colorectal Surgery, Cleveland Clinic Florida, Weston, FL, USA

G. Willy Davila, MD
Section of Urogynecology and Reconstructive Pelvic Surgery, Department of Gynecology, Cleveland Clinic Florida, Weston, FL, USA

James Doty, MD
Department of Colorectal Surgery, Cleveland Clinic Florida, Naples, FL, USA

Jonathan E. Efron, MD, FACS, FASCRS
Department of Colorectal Surgery, Cleveland Clinic Florida, Naples, FL, USA

Ahmed Elazab, MD
Glickman Urological Institute, The Cleveland Clinic Foundation, Cleveland, OH, USA

Lawrence P. Frank, MD
Department of Anesthesiology, Cleveland Clinic Florida, Weston, FL, USA

Tara L. Frenkl, MD, MPH
Glickman Urological Institute, The Cleveland Clinic Foundation, Cleveland, OH, USA

Gamal M. Ghoniem, MD, FACS
Section of Voiding Dysfunction, Female Urology and Reconstruction, Department of Urology, Cleveland Clinic Florida, Weston, FL, USA

Wagih W. Gobrial, MD
Section of Head Pain Management, Division of Anesthesiology, Cleveland Clinic Florida, Weston, FL, USA

Nathan Guerette, MD
Section of Urogynecology and Reconstructive Pelvic Surgery, Department of Gynecology, Cleveland Clinic Florida, Weston, FL, USA

Gassan Haddad, MD
Department of Internal Medicine, Cleveland Clinic Florida, Weston, FL, USA

John C. Hairston, MD
Section of Female Urology, Voiding Dysfunction and Pelvic Reconstruction, Cleveland Clinic Florida, Department of Urology, Weston, FL, USA

Lawrence S. Hakim, MD, FACS
Section of Sexual Medicine, Infertility and Prosthetics, Department of Urology, Cleveland Clinic Florida, Weston, FL, USA

Tracy L. Hull, MD
Department of Colorectal Surgery, The Cleveland Clinic Foundation, Cleveland, OH, USA

Jonathan Jay, MD, FACS
Department of Urology, Cleveland Clinic Florida, Naples, FL, USA, USA

Robert Kay, MD
Vice Chairman, Board of Governors, Chief of Staff, The Cleveland Clinic Foundation, Cleveland, OH, USA

Usama M. Khater, MD
Section of Female Urology, Voiding Dysfunction and Pelvic Reconstruction, Cleveland Clinic Florida, Department of Urology, Weston, FL, USA

M. Louis Moy, MD
Glickman Urological Institute, The Cleveland Clinic Foundation, Cleveland, OH, USA

Minda Neimark, MD
Section of Urogynecology and Reconstructive Pelvic Surgery, Department of Gynecology, Cleveland Clinic Florida, Weston, FL, USA

Juan J. Nogueras, MD
Department of Colorectal Surgery, Cleveland Clinic Florida, Weston, FL, USA

Marie Fidela R. Paraiso, MD
Department of Obstetrics and Gynecology, Urological Institute and Women's Health Center, Section of Urogynecology and Reconstructive Pelvic Surgery, The Cleveland Clinic Foundation, Cleveland, OH, USA

Benjamin Person, MD
Department of Colorectal Surgery, Cleveland Clinic Florida, Weston, FL, USA

Ronnie R. Pimentel, MD, FACP, FACG
Department of Gastroenterology, Cleveland Clinic Florida, Weston, FL, USA

Jennifer T. Pollak, MD
Florida Center for Urogynecology, Hollywood, FL, USA

Raymond R. Rackley, MD
Glickman Urological Institute, The Cleveland Clinic Foundation, Cleveland, OH, USA

Virgilio Salanga, MD, MS
Department of Neurology, Cleveland Clinic Florida, Weston, FL, USA

Dana R. Sands, MD
Department of Colorectal Surgery, Cleveland Clinic Florida, Weston, FL, USA

T. Cristina Sardinha, MD
Department of Colorectal Surgery, Cleveland Clinic Florida, Weston, FL, USA

Andrew I. Sokol, MD
Section of Urogynecology and Reconstructive Pelvic Surgery, Department of Obstetrics and Gynecology, The Cleveland Clinic Foundation, Cleveland, OH, USA

Wael Solh, MD
Department of Colorectal Surgery, Cleveland Clinic Florida, Weston, FL, USA

Kevin J. Stepp, MD
Section of Female Pelvic Medicine and Reconstructive Pelvic Surgery, Department of Obstetrics and Gynecology, The Cleveland Clinic Foundation, Cleveland, OH, USA

Joanna M. Togami, MD
Section of Female Urology, Voiding Dysfunction and Pelvic Reconstruction, Cleveland Clinic Florida, Department of Urology, Weston, FL, USA

Sandip P. Vasavada, MD
Section of Voiding Dysfunction and Female Urology, Glickman Urological Institute, The Cleveland Clinic Foundation, Cleveland, OH, USA

Dawn Vickers, RN
Department of Gastroenterology, Cleveland Clinic Florida, Weston, FL, USA

Mark D. Walters, MD
Section of Female Pelvic Medicine and Reconstructive Pelvic Surgery, Department of Obstetrics and Gynecology, The Cleveland Clinic Foundation, Cleveland, OH, USA

Eric G. Weiss, MD
Department of Colorectal Surgery, Cleveland Clinic Florida, Weston, FL, USA

Steven D. Wexner, MD, FACS, FRCS, FRCSED
Department of Colorectal Surgery, Cleveland Clinic Florida, Weston, FL, USA

Samir M. Yebara, MD
Department of Colorectal Surgery, Cleveland Clinic Florida, Weston, FL, USA

Section I

Concept of the Pelvic Floor as a Unit

1-1

Concept of the Pelvic Floor as a Unit

G. Willy Davila

Disorders of urinary and fecal continence, as well as genital and rectal support, are common in adult women. Clinicians who address these problems include urologists, gynecologists, and colorectal surgeons. Great advances have been made in the treatment of continence and support disorders. Unfortunately, our current model of vertical integration for care of disease entities has limited a clinician's understanding to the vertical unit in which the clinician has an expertise. This frequently leads to serial surgeries because of the lack of identification of dysfunction in an adjacent organ system of the pelvic floor. There has been little overlap in the verticalized spectrum of care provided by urologists (kidneys, ureters, bladder, urethra), gynecologists (uterus, vagina, perineum), and colorectal surgeons (colon, rectum, anus). Coexistence of dysfunction of urinary and bowel control is high and is well established.[1] Unfortunately, clinicians who address pelvic floor dysfunction syndromes have been slow to adapt their practice styles to address all pelvic floor dysfunctions in one setting. Our concept of horizontal integration of pelvic floor dysfunction evaluation and management (Figure 1-1.1) is not new.[2] This concept is expanding as clinicians realize the importance of a team approach to evaluation and treatment of pelvic floor dysfunction. We hoped to emphasize these principles as we prepared this textbook.

Defining the Female Pelvic Floor

The term "pelvic floor dysfunction" has different meanings for different clinicians. From our viewpoint, the female pelvic floor includes all of the structures within the bony pelvis: from pubic symphysis to coccyx and from lateral pelvic sidewall to lateral pelvic sidewall. It thus includes not only the lower urinary tract, reproductive tract, and lower gastrointestinal tract, but also the neuromuscular components of their support. The neuro-musculo-visceral anatomy of the pelvic floor is described further in Chapter 3-6. The musculature of the pelvic floor includes the levator musculature and the perineal musculature. The levator musculature provides support to all of the pelvic floor organs and is transversed by the urethra, vagina, and anus. Because the levator muscle complex provides support to all three organ systems, its weakness will result in impaired function of any, or all, of the structures that the muscles support. Muscular dysfunction can result from stretch or tear injuries to the pelvic floor muscles. However, the most common etiology for muscular dysfunction is a denervation injury from childbirth or lower back trauma. A denervation injury will result in partial paralysis of the supplied muscle groups. As a consequence, any lower back injury can result in weakness of pelvic floor support. This is especially true of the lower components of the pelvic floor. Injury to the pudendal nerve can result in dysfunction of the urethral sphincter, anal sphincter, and motor or sensory dysfunction of the perineum. Significant injury to the pudendal innervation will typically result in multisystem dysfunction, such as urinary and fecal incontinence.

The organ systems of the pelvic floor are enveloped in moderately thick layers of connective neuromuscular tissue. Labeled as endopelvic fascia, this neuromuscular tissue provides circumferential support to the three cavities that transverse the pelvic floor muscles. In addition, they constitute separating structures between the organ systems: the vesicovaginal septum and the rectovaginal septum. Lack of integrity of the fibromuscular layer between two organ systems will lead to herniation of one organ system into another. The resultant cystocele, enterocele, or rectocele may then result in dysfunction of the underlying visceral organ including disorders of urinary continence and storage, or dysfunction of fecal continence or storage. Frequently, multiple sites of fibromuscular layer damage are found. This is represented by the frequent coexistence of prolapse of the anterior and posterior vaginal walls requiring repair of both.

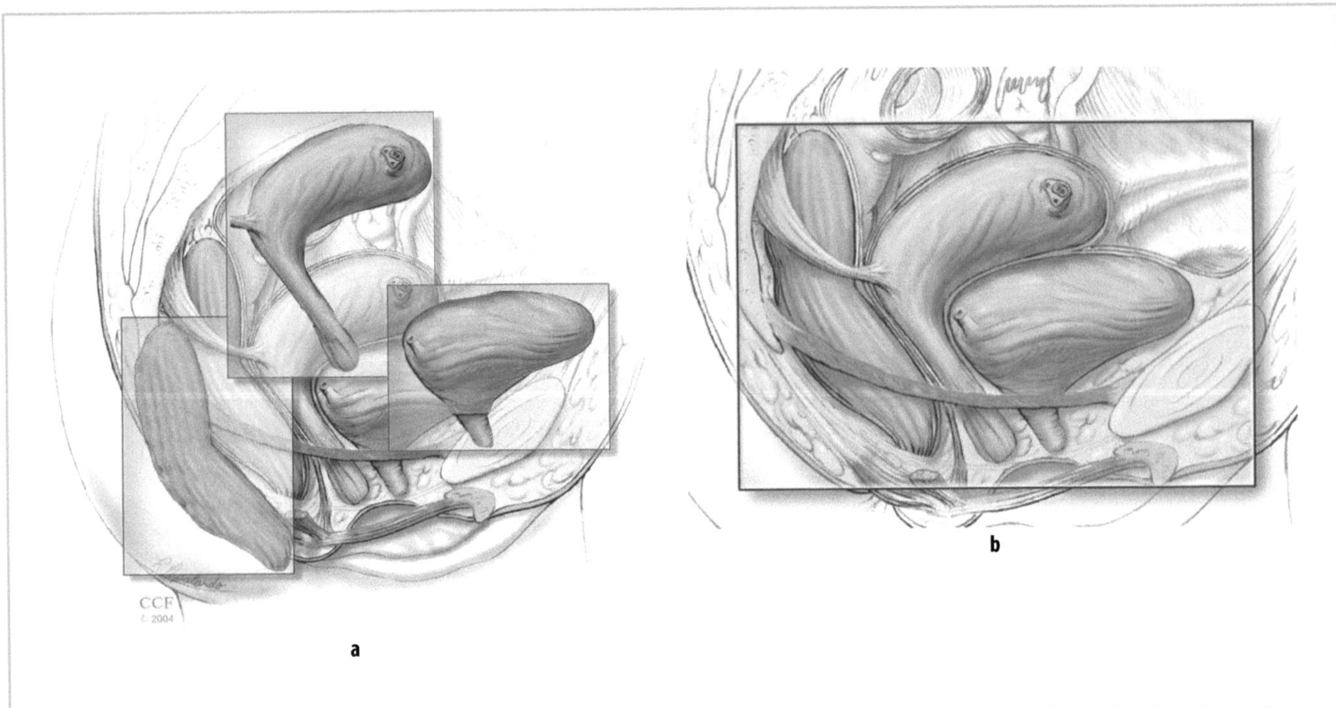

Figure 1-1.1. **a**, Traditional vertical view of pelvic floor dysfunction. **b**, Horizontal integration of pelvic floor dysfunction. (Reprinted with the permission of The Cleveland Clinic Foundation.)

Embryologic Origins

Our consideration of the pelvic floor organs as a single function unit is further augmented by understanding the common embryologic origin of the structures that compose the pelvic floor. The urogenital sinus and endodermal cloaca lead to the formation of the structures comprising the pelvic floor. It is therefore not surprising that symptomatic conditions can affect adjacent organ systems, a clear example being urogenital sinus irritative symptoms (Chapter 9-2). Abnormal embryologic development can thus also lead to a predisposition for pelvic floor dysfunction in certain women, including short urethra syndrome, shortened perineal body, and altered mucosal sensitivity. These factors are obviously immutable.

Obstetric Factors

The vaginal birth process represents the common denominator for many pelvic floor dysfunction symptoms. Our understanding of neuromuscular damage, which occurs during the vaginal birth process, has been enhanced with the use of neurophysiologic testing and endoanal ultrasound. As such, the vaginal birth process not only results in significant stretching of the levator musculatures in a vertical direction, but, more importantly, in stretching of the pudendal nerves in the vertical direction. Any stretch of a somatic nerve of more than 12% has been reported to lead to a degree of permanent injury to that nerve. During the vaginal birth process, perineal descent results in stretching of the pudendal nerve to a degree in which permanent injury can result. Beyond direct injury to muscles and nerves of the pelvic floor, the vaginal birth process also results in significant symptomatic as well as occult injury to the anal sphincteric mechanism. Many anal sphincter tears are symptomatic. However, it is unknown whether occult anal tears will result in subsequent fecal incontinence a few years postpartum.[3] It is thus not surprising that there is a high incidence of postpartum flatal and fecal incontinence.

An aside should probably be made at this point. Our increased understanding of the neuromuscular consequences of the vaginal delivery process has been the subject of much debate in the recent past, especially as relating to the acceptance of elective, on-demand cesarean delivery. Many cultures around the world have already espoused and accepted this concept for various reasons, including resultant pelvic floor dysfunction. In the United States, this topic has been the subject of significant debate among obstetricians and gynecologists. The urogynecologic stance on this has been one of providing patients information regarding potential consequences of pelvic floor dysfunction and allowing patients to make a decision whether they wish to undergo an elective cesarean delivery.[4] In a preliminary survey of pregnant women, it seems that providing women with information regarding potential pelvic floor consequences of a vaginal delivery does not alter their decision-making process significantly.[5] Nevertheless, providing patients information regarding the risks and benefits of the vaginal delivery process, as with any other invasive procedure informed-consent process, will likely become a part of antepartum obstetric care.

Coexistence of Pelvic Floor Dysfunction Symptoms

Various surveys have been performed to obtain a better understanding of the coexistence of symptoms of urinary, genital, and fecal dysfunction. Not surprisingly, there is a high incidence of coexistence of incontinence and support defects (Table 1-1.1). It has been reported that in patients seen with fecal incontinence, 24% to 53% also complained of urinary incontinence, and 7% to 22% complained of genital prolapse. Of patients who presented with rectal prolapse, 66% also complained of urinary incontinence and 34% complained of genital prolapse.[6–8] In a survey of patients who had undergone surgery for rectal prolapse and fecal incontinence at our institution, with an included control group of clinic patients, the incidence of urinary incontinence was 53% in those who had previous surgery for fecal incontinence and 65% in those who had previous surgery for rectal prolapse. Genital prolapse was found in 18% of patients with fecal incontinence and 34% of patients with rectal prolapse. The control group had a urinary incontinence incidence of 30% and genital prolapse incidence of 12.5%; both incidence rates are consistent with a normal population.[6]

The frequent coexistence of symptoms of urinary and colorectal dysfunction provides further emphasis on the need for a team approach to the evaluation and management of women with any of these conditions. In addition, the fact that rectal prolapse patients have a higher incidence of urinary incontinence and genital prolapse suggests that rectal prolapse may represent a more advanced degree of pelvic floor dysfunction. Anecdotally, in patients we have followed over time with urinary incontinence or other forms of pelvic floor dysfunction who were treated either surgically or medically, the progression of pelvic floor dysfunction was seen with subsequent development of rectal prolapse not being an unusual finding. There is a great need to increase our understanding of the natural history of pelvic floor dysfunction, as its progression may not be possible to stop, but only slowed by our interventions.

Table 1-1.1. Population analysis of the coexistence of pelvic floor dysfunction symptoms

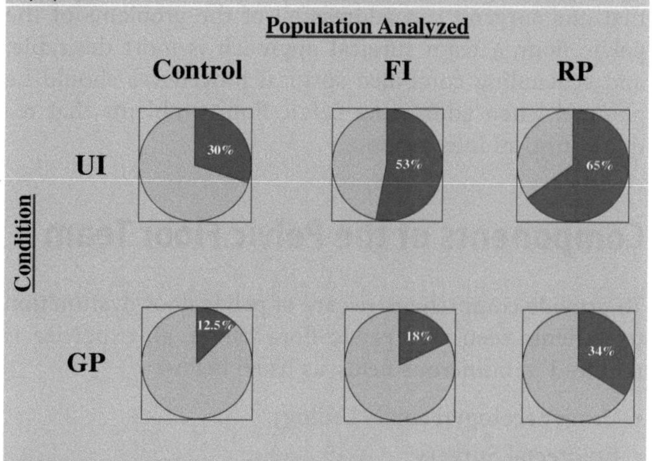

UI, urinary incontinence; GP, genital prolapse; FI, fecal incontinence; RP, rectal prolapse.[6]

Current Needs and Future Perspectives

A very strong case can be made for a multidisciplinary approach to treatment of pelvic floor dysfunction. Certainly, the concept of serial surgeries in a patient with pelvic floor dysfunction cannot be beneficial to the patient. Therefore, correction of all pelvic floor defects at one setting and postoperative physiotherapeutic pelvic floor rehabilitation may represent the most desirable means of treating asymptomatic patients. Less-severe degrees of pelvic floor dysfunction and syndromes not associated with anatomic alterations are amenable to pelvic floor rehabilitative interventions. Physiotherapeutic modalities and their indications will be discussed in Chapter 12. An emphasis must be made on the fact that pelvic floor physiotherapy will benefit both urinary and fecal continence disorders. In fact, patients with concomitant urinary and fecal incontinence, especially associated with urgency symptoms, are optimal candidates for pelvic floor physiotherapy/rehabilitation. Although current reimbursement practices in the United States may pose a barrier to accessibility to pelvic floor rehabilitation, the value of this conservative treatment modality has been demonstrated in many studies involving various types of pelvic floor dysfunction. We consider physiotherapy an integral part of our treatment plan for many patients with pelvic floor dysfunction.

If it is determined that a patient has various forms of pelvic floor dysfunction that are amenable to surgical therapy, a combined surgical approach is indicated. Our previous experience has demonstrated that morbidity is not increased by performing urogynecologic and colorectal surgeries in one setting. In fact, there is no significant difference in morbidity, length of stay, or recuperation phase (Table 1-1.2). The longer operative time required for

Table 1-1.2. Colorectal (CRS) and combined colorectal and urogynecologic surgery (CRS + UG)[9,10]

	Hospital Stay (d)	Complications (%)	Return to Work (wk)	Operating Room Time (h)
CRS	3.55	5–6	6.8	1.5–1.8
CRS + UG	3.85	5–8	6.2	4.5–5

a combined surgical procedure does not seem to be associated with increased morbidity. Because we do not believe that one surgeon can address all of the problems of the pelvic floor, a team surgical approach is most desirable, and scheduling combined surgical procedures should be planned when addressing pelvic floor problems that require surgical intervention.

Components of the Pelvic Floor Team

To provide comprehensive care of pelvic floor dysfunction to patients seen at a pelvic floor center, an expertise is required in numerous fields, as listed below:

- Urogynecology/Female Urology
- Colorectal Surgery
- Gastroenterology
- Physiotherapy
- Sexual Dysfunction Specialist
- Mental Health Specialist
- Nursing Personnel

Communication between the various team members is very important. At Cleveland Clinic Florida, each service obtains information regarding dysfunction in the other compartments and then schedules the patient for appropriate evaluation. Our plan is to have a triage nurse to perform the initial history-taking for patients seen with any form of pelvic floor dysfunction. This would then allow scheduling of appropriate consultation and evaluation services. Constant and accurate communication between team members is very important, especially as relates to planning combined surgical interventions. Although we do not have a combined database at this time, our clinical databases cross-reference each other in order to be able to identify patients who are being seen by multiple services. In addition, joint pelvic floor conferences are necessary and important to discuss complex cases. We hold a pelvic floor conference once a month in which joint cases are discussed and outcomes are provided. We have been able to achieve an excellent working relationship among the various subspecialists of the pelvic floor. This is crucial in the development of a close working relationship, to avoid any "turf" issues. As such, rectocele repairs are typically performed via the vaginal approach by the urogynecology service whereas anal sphincteroplasties are performed by the colorectal service. Our decisions regarding delegation of surgical duties to different services are based on existing literature. Postgraduate fellows are encouraged to participate in the surgical management of problems – especially when operated on by another service – to allow for continuity of patient care.

Summary

The Cleveland Clinic pelvic floor dysfunction team approach is based on a close and cooperative relationship between clinicians of the various disciplines involved in the management of pelvic floor dysfunction. Ultimately, the patient benefits from these cooperative efforts, because serial surgeries are avoided. We hope to be able to demonstrate that outcomes are improved, especially on a long-term basis. We encourage other clinicians to adopt our philosophy and team approach, and welcome any visitors to our institutions as observers or research fellows.

References

1. Swash M, Henry MM, Snooks SJ. Unifying concept of pelvic floor disorders and incontinence. J R Soc Med 1985;78:906–911.
2. Davila GW, Ghoniem GM. Pelvic floor dysfunction: the importance of a multidisciplinary approach. Clin Colon Rectal Surg 2003;16:3–4.
3. Sultan AH, Kamm MA, Hudson CN, Thomas JM, Bartram CI. Anal-sphincter disruption during vaginal delivery. N Engl J Med 1993; 329:1905–1911.
4. Davila GW. Informed consent for obstetrics management: a urogynecologic perspective. Int Urogynecol J Pelvic Floor Dysfunct 2001; 12:289–291.
5. Pollak JT, Chong D, Bratter J, Pietro P, Davila GW. C-section versus vaginal delivery: what do patients want? Int Urogynecol J Pelvic Floor Dysfunct 2003;14(suppl):S53.
6. Gonzalez-Argente FX, Jain A, Nogueras JJ, Davila GW, Weiss EG, Wexner SD. Prevalence and severity of urinary incontinence and pelvic genital prolapse in females with anal incontinence or rectal prolapse. Dis Colon Rectum 2001;44:920–926.
7. Khullar V, Damiano R, Toozs-Hobson P, Cardozo L. Prevalence of faecal incontinence among women with urinary incontinence. Br J Obstet Gynaecol 1998;105:1211–1213.
8. Meschia M, Buonaguidi A, Pifarotti P, Somigliana E, Spennacchio M, Amicarelli F. Prevalence of anal incontinence in women with symptoms of urinary incontinence and genital prolapse. Obstet Gynecol 2002;100:719–723.
9. Halverson AL, Hull TL, Paraiso MF, Floruta C. Outcome of sphincteroplasty combined with surgery for urinary incontinence and pelvic organ prolapse. Dis Colon Rectum 2001;44:1421–1426.
10. Sun JH, Aguirre O, Davila GW. Team approach to pelvic floor dysfunction benefits the patient. Dis Colon Rectum 1999;42:A15.

Section II

Epidemiology and Prevalence

2-1

Epidemiology of Non-Neurogenic Urinary Dysfunction

Usama M. Khater, Gassan Haddad, and Gamal M. Ghoniem

Millions of the general population, particularly women and the elderly, have bladder dysfunction. Although most studies have mainly focused on incontinence, more recently, attention has also included overactive bladder (OAB). Therefore, this chapter will be focused on urinary incontinence (UI) and OAB, because both of them constitute the most common causes of urinary dysfunction.

In 1988, the International Continence Society (ICS) defined UI as the involuntary loss of urine that is objectively demonstrable and is a social or hygienic problem. Recently, the ICS redefined it as the complaint of any involuntary leakage of urine.[1] The impact of the new definition is that more cases of UI will be discovered and hopefully treated. Urinary incontinence symptoms mainly are stress (SUI), which is leakage with physical exertion, urge (UUI), which is leakage with a strong desire to void, and mixed, which is a combination of SUI and UUI. When SUI is urodynamically proven, it is also called genuine stress incontinence (GSI). Urinary urgency and frequency with urge incontinence has recently been defined as OAB/wet. There are other types of incontinence, although less frequent, including overflow incontinence, functional, and extraurethral incontinence.

Urinary Incontinence

Potential Risk Factors

Epidemiologic studies conducted in various populations reveal a number of variables related to the development of UI including several possible risk factors.[2]

Age

Most studies have found that the prevalence of UI tends to increase with advancing age.[3–5] Prevalence has always been higher in institutionalized adults because residents in institutions tend to be older and more impaired than community dwellers. Some studies have found that the proportions of types of UI differ by age. A survey of young and middle-aged women suggests that pure stress incontinence predominates in that age group. Other surveys of older women suggest that urge and mixed incontinence predominate.[6–8]

Race

Ethnicity may be associated with UI. There are several studies of non-Caucasians showing a wide variation in prevalence.[9–11] These studies have used different methods and definitions. Therefore, the results are difficult to compare. Mattox and Bhattia[12] evaluated the prevalence of incontinence among white and Hispanic women with similar age, gravity, and parity who had undergone urodynamic evaluation. The prevalence of OAB/wet (urge incontinence) was 18% in the white population and 9% in the Hispanic population. In another study of more than 4000 women 70 years of age and older, the prevalence of involuntary urine loss was significantly higher in white women (23%) than in black women (16%).[13] There are reports stating that black South Africans rarely develop stress incontinence, and they develop the related disorder of genital prolapse at a rate 80 times lower than whites. The authors explained the rarity of stress incontinence among blacks as a function of differing urethral pressures and length as well as pubococcygeal muscle strength.[14] In the United States, Howard et al.[15] reported clinical data suggesting that African American women have higher urethral closure pressure, larger urethral volume, and greater vesical mobility.

Sex

The prevalence rates of UI are higher in women than in men. The prevalence range for UI is 4.5% to 53% in women and 1.6% to 24% in men.[16]

Pregnancy and Childbirth

Urinary incontinence in women is often assumed to be attributable to the effects of pregnancy and childbirth. The literature shows that UI is a more common occurrence among pregnant women compared with other groups of women, with reported prevalence rates of 31% and 60%.[17,18] Urinary incontinence during pregnancy is a self-limited condition. Viktrup et al.[19] found a 28% prevalence rate of SUI during pregnancy, with 16% becoming free of symptoms in the puerperium. It is still questionable whether pregnancy itself is a risk factor for UI in later life or if it is the vaginal delivery that is the main risk factor. The authors also compared continent women having delivered vaginally with women who underwent a cesarean delivery and found a difference in favor of cesarean delivery. However, 3 months after delivery, the difference became statistically insignificant.

A link between UI and parity has been demonstrated, and several explanations have been offered. First, childbirth may result in pelvic floor laxity as a consequence of weakening and stretching of the muscles and connective tissue during delivery. Second, damage may occur as a result of spontaneous lacerations and episiotomies during delivery. Both may result in impairment of the position and support of the pelvic organs. Third, the stretching of the pelvic tissues during vaginal delivery may damage the pudendal and pelvic nerves, as well as the muscles and connective tissue of the pelvic floor.[20]

Menopause

Postmenopausal women are more likely to have UI than premenopausal women. Evidence that atrophy of the urogenital mucosa can be reversed with estrogen suggests that estrogen loss contributes to the problem. However, the literature is inconsistent in describing the role of menopause and estrogen loss as significant contributors. Rekers et al.[21] compared premenopausal women with postmenopausal women and found no significant difference in the prevalence of UI between the two groups. Another study found that women who experience surgical menopause had a higher rate of UI (36%) compared with those who experience natural menopause (22%).[22]

Hysterectomy

The role of hysterectomy is controversial. Some studies have shown significant association between UI and hysterectomy, as well as oophorectomy. Brown et al.[23] explained this association by the fact that hysterectomy may disturb the musculofascial attachments of the bladder to the surrounding pelvic wall, and oophorectomy results in surgical menopause.

Obesity

Obesity and increased body mass index have been shown to be frequently associated with UI. Obesity may cause stretching and weakening of the muscles, nerves, and other structures of the pelvic floor. There is clear epidemiologic support for the role of obesity in UI.[24]

Other Factors

There are many other risk factors that may contribute to the development of UI. Patients with dementia, Parkinson's disease, multiple sclerosis, and stroke are at a high risk of manifesting bladder dysfunction during the course of their disease. Patients lacking mental orientation have a greater risk of being incontinent than those with normal mental status. Studies in nursing homes have suggested a link between dementia and UI.[25,26]

There are other factors correlated with urinary dysfunction, including chronic obstructive lung diseases,[27] smoking,[28] diabetes,[3] constipation, fecal incontinence,[25] impaired function of levator muscles, genital prolapse,[29] previous gynecologic surgery, perineal suturing,[30] and history of childhood enuresis.[31]

Overactive Bladder

Epidemiology of Overactive Bladder

Overactive bladder is defined by the ICS as a medical condition referring to the urinary symptoms of frequency and urgency, with or without urge incontinence, when appearing in the absence of local pathologic factors.[1]

Spectrum of Overactive Bladder

Patients with OAB experience a wide range of symptoms, including frequency, both daytime and nighttime (nocturia), as well as urgency. A continuum of symptomatology is necessary to accurately describe OAB. Approximately one-third of patients with OAB have urge incontinence.[32]

National Telephone Survey

The National Overactive Bladder Evaluation (NOBLE) program is a nationwide, population-based survey of the United States adult population. The goals of the program are to describe the epidemiology of OAB symptoms, assess the use of coping strategies among people with OAB, and to estimate the burden of the disease. A validated computer-assisted telephone interview was conducted to assess bladder symptoms and coping behavior. In all, 17231 households were contacted by telephone from November

2000 to January 2001. A total of 5204 individuals completed the telephone interview. Eight hundred sixty-four cases of OAB were identified: 545 were classified as OAB without incontinence and 319 were classified as OAB with incontinence. Overactive bladder without incontinence was defined as four or more episodes of urgency within 4 weeks preceding the interview plus a frequency of eight or more voids daily or at least one bladder-control coping behavior. Overactive bladder with incontinence was defined using the same criteria as OAB without incontinence plus three or more episodes of urge UI within the 4 weeks preceding the interview. According to data derived from the NOBLE program, the overall prevalence of OAB in the United States is 16.6%. This means that approximately 33 million American adults may have OAB, although not all of these individuals will require treatment.

The results of the NOBLE program, in addition to indicating that the prevalence of OAB is 16.6%, have demonstrated that the prevalence of OAB is similar between men (16.0%) and women (16.9%). The incidence increases with age in both men and women.[32]

Milsom et al.[33] assessed the prevalence of OAB symptoms in men and women 40 or more years of age from six European countries. There were 16776 interviews conducted in person or by telephone. The overall prevalence of OAB symptoms in European men and women was 16.6%, and the prevalence increased with age.

Prevalence of Overactive Bladder: Incontinent Versus Continent

Of the 33.3 million adults with OAB in the United States, the NOBLE program found that OAB without incontinence (OAB/dry) was more prevalent in men, whereas OAB with incontinence (OAB/wet) was more prevalent in women. Thirty-seven percent of people with OAB have OAB/wet; the remaining 63% have OAB/dry.[32]

Prevalence of Symptoms in the United States

Among the representative population, the NOBLE program assessed the prevalence of the individual bothersome symptoms that characterize OAB. The prevalence of urge incontinence was 6%, frequency was 15%, urgency was 22%, and nocturia was 29%.[32]

Prevalence of OAB and Chronic Conditions in the United States

Overactive bladder is as prevalent a condition as chronic diseases such as arthritis, allergic rhinitis, and sinusitis. Overactive bladder is more prevalent than heart disease, asthma, diabetes, or Alzheimer's disease.

Prevalence of Overactive Bladder with Selected Comorbidities

Overactive bladder with and without incontinence were more prevalent among men and women with certain comorbid conditions, including cancer, diabetes, congestive heart failure, neurologic disorder, and, in men, enlarged prostate.

The prevalence ratio is greater than 1 for most comorbid conditions in men and women with OAB with incontinence. Persons less than 45 years of age have higher prevalence ratios for most comorbid conditions compared with persons 45 years of age and older. This suggests that the influence of aging on the prevalence of OAB is a predominant factor.

Falls and Fractures

Brown et al.[34] administered a questionnaire to assess the risk of falls and fractures in 6049 women with UI. The rationale for assessing women specifically with UUI was that these women have increased diurnal and nocturnal voiding. Additionally, women with UI often rush to the bathroom to avoid an episode of incontinence. Therefore, urgency in these women may predispose them to an increased risk of falls and fractures. Brown et al. reported that 19% to 42% of women with UI sustain falls and that fractures occur in 4% to 9% of falls. Hip fractures are associated with a high rate of morbidity and mortality. Weekly or more frequent episodes of urge incontinence were independently associated with a 30% increased risk of falls and fractures.

Association Between Overactive Bladder and Depression

Overactive bladder may be associated with an increased risk of depression. Zorn et al.[35] assessed 115 incontinent patients to determine whether there was an association between incontinence and depression. Thirty percent of patients with incontinence had an abnormal Beck Depression Inventory (BDI) score compared with only 17% of controls. They found evidence of depression in 60%, 44%, and 14% of patients with idiopathic urge, mixed, and stress incontinence, respectively. The authors hypothesized that OAB and depression may share a common etiology involving the neurotransmitter serotonin. Whereas it seems that there is a positive correlation between depression and idiopathic urge incontinence, the association between depression and stress incontinence is less clear.

Economic Costs of Overactive Bladder

The total economic costs of OAB in the United States in the year 2000 were $18.2 million dollars.[32] This estimate includes direct and indirect costs at the community and institutional levels but does not consider intangible costs; therefore, although this current cost of OAB is higher than the cost of disease states such as osteoporosis and Parkinson's disease, the economic burden of OAB in the United States may be underestimated.

Impact of Cost to the Patient

To appropriately and accurately assess the economic burden of OAB, it is necessary to consider three types of costs: direct, indirect, and intangible costs. Direct costs include routine care costs (e.g., catheter, pads), treatment costs (e.g., drugs, behavior therapy), diagnostic costs (e.g., physical examinations, urodynamic evaluations), and consequence costs (e.g., treatment for sequels such as falls or infections). Indirect costs include money, time, and productivity that are lost by patients and caregivers because of the disease. Intangible costs are the most difficult to estimate. These costs are derived from subjectively measured parameters such as effects on quality of life, welfare and well-being, and productivity.

Prevalence and Impact

The short form, SF-36 scale is a validated instrument that provides measurements of various domains reflecting quality of life. Higher scores represent higher quality of life. On the SF-36, scores in the physical and mental domains are significantly lower in men and women with dry or wet OAB compared with controls.

Affected individuals often limit or even cease participation in physical activities. Sexual contact and intimacy may also be avoided. Both work and home life are affected by OAB. Social interactions may be reduced because of the need to plan activities and travel around the availability of toilet facilities.

References

1. Abrams P, Cardozo L, Fall M, et al. The standardisation of terminology of lower urinary tract function: Report from the Standardisation Sub-committee of International Continent Society. Neurourol Urodyn 2002;21:167–178.
2. Fantl A, Newman K. Urinary Incontinence in Adults, Clinical Practice Guideline Update. Rockville, MD: US Department of Health Services, Public Health Agency for Health Care Policy and Research; March 1996. AHCPR publication 96-0682.
3. Chiarelli P, Brown W, McElduff P. Leaking urine: prevalence and associated factors in Australian women. Neurourol Urodyn 1999;18: 567–577.
4. Nygaard IE, Lemke JH. Urinary incontinence in rural older women: prevalence, incidence and remission. J Am Geriatr Soc 1996;44:1049.
5. Yarnell JW, Voyle GJ, Richards CJ, Stephenson TP. The prevalence and severity of urinary incontinence in women. J Epidemiol Community Health 1981;35:17.
6. Cheater M, Castleden M. Epidemiology and classification of urinary incontinence. Clin Obstet Gynecol 2000;14:183.
7. Diokno C, Broke M, Herzog R. Prevalence of urinary incontinence and other urological symptoms in the noninstitutional elderly. J Urol 1986;136:1022.
8. Peet SC, Castleden CM. The prevalence of urinary and fecal incontinence in hospitals and residential and nursing homes for older people. BMJ 1995;311:1063.
9. Brieger GM, Mongelli M, Hint LY, et al. The epidemiology of urinary dysfunction in Chinese women. Int Urogynecol J Pelvic Floor Dysfunct 1997;8:191.
10. Lara C, Nacey J. Ethnic differences between Mori, Pacific Island and European New Zealand women in prevalence and attitudes to urinary incontinence. N Z Med J 1994;107:374.
11. Nakanishi N, Atar K, Naramura H, et al. Urinary and fecal incontinence in a community-residing older population in Japan. J Am Geriatr Soc 1997;45:215.
12. Mattox F, Battia N. The prevalence of urinary incontinence or prolapse among white and Hispanic women. Am J Obstet Gynecol 1993; 81:421–425.
13. Fultz H, Herzog R, Raghunathan E, Wallace B, Diokno C. Prevalence and severity of urinary incontinence in older African American and Caucasian women. J Gerontol A Biol Sci Med Sci 1999;54:M299–M303.
14. Knobel J. Stress incontinence in black female. S Afr J Obstet Gynecol 1975;49:430.
15. Howard D, Dvies S, Delancey O, Small Y. Differences in perineal lacerations in black and white primiparas. Obstet Gynecol 2000; 96:622.
16. Davila G, Neimark M. The overactive bladder: prevalence and effects on quality of life. Clin Obstet Gynecol 2002;45:173–181.
17. Burgio L, Locher L, Zyczynski H, Hardin M, Singh K. Urinary incontinence during pregnancy in racially mixed sample: characteristics and predisposing factors. Int Urogynecol J Pelvic Floor Dysfunct 1996;7:69.
18. Mellier G, Delille MA. Urinary disorders during pregnancy and postpartum. Rev Fr Gynecol Obstet 1990;85:525–528.
19. Viktrup L, Lose G, Rolff M, Farfoed K. The symptom of stress incontinence caused by pregnancy or delivery in primiparas. Obstet Gynecol 1992;79:945.
20. Foldspang A, Mommsen S, Lam GW, Elving L. Parity as a correlate of adult female urinary incontinence prevalence. J Epidemiol Community Health 1992;46:595–600.
21. Rekers H, Drogendijk C, Valkenburg H, Riphagen F. Urinary incontinence in women from 35 to 79 years of age: prevalence and consequences. Eur J Obstet Gynecol Reprod Biol 1992;43: 229.
22. Milsom I, Ekelund P, Molander U, Arvidsson L, Areskoug B. The influence of age, parity, oral contraception, hysterectomy and menopause on prevalence of urinary incontinence in women. J Urol 1993; 149:1459.
23. Brown S, Sawaya G, Thom H, Grady D. Hysterectomy and urinary incontinence: a systemic review. Lancet 2000;356:535.
24. Mommsen S, Foldspang A. Body mass index and adult female urinary incontinence. World J Urol 1994;12:319.
25. Aggazzotti G, Pesce F, Grassi D, et al. Prevalence of urinary incontinence among institutionalized patients: a cross sectional epidemiologic study in a midsized city in northern Italy. Urology 2000;56: 245.
26. Brandeis H, Baumann M, Hossain M, Morris JN, Resnick NM. The prevalence of potentially remediable urinary incontinence in frail older people: a study using the minimum data set. J Am Geriatr Soc 1997;45:179–184.
27. Maggi S, Minicuci N, Langlois J, Pavan M, Enzi G, Crepalid G. Prevalence rate of urinary incontinence in community-dwelling elderly individuals: the Veneto study. J Gerontol A Biol Sci Med Sci 2001; 56:M14–M18.

28. Bump C, Mcclish K. Cigarette smoking and urinary incontinence in women. Am J Obstet Gynecol 1992;167:1213.
29. Hording U, Pendersen H, Sidenius K, Hedegaard L. Urinary incontinence in 45-year-old women. An epidemiological survey. Scand J Urol Nephrol 1986;20:183.
30. Jolleys V. Reported prevalence of urinary incontinence in women in a general practice. BMJ 1988;296:1300.
31. Foldspang A, Mommsen S. Adult female urinary incontinence and childhood bedwetting. J Urol 1994;152:85.
32. Stewart W, Herzog R, Wein A, et al. Prevalence and impact of overactive bladder in the USA: results from the NOBLE program. Neurourol Urodyn 2001;20:406.
33. Milsom I, Abrams P, Cardozo L, Roberts RG, Thuroff J, Wein A. How widespread are the symptoms of an overactive bladder and how are they managed? A population-based prevalence study. BJU Int 2001; 87:760–766.
34. Brown JS, Vittinghom E, Wyman JF, et al. Urinary incontinence: does it increase risk for falls and fractures? Study of Osteoporotic Fractures Research Group. J Am Geriatri Soc 2000;48(7):721–725.
35. Zorn H, Montgomery H, Pieper K, Gray M, Steers D. Urinary incontinence and depression. J Urol 1999;162(1):82–84.

2-2

Bowel Dysfunction

James Doty and Jonathan E. Efron

Pelvic floor disorders involving bowel dysfunction include several different clinical problems such as fecal incontinence, constipation, and prolapse of the rectum. These disorders are often complex, involving the functions of smooth and skeletal muscles, their nerves, and connective tissues. Etiologies of these disorders are either posttraumatic (such as from vaginal birth or prior anorectal surgery) or acquired, developing from chronic evacuation problems or with age. In some cases, the cause is idiopathic. Management strategies for the treatment of these disorders have evolved and continue to evolve as we better understand the nature of these problems.

Evacuation Disorders

One of the most common complaints of patients seeing colorectal surgeons is constipation. The prevalence of chronic constipation in the United States varies from 2% to 34%.[1] Constipation is caused by several etiologies. Pelvic outlet obstruction (POO) is a common cause of constipation and is attributed to muscular dysfunction of the pelvic floor. The reported incidence of POO as a cause of constipation was reported as high as 50% at a tertiary referral center.[2] Population studies have demonstrated the prevalence of POO in the elderly (age 65 years and older) as 20%.[3] In a population study performed by Talley et al.,[1] the prevalence of POO was 16.5% in females and 5.2% in males. In another study that compared age-related prevalence in the same community of patients, aged less than 60 to more than 65 years, there was an increasing prevalence of POO with increasing age.[1,3]

Multiple risk factors suspected to be associated with POO have been reviewed. One such risk factor is gender. Constipation in young and middle-aged adults is approximately three times more prevalent in women, and the prevalence increases in both genders with age.[4] Nonsteroidal antiinflammatory drug usage has been shown to be associated with POO in some studies but not in others.[1,3] Other factors such as aspirin, Tylenol, alcohol, and tobacco use, marriage and employment status, level of education, and depression have not borne out as significant risk factors for POO.[1,3,5] Laxative and enema abuse have been associated with POO, but have not been shown to be significant. There are limited reports of other associated pelvic floor anomalies with POO. For example, Talley et al.[1] reported up to a 28% incidence of proctalgia fugax in patients with POO.

Among patients with POO, there are several neuromuscular-associated or etiologic syndromes, including: 1) nonrelaxing puborectalis syndrome, 2) rectocele, and 3) descending perineum syndrome. Nonrelaxing puborectalis syndrome occurs when there is failure of relaxation or paradoxical contraction of the puborectalis muscle at the time of defecatory effort. Paradoxical or nonrelaxation of the puborectalis muscle is the cause in 31% to 42% of patients with POO constipation.[2,6] As a result of the chronic straining associated with this syndrome, many other associated disorders may develop including rectal prolapse and rectocele.

The descending perineum syndrome occurs as a result of either injury of the sacral or pudendal nerves or damage to the pelvic floor muscles. Most often, this injury occurs secondary to childbirth or chronic straining at stool. The descending perineum syndrome is frequently associated with constipation and, later, development of fecal incontinence. Its incidence increases with age and it is more common in women than men.[7]

A rectocele is defined as a herniation or protrusion of the anterior rectal wall into the vagina and is associated with pelvic laxity. Rectoceles are found in 1% of patients presenting to specialists with complaints of chronic constipation.[2] They are very common and not often symptomatic, and so they are vastly underreported. Many rectoceles will remain asymptomatic until the fourth or fifth decade of life.[8] The majority of patients with rectoceles are multiparous and/or have chronic constipation with a history of straining.[9] During vaginal delivery or chronic straining, damage to the rectovaginal septum, pelvic floor muscles, and the pudendal nerves may be related to rectocele development.[9] Rectoceles are also associated with hysterectomies.[9]

Prolapse Syndromes

Rectal prolapse occurs when the full thickness of the rectal wall protrudes through the anus. It is the most common type of distal digestive tract prolapse. In mucosal prolapse, only the mucosa of the rectal wall protrudes through the anus. In contrast, internal intussusception can be full or partial thickness but the prolapse does not pass beyond the anus. Intussusception is common in normal, healthy volunteers (up to 50%) and may be normal. It seldom leads to full prolapse.[9]

The first description of rectal prolapse is said to be in the Ebers papyrus 1500 BC. The first treatment as outlined by Hippocrates involved hanging patients by their heels and shaking them.[10] Obviously, this was rarely successful in the long term. The true incidence of rectal prolapse (mucosal or complete) is unknown mostly because of underreporting. It is associated with long-standing constipation, chronic straining, pregnancy, prior surgery, female gender, aging, neurologic disease, mental illness (up to 53% in a study by Vongsangnak et al.), and other pelvic floor disorders.[11,12]

The peak incidence of prolapse is seen in the fourth and seventh decades of life, and women are affected more often, outnumbering men 10 to 1.[13] In fact, the incidence increases with age in women, but not in men. Approximately two-thirds of women with rectal prolapse have neurogenic weakness of the pelvic floor, which is usually a consequence of childbirth.[14] The incidence of prolapse, however, does not seem to be lower in nulliparous women, although they are less likely to have associated fecal incontinence.[14] In a study by Karulf et al.,[15] 35% of patients with rectal prolapse were nulliparous. Other associated conditions with rectal prolapse include rectosigmoid cancer, systemic sclerosis, laxative abuse, malnutrition, and increased intraabdominal pressure. Up to 40% of patients with rectal prolapse have fecal incontinence, and up to 60% have constipation.[16]

Solitary rectal ulcer syndrome (SRUS) is a condition described classically as a single ulceration of the rectal mucosa usually anteriorly between 5 and 12 cm from the anal verge.[17,18] The name of the syndrome, however, can be misleading, because patients do not need to have ulceration to be diagnosed with this condition. Typically, patients present with symptoms of bleeding (56%), passage of mucous, straining, or a sense of incomplete evacuation. Endoscopic findings range from mucosal ulcerations to mild erythema. It is thought that years of straining because of chronic constipation or birth-related injuries may weaken the pelvic floor with resultant pudendal nerve damage. Pudendal nerve damage leads to weakened sphincters and thus prolapse. Intussuscepted mucosa may lead to direct mucosal injury or ischemia, which leads to rectal mucosal injury or ulceration. Associated conditions include rectal prolapse, failure of puborectalis relaxation, increased anorectal angle, abnormal perineal descent, and decreased anorectal sensitivity to balloon inflation.[19] Abnormal perineal descent has been seen in patients with SRUS, and behaviors such as digitization have been related. Other disorders such as defecation disorders, constipation, fecal impaction, erythema nodosum, and lupus have also been associated with SRUS.[20]

The annual incidence of SRUS is 1 to 4 per 100 000. Eighty percent of patients are less than 50 years old. Females are affected slightly more often than males.[9] The mean age of presentation is 49, but 25% of patients present after age 60.[9] As noted above, many authors believe rectal prolapse is the etiology. It is proposed that prolapse causes ischemia, which leads to ulceration; however, only 30% of patients with SRUS have rectal prolapse.[19]

Continence Abnormalities

Fecal incontinence, similar to other pelvic floor disorders, is vastly underreported by patients, and thus the true prevalence is unknown. In a telephone survey in Wisconsin, 2.2% of the population reported incontinence, half of whom were incontinent to solid or liquid stool.[21] Studies using anonymous questionnaires have shown higher rates, with 4.8% having incontinence to solid stool and 6.7% with incontinence to liquid stool.[22] Women are 50% more likely to report incontinence than men.[22]

Obstetric trauma is the most important etiologic factor in the pathogenesis of fecal incontinence in women. There is evidence that hormonal changes during pregnancy lead to smooth muscle relaxation attributed to progesterone. Relaxin is an ovarian hormone that peaks late during pregnancy and leads to connective tissue remodeling in the pelvic floor.[23] With parturition, there is stretching of the levators, stretching and tearing of the rectovaginal septum, stretching of the vaginal wall, and compression of the pudendal nerves against the pelvic side wall. All these factors may contribute to fecal incontinence.

A published study by Sultan et al.[24] revealed anal sphincter defects in 30% to 40% of asymptomatic postpartum females. Fortunately, the minority of these patients were symptomatic (32%). However, these patients may become symptomatic later in life or with subsequent vaginal deliveries. In addition, pudendal nerve injury documented by electromyography has been demonstrated in 42% of postpartum females by Snooks et al.[25,26] Sixty percent of these patients recovered nerve function 2 months after delivery, but 40% did not. Four percent of 906 postpartum women in a study by MacArthur et al.[27] reported new symptoms of incontinence after childbirth. Sultan et al.[28] showed a 1% incidence of frank fecal incontinence and a 25% incidence of decreased flatal control at 9 months' follow-up after vaginal delivery.

The incidence of sphincter injury is higher in patients with perineal tears. Up to 25% of patients developed fecal incontinence symptoms after a third degree tear in a study by Wood et al.[29] Third degree tears, involving the sphincter muscle, occur in approximately 0.6% of all vaginal deliveries.[30] Repairs of sphincter defects by obstetricians at the time of injury have shown variable results with up to 85% of patients having persistent sphincter defects after repair. These repairs were performed with figure-of-8 in an end-to-end manner.[30] Overlapping repairs have shown better results with only a 15% incidence of persistent sphincter defects.[31] Other studies, however, have not supported these results. A study by Fitzpatrick et al.[32] showed no significant difference in outcome by either method of repair.

Episiotomies, similar to tears, are associated with incontinence. Sultan et al.[31] found episiotomy to be associated with an increased risk of sphincter injury. Signorello et al.[33] showed a threefold increase in fecal incontinence after midline episiotomy as compared with spontaneous laceration; therefore, a mediolateral episiotomy is recommended.

Menopause is noted to be a risk factor for the development of altered fecal continence. Some preliminary studies have shown a plethora of estrogen receptors in the sphincters of women, and have shown a subjective improvement in fecal incontinence with estrogen therapy.[34,35] Aging is also a significant risk factor for fecal incontinence. The prevalence of fecal incontinence is between 9% and 17% in women 75 years or older. The incidence is higher in women (8 : 1) probably because of a history of obstetric injury and the loss of estrogen.[36]

Although vaginal delivery is probably the most common cause of fecal incontinence, there are several other less common causes. Other traumatic etiologies include surgery for perianal fistula (with an incidence up to 34%), sphincterotomy for fissure, or hemorrhoids.[37] Pelvic fractures or direct anal trauma are other traumatic causes. Infiltration of the nerves and/or muscles by neoplastic processes can also result in incontinence. Congenital anorectal malformations or the surgery performed to correct them account for a small percentage of fecal incontinence. Patients with a postsurgery for imperforate anus have an incidence of incontinence up to 47% in some series.[38] Idiopathic causes in which there is pelvic floor denervation or dysfunctional, yet intact, musculature account for a significant proportion of fecal incontinence, especially in the elderly population. Etiologies include incontinence associated with rectal prolapse, chronic straining at stool, or other neurologic dysfunction. Specific neurologic causes may include multiple sclerosis, spinal cord injury, or diabetes in which there may be up to a 9% incidence of fecal incontinence secondary to autonomic neuropathy.[39]

Regardless of the etiology, the prevalence of bowel dysfunction in pelvic floor disorders can be significant. For many of these problems, treatment is available but not always satisfactory. As our understanding of these disorders improves, more successful and more durable treatments will likely develop. Patient and physician education, as well as early intervention, can often make a sizable difference in outcome resulting in a true understanding of the incidence of these problems as well as improved results of therapy.

References

1. Talley NJ, Weaver AL, Zinsmeister AR, Melton LJ. Functional constipation and outlet delay: a population-based study. Gastroenterology 1993;105:781–790.
2. Surrenti E, Rath DM, Pemberton JH, Camilleri M. Audit of constipation in a tertiary referral gastroenterology practice. Am J Gastroenterol 1995;90(9):1471–1475.
3. Talley NJ, Fleming KC, Evans JM, et al. Constipation in an elderly community: a study of prevalence and potential risk factors. Am J Gastroenterol 1996;91(1):19–25.
4. Johnson JF, Sonnenberg A, Koch TR. Clinical epidemiology of chronic constipation. J Clin Gastroenterol 1989; 11:516–536.
5. Grozt RL, Pemberton JH, Talley NJ, Rath DM, Zinsmeister AR. Discriminant value of psychological distress, symptoms profiles and segmental colonic dysfunction in outpatients with severe idiopathic constipation. Gut 1994;35(6):798–802.
6. Glia A, Lindberg G, Nilsson LH, Mihocsa L, Akerlund JE. Constipation assessed on the basis of colorectal physiology. Scand J Gastroenterol 1998;33(12):1273–1279.
7. Bannister JJ, Abouzekry L, Read NW. Effect of aging on anorectal function. Scand J Gastroenterol 1997;28:353–357.
8. Pfeifer J, Agachan F, Wexner SD. Surgery for constipation. Dis Colon Rectum 1996;39(4):444–460.
9. Felt-Bersma RJF, Cuesta MA. Rectal prolapse, rectal intussusception, rectocele, and solitary rectal ulcer syndrome. Gastro Clin North Am 2001;30(1):199–218.
10. Pemberton JH, Swash M, Henry M. The Pelvic Floor: Its Function and Disorders. Philadelphia: WB Saunders; 2002:265.
11. Altemeier WA, Culbertson WR, Schowengerdt C, Hunt J. Nineteen years' experience with the one-stage perineal repair of rectal prolapse. Ann Surg 1971;173(6):993–1006.
12. Vongsangnak V, Varma JS, Smith AN. Reappraisal of Thiersch's operation for complete rectal prolapse. J R Coll Surg Edinb 1985; 30(3):185–187.
13. Poritz LS. Rectal prolapse. eMedicine February 2003.
14. Neil ME, Parks AG, Swash M. Physiological studies of the anal sphincter musculature in faecal incontinence and rectal prolapse. Br J Surg 1981;68(8):531–536.
15. Karulf RE, Madoff RD, Goldberg SM. Rectal prolapse. Curr Probl Surg 2001;38(10):771–832.
16. Rashid Z, Basson MD. Association of rectal prolapse with colorectal cancer. Surgery 1996;119(1):51–55.
17. Madigan MR, Morson BC. Solitary ulcer of the rectum. Gut 1969;10: 871–881.
18. Womack NR, Williams NS, Holmfield JH, Morrison JF. Pressure and prolapse: the cause of solitary rectal ulceration. Gut 1987;28:1228–1233.
19. Keighley MRB, Shouler P. Clinical and manometric features of the solitary rectal ulcer syndrome. Dis Colon Rectum 1984;27:507–512.
20. Tandon RK, Atmakuri SP, Mehra NK, Malaviya AN, Tandon HD, Chopra P. Is solitary rectal ulcer syndrome a manifestation of a systemic disease? J Clin Gastroenterol 1990;12:286–290.
21. Nelson R, Norton N, Cautley E, Furner S. Community based prevalence of anal incontinence. JAMA 1995;274:559–561.
22. Giebel GD, Lefering R, Troidl H, Blochl H. Prevalence of fecal incontinence: what can be expected? Int J Colorectal Dis 1998: 13:73–77.
23. Lepert PC. Anatomy and physiology of cervical ripening. Clin Obstet Gynecol 1995;38:267–279.

24. Sultan A, Kamm M, Hudson C, Thomas J, Bartram C. Anal sphincter disruption during vaginal delivery. N Engl J Med 1993;329;1905–1911.
25. Snooks SJ, Swash M, Mathers SE, Henry MM. Effect of vaginal delivery on the pelvic floor: a five year follow-up. Br J Surg 1990;77:1358–1360.
26. Snooks SJ, Swash M, Setchel M, Henry MM. Injury to innervation of the pelvic floor sphincter musculature in childbirth. Lancet 1984;ii:546–550.
27. MacArthur C, Bick DE, Keighley MRB. Faecal incontinence after childbirth. Br J Obstet Gynaecol 1997;104:46–50.
28. Sultan AH, Kamm MA, Bartram CI, Hudson CN. Anal sphincter trauma during instrumental delivery. Int J Gynecol Obstet 1993;43:263–270.
29. Wood J, Amos L, Rieger N. Third degree anal sphincter tears: risk factors and outcome. Aust N Z J Obstet Gynaecol 1998;38:414–417.
30. Sultan AH, Kamm MA, Hudson CN, Bartram CI. Third degree obstetric anal sphincter tears: risk factors and outcomes of primary repair. BMJ 1994;38:887–891.
31. Sultan AH, Monga AK, Kumar D, Stanton SL. Primary repair of obstetric anal sphincter rupture using the overlap technique. Br J Obstet Gynaecol 1999;106:318–323.
32. Fitzpatrick M, Fynes M, Cassidy M, Bechan M, O'Connel PR, O'Herlihy C. Prospective study of the influence of parity and operative technique on the outcome of primary anal sphincter repair following obstetric injury. Eur J Obstet Gynecol Reprod Biol 2000;89:159–163.
33. Signorello LB, Harlow BL, Chekos AK, Repke JT. Midline episiotomy and anal incontinence: retrospective cohort study. BMJ 2000;320(8):86–90.
34. Donnelly V, O'Connell PR, O'Herlihy C. The influence of oestrogen replacement on faecal incontinence in postmenopausal women. Br J Obstet Gynaecol 1997;104:311–315.
35. Khullar V, Cardozo L. Incontinence in the elderly. Curr Opin Obstet Gynecol 1998;10:391–394.
36. Brocklehurst JC. Urinary incontinence in the community-analysis of a MORI poll. BMJ 1993;306:832–834.
37. Shouler PJ. Grimley RP, Keighley MR, et al. Fistula in ano is usually simple to manage surgically. Int J Colorectal Dis 1986;1:113–115.
38. Javid PJ, Barnhart DC, Hirschl RB, et al. Immediate and long-term results of surgical management of low imperforate anus in girls. J Pediatr Surg 1998;33(2):198–203.
39. Bytzer P, Talley NJ, Young LJ, et al. GI symptoms in diabetes mellitus are associated with both poor glycemic control and diabetic complications. Am J Gastroenterol 2002;97(3):604–611.

2-3

Genital Prolapse, Urogenital Atrophy, and Sexual Dysfunction

Minda Neimark

This chapter will review the epidemiology, incidence, prevalence, and risk factors involved with genital prolapse. In addition, it will also review urogenital atrophy and sexual dysfunction related to prolapse and its incidence after pelvic reconstructive surgery.

Genital Prolapse

Genital prolapse is a common condition that can affect women of all ages. Multiple risk factors have been identified and studied that are thought to contribute to the development of genital prolapse, including age, parity, menopausal status, body mass index (>30 kg/m^2), race, genetics, connective tissue disease, tobacco use, chronic lung disease, chronic constipation, occupational straining, and previous surgery.

Genital prolapse is a general term for weakening or loss of support to the pelvic organs (bladder, vagina, uterus, and rectum) resulting in a herniation of those pelvic organs. Site-specific pelvic examinations are used to determine the exact location of pelvic floor weakness, and the current grading systems [Baden-Walker, pelvic organ prolapse quantification (POP-Q) system] allow documentation of the degree of prolapse, which can be communicated to other clinicians. The types of genital prolapse include:

- Vaginal vault prolapse – herniation of the vaginal vault caused by loss of support or weakening of the uterosacral ligaments, cardinal ligaments, and loss of attachment of the endopelvic fascia to the white line at the level of the sacrospinous ligament, or a combination of the above mentioned.
- Uterine prolapse – herniation of the uterus caused by loss of support of the uterosacral and/or cardinal ligaments.
- Cystocele – herniation of the anterior vaginal wall and bladder caused by tearing, stretching, or a combination of the two, of the anterior wall endopelvic fascia. The insult may either be midline, resulting in a central cystocele, or a lateral insult resulting in a paravaginal defect.
- Enterocele – herniation of the superior portion of the posterior vaginal wall caused by tearing, stretching, or a combination of the two, of the posterior vaginal wall endopelvic fascia.
- Rectocele – herniation of the inferior portion of the posterior vaginal wall and rectum caused by tearing, stretching, or a combination of the two, of the posterior vaginal wall endopelvic fascia.

Epidemiology

The overall incidence of genital prolapse is underestimated because of limited patient access to medical care, limited number of clinicians trained to recognize prolapse, mild cases of prolapse that may be asymptomatic and therefore unknown to the patient and ignored by the physician, as well as the patient's desire to not disclose her problem to the clinician. Therefore, the studies published to date that cite incidence and prevalence of genital prolapse can only be generalized to similar community settings rather than to worldwide populations. In addition, the incidence of grades of prolapse severity are unattainable because, to date, not all clinicians who treat genital prolapse use the POP-Q system, which is an objective tool used to grade genital prolapse. Typically, patients with genital prolapse have more than one area of weakness, with more severe prolapse in some areas, depending on the initial damage to the pelvic floor as well as the length of time the prolapse has gone untreated.

Surgery for Prolapse

Despite the fact that pelvic organ prolapse is one of the most common indications for gynecologic surgery, there is little epidemiologic information regarding surgical procedures for prolapse. In a 1997 survey, approximately 226 000 women underwent surgery for pelvic organ prolapse in the United States, making prolapse one of the most common indications for gynecologic surgery.[1] The average age of women undergoing prolapse surgery was 55 years. The overall rate of pelvic organ prolapse surgery was 22.7 per 10 000 women in the United States. This increased with age to a peak rate of 42.1 per 10 000 women aged 60 to 69 years. Vaginal hysterectomy with rectocele and cystocele repair was the most common surgery performed. There was a higher percentage of women aged 70 to 79 years compared with the other age groups that underwent vault fixation or suspension. Women in the southern part of the United States were more likely to undergo surgery for prolapse than women in the midwest or northern states (29.3 per 10 000 vs. 21.7 per 10 000 vs. 19.1 per 10 000, respectively). In addition, women aged 30 to 39 and 70 to 79 years in the South were six times more likely to have prolapse surgery than women in the Northeast.[1] The Oxford Family Planning Association followed up more than 17 000 women aged 25 to 39 years for approximately 26 years. The annual incidence of hospital admission with a diagnosis of prolapse was 0.204%, whereas the annual incidence of surgery for prolapse was 0.162%.[2] Depending on the population studied, up to 14% of all hysterectomies performed are for genital prolapse.

In a study to review the trends in surgery performed for pelvic organ prolapse from 1979 to 1997, there was a decrease in the number of surgeries performed for prolapse, including hysterectomy (abdominal and vaginal), as well as rectocele and cystocele procedures. However, there was an increase in vaginal vault suspensions and fixations.[3]

Age and Menopause

The incidence of prolapse increases with age, as does the rate of surgery for prolapse. Luber and colleagues[4] found that women seeking help at a pelvic floor disorders clinic had a mean age of 61.5 years and that more than half of the women (57%) were aged 60 years and older. It has been demonstrated that the rate of pelvic organ prolapse surgery increases with age, peaking in the sixth decade. The lifetime risk of undergoing a single operation for pelvic organ prolapse by the age of 80 is 11.1%. This incidence will be influenced by variables including access to a health care provider trained to treat these problems, financial concerns, and overall medical care access. Surgically managed patients represent only a small fraction of those affected, because many women, especially the elderly, are managed conservatively or never present for evaluation.[5]

To date, there are no studies evaluating the role of estrogen replacement initiated at the start of menopause in prevention of genital prolapse. However, because age has been shown to be a strong risk factor in the development of prolapse, it can be assumed that estrogen deficiency, which will occur in all women in the menopausal years, may contribute to weakening of the supports (epithelium, connective tissue, muscle). This may be a key factor in the development and progression of prolapse, explaining the increased incidence of prolapse in the postmenopausal years. Estrogen receptors have been identified throughout the nuclei of the connective tissue and smooth muscle cells of the bladder trigone, urethra, vaginal mucosa, levator ani muscle stroma, arcus tendineus, and the uterosacral ligaments. The collagen content of the pelvic floor is also estrogen dependent. In biochemical analysis of pelvic floor tissue, it has been demonstrated that there is a reduction in total collagen content when estrogen concentrations decrease. In addition, there is a reduction in collagen content in genitourinary tissue of patients with genital prolapse when compared with patients without prolapse, regardless of menopausal states.[6] Hormone replacement use for longer than 5 years may have a protective role in pelvic floor dysfunction development.[7]

Previous Surgery

Reoperation rates for recurrent prolapse have been estimated to be approximately 30%. The time interval between repeat procedures for recurrent prolapse has been shown to decrease with each successive repair.[5] Theoretically, previous surgical treatment will cause further damage to the nerves and surrounding support systems and is thus increasing potential risks for recurrent prolapse. Vaginal vault prolapse is a late and common complication after an abdominal or vaginal hysterectomy (Table 2-3.1), with a reported incidence as high as 43%.[8] Marchionni and colleagues[9] found that vault prolapse was higher when a hysterectomy had been performed for relaxation compared with a hysterectomy for benign disease (11.6% vs. 1.8%). For patients who underwent previous abdominal hysterectomy for benign disease, the incidence of vault prolapse was 2%, and, for those with a previous vaginal hysterectomy, the incidence was essentially zero.

Development of prolapse after retropubic urethropexy has been described in the literature with a reported occurrence rate of anywhere from 4% to 26%, depending on the surgical technique used, severity of prolapse, and duration of follow-up.[10] In theory, prolapse occurs after a urethropexy because of the anterior displacement of the infe-

Table 2-3.1. Common sites for genital prolapse

Before Hysterectomy	After Hysterectomy
Cystocele	Vault prolapse
Uterine prolapse	Enterocele
Rectocele	Cystocele
Enterocele	Rectocele
Vault prolapse	

rior portion of the vagina, which creates a change in vector forces, therefore making the upper portion of the posterior vaginal wall susceptible to increases in intraabdominal pressure.

The vaginal retroversion that is seen after sacrospinous fixation of the vaginal vault has been suggested to be a predisposing factor for the development of anterior vaginal wall defects. This is the result of excess intraabdominal pressure on the anterior vaginal wall from the exaggeration in the posterior vaginal wall axis deviation. Studies have indicated that the risk of development of a cystocele after sacrospinous ligament fixation is up to 20%.[11,12]

Obstetric Factors

Obstetric injury is thought to be a principal factor contributing to the development of pelvic floor dysfunction (prolapse, urinary and fecal incontinence). This is attributed to direct muscular and connective tissue injury or to denervation injury to the pudendal nerve. There is question as to whether the mode of obstetric delivery, or just the event of being pregnant, has a role in developing prolapse. In a case-control study, patients who delivered their first child before the age of 25 years had an almost fourfold increase of developing prolapse.[7] Up to 50% of all parous women have some degree of prolapse, with 10% to 20% being symptomatic.[13] Elective cesarean delivery in a nonlabored patient has been speculated to protect the pelvic floor from damage leading to prolapse; however, to date there are no prospective randomized studies to support this theory. It has been well established that vaginal parity, notably the first, is an independent risk factor for the future development of genital prolapse. It was demonstrated that the odds ratio was 3.0 for women who had one vaginal delivery and 4.5 for women with a history of two or more vaginal deliveries.[14] Mant et al.[2] analyzed risk factors for genital prolapse in 17 032 women in family planning clinics across England and Scotland. In those patients who were diagnosed with prolapse, parity was shown to have the strongest relationship to the development of prolapse. The risk of prolapse increased with each child, up to two children; that is, a woman with two children was eight times more likely to develop prolapse compared with a nulliparous patient.

The event of a vaginal delivery results in dilation and stretching of the pelvic floor – nerves, connective tissue supports including the uterosacral and cardinal ligaments, endopelvic fascia, and the perineal body. Stretching of the pudendal nerve during vaginal delivery is thought to be a major cause of nerve damage leading to pelvic floor dysfunction including prolapse as well as incontinence. The type of obstetric event, which results in a neuropathy, has been evaluated in studies assessing pudendal nerve function. Tetzschner and colleagues[15] demonstrated an increase in terminal motor latency of the pudendal nerve (PTML) in women who underwent a vaginal delivery. In a study by Sultan et al.,[16] PTML was compared in women beyond 34 weeks gestation who either delivered vaginally or by elective cesarean delivery. Terminal motor latency of the pudendal nerve was significantly more prolonged in those women who had a vaginal delivery compared with women who had an elective cesarean delivery either primiparous or multiparous. Women who had a cesarean delivery after the onset of labor also had prolongation of PTML.[16] Instrumental vaginal delivery with forceps has been shown to be responsible for a greater incidence (fourfold increase) of pelvic floor dysfunction compared with vaginal deliveries without forceps. This determination is based on objective data including physical examination.[17]

Race and Genetics

Population studies have shown that prolapse is more common in whites, less common in Asians, and uncommon in blacks. Differences in the incidence of prolapse between racial groups suggest congenital and cultural factors. Genetics may also have a role in those women who have collagen or connective tissue diseases contributing to the development of prolapse.

Constipation

Chronic constipation and repeated prolonged defecatory straining efforts have been shown to contribute to progressive neuropathy and dysfunction.[18] It has also been demonstrated that a history of constipation and extensive straining during defecation as a young adult, before the onset of pelvic floor dysfunction, was significantly more common in women who developed genital prolapse compared with those women who did not develop prolapse (61% vs. 4%).[19]

Female Sexual Dysfunction

Female sexual dysfunction is a generalized term for abnormalities in the normal female sexual cycle. It is a highly prevalent problem affecting anywhere from 30% to 50% of women in all age groups. The National Health and Social Life survey of 1749 women found that 43% complained of sexual dysfunction.[20] Only recently have the problems involved with female sexuality gained recognition by clinicians. There are risk factors associated with female sexual dysfunction including age, menopause, previous surgery, mental and physical health, and availability of a partner.

In 1966, Masters and Johnson[21] first described the female sexual response consisting of four successive phases – excitement, plateau, orgasmic, and resolution. This was later modified to the phases of desire, arousal, orgasm, and resolution. It is well recognized today that the female sexual cycle is not just a sequence of events, but a very complex entity including biological, psychological, sociocultural, and interpersonal factors that affect each part of the sexual cycle.

Sexual Function after Pelvic Surgery

Pelvic surgery to correct prolapse may affect sexual function for a number of reasons including narrowed vaginal canal, poor lubrication, and fear of urinary incontinence. Some studies suggest that sexual dysfunction can occur up to 20% of the time after surgery for prolapse or incontinence. It is thought that the vaginal dissection may lead to pelvic floor neuropathy affecting the pudendal nerve, which subsequently affects vaginal sensation and orgasm. Hysterectomy has been associated with sexual dysfunction. It is thought that removal of the cervix alters the upper portion of the vaginal canal and causes a neuropathy, which is the source for dyspareunia as well as anorgasmia. However, studies in the literature are conflicting as to the exact cause of sexual dysfunction after hysterectomy.[22] Another possible cause for dyspareunia is poor estrogenization of the vaginal mucosa in those women who undergo surgical menopause at the time of a hysterectomy. Techniques used to repair the rectocele have been implicated in the cause of postoperative dyspareunia. Posterior colporrhaphy using levator plication has been implicated as a cause for postoperative dyspareunia. Kahn and Stanton[23] evaluated 209 women who underwent a posterior colporrhaphy. They reported an increase in sexual dysfunction in up to 27% of the women included in the study. Site-specific repair of rectoceles has been reported to result in less dyspareunia. In a retrospective study, 73 women who underwent a site-specific rectocele repair were followed 6 months postoperatively to evaluate quality of life, sexual function, and bowel function. It was demonstrated that sexual function was not affected, but dyspareunia was significantly improved or cured after operation in 73% of patients and only worsened in 19% of patients.[24] Kenton and colleagues[25] also reported a significant reduction in dyspareunia, 92% after site-specific repair of rectoceles at 1 year. However, to date, there are no prospective randomized controlled trials evaluating the two techniques and their impact on sexual dysfunction.

New-onset dyspareunia after sacrospinous fixation has been reported in up to 50% of women. This was attributed to vaginal dryness, vaginal constriction, or a shortened vaginal length.[26]

Prolapse

Problems with sexual function have been reported to occur in women with urinary incontinence and uterovaginal prolapse.[27] However, the precise relationship between prolapse and sexual function has not been well documented. In a study that compared sexual function in women with and without uterovaginal prolapse, the authors found that overall sexual dysfunction in women without prolapse did not differ from women who had prolapse. In addition, the degree of prolapse did not have a role in whether sexual dysfunction occurred among those with prolapse. Psychological and relationship conflicts tended to have a greater impact on sexual function for both groups of women.[28]

Menopause

Several menopause-related changes in sexual function occur that have been described in the literature: diminished sexual responsiveness, dyspareunia, decreased sexual activity, decrease in sexual desire, and a dysfunctional male partner. Sarrel[29] described the underlying cause of biological changes that occurred with sexual dysfunction to be estrogen deficiency. The postmenopausal ovary has been shown to be responsible for up to 50% of the testosterone believed to be associated with libido. Many clinicians believe that a combination of both estrogen and testosterone is required to improve female sexual function.

The effect of menopause on vaginal physiology has been described. There is a pronounced decrease in pelvic blood flow. This in turn results in a thinning of the vaginal mucosa and loss of the normal microbial environment. With the loss of estrogen, there is change in the integrity of the vaginal tube, making it less compliant for coitus. Lubrication is also compromised because of loss of estrogen. This has been confirmed with studies demonstrating that these factors will improve with local estrogen cream over time.

Atrophy

As modern medicine continues to improve, life expectancy will continue to increase. It has been estimated that, there are currently more than 25 million women who will live 30 years beyond menopause. Menopause by definition is an estrogen-deficient state resulting in physiological changes to many female organs. The urogenital organs, including the urethra, bladder, vagina, and vulva, are highly estrogen dependent. Blood flow to the pelvis is also estrogen dependent. Without an appropriate estrogen level, urogenital atrophy will develop. This means that 100% of women who enter into the menopausal state will develop atrophy of the urogenital system.

When vulvovaginal atrophy occurs, there is a loss of the vaginal epithelial integrity and loss of normal vaginal microflora. This results in a thinning of the vaginal mucosa and bacterial overgrowth caused by the loss of the acidic state provided by the normal flora. Loss of blood flow to the pelvis results in loss of the transudation of fluid across the vaginal mucosa required for vaginal lubrication during the normal sex response cycle, leading to dyspareunia. In addition, there is increased incidence of vaginal and urinary tract infections. The signs as well as the symptoms of atrophy probably begin before a woman completes the transition into menopause. However, the degree of estrogen deficiency that results in an atrophic state is unknown to date, and is likely individually dependent.

The symptomatology from urogenital atrophy varies among women. Davila et al.[30] demonstrated that the degree of atrophy evidenced on physical examination by vaginal pH and maturation index did not correlate well with symptoms of urogenital atrophy. The treatment for an estrogen-deficient state is hormone replacement therapy. However, approximately 40% of women on systemic hormone therapy will still develop urogenital atrophy requiring further therapy. Local vaginal estrogen therapy will improve the vaginal mucosa by reestablishing the normal microflora and pH, as well as enhancing vaginal epithelial integrity.

References

1. Brown JS, Waetjen LE, Subak LL, Thom DH, Van den Eeden S, Vittinghoff E. Pelvic organ prolapse surgery in the United States, 1997. Am J Obstet Gynecol 2002;186:712–716.
2. Mant J, Painter R, Vessey M. Epidemiology of genital prolapse: observations from the Oxford Family Planning Association study. Br J Obstet Gynaecol 1997;104:579–585.
3. Boyles SH, Weber A, Meyn L. Procedures for pelvic organ prolapse in the United States 1979–1997. Am J Obstet Gynecol 2003;188: 108–115.
4. Luber KM, Boero S, Choe JY. The demographics of pelvic floor disorders: current observations and future projections. Am J Obstet Gynecol 2001;184:1496–1503.
5. Olsen AL, Smith VJ, Bergstrom JO, et al. Epidemiology of surgically managed pelvic organ prolapse and urinary incontinence. Obstet Gynecol 1997;89:501–506.
6. Jackson SR, Avery NC, Tarlton JF, Eckford SD, Abrams P, Bailey AJ. Changes in metabolism of collagen in genitourinary prolapse. Lancet 1996;347:1658–1661.
7. Moalli PA, Ivy AJ, Meyn LA, et al. Risk factors associated with pelvic floor disorders in women undergoing surgical repair. Obstet Gynecol 2003;101:869–874.
8. Virtanen HS, Makinen JI. Retrospective analysis of 711 patients operated on for pelvic relaxation in 1983–1989. Int J Gynaecol Obstet 1993;42:109–115.
9. Marchionni M, Bracco GL, Checcucci V, et al. True incidence of vaginal vault prolapse. J Reprod Med 1999;44:679–684.
10. Langer R, Lipshitz Y, Halperin R, Pansky M, Bukovsky I, Sherman D. Prevention of genital prolapse following Burch colposuspension: comparison between two surgical procedures. Int Urogynecol J Pelvic Floor Dysfunct 2003;14:13–16.
11. Meschia M, Bruschi F, Amicarelli F, Pifarotti P, Marchini M, Crosignani PG. The sacrospinous vaginal vault suspension: critical analysis of outcomes. Int Urogynecol J Pelvic Floor Dysfunct 1999;10:155–159.
12. Smilen SW, Saini J, Wallack SJ, et al. The risk of cystocele after sacrospinous ligament fixation. Am J Obstet Gynecol 1998;179: 1465–1472.
13. Bidmead J, Cardozo LD. Pelvic floor changes in the older woman. Br J Urol 1998;82(suppl):18–25.
14. Chiaffarino F, Chatenoud L, Dindelli M, et al. Reproductive factors, family history, occupation and risk of urogenital prolapse. Eur J Obstet Gynecol Reprod Biol 1999;82:63–67.
15. Tetzschner T, Sorensen M, Jonsson L, Lose G, Christiansen J. Delivery and pudendal nerve function. Acta Obstet Gynecol Scand 1997;76:324–331.
16. Sultan AH, Kamm MA, Hudson CN. Pudendal nerve damage during labor prospective study before and after childbirth. Br J Obstet Gynaecol 1994;101:22–28.
17. Meyer A, Hohfeld P, Actari C, et al. Birth trauma: short and long term effects of forceps delivery compared with spontaneous delivery on various pelvic floor parameters. Br J Obstet Gynaecol 2000;107: 1360–1365.
18. Lubowski DZ, Swash M, Nicholls RJ, Henry MM. Increases in pudendal nerve terminal motor latency with defaecation straining. Br J Surg 1988;75:1095–1097.
19. Spence-Jones C, Kamm MA, Henry MM, et al. Bowel dysfunction: pathogenic factor in uterovaginal prolapse and urinary stress incontinence. Br J Obstet Gynaecol 1994;101:147–152.
20. Laumann E, Paik A, Rosen R. Sexual dysfunction in the United States: prevalence and predictors. JAMA 1999;281:537–544.
21. Masters EH, Johnson VE. Human Sexual Response. Boston: Little, Brown; 1966.
22. Rhodes JC, Kjerulff KH, Langenberg PW, et al. Hysterectomy and sexual functioning. JAMA 1999;282:1934–1947.
23. Kahn MA, Stanton SL. Posterior colporrhaphy: its effects on bowel and sexual function. Br J Obstet Gynaecol 1997;104:82–86.
24. Porter WE, Steele A, Walsh P, et al. The anatomic and functional outcomes of defect-specific rectocele repairs. Am J Obstet Gynecol 1999; 181:1353–1359.
25. Kenton K, Shott S, Brubaker L. Outcome after rectovaginal fascia reattachment for rectocele repair. Am J Obstet Gynecol 1999;181: 1360–1364.
26. Paraiso MFR, Ballard LA, Walters M, et al. Pelvic support defects and visceral and sexual function in women treated with sacrospinous ligament suspension and pelvic reconstruction. Am J Obstet Gynecol 1996;175:1423–1431.
27. Field SM, Hilton P. The prevalence of sexual problems in women attending for urodynamic investigation. Int Urogynecol J Pelvic Floor Dysfunct 1993;4:212–215.
28. Weber AM, Walters MD, Schover LR, Mitchinson A. Sexual function in women with uterovaginal prolapse and urinary incontinence. Obstet Gynecol 1995;85:483–487.
29. Sarrel PM. Sexuality and menopause. Obstet Gynecol 1990;75:26S–30S.
30. Davila GW, Singh A, Karapanagiotou I, et al. Are women with urogenital atrophy symptomatic? Am J Obstet Gynecol 2003;188: 382–388.

Section III

Evaluation

Section III

Evaluation

Gamal M. Ghoniem

It is important for a physician evaluating pelvic floor disorders to differentiate between functional and anatomical abnormalities. Impressive genitourinary bulges very often do not explain a patient's symptoms. Many women with a large prolapse may not complain of a bulge, but of other symptoms such as urinary or fecal incontinence.

The goal of physical examination and subsequent investigation is to explain the symptoms, identify the causative mechanism and its risk factors, and finally propose treatment. Using sophisticated testing, it may be possible to even predict outcomes. For example, a patient with impaired bladder contractility on preoperative urodynamics, who is undergoing a sling procedure for stress urinary incontinence, may need to be taught clean self-catheterization after surgery. Preoperative counseling to predict such specific outcomes can help to avoid unsatisfied patients. Because pelvic floor disorders are mainly quality-of-life issues, it is imperative that physicians discuss with their patients their goals and expectations for the treatments offered to them.

The following chapters cover multitudes of evaluation. The history and physical examination are the starting building blocks to what investigation is needed next.

3-1

Primary Evaluation of the Pelvic Floor

Nathan Guerette, Dana R. Sands, and G. Willy Davila

Evaluation of the Urogynecology Patient

Proper evaluation of women with symptomatic pelvic organ prolapse and urinary incontinence requires an organized approach. A thorough history and physical examination along with basic office testing will often provide ample evidence to make a diagnosis and develop an effective treatment plan.

Urogynecology History

Urinary incontinence and pelvic organ prolapse are problems that cause social embarrassment and inconvenience. It is, therefore, critical to fully understand the patient's chief complaint and its effect on quality of life. Detailed questions should be asked regarding the degree to which the problem has impacted social functioning and living situations.

Specific questions will often help clarify the type of incontinence distressing the patient. Women with stress urinary incontinence (SUI), or urine loss associated with exertion, report urine loss with coughing, sneezing, or laughing. Urine loss during physical activity that increases intraabdominal pressure, such as running and sit-ups, also occurs frequently. Women with urge urinary incontinence (UUI), or urine loss associated with uninhibited detrusor muscle contractions, will typically also complain of urinary urgency, nocturia, and symptoms of frequency. These patients often report an urgency to void that occurs with little warning and cannot be repressed, with leakage often occurring on the way to the toilet. They should be asked how many daytime and nocturnal voids they average. More than eight voids per day and two voids per night is considered abnormal. Episodes of enuresis should be specifically elicited. Urge incontinence will often have triggers such as running water, handwashing, sexual intercourse, or emotional stress. Inquiry into these contributory factors will frequently be helpful in making a diagnosis. Women with mixed incontinence (MI) will have a combination of stress and urge symptoms. Elucidating which type of symptoms predominates will be helpful in later making therapeutic recommendations. A bladder diary, kept over 3 days, reporting voids, urgency events, incontinence events, and fluid intake can be very useful in clarifying often unclear and complex symptoms (Figure 3-1.1).

The degree of urine loss should also be quantified. The frequency of accidents should be noted. Pad use will often indicate symptom severity. The patient should be asked if pads are used and, if so, how many per day.

Voiding dysfunction is common in this population of patients. Questions about difficulty initiating the void, interrupted or slow voiding, needing to strain with voiding, a feeling of incomplete emptying, frequent urinary tract infections, postvoid dribbling, and double voids will aid in diagnosis and dictate need for a voiding study during urodynamics.

Patients with significant prolapse will typically complain of pelvic pressure or low back pain that worsens with prolonged standing. If the prolapse is exteriorized, they will often note a vaginal mass or bulge. Patients with prolapse involving the posterior vaginal wall often describe the need to digitally splint their vagina or manually extract stool in order to evacuate the rectum.

Sexual dysfunction frequently occurs in conjunction with incontinence and prolapse, and it should not be ignored. Although some symptoms may be primarily hormonal in nature (i.e., decreased libido, vaginal dryness) and others primarily neurologic (i.e., decreased vaginal sensation, secondary anorgasmic), there is frequently a combination of etiologic factors. In addition, parameters defining normal sexual function have not yet been determined because of varying individual characteristics, as well as cultural factors.

Symptoms related to fecal incontinence and defecatory dysfunction are discussed later in this chapter.

After the symptoms are well characterized, a complete medical history should be obtained. Focused questions to

Time	Urinations in toilet	Urge	Leakage accidents			Fluid intake 1 = teacup 2 = glass/can 3 = water bottle
			Urge	Stress	Other	
12 midnight - 2 a.m.	☐ ☐ ☐ ☐	☐ ☐	☐	☐	☐	☐1 ☐2 ☐3
2 - 4 a.m.	☐ ☐ ☐ ☐	☐ ☐	☐	☐	☐	☐1 ☐2 ☐3
4 - 6 a.m.	☐ ☐ ☐ ☐	☐ ☐	☐	☐	☐	☐1 ☐2 ☐3
6 - 7 a.m.	☐ ☐ ☐ ☐	☐ ☐	☐	☐	☐	☐1 ☐2 ☐3
7 - 8 a.m.	☐ ☐ ☐ ☐	☐ ☐	☐	☐	☐	☐1 ☐2 ☐3
8 - 9 a.m.	☐ ☐ ☐ ☐	☐ ☐	☐	☐	☐	☐1 ☐2 ☐3
9 - 10 a.m.	☐ ☐ ☐ ☐	☐ ☐	☐	☐	☐	☐1 ☐2 ☐3
10 - 11 a.m.	☐ ☐ ☐ ☐	☐ ☐	☐	☐	☐	☐1 ☐2 ☐3
11 a.m. - 12 noon	☐ ☐ ☐ ☐	☐ ☐	☐	☐	☐	☐1 ☐2 ☐3
12 noon - 1 p.m.	☐ ☐ ☐ ☐	☐ ☐	☐	☐	☐	☐1 ☐2 ☐3
1 - 2 p.m.	☐ ☐ ☐ ☐	☐ ☐	☐	☐	☐	☐1 ☐2 ☐3
2 - 3 p.m.	☐ ☐ ☐ ☐	☐ ☐	☐	☐	☐	☐1 ☐2 ☐3
3 - 4 p.m.	☐ ☐ ☐ ☐	☐ ☐	☐	☐	☐	☐1 ☐2 ☐3
4 - 5 p.m.	☐ ☐ ☐ ☐	☐ ☐	☐	☐	☐	☐1 ☐2 ☐3
5 - 6 p.m.	☐ ☐ ☐ ☐	☐ ☐	☐	☐	☐	☐1 ☐2 ☐3
6 - 7 p.m.	☐ ☐ ☐ ☐	☐ ☐	☐	☐	☐	☐1 ☐2 ☐3
7 - 8 p.m.	☐ ☐ ☐ ☐	☐ ☐	☐	☐	☐	☐1 ☐2 ☐3
8 - 9 p.m.	☐ ☐ ☐ ☐	☐ ☐	☐	☐	☐	☐1 ☐2 ☐3
9 - 10 p.m.	☐ ☐ ☐ ☐	☐ ☐	☐	☐	☐	☐1 ☐2 ☐3
10 - 11 p.m.	☐ ☐ ☐ ☐	☐ ☐	☐	☐	☐	☐1 ☐2 ☐3
11 p.m. - 12 midnight	☐ ☐ ☐ ☐	☐ ☐	☐	☐	☐	☐1 ☐2 ☐3

Figure 3-1.1. Bladder diary.

uncover contributing factors should be asked. Past urologic and gynecologic procedures such as incontinence surgery, urethral dilatations, radical hysterectomy, or radiation treatment might help explain the etiology of the patient's symptoms. A careful obstetric history including modes of delivery, fetal weights, use of forceps, large tears, or other indications of significant urogenital trauma are important. Neurologic conditions such as multiple sclerosis, Parkinson's disease, stroke, or previous back trauma also often contribute to incontinence and prolapse symptoms. Pulmonary pathology associated with a chronic cough is a risk factor for SUI and prolapse and will also challenge a repair. Moreover, the risk of all forms of urinary incontinence is increased by up to three times with a history of smoking.[1] Therefore, a smoking history is also important to obtain.

Other conditions associated with repetitive significant increases in intraabdominal pressure, such as chronic constipation and heavy lifting, are also important to inquire about because they can be associated with voiding dysfunction, pelvic neuropathy, and pelvic organ prolapse.

The patient's medications should be thoroughly reviewed to identify drugs that may be contributing to symptoms and to prevent untoward interactions. In particular, usage of medications with α-adrenergic activity (agonists such as phenylephrine enhance urethral sphincteric function; blockers such as prazosin relax the urethral sphincter) and cholinergic effects (procholinergic effects, from bethanechol, increase detrusor form; anticholinergic effects, from oxybutynin and tolterodine, relax the detrusor muscle) should be recorded. The patient's menopausal and estrogenation status should be addressed, because of the marked effect female hormones may have on urogenital function (Chapter 11-1). Likewise, the patient's diet and dietary supplements should be examined to identify bladder irritants that may worsen possible symptoms.

Colorectal History

Patients with complex disorders of evacuation will often have a confusing list of complaints, which can include both constipation and fecal incontinence. It is important, in the history of the present illness, to determine the patient's baseline bowel function. For patients with constipation, frequency and consistency of bowel movements should be ascertained. The sensation of the need to evacuate with significant straining leads the physician to consider outlet obstruction in the differential diagnosis. Infrequent bowel movements with no sensation of rectal fullness may lead to consideration of colonic inertia as a cause for constipation. It is always important to ask the patient if there is any prolapse of tissue from the anal canal. Often, patients will complain of severe hemorrhoidal prolapse, when in fact, the patient has full-thickness rectal prolapse. It is always important to determine whether there is any associated rectal bleeding. Obviously, occult malignancy must be ruled out in any patient with changes in bowel habits. For patients complaining of incontinence, the duration and severity of the symptoms are important initial questions. A detailed obstetric history as well as any history of anorectal surgery or inflammatory bowel disease can provide important insight into the etiology of fecal incontinence.

The independently validated Cleveland Clinic Florida Fecal Incontinence Score (0–20) provides an objective measurement of severity of incontinence both before and after intervention[2,3] (Table 3-1.1). This score takes into account the frequency of incontinence to solid, liquid, and gas, as well as the need to wear protective undergarments and the impact on the patient's lifestyle. A detailed bowel history can also lead the physician to a better understanding of the etiology of the incontinence. Passive soiling, the unwanted loss of stool without the patient's awareness, may indicate internal sphincter pathology. Urge fecal incontinence, the unwanted loss of stool despite attempts to inhibit defecation, may signify external sphincter pathology. Post-defecation soiling, the unwanted loss of stool immediately after bowel movement with normal continence other times is often a sign of incomplete emptying or perianal disease. Stool consistency is an important component of continence. Frequent loose bowel movements can pose a particular challenge for those patients with borderline continence. Oftentimes, improvement in the consistency of the stool in these patients can provide great symptomatic relief. In any patient with pelvic floor dysfunction, it is important to assess the presence of anterior compartment symptoms, such as urinary incontinence or prolapse.

A detailed medical and surgical history is important in the evaluation of any patient. History of metabolic disturbances such as hypothyroidism or hypercalcemia may provide insight into the etiology of constipation. History of inflammatory bowel disease and proctitis will often contribute to a patient's symptoms of incontinence. Diabetes and peripheral neuropathy can also have an important role in incontinence. As mentioned previously, a detailed birth history with particular attention to previous episiotomy, high-birth-weight children, prolonged labor, and the use of forceps for delivery is of paramount importance. Surgical history, with a focus on previous anorectal surgery or bowel resection, will obviously contribute to the evaluation of these patients. Medications can also have a role in a patient's symptoms. Anticholinergic medications, narcotics, and calcium supplements will often contribute to constipation symptoms. Any medications that cause diarrhea will often worsen the severity of incontinence.

Table 3-1.1. Fecal incontinence scoring system

Type of Incontinence	Never	Rarely <1/mo	Sometimes <1/wk >1/mo	Usually <1/d >1/wk	Always >1/d
Solid	0	1	2	3	4
Liquid	0	1	2	3	4
Gas	0	1	2	3	4
Wears pad	0	1	2	3	4
Lifestyle alteration	0	1	2	3	4

Urogynecology Physical Evaluation

A complete physical examination, including a thorough lower neurologic examination, and gynecologic examination should be performed for every patient presenting with complaints of prolapse or incontinence. The general examination should look for evidence of pulmonary conditions that might contribute to the symptom profile. Hernias and masses should be excluded on abdominal examination, and costovertebral angle tenderness must be ruled out.

Neurologic Examination

Multiple sclerosis, Parkinson's disease, dementia, minor cerebrovascular accidents, and numerous additional neurologic processes can have genitourinary dysfunction as the presenting symptom. Back trauma and surgery, spinal stenosis, peripheral neuropathy, and damage to pelvic nerves also predispose women to incontinence and prolapse. Ambulation and mobility problems may jeopardize an individual's ability to reach a toilet, and should be identified. During the examination, the patient's mental status should be assessed. Cognitive impairment is a major impediment to behavioral modification and other forms of therapy. Realistic expectations regarding the ability to achieve continence in a demented patient should be addressed with family members and care providers.

Sensory and motor function of L1–S4 are of particular importance. Testing for light touch and pinprick over the perineum, inner and anterior thighs, and pretibial area will adequately test sensory function. The motor examination should include extension and flexion of the hip, knee, and ankle. The intact sacral reflexes are confirmed by checking the bulbocavernosus reflex and anal wink. The bulbocavernosus reflex is tested by stroking the labia or tapping the clitoris and looking for contraction of the bulbocavernosus and ischiocavernosus muscles. The anal wink is performed by stroking the perianal skin and noting contraction of the external anal sphincter. These tests should be performed bilaterally. Sacral reflexes can be difficult to evaluate and are often absent in neurologically intact women.

Pelvic Examination

A complete pelvic examination is necessary for proper evaluation. The labia and vagina should be inspected for signs of atrophy, because hypoestrogenemia may contribute to incontinence and prolapse symptomatology and impair proper healing after surgery. A vaginal mucosal maturation index can often help with this assessment. Any ulcerations of the mucosa should be noted, particularly in women who have used a pessary. A bimanual examination is important to rule out pelvic masses and assess for pain. A rectal examination should be performed, and the resting tone and power of the rectal squeeze should be recorded. Any rectal mucosal prolapse should prompt further investigation. Impaction should be ruled out, because it is strongly associated with voiding dysfunction in elderly women.

Prolapse Examination

The prolapse examination should be site specific, carefully looking for support defects of the uterus, vaginal apex, anterior vaginal wall, and posterior vaginal wall. This is performed using a standard Grave's speculum and needs to be done in a methodical manner. Typically, we evaluate uterine and vault support first by having the patient Valsalva with either the cervix or vaginal apex isolated with the speculum (Figure 3-1.2). The speculum is then split and the anterior and posterior vaginal compartments are evaluated independently (Figure 3-1.3). The vault should be supported during anterior and posterior wall evaluation to accurately develop an understanding of the patient's specific pelvic floor weaknesses. It is often necessary to have the patient stand to accurately assess their degree of prolapse. The prolapse should be documented with a standardized system, as described below.

Classification of Pelvic Organ Prolapse

A standardized description and classification of pelvic organ prolapse is critical for accurate planning of treatment, improving physician communication, and advancing research. Two systems are currently accepted, the Baden–Walker Halfway system and the Pelvic Organ Prolapse Quantification (POP-Q) system. The Baden–Walker system has good interobserver validity and is simpler and more commonly understood.[4] In this classification system,

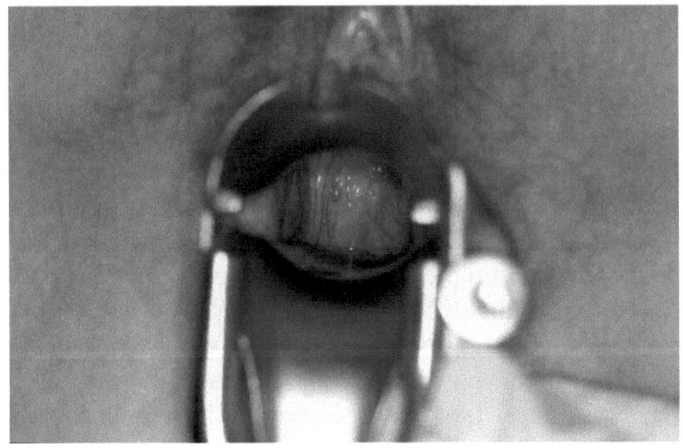

Figure 3-1.2. Prolapse evaluation with speculum.

prolapse of each vaginal compartment (cystocele, enterocele, rectocele, uterus, and vaginal vault) is isolated and the most dependent position during maximum straining is recorded. Each compartment is recorded in accordance to its relationship with the hymen and the mid-vaginal plane (Figure 3-1.4).

The POP-Q system is the currently established International Continence Society standard for description of female pelvic organ prolapse.[5] In this system, six points are used to make site-specific measurements during maximal straining (Figure 3-1.5). This replaces previously used descriptive terms such as cystocele, rectocele, and enterocele. The location of the points is recorded in centimeters above (negative numbers) or below (positive numbers) the hymen. The points are defined as follows:

Anterior vagina

Point Aa: anterior midline vaginal wall 3 cm proximal to the external urethral meatus (range, −3 to +3 cm).

Point Ba: the most dependent position of the anterior vagina up to and including the anterior vaginal fornix or vaginal cuff.

Posterior vagina

Point Ap: posterior midline vaginal wall 3 cm proximal to the hymen (range, −3 to +3 cm).

Point Bp: the most dependent position of any part of the posterior vagina up to and including the posterior vaginal fornix or the vaginal cuff.

Apical vagina

Point C: most dependent edge of the cervix or vaginal cuff after total hysterectomy.

Point D: apex of the posterior fornix in a woman who has a cervix or the level of attachment of the uterosacral ligaments to the posterior cervix. It is useful to differentiate between loss of apical vaginal and uterine support and cervical hypertrophy. This point is not used in a woman who has had a hysterectomy.

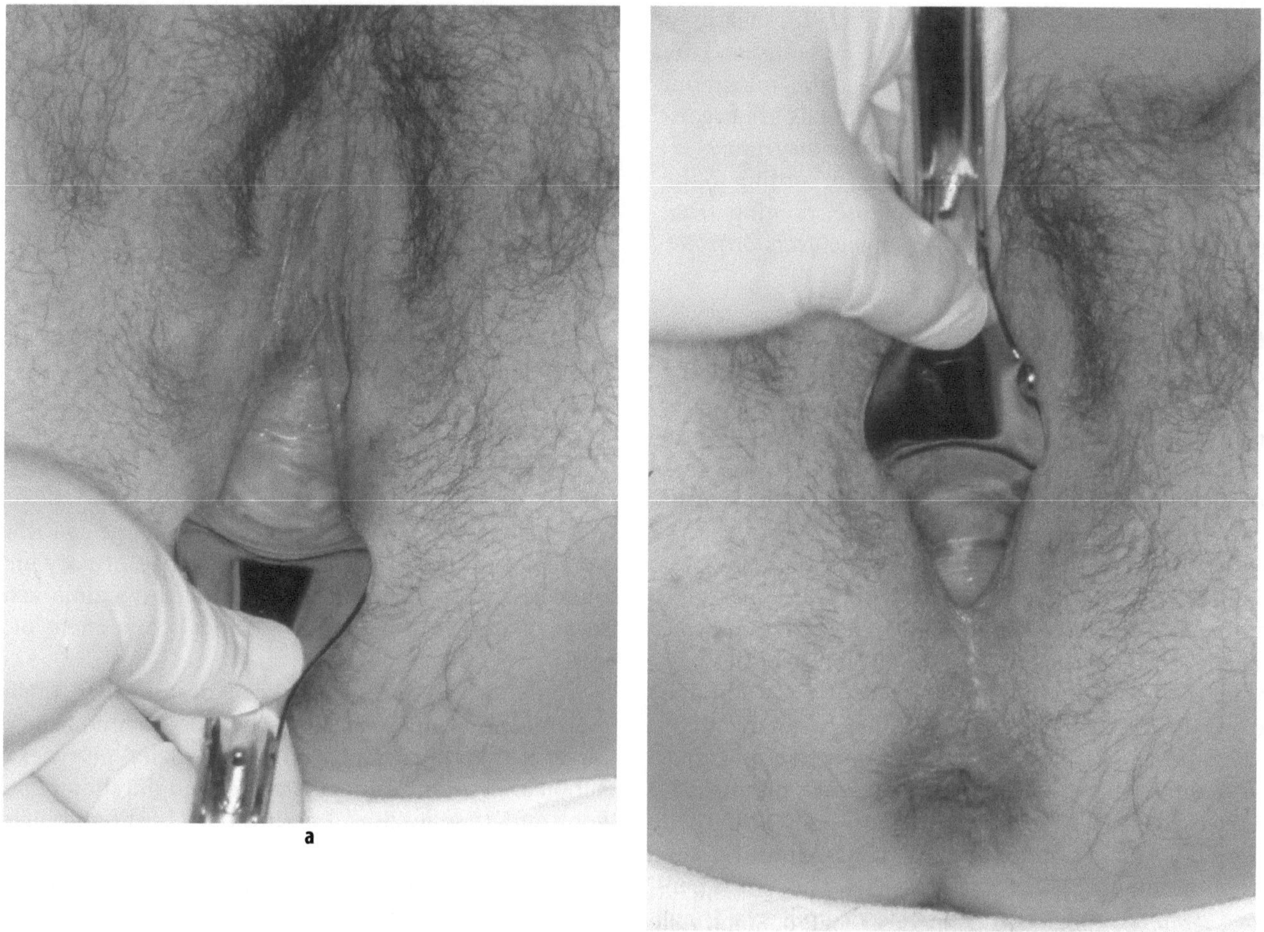

Figure 3-1.3. a, Anterior vaginal compartment evaluation. b, Posterior vaginal compartment evaluation.

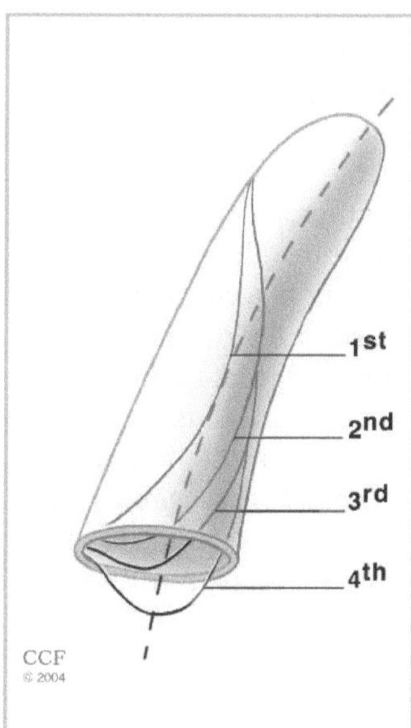

Figure 3-1.4. Cystocele Baden–Walker vaginal profile. (Reprinted with the permission of The Cleveland Clinic Foundation.)

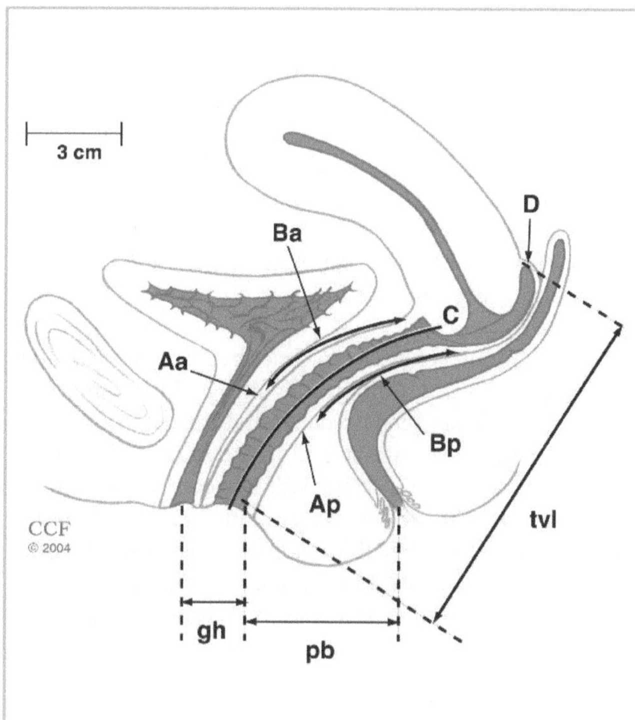

Figure 3-1.5. POP-Q Scoring System. (Reprinted with the permission of The Cleveland Clinic Foundation.)

Additional measurements include the genital hiatus (GH), measured from the urethral meatus to the posterior hymen in the midline, the perineal body (PB), measured from the midline posterior hymen to the midanus, and the total vaginal length (TVL), recorded as the maximum vaginal length from the hymen. Measurements are organized as a 3 × 3 grid and/or a vaginal profile diagram.

The overall POP-Q stage is determined according to the most dependent portion of the prolapse, ranging from stage 0 (no prolapse) to stage IV (complete eversion).

Table 3-1.2. Treatment of stress urinary incontinence algorithm based on urethral evaluation

		Function (UPP, VLPP)	
		Normal	Low
Mobility (Q-tip test)	>30	Kegels, physio, pessary, TVT, Burch, TOS	Sling (traditional),
	<30	Kegels, physio	Bulking agents

UPP, urethral pressure profilometry; VLPP, Valsalva leak point pressure; TVT, tension-free vaginal tape; TOS, transobturator sling; physio, pelvic floor rehabilitation.

Urethral Examination

Assessment of urethral function should be included as part of the initial examination. The anterior vagina and urethra should be palpated to evaluate for tenderness, masses, or a urethral discharge that might suggest a diverticulum or other urethral pathology.

An empty supine stress test (ESST) should be performed. This is done with the patient in the lithotomy position after a recent void. The labia are separated and the woman is asked to cough or Valsalva, and any urine leakage is noted. Lobel and Sand[6] conducted a 1996 study and reported that the ESST had a 98% positive predictive value for diagnosing SUI documented on urodynamic testing. We use this test to screen for the presence of intrinsic sphincteric deficiency.

Another examination technique that is useful for evaluation of urethral support is the Q-tip test. This test is performed by placing a sterile, lubricated cotton swab into the urethra to the level of the bladder neck with the patient in the flat dorsal lithotomy position. She performs a Valsalva maneuver and the maximum angle of deflection is recorded with a protractor. An angle of deflection of positive 30 degrees or more from the horizontal is considered indicative of urethral hypermobility. Although a high strain angle is associated with the presence of SUI, there is a large overlap between incontinent and continent women. Because of this, the Q-tip test should never be used as a single factor in making a diagnosis, but rather as an adjunct to the physical examination. It is often useful in choosing a treatment regimen. We use the Q-tip test along with testing of urethral sphincteric function to make therapeutic recommendations in women with SUI (Table 3-1.2).

Other techniques have been described to assess bladder neck support. Cystourethrography using contrast material or a bead-chain has been used. Sagittal images are taken at rest and with maximum straining. Rotational descent of the urethra can thus be calculated. The contrast technique also allows visualization of the proximal urethra. This can help determine whether there is funneling of the bladder neck and may diagnose a urethral diverticulum.

Perineal and vaginal ultrasound techniques have also been described to obtain similar information. In the hands of experienced operators, sonography seems to produce similar results without the risk and complication of fluoroscopy and contrast materials.

Assessment of Pelvic Floor Function

Weakness of the pelvic floor muscles is a large contributor to the occurrence of pelvic organ prolapse, urinary incontinence, and fecal incontinence. Proper dynamic assessment of muscle function is a critical component of the initial evaluation.

Numerous methods have been described to quantify dynamic pelvic floor function. The most common and simplest techniques currently in use are digital pelvic examination scales, which note the power and duration of pelvic floor muscle squeeze and perineal lift. Intra- and interobserver reliability have been shown to be good, using the digital assessment. This method's scientific usefulness is limited, however, because of its subjectivity.

Perineometry using either electromyography pads or pressure sensors on a vaginal plug are also frequently used methods. These are also limited because they are indirect assessments of muscle activity and subject to numerous sources of artifact such as contraction of the rectus abdominus and/or adductor muscles.

Weighted cones are also useful for evaluation of the pelvic floor. Pelvic floor strength is calculated by the ability of the patient to retain the progressively weighted cones. Cone retention, however, is often poorly correlated with physical findings. This is attributed to retention of the cone in the posterior pelvis of women with severe prolapse.

The Kolpexin® Pull Test, a new and recently validated dynamic pelvic floor assessment tool in use at Cleveland Clinic Florida, attempts to overcome the shortcomings of the previously described methods and give a more direct, objective measurement of pelvic floor strength.[7] To perform the pull test, the patient is placed in the dorsal lithotomy position. A lightly lubricated Kolpexin® sphere with attached force-gauge is then placed in the vagina above the level of the levators. The patient is then asked to perform a maximum pelvic floor contraction and the sphere is withdrawn. The maximum force generated is recorded in pounds or kilograms to three significant digits. This test is rapid, inexpensive, and the most direct evaluation tool for pelvic floor strength currently available.

Diagnostic Testing

Laboratories

All patients should have a clean-catch urinalysis at their initial visit. If the results are suspicious for infection, the urine should be sent for culture and sensitivity. Urine cytology is obtained in patients with persistent hematuria in the absence of infection. The utility of urine cytology as a simple test, however, has recently been called into question. Additional laboratory tests are rarely needed.

Simple Cystometry and Uroflowmetry

To objectively confirm a diagnosis, simple cystometry and uroflowmetry can be performed in most gynecology offices. The patient begins the testing with a full bladder. She voids in the most natural way possible, and the voiding time and volume are documented. She should be asked if this represents a typical void. If she indicates it was not, the test should be repeated.

The patient is then placed in the lithotomy position, a sterile preparation is performed, and a catheterized postvoid residual (PVR) is obtained. We consider a PVR of more than 100 mL to be abnormal, but it should be kept in mind that standardized normal values have not been established and a "normal" PVR in elderly individuals could be more than 100 mL. A large syringe is then attached to the catheter and the patient is asked to sit up or stand. The bladder is slowly filled with sterile water. Care must be taken not to fill the bladder too quickly or with cold water because these cause artifactual detrusor contractions even in normal women. A typical filling rate is 50 to 70 mL/min. Initial filling sensation, normal desire (when the patient would first consider voiding), and maximum cystometric capacity is recorded. Normal bladder capacity is 400 to 650 mL although asymptomatic women are not infrequently outside this range. The water level in the syringe should be closely monitored. A bladder contraction will cause the level to increase and may induce urgency and leakage. At capacity the catheter is removed, and provocative maneuvers, such as coughing, heel-bouncing, or running water are done to induce an incontinence episode. Communication throughout testing is important to determine whether the patient's symptoms are replicated, especially during a urine-loss event.

Cystoscopy

Irritative voiding, painful bladder symptoms, and persistent hematuria should be evaluated with cystoscopy. Evidence of neoplasms, inflammation, stones, or trabeculations should be carefully noted. A zero-degree scope can be used to assess urethral coaptation, presence of a urethral mass or diverticulum, and bladder neck mobility.

Perineal Pad Tests

Perineal pad testing can be helpful in evaluating a patient complaining of urinary incontinence, when objective urine loss has not been demonstrated on routine office testing. The patient should be observed over a long enough period of time to witness their symptoms, with most common time intervals of either 1 hour or 24 hours.

The 1-hour pad test is usually performed in the office. The patient's bladder is prefilled, either with a catheter or by having the patient drink water, until it feels full or a set volume is reached (i.e., 250 mL). The patient then performs a standard set of activities for the hour. These activities must be representative of routine daily life such as bending, coughing, and climbing stairs. The pad must be weighed before and after the test and the weight gain represents the volume of fluid loss. Every effort should be made to prevent drying of the pad. Ideally, waterproof underclothing should be worn during the test period and, once removed, the pad should be sealed in a plastic bag until it is weighed. It is important to note that a weight gain of up to 1 g over an hour can be accounted for by sweat and vaginal discharge.

The 24-hour pad test is performed at home. This test is not as standardized as the 1-hour test. The patient is sent home with a preweighed pad and is instructed to wear it for a set period of time (usually 4–6 hours) and then seal it in a plastic bag. The pad is then reweighed to estimate fluid loss.

Changes in weight often do not correspond well to patient and urodynamic assessment of incontinence severity. A 1998 study by Ryhammer et al.[8] showed no significant difference between increases in pad weight between self-reported incontinent and continent women over a 24-hour period. This and the time required to perform the tests have relegated pad weights to a second-line diagnostic option. They can be useful, however, as an assessment of severity or to assess improvement after treatment.

In women in whom office evaluation fails to demonstrate urine loss, Pyridium or another urine-coloring agent can be used to make objective urine loss on a pad more obvious. The woman ingests a Pyridium tablet and then wears a pad for a set period of time. The major shortcoming of the Pyridium pad test is the high false-positive rate. Fifty-two percent of healthy, continent women in a study by Wall et al.[9] had a positive test.

Indications for Complex Testing

Many women with mild to moderate incontinence and/or prolapse can be satisfactorily treated following the evaluation techniques described above. The impact of the patient's symptoms on her lifestyle can often be ameliorated with low-risk interventions such as medications, physical therapy, and behavioral changes. If, however, the patient does not acceptably improve with conservative therapy, surgery is being considered, or the diagnosis is uncertain, referral to a center that has the capability of performing multichannel urodynamics and other complex lower urinary tract evaluation testing is warranted. Other indications for multichannel urodynamics include the following:

- Previous failed surgery
- Exteriorized prolapse
- Complex urinary incontinence symptoms
- Intrinsic sphincter deficiency suspected (severe SUI)
- Urinary retention/abnormal voiding
- Complex medical problems
- Pelvic reconstructive surgery planned

Colorectal Physical Evaluation

A detailed physical examination is conducted on all patients. Careful abdominal examination, looking for any evidence of tenderness, masses, incisions, or hernias is an important initial component. Anorectal examination is first conducted in the prone jackknife position. A proper light source is essential. Initial inspection of the perianal skin can reveal any dermatologic conditions such as excoriation or dermatitis. Previous scars from either episiotomy or anorectal surgery are identified. The anus is assessed, and it is noted if the anus is closed or patulous in the resting state. Assessment of sensation is the next step in the examination. Any loss of either sharp or dull sensation is noted in the perianal area. The presence or absence of the anal wink reflex is also ascertained. Absence of the anal wink may signify pudendal neuropathy.

Digital rectal examination allows gross estimations of the patient's resting anal tone and voluntary squeeze pressure. It is, however, important to exclude accessory muscle contraction when assessing squeeze pressure. Bidigital anovaginal examination is used to estimate perineal bulk and grossly identify sphincter defects. It is with this technique that the rectovaginal septum is also evaluated for any defects or masses. When evaluating any patient for pelvic floor pathology, it is important to conduct a portion of the examination with the patient straining or performing a Valsalva maneuver. This will help to identify increased perineal descent, or the presence of a rectocele, cystocele, or perineal hernia. Frequently, it will also be necessary to repeat this portion of the examination, with the patient in the squatting position or sitting on the toilet. Although the prone jackknife position allows for excellent visualization of the anal canal, pelvic organ prolapse can be underestimated with this technique.

After a careful history and physical examination, the physician is able to diagnose most posterior compartment pelvic floor disorders. Focused anorectal physiologic evaluation and radiographic studies are then obtained to complete the evaluation and determine a treatment plan.

References

1. Bump RC, McClish DK. Cigarette smoking and urinary incontinence in women. Am J Obstet Gynecol 1992;167:1213–1218.
2. Jorge JMN, Wexner SD. Etiology and management of fecal incontinence. Dis Colon Rectum 1993;36:77–97.
3. Rothbarth J, Bemelman WA, Meijerink WJ, et al. What is the impact of fecal incontinence on quality of life? Dis Colon Rectum 2001; 44:67–71.
4. Baden WF, Walker TA, Lindsey JH. The vaginal profile. Tex Med 1968; 64:56–58.
5. Bump RC, Mattiasson A, Bo K, et al. The standardization of terminology of female pelvic organ prolapse and pelvic floor dysfunction. Am J Obstet Gynecol 1996;175:10–17.
6. Lobel RW, Sand PK. The empty supine stress test as a predictor of intrinsic urethral sphincter dysfunction. Obstet Gynecol 1996;88:128–132.
7. Guerette N, Neimark M, Kopka SL, Jones JE, Davila GW. Initial experience with a new method for the dynamic assessment of pelvic floor function in women: the Kolpexin Pull Test. Int Urogynecol J Pelvic Floor Dysfunct 2004;15:39–43.
8. Ryhammer AM, Laurberg S, Djurhuus JC, Hermann AP. No relationship between subjective assessment of urinary incontinence and pad test weight gain in a random population sample of menopausal women. J Urol 1998;159:800–803.
9. Wall LL, Want K, Robson I, Stanton SL. The Pyridium pad test for diagnosing urinary incontinence. J Reprod Med 1990;35:682–684.

3-2

Urodynamics

Gamal M. Ghoniem and Usama M. Khater

Urodynamic investigation is a functional assessment of the lower urinary tract, which is performed to provide an objective pathophysiologic explanation of urinary tract dysfunction symptoms. Urodynamic studies comprise a series of tests, and the information provided by these tests is useful in establishing the etiology of the dysfunction and selecting the most appropriate intervention. The appropriate test should be selected and performed in an attempt to answer a specific question regarding the function to be evaluated. Urodynamic evaluation is an integral part of the evaluation of patients with voiding dysfunction. Before urodynamic investigation, a medical history, physical examination, and voiding diary should be completed. This information is necessary to select the appropriate studies and anticipate which events might occur during the urodynamic investigation.

Uroflowmetry

Uroflowmetry is the measurement of urine flow rate over time, and is the most commonly used urodynamic study. It is performed as an initial screening test when voiding dysfunction is suspected, and the results may prompt further investigations. Uroflowmetry is performed by having the patient void into a specially designed commode that allows the volume voided to be measured over time. Uroflowmeters use various measurement techniques, such as the rotating disk method, electronic dip, weight transducers, and the gravimetric method. The rotating disk method is the most commonly used; it directs the voided urine onto a rotating disk, which produces a proportionate increase in the disk's inertia. The power required to keep the disk rotating at a constant rate is measured, thus allowing calculation of the flow rate. The patient is instructed to void normally in either a sitting or standing position with a comfortably full bladder, in as much privacy as possible to remove the inhibitory effect of the test environment. Privacy, voided volume, and patient's age and sex are factors affecting the parameters of uroflowmetry. Reported uroflowmetric parameters include total voided volume (V), maximum flow rate (Qmax), time to Qmax, voiding time (Vt), flow time, and average flow rate (Qave). The pattern of the flow curve is also observed. Voided volume represents the total voided volume as a single micturition event. The maximum flow rate is measured during a single void, and the time to reach this value is the time to Qmax. Flow time is the time over which measurable urine flow occurs. Average flow rate is calculated by dividing the voided volume over the flow time. Voiding time is the total duration of micturition including interruptions. When voiding is completed without interruption, voiding time is equal to flow time.[1] The voided volume should be at least 150 mL and preferably 200 mL (50 mL for a child), because Qmax increases by 5.6 mL/s with each 100-mL increase of voided volume, but only up to a voided volume of 200 mL. In young males, Qmax should be 15 to 25 mL/s. Women typically void with a slightly higher Qmax (5–10 mL/s) for a given volume than men.[2] A normal flow pattern is a continuous bell-shaped smooth curve with rapidly increasing flow rate reaching Qmax within one-third of the total voiding time (Figure 3-2.1). The first 45% of voided volume is reached before Qmax is reached. The overall appearance of the flow pattern may disclose abnormalities. An irregular pattern or curve with multiple peaks represents intermittent flow in which downward deflections do not reach 2 mL/s. This is often attributable to abnormal straining superimposed on a detrusor contraction. An interrupted pattern is consistent with intermittent flow in which downward deflection reaches 2 mL/s or less, leading to several micturitions separated by 2 to 20 seconds, often caused by straining in the absence of a detrusor contraction. Obstructed flow is characterized by prolonged flow time, sustained low flow rate, low Qmax and Qave, and a plateau-shaped curve (Figure 3-2.2). There is slow initial rise with an increased time to achieve Qmax, but because the voiding time is longer, Qmax may be seen relatively earlier.[3] The flow pattern of a patient with urethral obstruction is a flat-topped curve with decreased Qmax, which is reached quickly, but remains at the same level for most of micturi-

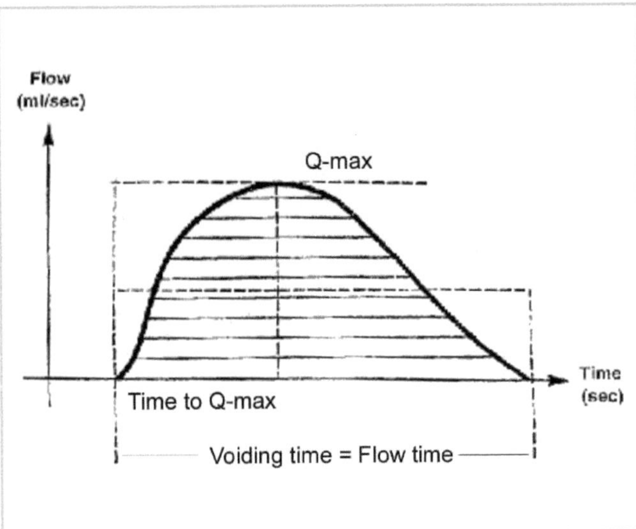

Figure 3-2.1. Normal flow pattern.

tion. The curve is flat and unbroken, with a large part of the volume voided at a constant Qmax.[4] Urinary flow rate provides useful information about whether there is outflow obstruction, especially in males. A flow rate greater than 40 mL/s is considered superflow. It may be attributable to decreased outlet resistance,[5] and is common in women, particularly those with genuine stress urinary incontinence (SUI) in which outlet resistance is reduced and in those with marked bladder activity. Although low Qmax may indicate urinary outlet obstruction, measurement of the flow rate alone has limited value when determining whether obstruction is present in a particular patient. To provide detailed information, uroflowmetry can be combined with a measurement of the postvoid residual volume (PVR). Postvoid residual volume is the bladder volume immediately after voiding. It reflects bladder contractility, and is an excellent assessment of bladder emptying. One method for measuring PVR is by ultrasound, with a large stationary machine, or, more commonly, using a small portable ultrasound unit. Transurethral catheterization immediately after a voluntary void is the most accurate method of measuring PVR; however, it is invasive and associated with a small risk of infection. Residual volume in an adult can be up to 25 mL; a residual volume of more than

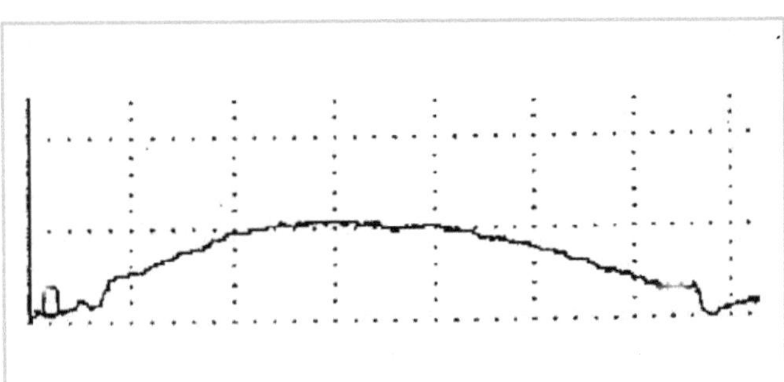

Figure 3-2.2. Obstructed flow pattern on uroflowmetry.

100 mL warrants surveillance or treatment. In normal children, residual volume is less than 10% of bladder capacity.[6] Residual volume may be overestimated in patients with vesicoureteral reflux, hydroureteronephrosis, or bladder diverticulum.

When a normal Qmax and normal voided volume without residual urine are present, infravesical obstruction or reduced contractility is unlikely. Several nomograms have been used to interpret measured flow rates. Siroky's nomogram is used for males; the Liverpool nomograms have scales for males and females. A peak flow more than the 90th percentile on the Liverpool nomogram may indicate detrusor instability.[7] A low Qmax in the presence of significant residual volume indicates obstruction. A low Qmax without significant residual volume indicates reduced contractility. However, a pressure flow study can provide more accurate diagnosis and differentiation of the two conditions.

Cystometry

Cystometry is a measure of bladder pressure in response to filling. A cystometrogram (CMG) is a graphic representation of pressure as a function volume: the *x* axis represents volume and the *y* axis represents pressure. Filling CMG is an excellent representation of the passive properties of the detrusor (i.e., viscoelastic properties). Several parameters are measured by this investigation: bladder filling sensation, stability, compliance, capacity, control over micturition, detrusor contractility, and emptying.

Single-channel Cystometry

Cystometry can be performed as a single-channel study in which the intravesical pressure [the pressure within the bladder (Pves)] is measured during filling and storage. It consists of a Foley catheter in the bladder attached to a wide-mouthed syringe without plunger and held at the level of the superior edge of the pubic symphysis. Saline is incrementally poured into the syringe and thus by retrograde flow into the bladder. The height of the saline column indicates the pressure of the tonic segment of the cystometric curve. A detrusor contraction during bladder filling results in a rapid increase in the saline column. With this technique, the monitored pressure is the intravesical pressure (Pves). This reflects the cumulative effects of all sources of pressure on the bladder, namely, intraabdominal pressure (i.e., the pressure surrounding the bladder) and detrusor pressure (i.e., the pressure derived from active and passive bladder wall smooth muscle activity). With single-channel recording, it is difficult to ascertain the exact source of the increase in Pves, caused by either detrusor contraction or increased intraabdominal pressure (e.g., cough or patient movement). To distinguish the source of the increased Pves, the patient should be asked

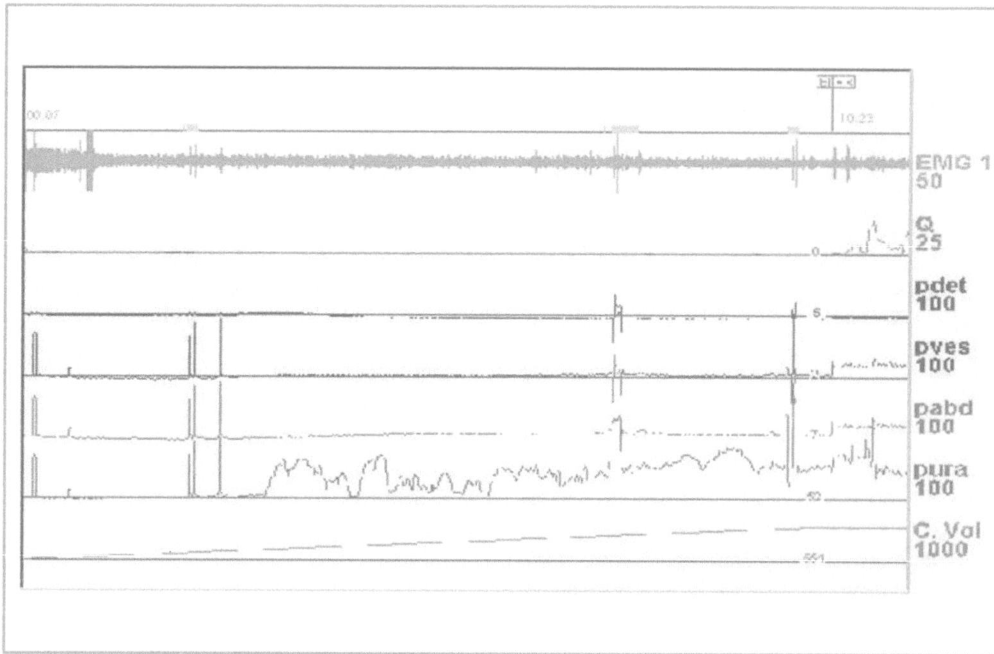

Figure 3-2.3. Multichannel cystometry showing normal CMG.

to hold his or her breath, or multichannel cystometry should be performed.

Multichannel Cystometry

During multichannel cystometry, concurrent measurements of intravesical (Pves), abdominal (Pabd), and urethral pressures (Pura) are performed (Figure 3-2.3). Detrusor pressure (Pdet) is estimated by subtracting Pabd from Pves. Electromyography (EMG) is also monitored.

Intravesical pressure represents the action of two forces acting upon the bladder: Pdet and Pabd (Figure 3-2.4). Detrusor pressure is a component of Pves that is created by passive and active forces in the bladder wall. Passive forces are produced by the viscoelastic property of the bladder wall, and active forces are created by detrusor contraction and tone. Exterior forces act upon the bladder as Pabd. Therefore, Pves = Pdet + Pabd, and Pdet = Pves − Pabd. In a nearly empty bladder, Pdet = 0 as in the situation Pves = Pabd (Pabd may be slightly greater than Pves because of rectal balloon compliance).

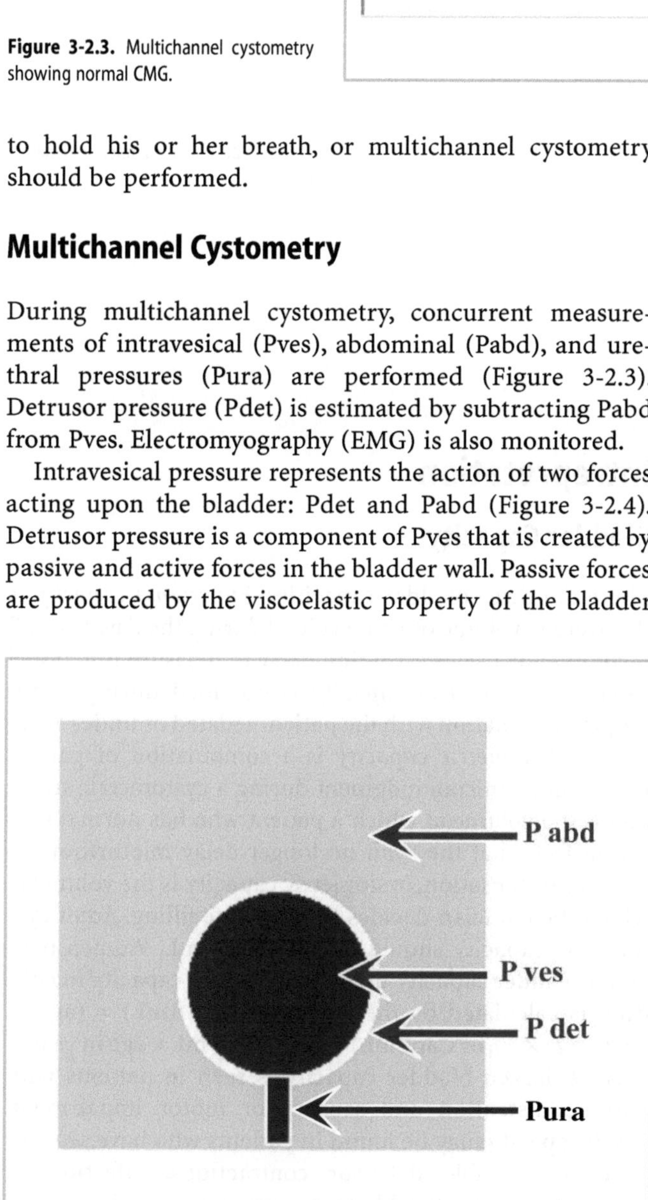

Figure 3-2.4. Urodynamic algebra.

Technical Aspects

Urodynamic laboratories use various instruments and techniques (Figure 3-2.5). There are a variety of catheters, such as microtip, air-charged, fiberoptic, and water-infused. Our department uses water-infused catheters. Microtip catheters are primarily used by urogynecologists. Intravesical pressure is measured by placing a catheter into the bladder that both fills the bladder and detects the pressure. The catheter has two lumens: one for filling and another for monitoring pressure. A triple-lumen catheter is used when urethral pressure is simultaneously monitored. Because the rectum and bladder are in close proximity and the intraabdominal pressure experienced by both organs is equal, abdominal pressure is measured with a rectal balloon catheter. Vaginal pressure monitoring may be used in women; however, the vagina is subject to many anatomical changes that may alter pressure transmission. The vagina should be used if there are anatomical changes of the rectum. The rectal tube should be soft and moderately sized, with a maximally distensible balloon. After instrumentation, the bladder is filled. Historically, both liquid and gas have been used as the filling medium. Gas is inexpensive and allows rapid filling. However, it is unphysiologic, compressible, and easily provokes detrusor overactivity, and is not suitable for studying voiding and leakage. The liquid filling medium may be saline, water, or radi-

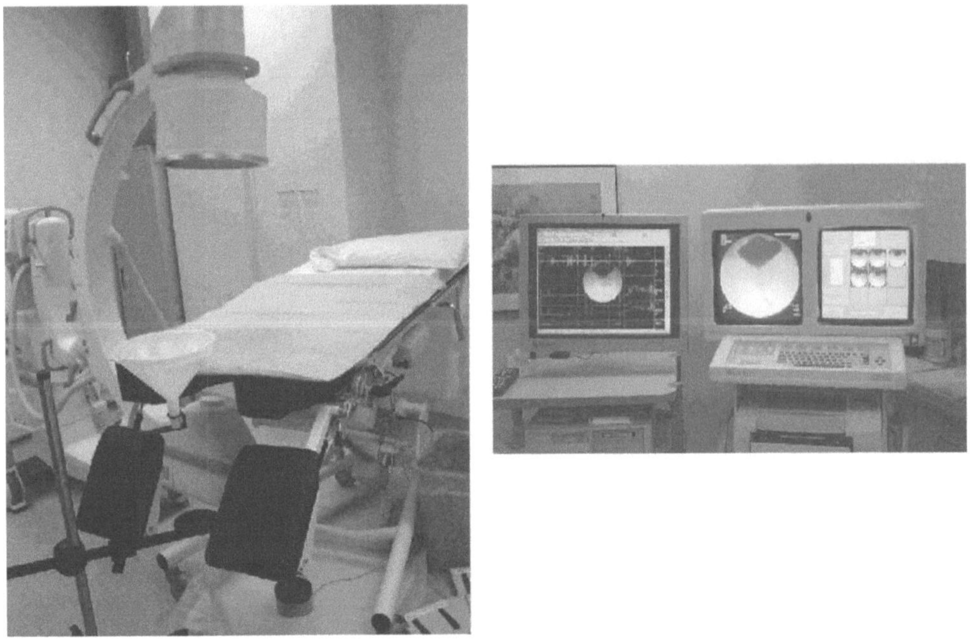

Figure 3-2.5. Multichannel urodynamic unit.

ographic contrast. The bladder can be filled either antegradely or retrogradely. The latter is used more often in the office setting because of time considerations. The temperature of the liquid should be either room or body temperature. According to the International Continence Society (ICS), the filling rate is considered fast (>100 mL/min), medium (10–100 mL/min), or slow (<10/min).[6] Natural bladder filling averages 1 to 2 mL/min. Rapid infusion may induce involuntary contractions or give the appearance of decreased compliance. A rapid infusion rate is used to provoke occult detrusor instability, to provoke contractions in a patient who did not demonstrate them at a slower infusion rate, or to determine the presence of sensation in a large bladder.[8] Slow infusion is used for children and adults with small functional capacity or unusual sensitivity to filling at medium filling rates. At Cleveland Clinic Florida, we use room-temperature sterile water as a filling medium, with a filling rate of 60 mL/min in adults and 30 mL/min in patients with suspected overactive (unstable) bladder. A triple-lumen 7-French catheter is used for filling and pressure measurements. Catheter sizes larger than 8 French are discouraged, because they may impede voiding. The patient's bladder is filled in the supine or sitting position, and the patient is then asked to void in either a standing or sitting position.

Intrepretation

Bladder Capacity

There are three bladder capacities. *Functional capacity* is the average volume of urine voided during the day. The ICS recommends use of a voiding diary to measure functional capacity. *Anatomical capacity* is obtained during endoscopic examination with the patient sedated or under anesthesia. *Cystometric capacity* is a combination of patient report and clinician judgment during a cystometric study, and is the volume at which a patient who has normal sensation feels that they can no longer delay micturition. In absence of sensation, cystometric capacity is the volume at which the clinician decides to terminate filling. Adult cystometric capacity should be 300 to 600 mL. Women have larger bladder capacity than men. Bladder capacity in children is calculated by the equations, $\text{Cap (mL)} = (\text{age in years} + 2) \times 30$, or $\text{Cap (mL)} = 30\ \text{mL} + 30\ \text{mL} \times \text{age in years}$.[9]

An enlarged bladder capacity is seen in patients with chronic distention and sensory or motor impairment. Small capacity may be found in patients who have sensory urgency, unstable detrusor contractions, infection or inflammation, or after bladder surgery or radiation, and is related to low compliance.

Compliance

Compliance is change in volume per change in pressure, and is expressed in milliliters/centimeter water. It represents bladder distensibility and is attributed to the viscoelastic properties and spherical shape of the bladder. Absence of involuntary contractions during filling also contributes to bladder compliance. As a result, the detrusor pressure remains low throughout bladder filling. Normally during bladder filling, the detrusor pressure does not exceed 5 to 10 cm H_2O greater than the starting pressure. Therefore, the kidneys are protected.

The CMG is divided into four phases. The first three phases occur during filling and the fourth occurs during voiding. Phase I reflects the bladder's initial response to filling. Phase II is the tonus limb and reflects the bladder pressure during the majority of the filling phase. Phase III occurs when the upper limit of viscoelastic properties is reached and the pressure begins to increase. Phase IV represents the voluntary contraction. Compliance is calculated at the end of filling as Δ volume/Δ Pdet.

According to Ghoniem,[10] compliance in the normal individual is ⩾20 mL/cm H_2O. Compliance less than 10 mL/cm H_2O is clinically significant (Figure 3-2.6). Low compliance is found in patients with severe outlet obstruction, meningomyelocele, chronic inflammation (tuberculosis and schistosomiasis), and after radiotherapy. High compliance may be associated with large capacity and overdistended bladders, and has little clinical significance as an isolated finding.[10]

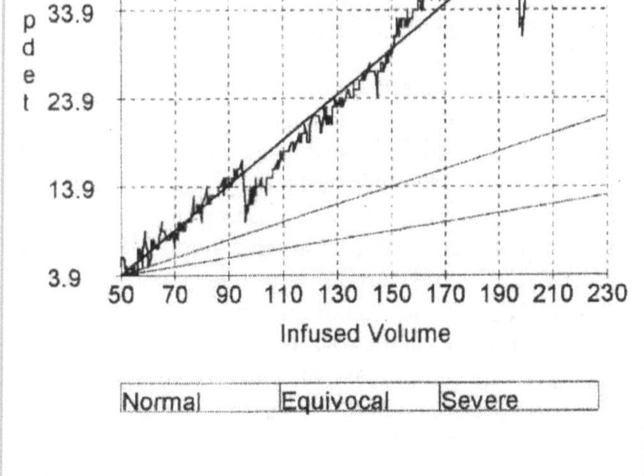

Figure 3-2.6. Impaired compliance.

Bladder Sensation

Measurement of sensation during cystometry is subjective and variable. Sensitivity and functional capacity are best determined with voiding diaries and not by CMG. The first sensation is the first awareness of filling, and is considerably variable. Many patients describe a filling sensation immediately. Typically, the first sensation to urinate is felt at 90 to 150 mL. The normal desire to void is the feeling that leads the patient to pass urine at the next convenient moment, but when voiding can be delayed if necessary. A strong desire to void is felt when there is a persistent urge to void without the fear of leakage. Urgency is a strong desire to void accompanied by fear of leakage or pain. Urgency is felt at 200 to 400 mL and bladder fullness occurs at 300 to 600 mL. Pain during bladder filling or micturition is abnormal.

Early sensations of filling (sensory urgency) may indicate inflammation, unstable detrusor contractions, or decreased bladder compliance. Delayed sensations of filling may represent a neurologic disorder such as peripheral neuropathy of the pelvic plexus, trauma, or disease affecting the sacral nerves.

Stability

Normally, the bladder stores increasing volumes of urine without a significant increase in pressure because of compliance and absence of involuntary detrusor contractions. The bladder should contract only during the voluntary act of voiding. Under certain circumstances, the bladder may show involuntary contractions, which may be associated with symptoms (Figure 3-2.7). These involuntary contractions are divided into two categories by the ICS: idiopathic and neurogenic detrusor overactivity. Idiopathic detrusor overactivity is defined as involuntary contractions not associated with an underlying neurologic lesion. The cause of this overactivity may be idiopathic, aging, infection, inflammation, or bladder outlet obstruction. Neurogenic detrusor overactivity is defined as involuntary contractions associated with a known neurologic lesion, typically an upper motor neuron lesion. Idiopathic and neurogenic detrusor overactivity may be indistinguishable on CMG.[1] The terms idiopathic and neurogenic overactivity are defined strictly by the patient's neurologic status rather than the appearance on CMG. While recording the contractions, the clinician must ensure that they are involuntary, because patients may become confused during the study. During the involuntary contractions, the patient's symptoms must be reproduced (e.g., urgency), and the bladder volume as well as any leakage should be recorded. Absence of involuntary contractions does not exclude the existence of contractions, especially if the patient's symptoms are not reproduced during testing.

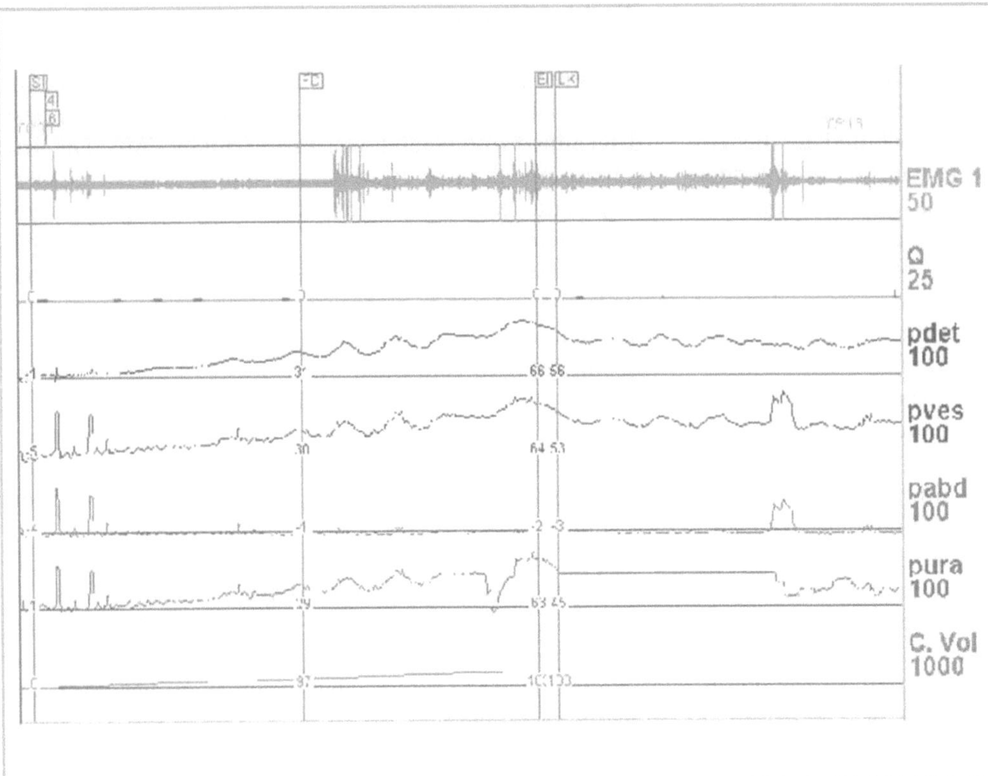

Figure 3-2.7. Neurogenic overactive bladder with low compliance and uninhibited contraction.

Rectal Contractions

Rectal contractions may occur during multichannel urodynamic studies. Although previously considered artifacts or variants of normal, Combs and Nitti[11] have reported that rectal contractions occur more frequently in patients with voiding dysfunction. Ghoniem[12] evaluated 71 patients with symptomatic benign prostatic hyperplasia (BPH) during 110 urodynamic studies. Thirty-three had detrusor instability. Rectal contractions were found in 40% of patients with symptomatic BPH and unstable bladder (detrusor overactivity), and in 10% of symptomatic BPH patients without bladder instability. This difference was statistically significant (Figure 3-2.8).

Stress-induced Detrusor Instability

Increased abdominal pressure during coughing may trigger uninhibited detrusor contractions in some patients. If the contractions are associated with incontinence, the leakage occurs because of the contractions and not the increased abdominal pressure itself. This phenomenon is responsible for 7% of 98 women with stress incontinence.[13] Identification of stress-induced instability can have important implications in the treatment of this type of incontinence. Anticholinergics and timed voiding are the treatment of choice in such situations. Skilled Kegel exercises (i.e., pelvic floor muscle contractions) before an anticipated cough may abort such contractions.

Pressure-flow Studies

Pressure-flow studies are simultaneous measurements of detrusor pressure and urine flow rate (Figure 3-2.9). They provide more accurate and useful information than uroflow rate alone. The major goal of these studies is to diagnose outlet obstruction and distinguish obstruction from impaired detrusor contraction. Approximately 20 to 35 cm H_2O of pressure is required to drive urine across the normal male urethra.[14] Women void with lower detrusor contraction and pressure, and higher flow rates than men.[15] Detrusor pressures more than 30 cm H_2O typically indicate the presence of some degree of obstruction. No generally accepted pressure-flow criteria for obstruction in women exist.

The results of pressure-flow studies can be classified according to a number of nomograms. Most of these nomograms have been described in men because of the high prevalence of bladder outlet obstruction in this population. The drawback of these nomograms is the broad boundary between obstruction and nonobstruction. In Abrams-Griffith (ICS) nomogram, the Qmax and corresponding Pdet at the maximum flow rate are plotted against each other (Figure 3-2.10). The detrusor pressure is expressed in centimeters of water on the y axis and uroflow is expressed in milliliters/second on the x axis. The location of the plotted maximal flow point on the graph determines the presence, absence, or an equivocal state of obstruction.[16]

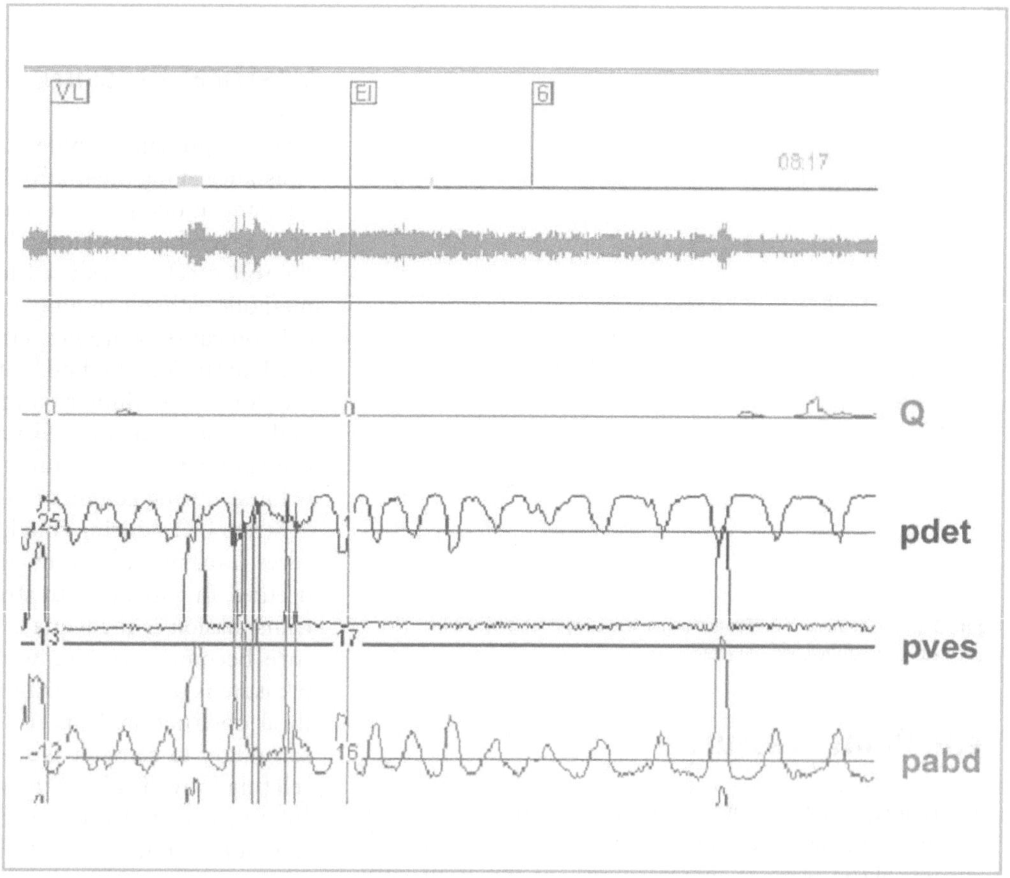

Figure 3-2.8. Rectal contractions may cause artifacts in detrusor pressure recording. Notice the stable vesical pressure during filling.

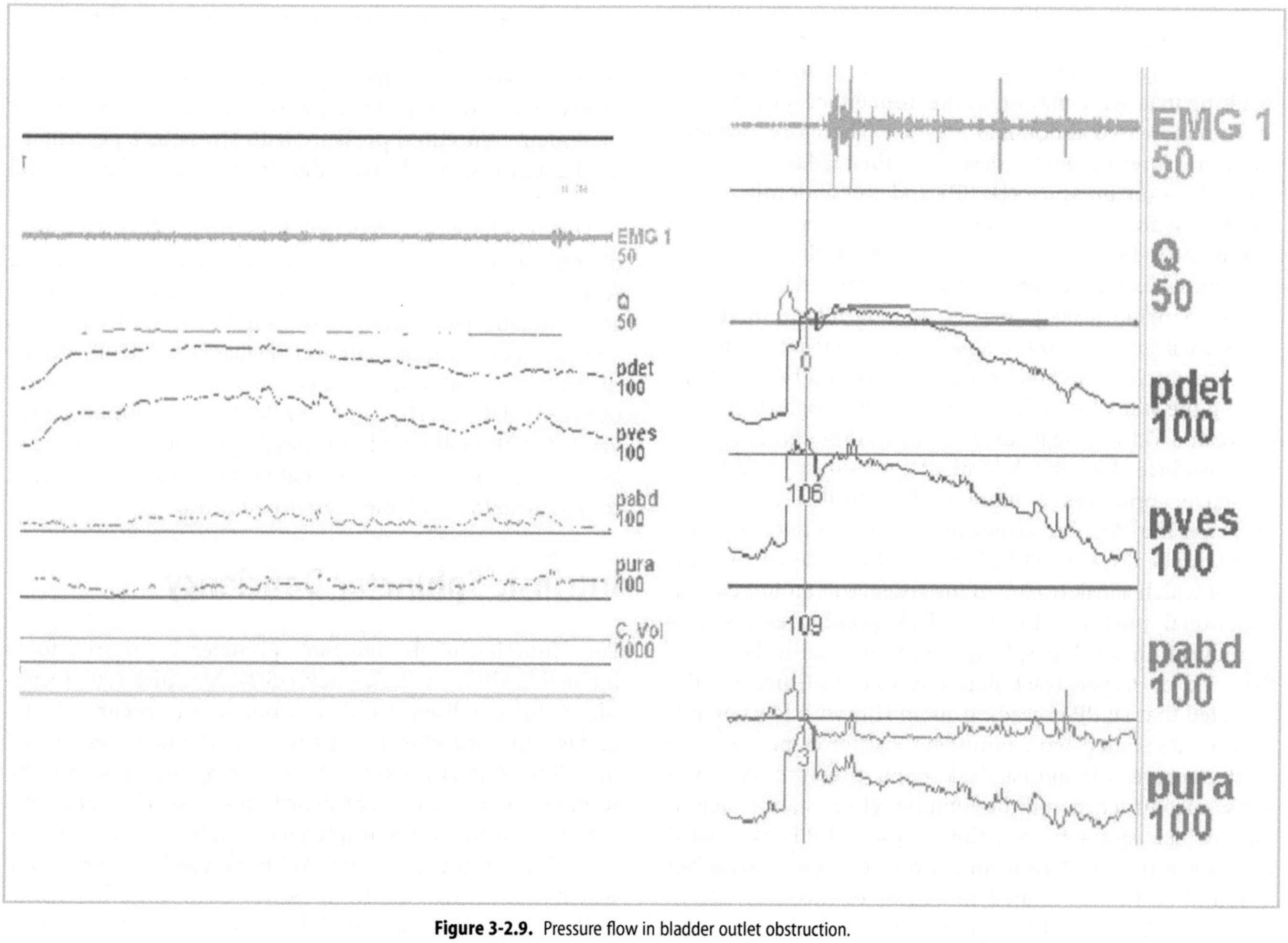

Figure 3-2.9. Pressure flow in bladder outlet obstruction.

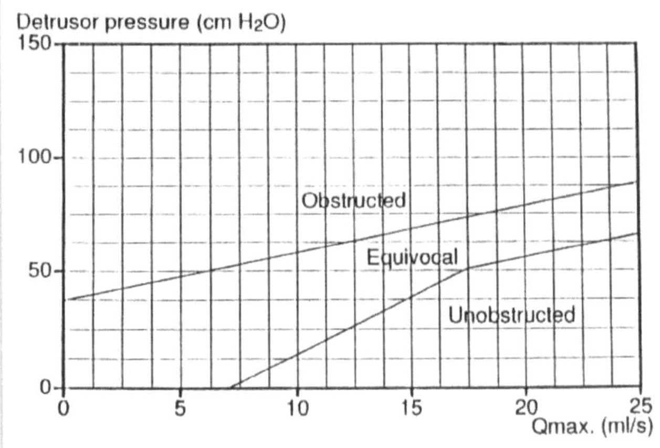

Figure 3-2.10. The Abrams-Griffith nomogram depicting three pressure flow zones: obstructed, nonobstructed, and equivocal.

Leak Point Pressure

Historically, leak point pressure was introduced by McGuire et al.[17] as a method to predict which children with myelodysplasia were at increased risk of upper urinary tract impairment because of chronically increased bladder pressure. The leak point pressure is the bladder pressure at which leakage occurs. The increase in bladder pressure resulting in leakage may originate from either the detrusor or from increased abdominal pressure. Therefore, two leak point pressures have been described: the bladder or detrusor leak point pressure (DLPP) and the abdominal leak point pressure (ALPP). Each measures the closure function of the entire bladder outlet under various circumstances. Abdominal leak point pressure is the intravesical pressure at which urine leakage occurs as a result of increased abdominal pressure in the absence of a detrusor contraction. The increase in abdominal pressure can be produced during the study by asking the patient to cough (CLPP) or perform a Valsalva maneuver (VLPP). Although coughing can produce higher abdominal pressures, a Valsalva maneuver produces a more well-controlled abdominal pressure. If a Valsalva maneuver fails to produce leakage, CLPP can be done. The DLPP is the value of detrusor pressure at which a leak occurs in the absence of an increase in abdominal pressure. Detrusor leak point pressure was introduced in myelodysplastic children as an indicator of the risk of upper tract deterioration. McGuire et al.[17] reported that DLPP exceeding 40 cm H_2O with low compliance is associated with imminent risk of upper urinary tract damage. Abdominal leak point pressure evaluates stress incontinence and the total outlet resistance including intrinsic sphincter function.[18] Low ALPP is associated with a severe degree of incontinence in some reports but is found to be not related to degree of incontinence in others.[19–21] Abdominal leak point pressure less than 90 cm H_2O suggests intrinsic sphincter deficiency (ISD).[22–23]

Urethral Pressure Profile

Urethral pressure profile (UPP) is a measure of urethral resistance along the urethral lumen. The measurement can be made at one point in the urethra over a period of time or at several points along the length of the urethra. There are three techniques used to measure UPP: the perfusion method, catheter-mounted transducers, and air-charged balloon catheter profilometry. We employ the most widely used perfusion method. Intravesical pressure should be measured to exclude a simultaneous detrusor contraction and to calculate the closure pressure. The intraluminal pressure can be determined at rest, at any given bladder volume, during cough, or during voiding.

Stress UPP (during cough) measurement evaluates pressure transmission from the abdominal cavity to the urethra. In patients with stress incontinence, this pressure transmission is inadequate and the urethral closure pressure becomes negative with coughing. Voiding urethral pressure is used to determine the pressure and site of urethral obstruction. This technique is plagued by artifacts secondary to catheter movement. Relevant measurements relating to resting UPP include maximal urethral pressure (MUP), maximal urethral closure pressure (MUCP), and functional urethral length. The maximum pressure measured is the MUP, and the maximum difference between urethral pressure and intravesical pressure is the MUCP. The functional profile length is the length of the urethra along which the urethral pressure exceeds intravesical pressure in women. Pressure transmission ratio is the increment in urethral pressure with stress, as a percentage of the simultaneously recorded increment in intravesical pressure.[1]

The drawback of UPP is that no normal values have been defined, and considerable variations have been reported. Although MUCP tends to be lower in women with genuine stress incontinence, there is overlap between SUI patients and normal subjects. Other drawbacks of UPP are its difficulty to standardize and perform. Also, the correlation between SUI severity and low MUCP has limited relevance.[24] Although UPP has limited applications because of broad inter- and intraindividual variations, it can be used to diagnose ISD, or "low pressure urethra."

Intrinsic Sphincter Deficiency

Poor function of the internal sphincter is referred to as intrinsic sphincter deficiency (ISD). McGuire (20) recognized this condition in some women with recurrent SUI and low urethral closure pressure, and defined it as type III SUI. Pure ISD is relatively easy to diagnose, because the urethra is fixed and incontinence occurs with low pressure. It is often seen after multiple failed incontinence surgeries or radiation therapy. The ALPP is used to determine whether ISD exists in the presence of urethral hypermobility. It is generally accepted that proximal urethral closure pressure less than 10 cm H_2O, and that ALPP of 60 cm H_2O

or less indicates ISD. A "gray zone" exists between 60 and 90 cm H_2O. Ghoniem et al.[25] classified ISD into three subgroups based on videofluorourodynamic findings and ALPP (Figure 3-2.11). *ISD-A* can be diagnosed only by videofluorourodynamic, because radiographic diagnosis cannot be performed when the bladder neck is not open at rest (Figure 3-2.11a). *ISD-B* is characterized by a beak-shaped open bladder neck at rest and an ALPP of less than 90 cm H_2O (Figure 3-2.11b). *ISD-C* is characterized by an open, fixed, nonfunctioning urethra (pipe stem) with the bladder neck in a high position, and an ALPP of less than 70 cm H_2O (Figure 3-2.11c) (see Table 3-2.1).

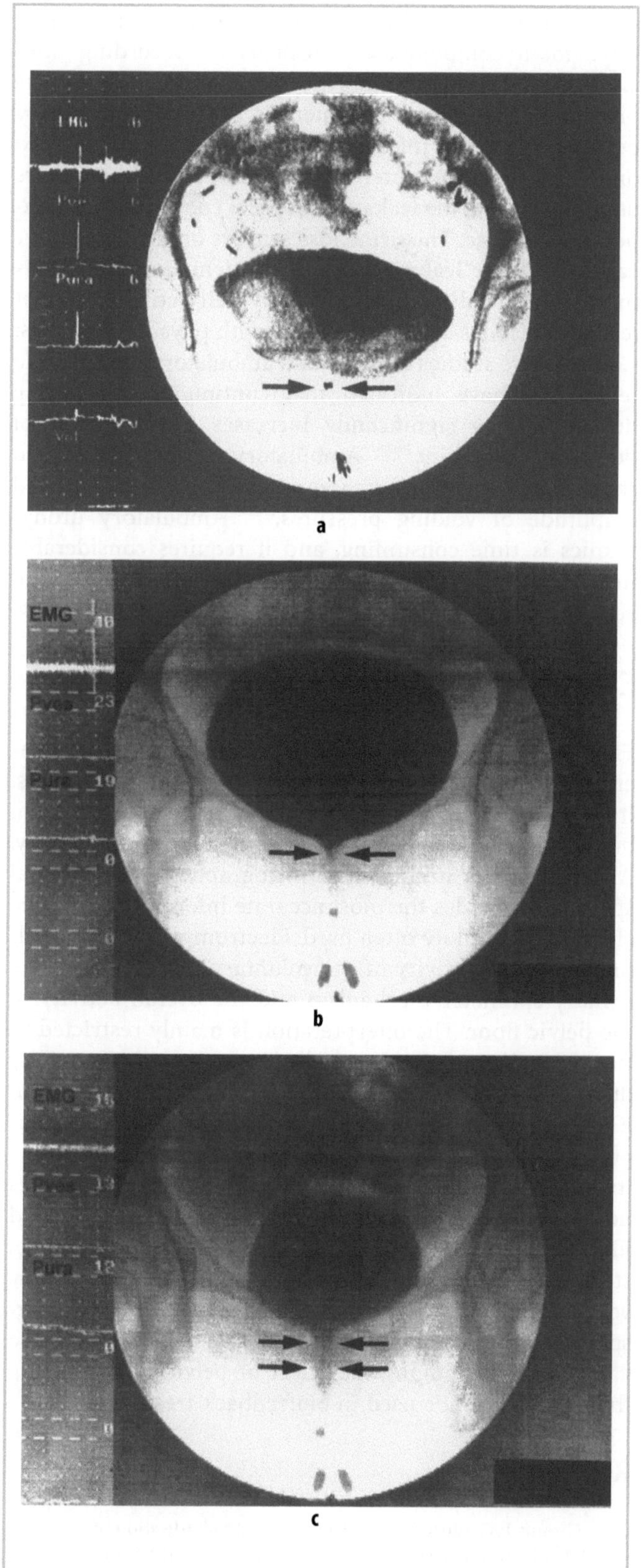

Figure 3-2.11. ISD classification (Ghoniem). ISD-A: Low poximal wetland pressure, high leak pressure. ISD-B: Beak-shaped bladder neck. ISD-C: Pipe stem urethra.

Table 3-2.1. Classification of ISD

	ISD-A	ISD-B	ISD-C
Stress test	+	Mostly +	Always
Bladder neck at FUDS	Closed at rest, open with stress	Beaked shaped, open at rest	Pipe stem open at rest
ALPP (cm H_2O)	<120	<90	<70

Source: Ghoniem et al.[24]
+ = positive

Videourodynamics

Videourodynamics combines a routine urodynamic study with X-ray or ultrasound imaging. Videourodynamics/fluorourodynamics (FUDS) is used for patients with complicated lower urinary tract dysfunction. In patients with neurologic conditions and lower urinary tract dysfunction, videourodynamics may enable a more accurate diagnosis. Because of the invasiveness, radiation exposure, and high initial and running costs, FUDS should not be considered as a first-line evaluation.

The indications for videourodynamics in incontinent patients are neurogenic lower urinary tract dysfunction and incontinence. In these patients, FUDS may help define the cause of incontinence when simple urodynamic studies do not lead to a definite diagnosis, or after failure of initial therapy based on less complicated methods of diagnosis.

The advantages of videourodynamics stem from the simultaneous measurement of pressure and visualization of the anatomy. Videourodynamics may reveal the following:

- Incompetent bladder neck
- Inadequate urethral closure during filling
- Location of urethral obstruction during voiding
- Descent of the bladder base
- Bladder base hypermobility (Figure 3-2.12)
- Intrinsic sphincter deficiency
- Incontinence as a leakage of contrast medium, which is demonstrated fluoroscopically
- Dyssynergia between detrusor and external sphincter or bladder neck
- Any sign of reflux during the filling or voiding phase
- Bladder or urethral diverticulum or other anatomical malformations

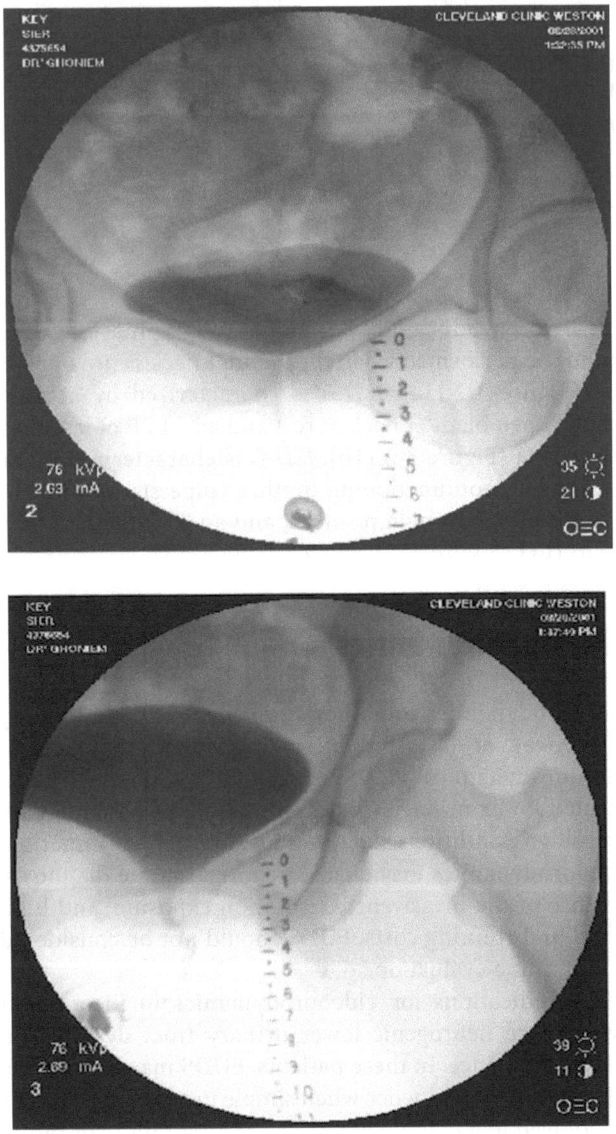

Figure 3-2.12. Bladder neck movement 2 cm during cough (hypermobility).

Technical Aspects

Images are captured during filling, and cine-loops are recorded during straining, coughing, and voiding. A postvoid image can show residual volume and may demonstrate urethral diverticulum. The CLPP should be recorded to check descent of the bladder base.

Ambulatory Urodynamics

Conventional or laboratory urodynamic studies are unphysiologic and conducted under unfavorable conditions for the patient and for a brief duration of time. Conventional studies typically involve artificial bladder filling with a specified liquid at a specific rate. Ambulatory urodynamics were developed to measure lower urinary tract function under more physiologic conditions (utilizing natural filling and during the subject's everyday activities), and to improve the sensitivity and accuracy of the urodynamic diagnosis. Ambulatory urodynamics systems have three main components: transducers, a recording unit, and an analyzing system. The transducers measure the pressures. The recording system is portable and battery powered, and the information collected is downloaded onto a personal computer for analysis. Two techniques are used to detect urine leakage: electrical conduction or temperature change. The urine loss sensors on the transducer catheter detect leakage, and the patient uses an events marker and voiding diary to record their activities so that leakage episodes can be correlated with physical activities. Comparative studies of office and ambulatory urodynamic techniques have indicated that continuous monitoring during activity significantly increases the diagnosis of detrusor instability.[26,27] Ambulatory urodynamics also shows lower pressure increases during filling and increased amplitude of voiding pressures.[28,29] Ambulatory urodynamics is time consuming, and it requires considerable interpretative skill. Its use is limited because there are no standardized parameters for normal or abnormal voiding.

Electromyography

Electromyography is the study of electrical activities generated by muscle, using either surface or needle electrodes. It provides useful information about sphincter function, but is most valuable when performed with cystometry. Although direct needle electromyography of the urethral sphincter provides the most accurate information, surface electrodes are more often used. Electromyography recording shows the activity of the voluntary component of the urinary sphincter mechanism and the overall activity of the pelvic floor. The interpretation is mainly restricted to the recording of progressively increasing EMG activity during bladder filling (guarding reflex). Reduced or silent EMG activity during bladder filling may indicate an incompetent urethral closure mechanism. Normally, EMG activity should decrease during voluntary voiding; if this does not occur, urethral function may be overactive. Increased EMG activity during voiding may indicate abdominal straining. Waxing and waning of the sphincter EMG during voiding suggests a functional obstruction (detrusor-sphincter dyssynergia). Surface EMG technique is also valuable when judging the effect of pelvic floor training therapy and can be used in biofeedback treatment.

References

1. Abrams P, Cardozo L, Fall M, et al. The standardisation of terminology of lower urinary tract function: report from the Standardization Sub-committee of the International Continence Society. Neurourol Urodyn 2002;21:167–178.

2. Drach G, Ignatoff J, Layton T. Peak urinary flow rate: observations in female subjects and comparison to male subjects. J Urol 1979;122: 215–219.
3. Abrams P, Torrens M. Urine flow studies. Urol Clin North Am 1979;6(1):71–79.
4. Jorgensen J, Jensen K. Uroflowmetry. Urol Clin North Am 1996;23: 237–242.
5. Thorup Andersen J, Jacobsen O, Gammelgaard PA, Hald T. Dysfunction of the bladder neck: a urodynamic study. Urol Int 1976; 31:78–86.
6. Abrams P, Khoury S, Wein A (eds.). Paris: Health Publication, 2004:317–372.
7. Webb R, Griffiths C, Ramsden P, Neal DE. Measurement of voiding pressures on ambulatory monitoring: comparison with conventional cystometry. Br J Urol 1990;65:152–154.
8. Susset JG, Ghoniem GM, Regnier CH. Clinical value of rapid cystometrogram in males. Neurourol Urodyn 1982;1:319–327.
9. Hjalmas K. Urodynamics in normal infants and children. Scand J Urol Nephrol Suppl 1989;114:20–27.
10. Ghoniem G. Disorders of bladder compliance. In: Krus E, McGuire E, eds. Female Urology. Philadelphia: Lippincott; 1994:83–94.
11. Combs A, Nitti V. Significance of rectal contractions noted on multichannel urodynamics. Neurourol Urodyn. 1995;14:73–80.
12. Ghoniem GM, Khater U, Elsergany R, Sokr M. The significance of rectal contractions in benign prostatic obstruction. Urodinomica 2005;15:33–38.
13. Nitti V. Cystometry and abdominal pressure monitoring. In: Nitti V, ed. Practical Urodynamics. Philadelphia: WB Saunders; 1998:47.
14. McGuire E. Urodynamic studies in prostatic obstruction. In: Fitzpatrick J, Krane R, eds. The Prostate. New York: Churchill Livingstone; 1989:103–109.
15. Ghoniem G. Urodynamics. In: Levy A, ed. Urology Pearls of Wisdom. Boston: Medical Publishing; 2001:113–118.
16. Abrams P, Griffiths D. The assessment of prostatic obstruction from urodynamic measurements and from residual urine. Br J Urol 1979; 51(2):129–134.
17. McGuire E, Woodside J, Borden T, Weiss RM. Prognostic value of urodynamic testing in myelodysplastic patients. J Urol 1981;126: 205–209.
18. Bump RC, Coates KW, Cundiff GW, Harris RL, Weidner AC. Diagnosing intrinsic sphincter deficiency: comparing urethral closure pressure, urethral axis, and Valsalva leak point pressures. Am J Obstet Gynecol 1997;177:303–310.
19. Cundiff GW, Harris RL, Coates KW, Bump RC. Clinical predictors of urinary incontinence in women. Am J Obstet Gynecol 1997;177: 266–267.
20. McGuire E. Urodynamic evaluation of stress urinary incontinence. Urol Clin North Am 1995;22:551–555.
21. Nitti V, Combs A. Correlation of Valsalva leak point pressure with subjective degree of stress urinary incontinence in women. J Urol 1996;155:281–285.
22. Lane T, Shah P. Leak-point pressures. BJU Int 2000;86:942–949.
23. McGuire E, Fitzpatrick C, Wan J, et al. Clinical assessment of urethral sphincter function. J Urol 1993;150:1452–1454.
24. Horbach NS, Ostegard DR. Predicting intrinsic sphincter dysfunction in women with stress incontinence. Obstet Gynecol 1994;84: 188–192.
25. Ghoniem G, Elgamasy A, El Sergany R, et al. Grades of intrinsic sphincter deficiency associated with female stress urinary incontinence. Int Urogynecol J 2002;13:99–105.
26. Brown K, Hilton P. The incidence of detrusor instability before and after colposuspension: a study using conventional and ambulatory urodynamic monitoring. BJU Int 1999;84:961–965.
27. Groen J, van Mastrigt R, Bosch R. Factors causing differences in voiding parameters between conventional and ambulatory urodynamics. Urol Res 2000;28:128–131.
28. Robertson A, Griffiths C, Ramsden P, Neal DE. Bladder function in healthy volunteers: ambulatory monitoring and conventional urodynamic studies. Br J Urol 1994;73:242–249.
29. Webb R, Griffiths C, Zachariah K, Neal DE. Filling and voiding pressures measured by ambulatory monitoring and conventional studies during natural and artificial bladder filling. J Urol 1991;146:815–818.

3-3

Imaging of the Genitourinary Tract in Females

M. Louis Moy and Sandip P. Vasavada

Pelvic organ prolapse and pelvic floor relaxation are common problems in older multiparous women, affecting approximately 16% of women aged 40 to 56 years.[1] A detailed knowledge of pelvic anatomy is paramount for the proper evaluation and management of such patients. Although a thorough pelvic examination is always indicated, even experienced clinicians may be misled by the physical findings, having difficulty differentiating among cystocele, enterocele, and high rectocele by physical examination alone. Depending on the position of the patient and strength of the Valsalva maneuver, the surgeon may be limited in his or her ability to accurately diagnose the components of pelvic prolapse. Furthermore, with uterine prolapse, the cervix and uterus may fill the entire introitus, making the diagnosis of concomitant anterior and posterior pelvic prolapse even more difficult. Regardless of the etiology of the support defect, the surgeon must identify all aspects of vaginal prolapse and pelvic floor relaxation for proper surgical planning. Accurate preoperative staging should reduce the risk of recurrent prolapse. Therefore, it must be understood that radiographic evaluation has an important role in the identification of these defects, and should be considered as an extension of the physical examination. One must realize that not all patients with complex pelvic floor pathology need imaging; however, it does enhance one's knowledge of the patient preoperatively, and accordingly finds its utility.

Imaging Techniques

Intravenous Urography

An intravenous urogram (IVU) is often used for the evaluation of hematuria, for the purpose of identifying a renal mass, renal or ureteral stone, upper tract urothelial filling defect, and hydroureteronephrosis. Furthermore, the functional nature of the IVU often allows the physician to distinguish between obstructive and nonobstructive hydronephrosis.

The most common reasons for obtaining an IVU in a woman with pelvic organ prolapse are to detect hydronephrosis and to evaluate for ureteral obstruction from previous pelvic surgeries. Moreover, although ureteral injuries during hysterectomy occur in 0.1% to 2.5% of cases, it has never been shown that routine preoperative IVU in patients with pelvic prolapse actually reduces the incidence of ureteral injury. Although the incidence of hydroureteronephrosis is low in the prolapse patient, it does increase with worsening pelvic prolapse, and is more common with uterine prolapse than with vault prolapse. As such, routine IVU in the setting of pelvic organ prolapse is not necessary unless one is concerned about urothelial pathology.

Fluoroscopy

Fluoroscopy is an excellent technique for investigating pelvic organ prolapse. By instilling contrast material into the bladder (anterior compartment), vagina (middle compartment), or rectum (posterior compartment), the dynamic relationships among the pelvic organs may be viewed in real time.

Cystography

Lateral phase cystography is still used quite often because it allows for a static view of the bladder and bladder outlet in relation to the remainder of the bony pelvis. Voiding cystourethrography is mainly used for demonstrating a cystocele, evaluating bladder neck hypermobility, and demonstrating an open bladder neck at rest (sphincteric incompetence). Dynamic lateral fluoroscopy at rest and during straining is an important adjunct to the urodynamic evaluation – useful for demonstrating the presence of, and degree of, urethrovesical hypermobility and cystocele formation. Although the radiographic findings do not always correlate well with urodynamic findings, in some patients, especially the incontinent ones, cystographic demonstration of an open bladder neck makes the diagnosis of sphincteric incompetence or intrinsic sphincter deficiency more certain. Other pathologic conditions that may be detected by voiding cystourethrography include vesicoureteral reflux, vesicovaginal fistula, and urethral diverticula, which further substantiates its utility.

Defecography

Defecography, or evacuation proctography, is used for evaluating the posterior (anorectal) compartment. Commonly measured variables include rectal volume, rectal emptying, perineal and pelvic floor muscle function, and anal sphincter function. Although the clinical value of defecography in the evaluation and management of constipation is not well proven, the presence of an obvious anatomic abnormality in a fecally incontinent patient such as a large rectocele, severe intussusception, or prolapse, supports surgical intervention. Furthermore, sigmoidoceles and upper rectal pathology and prolapse may lead to the decision to resect this portion of bowel in severe redundancy. To properly evaluate the sigmoid colon, one must specifically discuss this with the radiologist ahead of time or even perform a formal barium enema study. Concomitant therapy of a large sigmoidocele may allow one to effectively manage coexisting defecatory dysfunction with the prolapse repair.

Evacuation proctography relies on opacification of the rectal vault with barium paste. Instilling the paste is quite cumbersome, requiring a large-caliber enema. Typically, 80 to 300 mL of paste is instilled, often confounded by reflux into the sigmoid colon. Fluoroscopic images are recorded with the patient relaxed and while performing active contraction of the pelvic floor; this should result in elevation of the pelvic floor musculature. The patient is then examined during cough and during maximal straining maneuver, noting any pelvic floor descent or fecal incontinence. Finally, defecation is accomplished, and note is made of any rectocele, incomplete emptying, or need for digital assistance with evacuation (functional phase of the study).

In the cooperative patient, dynamic proctography allows precise identification and quantification of a rectocele, measured as the maximum extent of an anterior rectal bulge beyond the expected line of the rectum. Rectal intussusception may be visible as a circumferential invagination of the rectal wall, presenting as mucosal prolapse through the anus in its most severe form. Limitations of this examination are the cumbersome and potentially painful instillation of rectal barium paste, lack of correlation between the viscosity of the paste and the individual patient's stool, and the inability of many patients to defecate on command. Additionally, the presence of a rectocele in and of itself may be of limited concern, because previous studies have shown that an anterior rectal bulge is often demonstrated in nulliparous asymptomatic patients. Furthermore, the presence of, or size of, a rectocele does not correlate well with the completeness of barium evacuation.

Colpocystourethrography

First introduced in France in 1965, the colpocystourethrogram combines opacification of the bladder, urethra, and vagina.[2] Modified and made popular in the mid 1970s, the colpocystourethrogram is a dynamic study of pelvic support and function. The anatomical relationships among the bladder, urethra, and vagina may be demonstrated, and, when combined with proctography, may be even more useful in outlining the anatomy of the normal pelvis and of complex pelvic organ prolapse.

The accuracy of dynamic colpocystoproctography is even further enhanced by opacifying the small bowel. The patient drinks oral barium 2 hours before the examination. With the vagina, bladder, small intestine, and rectum opacified, the vaginal axis may be measured at rest and with straining, and any prolapse of the anterior, middle, or posterior vaginal compartment is evident. Moreover, the examiner should be able to distinguish among various organs that may prolapse into a widened rectovaginal space, differentiating enterocele from sigmoidocele, further refining the operative approach.

Magnetic Resonance Imaging

The development of fast-scanning magnetic resonance imaging (MRI) techniques has improved our ability to describe and quantify anatomical changes that may cause pelvic floor relaxation. Yang et al.[3] were the first to popularize dynamic fast MRI for the evaluation of pelvic organ prolapse. Since then, other investigators have shown that MRI is more sensitive than physical examination for defining pelvic prolapse.[4,5] Whereas some advocate the use of contrast opacification of the bladder, vagina, and rectum, others have shown that the vagina, rectum, bladder, urethra, and peritoneum are adequately visualized without any contrast administration. By avoiding instrumentation of the vagina or urethra, iatrogenic alteration of the anatomy is minimized.

Magnetic resonance imaging, however, can noninvasively survey the entire pelvis. The excellent differentiation among soft tissues and fluid-filled viscera provides visualization of the musculofascial support structures of the pelvic organs. Our group and others have previously demonstrated the clinical utility of MRI for evaluating bladder neck and urethral anatomy, and the utility of dynamic MRI for assessing pelvic floor descent and genital prolapse, and development of dynamic rapid sequencing has greatly improved the diagnostic utility of MRI by allowing exquisite anatomical detail during brief breath-holds (Figures 3-3.1 and 3-3.2). Magnetic resonance

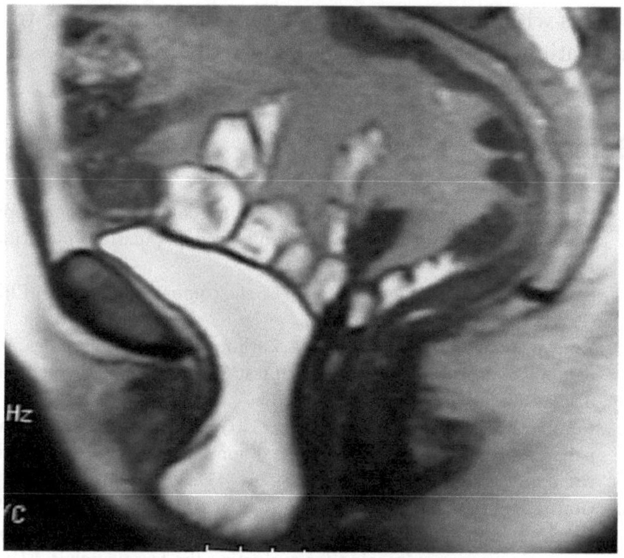

Figure 3-3.1. Magnetic resonance image of the pelvis demonstrating significant cystocele with descent of bladder base and trigone well below pubic symphysis.

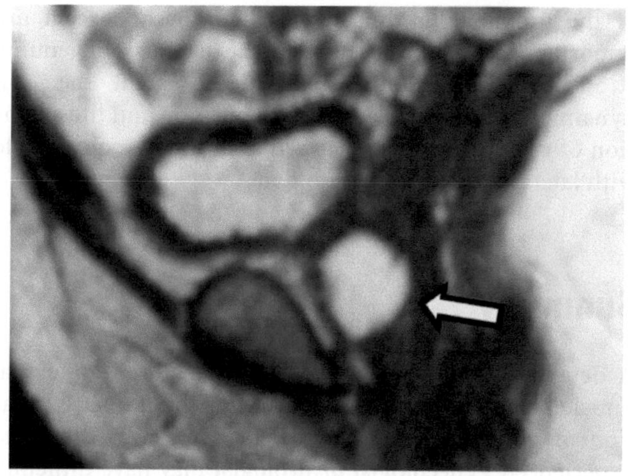

Figure 3-3.3. Magnetic resonance image of urethral diverticulum (see arrow).

imaging has become our choice for imaging patients with suspect urethral diverticula (Figure 3-3.3).

There are, however, several limitations of this technique. Defining normal values is difficult, because it would be quite expensive to perform dynamic MRIs on nulliparous women without any urologic complaints. Our choice for the nonprolapse group (normal patients) included patients in whom we desired an imaging study to evaluate pathology other than pelvic prolapse, such as recurrent infections, pelvic pain, and urethral pain. Another limitation is that a collapsed rectocele may not be visualized because of competition among prolapsing pelvic organs for limited introital space. In fact, physical examination has been shown to be slightly more accurate in demonstrating rectocele formation than MRI. Additionally, the study must be performed supine, simply because there are no upright MRI machines available at this time. However, dynamic MRI with relaxing and straining views has been shown to clearly demonstrate organ prolapse during straining in the supine position. An erect MRI is the next logical advancement. Finally, claustrophobic patients and those with cardiac pacemakers cannot go into the enclosed magnet. Despite these limitations, dynamic MRI has become the study of choice at our institution for evaluating high-grade pelvic prolapse and pelvic floor relaxation. Because this is a new technique, a standardized system for describing and quantifying organ prolapse and pelvic floor relaxation is important.

Magnetic resonance imaging findings were compared to physical examination and intraoperative findings. HASTE-sequence MRI was more accurate than physical examination in identifying cystoceles, enteroceles, vault prolapse, and pelvic organ pathology such as uterine fibroids, urethral diverticula, ovarian cysts, and Nabothian and Bartholin's gland cysts.[5] Comiter et al.[6] found that with dynamic MRI, surgical planning was altered in more than 30% of cases, most often because of occult enterocele not appreciated on physical examination.

In patients with severe prolapse, especially with renal insufficiency, the surgeon must rule out obstructive hydroureteronephrosis. This may be accomplished by magnetic resonance urogram, adding only minimally to overall examination time, and with no additional morbidity. Mag-

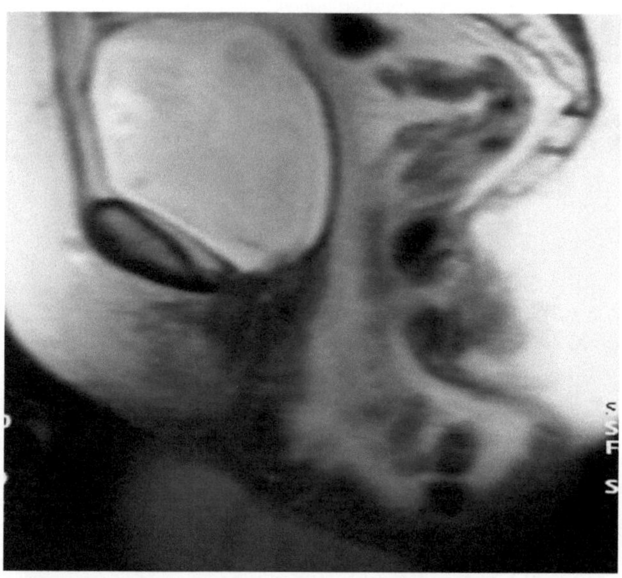

Figure 3-3.2. Magnetic resonance image demonstrating a well-supported bladder (white on T2-weighted study), but significant enterocele behind it.

netic resonance imaging may also be useful for the radiographic evaluation of stress incontinence. Hypermobility of the proximal urethra and bladder neck descent are important pathological features in the diagnosis of genuine stress urinary incontinence.[7,8] Measurement data on dynamic MRI for the bladder neck position and the extension of cystocele at maximal pelvic strain are comparable with lateral cystourethrogram data.[9]

Summary

A detailed working knowledge of normal and abnormal female pelvic anatomy is necessary for the proper evaluation of pelvic organ prolapse. However, even the most experienced gynecologist or urologist may have difficulty distinguishing among prolapsing organs competing for introital space. Accurate identification of all aspects of vaginal prolapse and pelvic floor relaxation are vital, not only to permit adequate surgical planning, but also to reduce the risk of recurrent prolapse. Radiographic evaluation of the woman with pelvic organ prolapse and pelvic floor relaxation should be viewed as a valuable extension of the physical examination. Urography, voiding cystography, dynamic colpocystodefecography, and MRI are each useful for the evaluation of pelvic prolapse and pelvic floor relaxation. Magnetic resonance imaging will continue to have an increasingly important role, because of its superior visualization of fluid-filled viscera and soft tissues, and the ability to simultaneously visualize all important pelvic organs without the need for patient preparation or instrumentation, or exposure to ionizing radiation.

References

1. Hagstad A, Janson PO, Lindstedt G. Gynaecological history, complaints, and examinations in a middle-aged population. Maturitas 1985;7:115–128.
2. Bethoux A, Bory S, Huguier M, Lan CS. Le colpocystogramme. J Chir (Paris) 1965;8:809–828.
3. Yang A, Mostwin JL, Rosenshein NB, Zerhouni CA. Pelvic floor descent in women: dynamic evaluation with fest MR imaging and cinematic display. Radiology 1991;179(1):25–33.
4. Comiter CV, Vasavada SP, Barbaric ZL, Gousse AE, Raz S. Grading pelvic prolapse and pelvic floor relaxation using dynamic magnetic resonance imaging. Urology 1999;54:454–457.
5. Gousse AE, Barbaric ZL, Safir MH, Madjar S, Marumoto AK, Raz S. Dynamic half Fourier acquisition, single shot turbo spin-echo magnetic resonance imaging for evaluating the female pelvis. J Urol 2000;164:1606–1613.
6. Comiter CV, Vasavada S, Raz S. Preoperative evaluation of pelvic prolapse using dynamic magnetic resonance imaging. 29th Annual International Continence Society; August, 1999; Denver, CO.
7. Enhorning G. Simultaneous recording of intravesical and intraurethral pressure. Acta Chir Scand 1956;276(suppl):1–68.
8. Jeffcoate TN, Roberts H. Observation on stress incontinence of urine. Am J Obstet Gynecol 1952;64:721–738.
9. Gufler H, DeGregorio G, Allmann KH, Kundt G, Dohnicht S. Comparison of cystourethrography and dynamic MRI in bladder neck descent. J Comput Assist Tomogr 2000;24:382–388.

3-4

Anorectal Physiology

T. Cristina Sardinha and Dana R. Sands

The last decade has brought a wealth of new information and methods for the evaluation of the structural and functional integrity of the anal sphincters and pelvic floor. Anal ultrasound, electromyography, pudendal nerve terminal motor latency assessment, anal manometry, and cinedefecography have all been used to evaluate patients with a plethora of disease states. The etiology and subsequent treatment plans for functional disorders such as fecal incontinence, constipation, and anal pain syndromes are often delineated in the anorectal physiology laboratory. This chapter will focus on two integral parts of any anorectal physiologic evaluation, anal manometry and defecography.

Anorectal Manometry

Anal manometry provides important information about the functional status of the anal sphincters and distal rectum. There is no standardized method of manometric evaluation, although several methods have been described.[1] Microtransducers can be used in the anal canal and are well tolerated by patients. Multichannel water perfused catheters are perhaps the most common tool used to perform anal manometry. Flow rates of 0.3 mL per channel per minute are required to adequately measure outflow pressure.[2] Higher flow rates can result in accumulation of fluid and distortion of the measurements.

The resistance of flow from the catheter determines pressure measurements, which can be taken in a continuous manner (continuous pull-through) or at set levels within the anal canal (station pull-through). The continuous pull-through technique requires the catheter to be removed at a continuous speed from the anal canal, using a computerized motor. This technique can provide a detailed recording of both the radial and longitudinal pressure profiles, while computer analysis can then generate a three-dimensional representation of the anal canal. The station pull-through technique measures anal canal pressures at 1-cm increments in the anal canal. It has been theorized that this method provides a more accurate assessment of anal pressures because there is a stabilization period before each reading, thereby reducing artifact.[3] This is the method used at the Cleveland Clinic Florida. The tip of the catheter is placed at 6 cm cephalad from the anal verge. Anal pressure measurements are taken at 1 cm intervals after stabilization of the catheter for 20 to 30 seconds. Rest and squeeze pressures are recorded at each centimeter up to 1 cm from the anal verge. After documentation of the pressures in the anal canal, rectal capacity, compliance, and rectoanal inhibitory reflex are tested. The evaluation of the rectal capacity will provide information regarding the rectal potential to accommodate the fecal matter if defecation must be delayed.

Measurements

Resting Pressure

The mean resting pressure in healthy volunteers ranges from 40 to 70 mm Hg. The internal anal sphincter generates the majority of the resting pressure.[4] This smooth muscle is in a continuous state of maximal contraction accounting for 55% to 60% of resting tone. The external anal sphincter contributes less to the resting anal tone. The final determinant of resting anal tone is the contribution of the hemorrhoidal plexi. Patients with alterations in continence related to the internal anal sphincter often have low baseline resting pressure.

Squeeze Pressure

The maximal squeeze pressure in healthy individuals is two to three times the baseline resting value. The external anal sphincter is the main contributor to the generation of these pressures. Traumatic defects of the external anal sphincter, whether obstetric or from surgery, often result in decreased squeeze pressures.

High-pressure Zone

The high-pressure zone is defined as the length of the internal anal sphincter through which pressures are greater than half of the maximal resting pressure. The high-pressure zone is approximately 2 to 3 cm in women and 2.5 to 3.5 cm in men.[4]

Rectoanal Inhibitory Reflex

The rectoanal inhibitory reflex is thought to have a role in the fine adjustments of continence. Rectal distension, usually with small volumes (10–30 mL), causes a contraction of the external anal sphincter followed by a pronounced internal anal sphincter relaxation. This reflex enables the sensory mucosa of the anal canal to "sample" the contents of the distal rectum and the patient to distinguish between gas, liquid, and solid stool. This reflex is absent or abnormal in patients with Hirschsprung's disease, Chagas' disease, dermatomyositis, and scleroderma.

Rectal Sensation

Alterations in rectal sensation can lead to decreased fecal continence. Rectal sensation is measured with an intrarectal balloon and incremental instillation of known volumes of air. Sensation is generally achieved with 20 mL of air. Overflow incontinence can result from a decrease in rectal sensation and subsequent fecal impaction.

Rectal Compliance

Compliance is determined by the change in pressure associated with a change in volume. This is calculated by subtracting the volume of first sensation from the maximum tolerable volume and dividing by the change in pressure at these two points. A noncompliant rectum can contribute to fecal incontinence if the patient is unable to accommodate the amount of stool presented to the rectum. This is common in conditions that cause proctitis.

Manometry is an important tool to evaluate patients with fecal incontinence. This test will determine functional weakness of the internal and/or external anal sphincter as well as identify the presence of sensory or muscular defects. Rectal capacity and compliance may also be significantly decreased in patients with fecal incontinence, proctitis, or loss of rectal reservoir. The role of manometric evaluation in patients with constipation is somewhat controversial. Manometry may have a role in the assessment of patients with paradoxical puborectalis contraction, lifelong history of constipation, especially in children and patients suspicious for Hirschsprung's or from areas endemic for Chagas' disease.

Cinedefecography

Cinedefecography is a radiologic evaluation of the functional status of the anus, rectum, and sigmoid. It is an important part of the physiologic evaluation of patients with constipation, pelvic organ prolapse, and rectal pain.

The dynamic evaluation of the defecation mechanism is an important tool in the overall investigation of evacuatory disorders. Defecography involves the instillation of a thick barium paste into the rectum as well as a water-soluble contrast into the vagina, with the subsequent radiographic imaging of evacuation. Both still and dynamic images are obtained with the patient seated on a radiolucent commode. Using standard fluoroscopic technique, the evacuation process is visualized and recorded with the aid of a videocassette recorder connected to the fluoroscopy device. Lateral X-ray films are taken during the different phases of the test, including resting, squeezing, and pushing. At the completion of the rectal emptying, static pictures are also taken to document the completion of evacuation of the contrast. From the still images, several measurements can be obtained.

Measurements

All measurements are taken from rest, squeeze, and push–static films:

Anorectal angle – angle between the axis of the anal canal and the posterior rectal wall (Figure 3-4.1).

Puborectalis length – distance between posterior anorectal angle and the anterior symphysis pubis.

Perineal plane distance – line drawn from tip of coccyx to anterior pubic symphysis (perineal plane). Perpendicular line drawn from perineal plane to the anorectal angle. This distance measures perineal descent. Increased perineal descent (>3 cm) in either the resting or squeeze phase correlates to fixed or dynamic descent, respectively (Figure 3-4.2).

Sigmoidocele – the degree of descent of the lowest portion of the sigmoid colon. First degree = above the pubococcygeal line; second degree = below the pubococcygeal line and above the ischiococcygeal line; third degree = below the ischiococcygeal line (Figure 3-4.3).

A wealth of information is gained from the dynamic images including the presence and significance of a rectocele or sigmoidocele (Figure 3-4.3), paradoxical puborectalis contraction (Figure 3-4.4), intussusception (Figure 3-4.5), prolapse, ability to straighten the anorectal angle, and the ability to empty the rectum. Seven hundred forty-four patients at the Cleveland Clinic Florida underwent defecography,[5] 60% of whom were constipated; 12.5% had "normal studies," 25.7% had rectoceles, 11% had sigmoidoceles, 12.6% intussusception, and 30% had a combination of findings. The number of significant findings

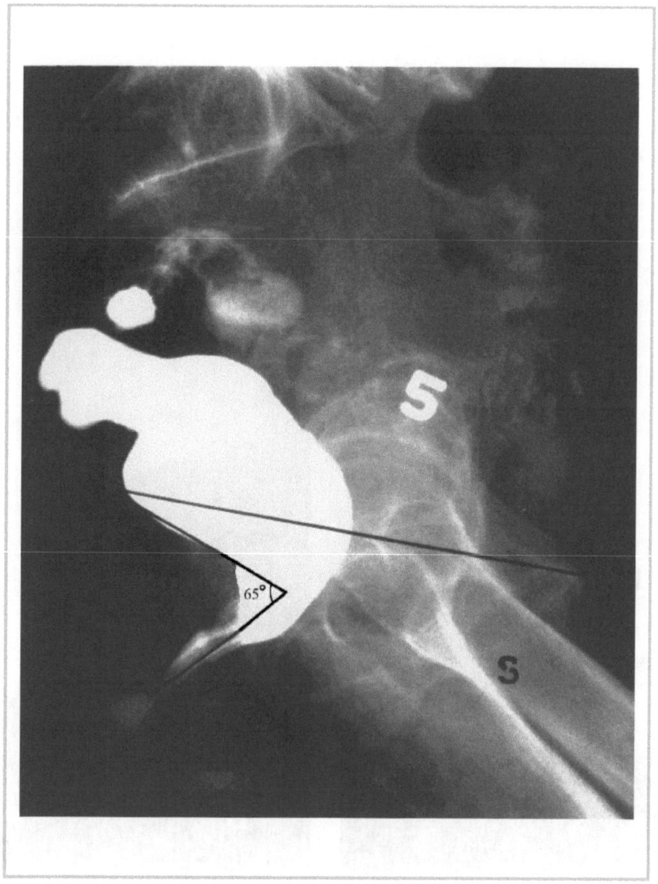

Figure 3-4.1. Anorectal angle on defogram.

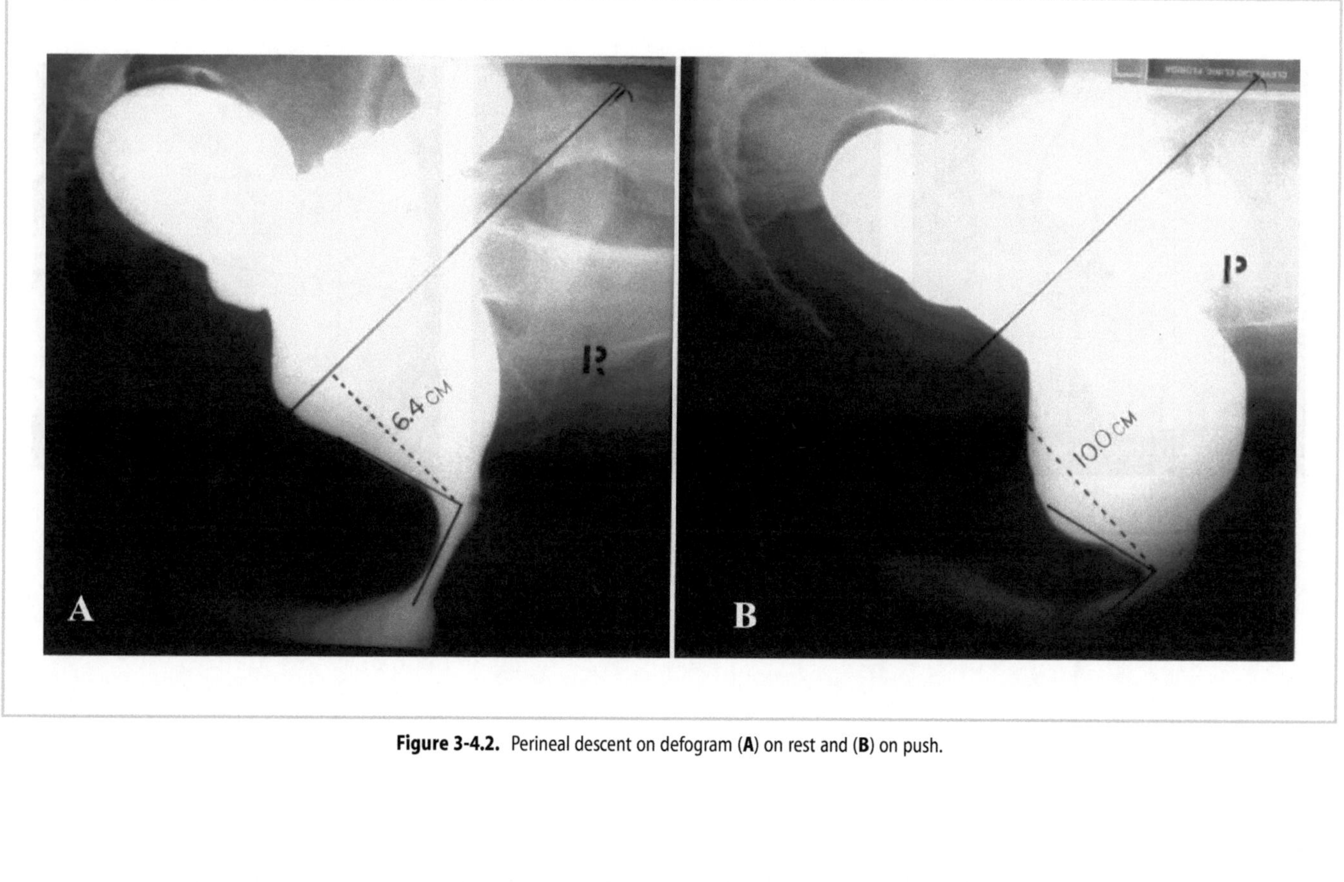

Figure 3-4.2. Perineal descent on defogram (**A**) on rest and (**B**) on push.

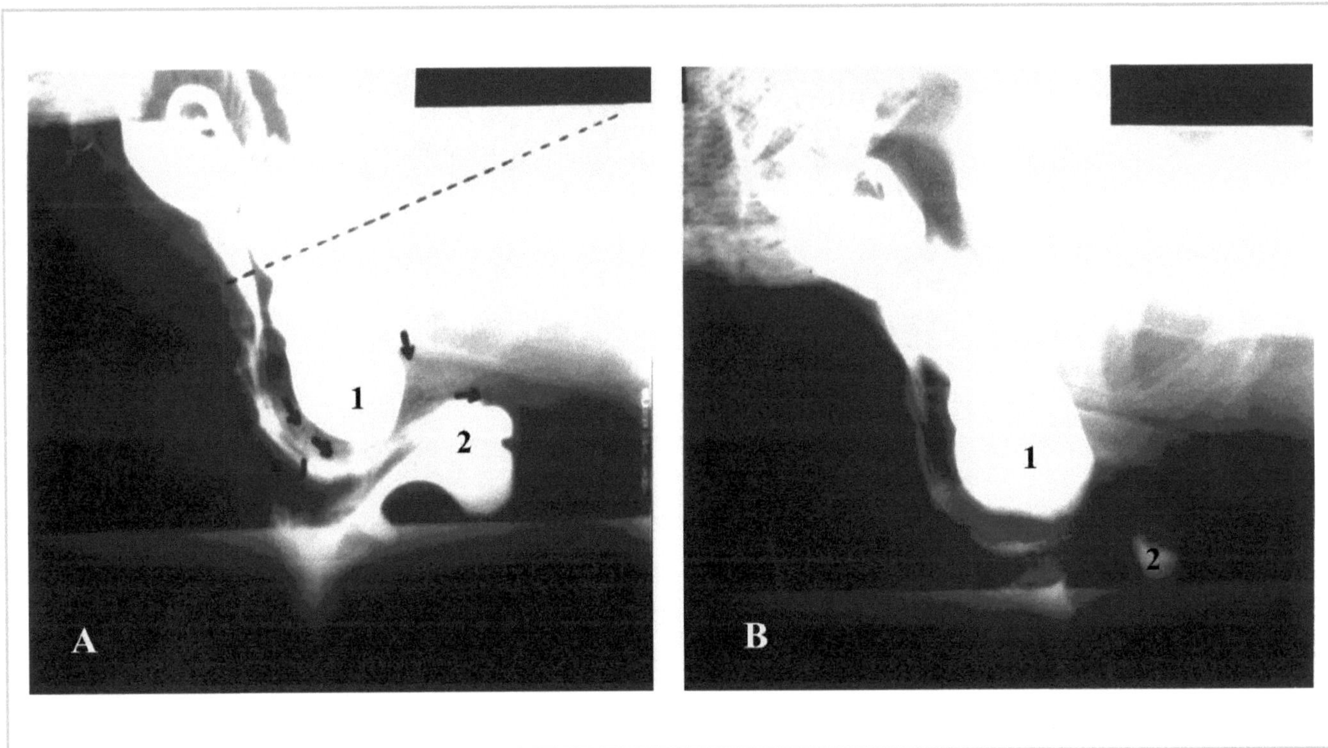

Figure 3-4.3. Sigmoidocele (1) and rectocele (2) on defogram, preevacuation (**A**) and postevacuation (**B**).

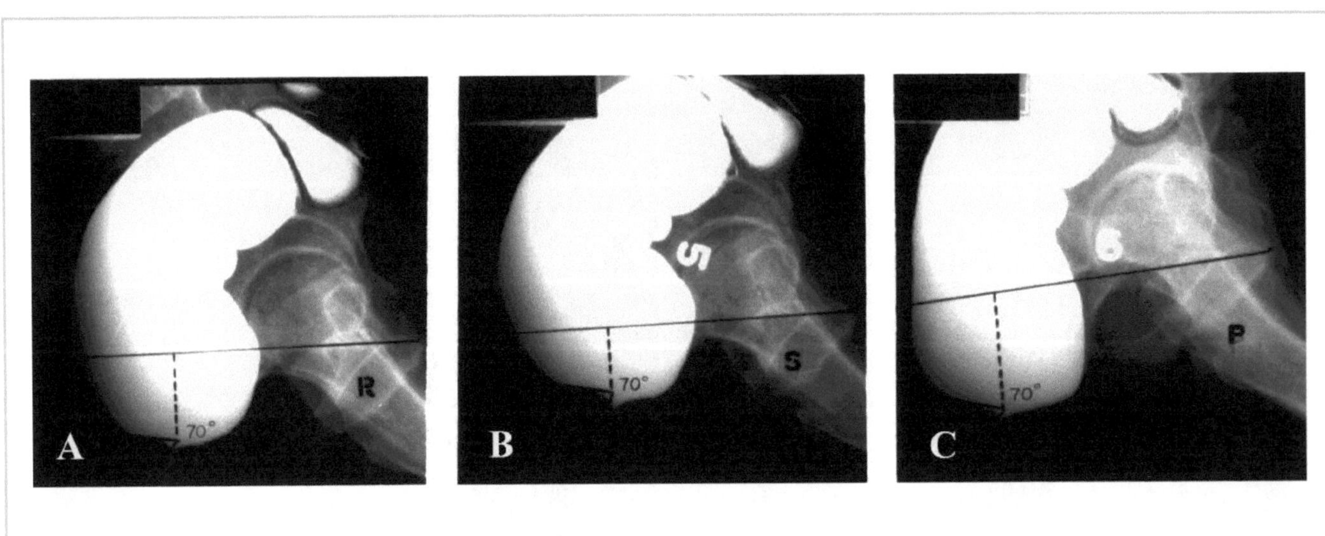

Figure 3-4.4. Paradoxical puborectalis contraction on defogram, on rest (**A**), on squeeze (**B**), and on push (**C**).

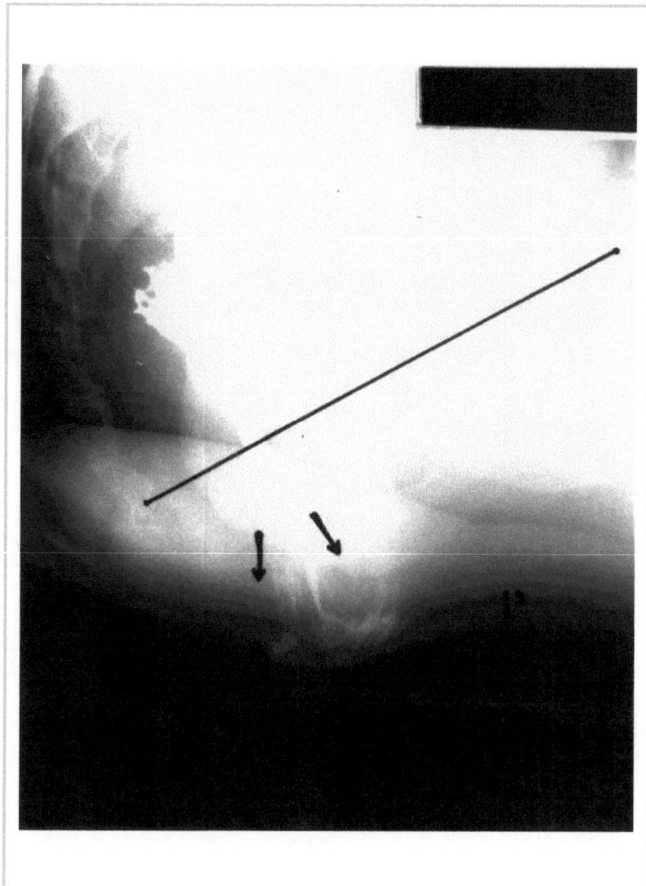

Figure 3-4.5. Intussusception on defogram.

within this group of patients underscores the need to always correlate radiographic findings with the clinical evaluation.

A normal cinedefecography is expected to demonstrate complete relaxation of the puborectalis muscle, a perineal descent less then 3 cm during rest, and complete evacuation of the rectal contrast.

Defecography is extremely helpful in the evaluation of patients with constipation. The documentation of the ability or inability to empty the rectum leads the physician to focus treatment plans on either the evacuatory mechanism or disorders of intestinal transit. The identification of pathology on the defogram allows the physician to correlate the anatomic abnormality with its impact on the evacuatory mechanism.

Conclusion

The anorectal physiology laboratory provides a wealth of information for the physician treating patients with evacuatory difficulties. The combination of anal ultrasound, electromyography, pudendal nerve terminal motor latency assessment, anal manometry, and cinedefecography allow for a complete evaluation of the posterior compartment of the pelvic floor. Proper interpretation of the studies and correlation with the clinical complaints of the patient are essential for the effective treatment of evacuatory dysfunction.

References

1. Roberts PL. Principles of manometry. Semin Colorectal Surg 1992; 3(2):64–67.
2. Jorge JM, Wexner SD. Anorectal manometry: techniques and clinical applications. South Med J 1993;86(8):924–931.
3. Coller JA. Clinical application of anorectal manometry. Gastrointest Clin North Am 1987;16:17–33.
4. Lestar B, Penninckx F, Kerremans R. The composition of anal basal pressure. An in vivo and in vitro study in man. Int J Colorect Dis 1989;4:118–122.
5. Agachan F, Pfeifer J, Wexner SD. Defecography and proctography. Results of 744 patients. Dis Colon Rectum 1996;39(8):899–905.

3-5

Anorectal Ultrasound

Juan J. Nogueras

During the past decade, anal imaging studies have become increasingly popular as diagnostic tools in the management of patients with both benign and malignant anorectal disorders. As surgeons become more comfortable with the application of these modalities, the indications for their use continue to expand. Anal ultrasonography is an accepted modality for the evaluation of several conditions, including fecal incontinence, anal sepsis, and anal cancer. Other ultrasonographic modalities are also used to image the anal canal, including transvaginal ultrasonography and transperineal ultrasonography. Magnetic resonance imaging (MRI) is yet another modality to image the anal region that seems to be increasing in popularity as its availability widens. Each of these imaging modalities and their indications for use in the anal canal will be discussed.

Anal Ultrasonography

Law and Bartram[1] described the technique of anal ultrasonography in 1989, and in the past 14 years, the technique has achieved widespread popularity. In recent years, endoanal ultrasonography has emerged as a popular diagnostic modality for anal sphincter mapping and imaging. The procedure is performed as an outpatient procedure, is relatively quick, is virtually painless, and therefore is preferred by patients over other more invasive tests, such as needle electromyography. Anal ultrasonography can provide a detailed image of the anal sphincter musculature.[2] The internal anal sphincter (IAS) appears endosonographically as a hypoechoic circular band, which is seen most prominently at the level of the mid anal canal. The range of thickness of this muscle is 1.5 to 4 mm for males and 2 to 4 mm for females. There does not seem to be a relationship between IAS thickness and body weight, gender, or height, but there is a direct correlation for thickness with advanced age.[3] The external anal sphincter (EAS) appears as a thicker circular mixed echogenic band outside of the hypoechoic internal sphincter. Other structures of the pelvic floor, such as the puborectalis, urethral sphincter, vagina, and outlines of the bony pelvic and ischiorectal fossae, are also demonstrated on endoanal ultrasonography.[4]

There is no special preparation necessary for anal ultrasonography. The patient is positioned in the left lateral decubitus position. A digital rectal examination is performed to evaluate for anal lesions, stenosis, and tone. We use a Bruel & Kjaer scanner with a 10-MHz transducer that has a sonolucent hard plastic cover filled with degassed water. A condom filled with ultrasound gel is placed over the probe and is lubricated with gel. This provides for effective acoustic coupling in the anal canal. Transducers vary in frequency; a probe with a higher frequency provides clearer sonographic images (resolution) but has a shorter focal range (penetration). A 7.0-MHz probe has a focal length of 2 to 5 cm, whereas a 10.0-MHz probe has a focal length of 1 to 4 cm.

Image Interpretation

The anal canal is divided into three compartments. The upper anal canal (Figure 3-5.1) is defined as the level where the puborectalis muscle sling is clearly seen. At this level, there is a normal separation of the muscle fibers anteriorly. It is important to recognize this anatomic level, or this anterior muscle separation can easily be misinterpreted as a sphincter defect.

The mid anal canal (Figure 3-5.2) represents the location where the IAS and EAS wrap around the anal canal circumferentially. We have arbitrarily chosen the site where the IAS muscle is at its maximum width as the mid anal canal; this is the level where sphincter defects are most often seen. The IAS is seen as a hypoechoic ring, which is surrounded by the mixed echogenic EAS.

The distal anal canal (Figure 3-5.3) represents the level where the IAS has tapered and is no longer visible, and the predominant muscle is the subcutaneous portion of the EAS. This mixed hyperechogenic band should be circumferentially intact.

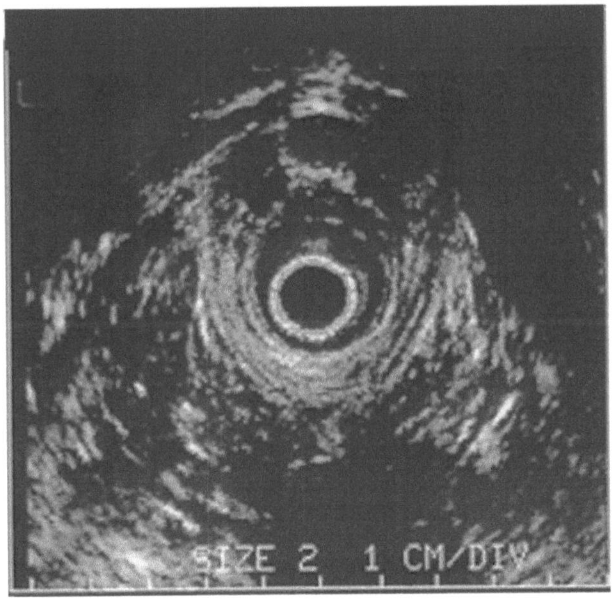

Figure 3-5.1. The upper anal canal.

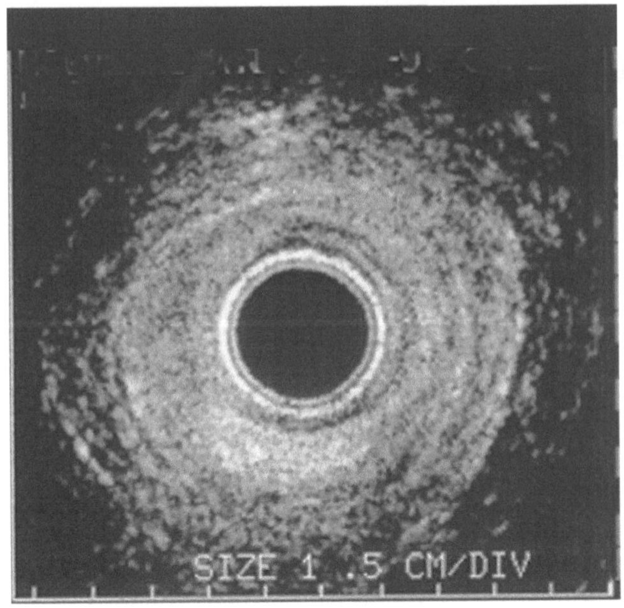

Figure 3-5.3. The distal anal canal.

Defects in the external or internal sphincters are defined as breaks in the continuity of the sphincter ring. The radial extents of the sphincter defects are measurable by the 360-degree view provided by the probe. Defects in the sphincters can be caused by obstetric trauma or prior operative procedure including hemorrhoidectomy, sphincterotomy, and fistula surgery. Endoanal sonography is reliable in assessing sphincter injury and is often used in the evaluation of patients with fecal incontinence.

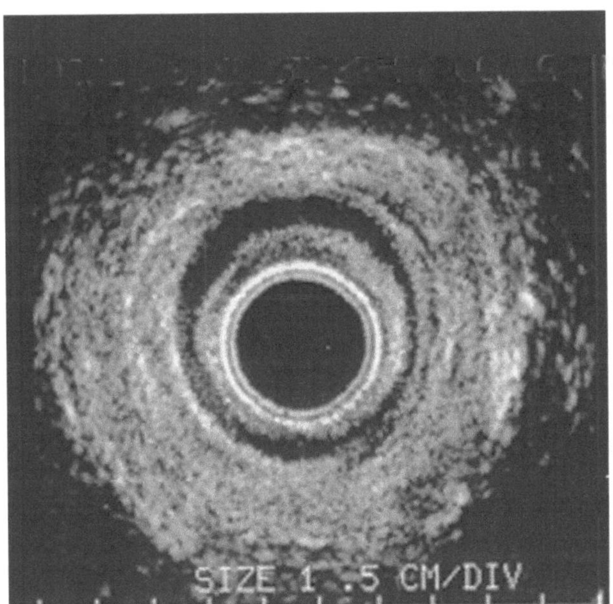

Figure 3-5.2. The mid anal canal.

Defects of the IAS (Figure 3-5.4) are easily recognized given the prominent appearance of the IAS in the mid anal canal. Typically, patients will present with a history of anorectal surgery or dilatation. In this scenario, there is a defect of the IAS in the left lateral quadrant with a thickening of the remaining IAS, suggesting a retraction phenomenon.

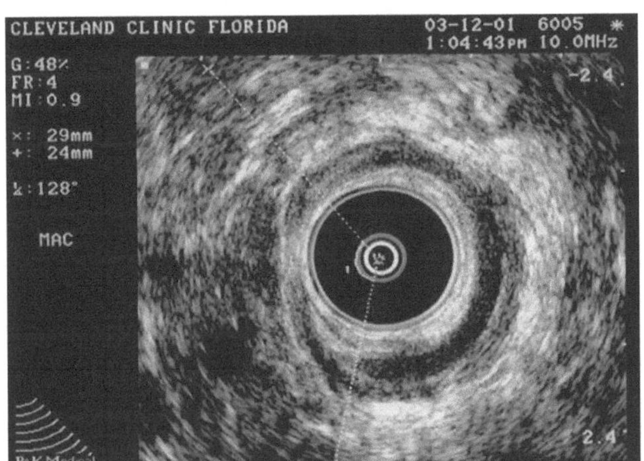

Figure 3-5.4. Defect of the internal anal sphincter.

Defects of the EAS occur for a variety of reasons such as birthing injuries, prior anal fistula surgery, and trauma. The appearance of an EAS defect is, by definition, a break in the circumferential integrity of the mixed hyperechoic density pattern. A defect can have either a hypoechoic or a hyperechoic density pattern. It can be difficult to differentiate between scarring from prior surgery and a true defect. In these difficult cases, there may be a role for electromyography single-fiber density studies. A good example of an anterior EAS defect is demonstrated on Figure 3-5.5, where the EAS is discontinuous anteriorly with a defect measuring 142 degrees.

To easily measure the perineal body thickness, a finger is inserted into the vagina, held gently against the posterior vaginal wall, and then the distance between the anal mucosa and the ultrasonographic reflection of the finger is measured (Figure 3-5.6). Zetterstrom et al.[5] determined that more than 90% of their patients with incontinence had perineal body thickness measurement less than 10 mm. Moreover, this maneuver improved the visualization of sphincter lesions in the majority of patients.

In patients with sphincter defects shown on endosonography, mean resting and squeeze pressures were significantly lower than in patients without anal sphincter defects in 46 subjects.[6] However, when the group was divided by clinical history (symptomatic incontinence vs. asymptomatic), the association of results between manometry and endosonography was not significant. There was a correlation between anal ultrasound findings and fecal incontinence.

The ability to delineate clinically occult injuries of the anal sphincter musculature remains one of the strongest arguments in support of the anal ultrasound for the investigation of the patient with incontinence. Some investigators have advocated that ultrasonography combined with pudendal nerve terminal motor latencies be recognized as the procedures of choice in the work-up of the patient with incontinence.[7–9] Each of the individual physiologic tests offers valuable information relevant to the continence mechanism, whereas ultrasonography yields results that are complementary to other tests. However, in this age of cost containment, ultrasound offers the advantage of providing a great deal of anatomic data in a single test. The patients clearly prefer this examination to other more painful procedures, such as needle mapping with electromyography. This fact also allows ultrasonography to serve as a surveillance tool to monitor the results after sphincteroplasty.[10]

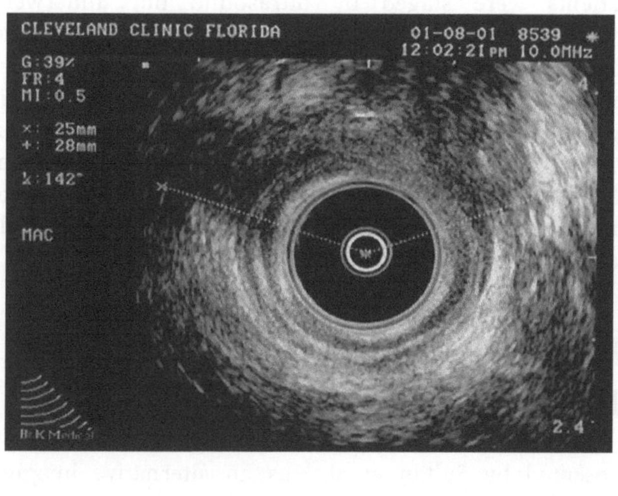

Figure 3-5.5. Defect of the anterior external anal sphincter.

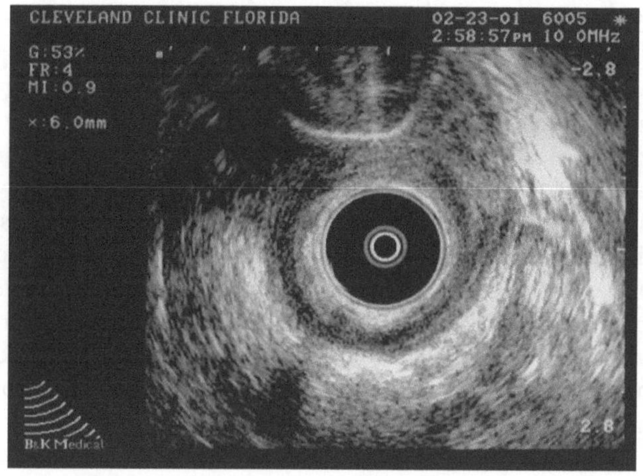

Figure 3-5.6. Measure of perineal body thickness.

Fistulas

Law et al.[11] performed ultrasonography in 22 patients with a diagnosis of recurrent perianal sepsis or fistula in ano. Internal openings were defined as breaks in the mucosal layer. Fistula tracts created hypoechoic bands within the intersphincteric plane and usually communicated with a cavity or scar that appeared as hypoechoic areas. Hypoechoic defects within the EAS were interpreted as transsphincteric fistula tracts. All sonographic data were prospectively collected, and compared with surgical findings. Ultrasonography correctly identified the internal opening in 8 of 12 patients, the primary tract in 11 of 12 patients, a superficial abscess in 2 of 3 patients, and an intersphincteric abscess in 10 of 12 patients. The ultrasound, however, failed to identify an infralevator abscess in three patients and a supralevator abscess in two patients.

Cho[12] defined the endosonographic criteria for an internal opening. He identified three criteria: intersphincteric tract, internal sphincter defect, and subepithelial breach.

The combination of all three findings yielded a 94% sensitivity, 87% specificity, 81% positive predictive value, and a 96% negative predictive value.

Choen et al.[13] performed a prospective trial comparing the accuracy of digital examination and anal ultrasonography in defining the anatomy of the fistula. They discovered no statistical difference between the consultant or research fellow's digital assessment and ultrasonography in identifying intersphincteric and transsphincteric tracts. These investigators found ultrasonography to be of no help in identifying suprasphincteric and extrasphincteric tracts compared with the clinical examination, which was accurate in 78% of these cases. These authors therefore believed that ultrasonography had limitations in the evaluation of the complex fistula.

Yang et al.[14] reported our initial experience with anal ultrasonography for anal fistulas. Sonographic data were compared with surgical findings in 11 patients with fistulas and 6 patients with a suspicion of abscess. In 82% of the patients, sonographic findings correlated with the operative findings. In one patient, a horseshoe fistula was incorrectly assessed as a lateral transsphincteric fistula, and in another patient with Crohn's disease, the primary tract was not visualized. We have since used hydrogen peroxide injection of the tract as an image-enhancement technique during anal ultrasonography for complex and recurrent fistulas.[15] Fistula tracts typically have a hypoechoic appearance. With the injection of hydrogen peroxide, the tract becomes hyperechoic as a result of the bubble-induced increased echogenicity. We believe this technique has helped us to identify tracts more easily. Poen et al.[16] also have found hydrogen peroxide injection to be useful in delineating the anatomic course of perianal fistulas. In a study of 86 patients, endoanal ultrasound was able to identify 74 anal fistulas (43 transsphincteric, 11 intersphincteric, 6 suprasphincteric, 3 superficial, and 11 anovaginal tracts) of which surgery confirmed the type of fistula in 86% and the location of the internal opening in 81%.[17]

In evaluating rectovaginal fistulas, Yee et al.[18] concluded that noncontrast endoanal ultrasound was not useful for the identification of the tract of the rectovaginal fistula. They found that the ultrasound study identified only seven of the rectovaginal fistulas. However, 23 sphincter injuries in 25 patients were identified, and they suggest that anal ultrasonography be used preoperatively in patients with rectovaginal fistulas to identify sphincteric defects.

The superiority of MRI over anal endosonography for the assessment of anal fistulae and abscesses has been controversial. Magnetic resonance imaging has been shown to be superior in some studies. For example, in a study of 39 patients conducted by Maier et al.,[19] MRI demonstrated a sensitivity of 84% and a specificity of 68% in identifying fistula, whereas endoanal ultrasound showed a sensitivity of 60% and specificity of 21%. Meanwhile, another study evaluating fistulas and abscesses complicating Crohn's disease by Orsoni et al.[20] demonstrated that anal ultrasound was more accurate than MRI with sensitivities of 89% and 48%, respectively, in identifying fistulas, and 100% and 55%, respectively, in identifying abscesses.

Pain

In the evaluation of anal pain, ultrasonography is useful in eliminating abscesses and fistulas as causes for acute pain. Intersphincteric abscesses are difficult to diagnose clinically but can easily be found with endoanal ultrasound. However, in a study of patients with chronic idiopathic anal pain, intra-anal ultrasound findings were abnormal (hyperechoic, hypoechoic, or thickened sphincters) in 12 patients.[21] However, these structural findings were nonspecific relative to the etiology of the anal pain. Endoanal ultrasonography has been useful in the guidance of anal sphincter biopsies and botulinum injections into the intersphincteric space in the treatment of chronic anal pain, but otherwise has little to contribute in the evaluation of chronic anal pain.

Anal Carcinoma

Tumors appear as hypoechoic areas and can be seen penetrating the different layers of the anal canal. Ultrasound can measure tumor size on screen and evaluate the invasion of surrounding structures. In imaging the anal canal, endoanal ultrasonography can stage an anal mass by evaluating the depth of infiltration in relation to the normal distinct layers of the anal canal. Typically, it is based on the penetration of the submucosa, the IAS, the EAS, the perianal tissues, or through and into adjacent structures. This is an accurate preoperative staging method that facilitates the choices for treatment.

Tarantino and Bernstein[22] correlated pathologic staging after abdominoperineal resection in five patients to the preoperative endoanal ultrasound staging. Seven other patients were staged by ultrasound but underwent chemoradiation. In this subgroup, they used the endoanal ultrasound to evaluate the patient after their medical therapy and on follow-up. All seven patients showed no evidence of residual tumor, and, on subsequent follow-up, showed no trace of the disease. Others have confirmed that ultrasonography can be used to determine the depth of tumor infiltration, to direct treatment options, and to evaluate tumor response to chemoradiation.[23]

Vaginal and Transperineal Ultrasonography

Presented by Sultan et al.[24] as an alternative imaging method for female patients, vaginal endosonography images the anal sphincters without distortion of the anal canal, submucosa, anal cushions, and sphincter muscles. The patients are placed in the left lateral position. A probe is inserted 3 cm into the vagina, using a transducer, covered by a water-filled balloon. A 7-MHz rotating endoprobe provides a 360-degree image in the same orientation as in anal ultrasonography. Comparisons between anal and vaginal ultrasonography of sphincter images in 20 patients demonstrated an underestimation of the sphincter thickness by the anal ultrasound. The authors concluded that the transanal hard probe resulted in luminal compression of the sphincter muscles. The significance of these observations is unknown.

Using conventional ultrasound equipment with a 5-MHz linear probe, images are obtained by positioning the probe on the midportion of the perineum, obtaining transverse images of the anal canal. The images obtained are comparable to those obtained from anal endosonography, with concentric muscular layers. Similar to transvaginal ultrasonography, the anal canal is imaged undisturbed, and the IAS cushions are visible. Unlike transvaginal ultrasonography, transperineal ultrasonography can be performed on men and women. Conventional ultrasound equipment is available in most hospitals, and because no special probe is necessary, transperineal ultrasonography may become more widely accessible.

Beer-Gabel et al.[25] have introduced dynamic transperineal ultrasound as a noninvasive method of assessing anatomy in patients with pelvic floor dysfunction. Defecography, the current method of dynamically evaluating the pelvic floor during evacuation, uses high levels of radiation and invasive injections of contrast dye into the anus and rectum. Transperineal ultrasound is noninvasive, and provides clear dynamic and still anatomical images of the pelvic floor, including the anal canal, the anal sphincters, the puborectalis sling, the urethra, and the bladder base. The anal sphincters can be seen in the transverse, sagittal, and longitudinal views, which may improve visualization of sphincteric defects or fistulous tracts.

Magnetic Resonance Imaging

Endoanal ultrasonography does have some limitations. Magnetic resonance imaging of the anorectal region was introduced as another imaging modality of this complex region in hopes of improving accuracy in image interpretation. The quality of the image has increased in recent years with the application of the endoanal receiver coil, which has only a 19-mm diameter, and compares favorably with the diameter of the 18-mm endoanal ultrasound probe.

Recent studies have defined MRI appearance of the anal sphincter musculature.[26–29] The IAS is seen as a ring of homogenously high-signal density with a low-signal intensity rim that is rich in collagen and contains neurovascular bundles. The EAS muscle can also be demonstrated with high resolution.

deSouza et al.[30] reported their experience with endoanal MRI in six women who developed incontinence immediately after childbirth. The MRI revealed the site and extent of the defect in all patients, and the findings were confirmed at the time of surgery. In another group of 16 patients who developed incontinence 15 to 30 years after childbirth, no demonstrable defect was found, but in all cases, atrophy of the external sphincter muscle was revealed. The degree of atrophy did not correlate with the degree of delay in pudendal nerve conduction velocity.

Magnetic Resonance Imaging and Endoanal Ultrasonography

These are two excellent imaging modalities that provide anatomic detail of the anorectal region. Each of these modalities has distinct advantages and disadvantages. Endoanal ultrasonography is rapid, less costly, portable, and accurate. However, there is a learning curve for image interpretation, interobserver interpretation is fair, there is a high false-positive rate for sphincter defects, and images are difficult to interpret when there is a great deal of fibrosis. This last point is particularly important for image interpretation in recurrent fistulas and incontinence after multiple prior sphincter surgeries. Magnetic resonance imaging offers clear images of the entire pelvis, and provides clear images of the anal sphincter musculature. Moreover, MRI has the capabilities for multiplanar images and computer manipulation of the images (T1- and T2-weighted images), which facilitate the interpretation of fibrosis from other tissue densities. However, MRI is expensive, requires specialized endoanal coils for optimal imaging of the anorectal region, and with current technology is time-consuming.

Nonetheless, these two different modalities should be regarded as complementary and not competitive to one another. The endoanal ultrasound is an office-based examination performed by the surgeon, and may be regarded as an extension of the physical examination. Some have labeled the ultrasound as the "surgeon's stethoscope." Many patients will be accurately diagnosed by ultrasonography and no further imaging modalities will be needed. However, for the patient in whom the ultrasound may not be accurate, as in the case of complex fistulas, additional information may be gained with an endoanal MRI. Further studies are currently underway to better delineate the role of each of these modalities in our diagnostic armamentarium.

References

1. Law PJ, Bartram CI. Anal endosonography: technique and normal anatomy. Gastrointest Radiol 1989;14:349–353.
2. Nielsen MB, Pederson JF, Hauge C, Rasmussen OO, Christiansen J. Endosonography of the anal sphincter: Findings in healthy volunteers. AJR Am J Roentgenol 1991;157(6):1199–1202.

3. Burnett SJD, Bartram CI. Endosonographic variations in the normal anal sphincter. Int J Colorectal Dis 1991;6:2–4.
4. Tjandra JJ, Milsom JW, Stolfi VM, et al. Endoluminal ultrasound defines anatomy of the anal canal and pelvic floor. Dis Colon Rectum 1992;35:465–470.
5. Zetterstrom JP, Mellgren A, Madoff RD, Kim DG, Wong WD. Perineal body measurement improves evaluation of anterior sphincter lesions during endoanal ultrasonography. Dis Colon Rectum 1998; 41:705–713.
6. de Leeuw JW, Vierhout ME, Struijk PC, Auwerda HJ, Bac DJ, Wallenburg HC. Anal sphincter damage after vaginal delivery: relationship of anal endosonography and manometry to anorectal complaints. Dis Colon Rectum 2002;45:1004–1010.
7. Burnett SJD, Speakman CTM, Kamm MA, Bartram CI. Confirmation of endosonographic detection of external anal sphincter defects by simultaneous electromyographic mapping. Br J Surg 1991;78: 448–450.
8. Emblem R, Dhaenens G, Stien R, Morkrid L, Aasen AO, Bergan A. The importance of anal endosonography in the evaluation of idiopathic fecal incontinence. Dis Colon Rectum 1994;37:42–48.
9. Sentovich SM, Wong WD, Blatchford GJ. Accuracy and reliability of transanal ultrasound for anterior anal sphincter injury. Dis Colon Rectum 1998;41:1000–1004.
10. Nielsen MB, Dammegaard L, Pedersen JF. Endosonographic assessment of the anal sphincter after surgical reconstruction. Dis Colon Rectum 1994;37:434–438.
11. Law PJ, Talbot RW, Bartram CI, Northover JMA. Anal endosonography in the evaluation of perianal sepsis and fistula in ano. Br J Surg 1989;76:752–755.
12. Cho DY. Endosonographic criteria for an internal opening of fistula-in-ano. Dis Colon Rectum 1999;42:515–518.
13. Choen S, Burnett S, Bartram CI, Nicholls RJ. Comparison between anal endosonography and digital examination in the evaluation of anal fistulae. Br J Surg 1991;78:445–447.
14. Yang YK, Wexner SD, Nogueras JJ, et al. The role of anal ultrasound in the assessment of benign anorectal disease. Coloproctology 1993; 5:260–264.
15. Cheong DMO, Nogueras JJ, Wexner SD, Jagelman DG. Anal endosonography for recurrent anal fistulas: image enhancement with hydrogen peroxide. Dis Colon Rectum 1993;36:1158–1160.
16. Poen AC, Felt-Bersma RJ, Eijsbouts QA, Cuesta MA, Meuwissen SG. Hydrogen peroxide-enhanced transanal ultrasound in the assessment of fistula-in-ano. Dis Colon Rectum 1998;41:1147–1152.
17. Sudol-Szopinska I, Gesla J, Jakubowski W, Noszczyk W, Szczepkowsi M, Sarti D. Reliability of endosonography in evaluation of anal fistulae and abscesses. Acta Radiol 2002;43:599–602.
18. Yee LF, Birnbaum EH, Read TE, Kodner IJ, Fleshman JW. Use of endoanal ultrasound in patients with rectovaginal fistulas. Dis Colon Rectum 1999;42:1057–1064.
19. Maier AG, Funovics MA, Kreuzer SH, et al. Evaluation of perianal sepsis: comparison of anal endosonography and magnetic resonance imaging. J Magn Reson Imaging 2001;14:254–260.
20. Orsoni P, Barthet M, Portier F, Panuel M, Desjeux A, Grimaud JC. Prospective comparison of endosonography, magnetic resonance imaging and surgical findings in anorectal fistula and abscess complicating Crohn's disease. Br J Surg 1999;86:1093–1094.
21. Christiansen J, Bruun E, Skjoldbye B, Hagen K. Chronic idiopathic anal pain: analysis of ultrasonography, pathology, and treatment. Dis Colon Rectum 2001;44:661–665.
22. Tarantino D, Bernstein MA. Endoanal ultrasound in the staging and management of squamous cell carcinoma of the anal canal: potential implications of a new ultrasound staging system. Dis Colon Rectum 2002;45:16–22.
23. Magdeburg B, Fried M, Meyenberger C. Endoscopic ultrasonography in the diagnosis, staging, and follow-up of anal carcinomas. Endoscopy 1999;31:359–364.
24. Sultan AH, Loder PB, Bartram CI, Kamm MA, Hudson CN. Vaginal endosonography: new approach to image the undisturbed anal sphincter. Dis Colon Rectum 1994;37:1296–1299.
25. Beer-Gabel M, Teshler M, Barzilai N, et al. Dynamic transperineal ultrasound in the diagnosis of pelvic floor disorders: pilot study. Dis Colon Rectum 2002;45:239–248.
26. deSouza NM, Gilderdale DJ, MacIver DK, Ward HC. High-resolution MR imaging of the anal sphincter in children: a pilot study using endoanal receiver coils. AJR Am J Roentgenol 1997;169:201–206.
27. deSouza NM, Puni R, Kmiot WA, Bartram CI, Hall AS, Bydder GM. MRI of the anal sphincter. J Comput Assist Tomogr 1995;19:745–751.
28. Hussain SM, Stoker J, Schutte HE, Lameris JS. Imaging of the anorectal region. Eur J Radiol 1996;22:116–122.
29. Strohbehn K, Ellis JH, Strohbehn JA, DeLancey JO. Magnetic resonance imaging of the levator ani with anatomic correlation. Obstet Gynecol 1996;87:277–285.
30. deSouza NM, Puni R, Zbar A, Gilderdale DJ, Coutts GA, Krausz T. MR imaging of the anal sphincter in multiparous women using an endoanal coil: correlation with in vitro anatomy and appearance in fecal incontinence. AJR Am J Roentgenol 1996;167:1465–1471.

3-6

Neurologic Evaluation of the Pelvic Floor

Virgilio Salanga

Several elegant methods of neurophysiologic assessment of the anal and urethral sphincters exist. Their purpose is to help localize the site of neurologic pathology leading to the disturbance of bowel and bladder continence and evacuation. Although the clinical neurologic examination may sufficiently determine if the lesion involves the muscle or the nerve control to the muscle, a quantitative assessment of the disturbance by neurophysiologic means is often needed for clinical monitoring and treatment planning.

Available neurophysiologic techniques include electromyography of the periurethral and anal sphincter muscles, perineal and pudendal motor conduction studies, sacral reflexes recording, somatosensory evoked response study from the pelvic floor to the cerebral sensory cortex, and transcutaneous spinal cord and cerebral motor cortex stimulation to record an evoked pelvic floor muscle response (Table 3-6.1).

The sacral reflex arc consists of an afferent limb originating in the bladder and urethra along visceral autonomic afferents to the T10-L2 spinal cord, and an efferent limb originating in the S2-S3-S4 ventral nerve roots to the pudendal nerve and its terminal branch, perineal nerve. In response to stimulation of the bladder wall or urethra, pelvic floor muscle contraction occurs. The urethral-anal reflex, bladder-anal reflex, and clitoral-anal reflex are neurophysiologically measurable. Abnormalities of the sacral reflexes may be seen in peripheral neuropathy, cauda equina and conus medullaris lesions, pelvic plexus, and pudendal nerve abnormalities.[1]

Pudendal nerve somatosensory evoked potential recordings provide information on lesions in the spinal cord and brain (central nervous system), but require interpretation in conjunction with pudendal nerve conduction studies.[1]

Transcutaneous spinal stimulation at two levels of the lumbar spine (L1 and L4) and recording the latencies of the evoked muscle responses from the anal sphincter or puborectalis muscle provides information on lesions involving the S2-S3-S4 motor nerve roots, and complements the information derived from pudendal nerve and perineal nerve conduction studies. Transcutaneous cervical spine (spinal cord) and cranial (motor cortex) stimulation and recording motor response from the pelvic floor muscles provide information on upper motor neuron disorders affecting the pelvic floor, when interpreted in conjunction with pudendal nerve or perineal nerve conduction studies.[2]

In our institution, my colleagues and I have considerable experience in measurement of pudendal nerve terminal motor latencies (PNTML) and external anal sphincter muscle concentric needle electromyography (AEMG), as part of the routine evaluation of bowel evacuatory disorders, especially fecal incontinence. In a 1991 survey of American and British colorectal surgeons, PNTML and AEMG were available only in 15% and 34% of anorectal physiology laboratories, respectively. Neurogenic injury in fecal incontinence was suggested by histometric and single fiber electromyogram (EMG) studies of the anal sphincter. Conditions that promote traction injury to the pudendal nerves (as in childbirth and prolapse) may lead to fecal incontinence. These neurophysiologic techniques provide objective assessment of nerve and/or muscle injury, and allow for optimal treatment approach.[3]

We perform PNTML studies according to the method described by Kiff and Swash[4] using a disposable St. Marks electrode (Dantec, Skovlunde, Denmark) with a Nicolet Viking II (Nicolet, Madison, WI) EMG machine. The St. Marks electrode is mounted onto the gloved index finger of the right hand, which is introduced into the rectum (Figure 3-6.1). The ischial spine and lateral edge of the sacrum are located so that the tip of the finger with the cathode-simulating electrode is in the vicinity of the pudendal nerve. Electrical stimuli are repeatedly delivered as the fingertip is methodically moved until the pudendal nerve is maximally stimulated based on the best amplitude of at least three reproducible and identical compound muscle action potentials recorded from the external anal sphincter muscle. The cathodal stimulating point is at the main trunk of the pudendal nerve; the terminal motor latency represents the conduction time along the inferior

Table 3-6.1. Potential usefulness of electrodiagnostic tests in certain pelvic floor disorders

	Pudendal Nerve (Motor Latency)	Perineal Nerve (Motor Latency)	Anal Needle (EMG)	Anal Surface (EMG)	Sacral Limb (Reflex)	NCT	SEP
Fecal incontinence	+	−	+	−	−	−	−
Constipation	+	−	+/−	+	−	−	−
Urinary incontinence	+	+	−	−	+	−	−
Pelvic organ prolapse	+	+	+	−	+	−	−
Pelvic trauma	+	+	+	−	+	−	−
Diabetic neuropathy	+	+	+	−	+	+	−
Cauda equine disorder	+	+	+	−	+	+	+
Myelopathy	−	−	−	−	+	−	+

NCT, nerve conduction test; SEP, somatosensory evoked potential.

rectal branch that innervates the external anal sphincter muscle. The other side is similarly done. Patients get a minimum of three recordings on each side to ensure consistency and accuracy in measurements. Based on previously published normative data, we consider PNTML of less than 2.3 ms as normal[5] (Figures 3-6.2 and 3-6.3).

Electromyography of the external anal sphincter is performed with a bipolar concentric needle electrode, 23-gauge, 25 to 75 mm long. The right and left halves or four quadrants (left, right, anterior, and posterior sectors) of the external anal sphincter muscle are routinely evaluated. Observations are made of muscle electrical activity at rest, during voluntary (squeeze) and reflexive (with a cough) contraction, and simulated defecation. Spontaneous potentials including positive sharp waves, fibrillations, fasciculations, and complex repetitive discharges are looked for (Figure 3-6.4). Alterations in motor unit potential (MUP) amplitude, duration, and polyphasia are observed (Figure 3-6.5). Reduction in MUP recruitment is noted and graded as minimal, moderate, severe, or complete absence of vol-

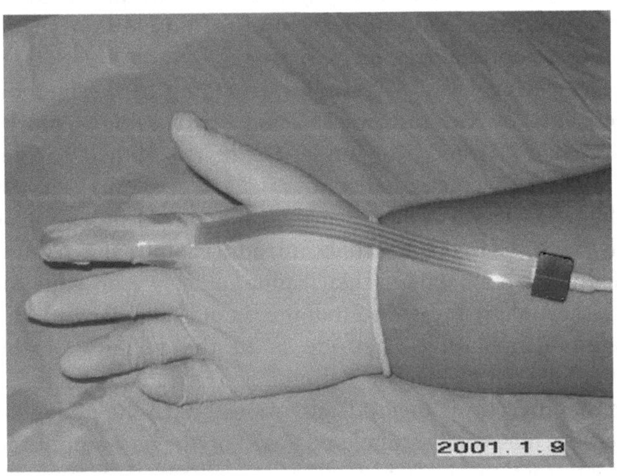

Figure 3-6.1. St. Mark's electrode.

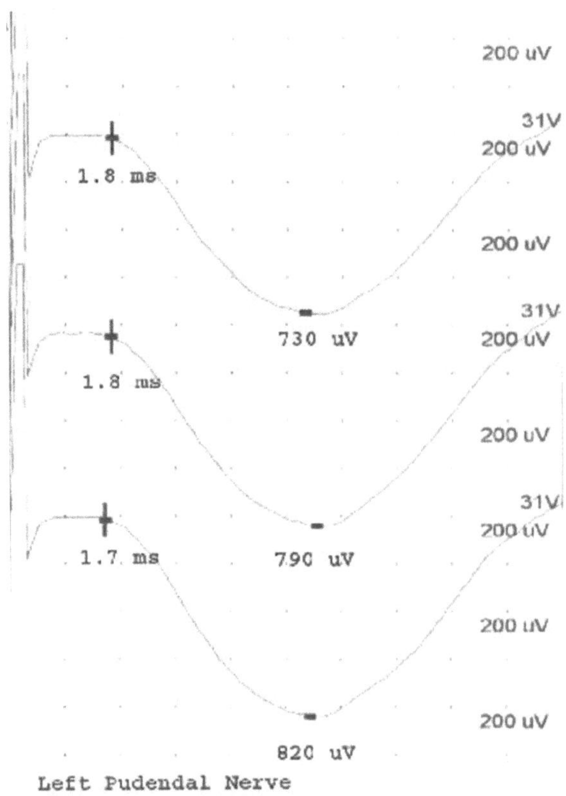

Figure 3-6.2. External anal sphincter compound muscle action potentials, left pudendal nerve stimulation.

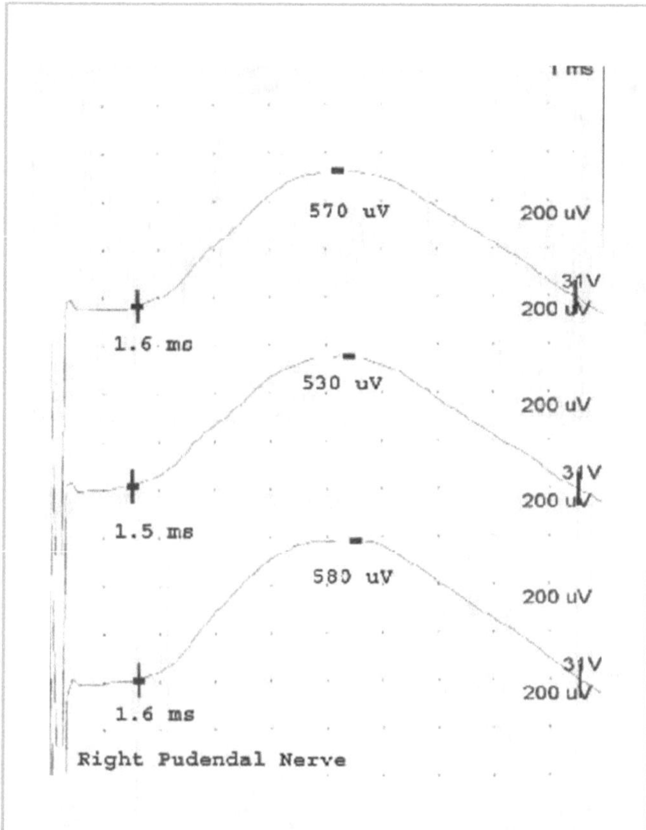

Figure 3-6.3. External anal sphincter compound muscle action potentials, right pudendal nerve stimulation.

untarily or reflexively activated MUPs (Figure 3-6.5). Sustained MUP recruitment during simulated defecation, which may indicate paradoxical pelvic floor muscle contraction, is noted.[5]

In patients with constipation, we use an intra-anal sponge electrode (Dantec) for surface EMG recording of the external anal sphincter muscle, instead of the concentric needle electrode. By avoiding needle-related discomfort, the kinetic activity of the anal sphincter during voluntary and reflexive contraction (squeeze and cough), and during simulated defecation is more accurately assessed, eliminating pain-induced contraction that may mimic paradoxical anal muscle contraction.[6]

In a prospective study of 225 consecutive patients with fecal incontinence (idiopathic 72, obstetric 45, prolapse 43, trauma 42, neurologic 23), PNTML and AEMG studies showed neurophysiologic evidence of neuromuscular abnormalities in 76% of patients: Reduced MUP recruitment during voluntary contraction (60%), prolonged PNTML (36%), paradoxical anal sphincter contraction during simulated defecation (19%), and assorted abnormalities of MUP morphology, fibrillations/positive waves, and complex repetitive discharges (46%). There was strong correlation between reduced voluntary MUP recruitment and decreased anal manometric pressures. Four-quadrant AEMG "mapping" showed good concordance with endoanal sonography in 35 of 41 patients (86%) tested with both techniques. Three patients were found with muscle defects by AEMG that were missed by sonography, and vice versa in another three patients. The two tests are complementary and not redundant. Our data in this group of patients did not clarify the exact role of pudendal neuropathy in the pathogenesis of fecal incontinence, because this was not present in 64%. Nonetheless, the identification of pudendal neuropathy is considered predictive of poor outcome of sphincter surgical repair. These neurophysiologic techniques assisted our colorectal surgeons in assigning the most appropriate therapy based on expected outcome.[5]

In urinary voiding abnormalities, the integrity of the perineal nerves is assessed by stimulation of the pudendal nerves through the vagina, using the St. Marks electrode, and recording the compound muscle action potential from the periurethral muscle with recording electrodes mounted on a Foley catheter. Simultaneous recording of the compound muscle action potentials from the left and right external anal muscle sphincter muscle is done with a pair of para-anal surface disc electrodes. The pudendal-perineal nerve terminal motor latencies are generally 0.2 to 0.5 msec longer than the pudendal-inferior rectal nerve terminal motor latencies.

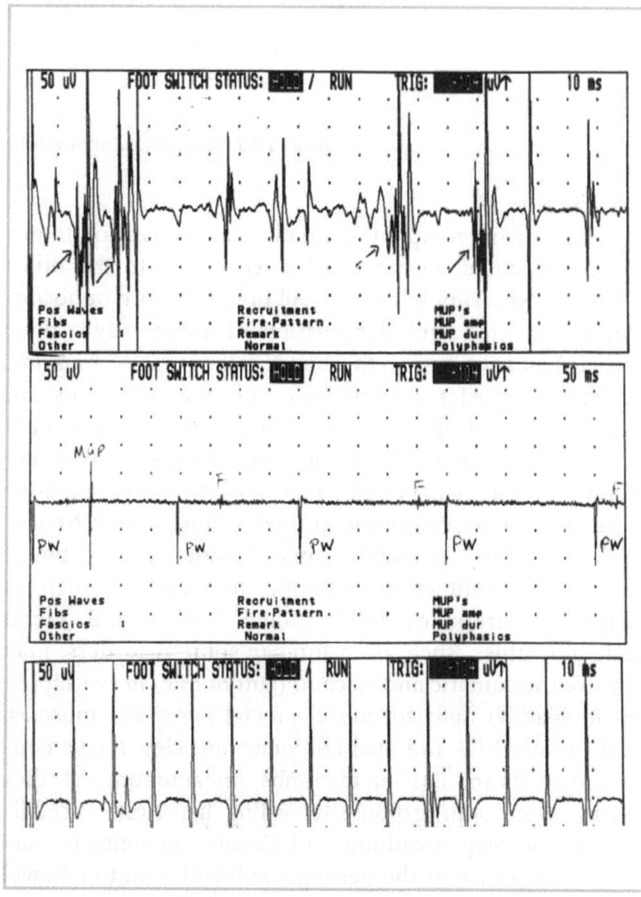

Figure 3-6.4. Needle EMG of external anal sphincter: polyphasic MUPs (arrows), positive waves (pw), fibrillations (f), high frequency repetitive discharges (bottom).

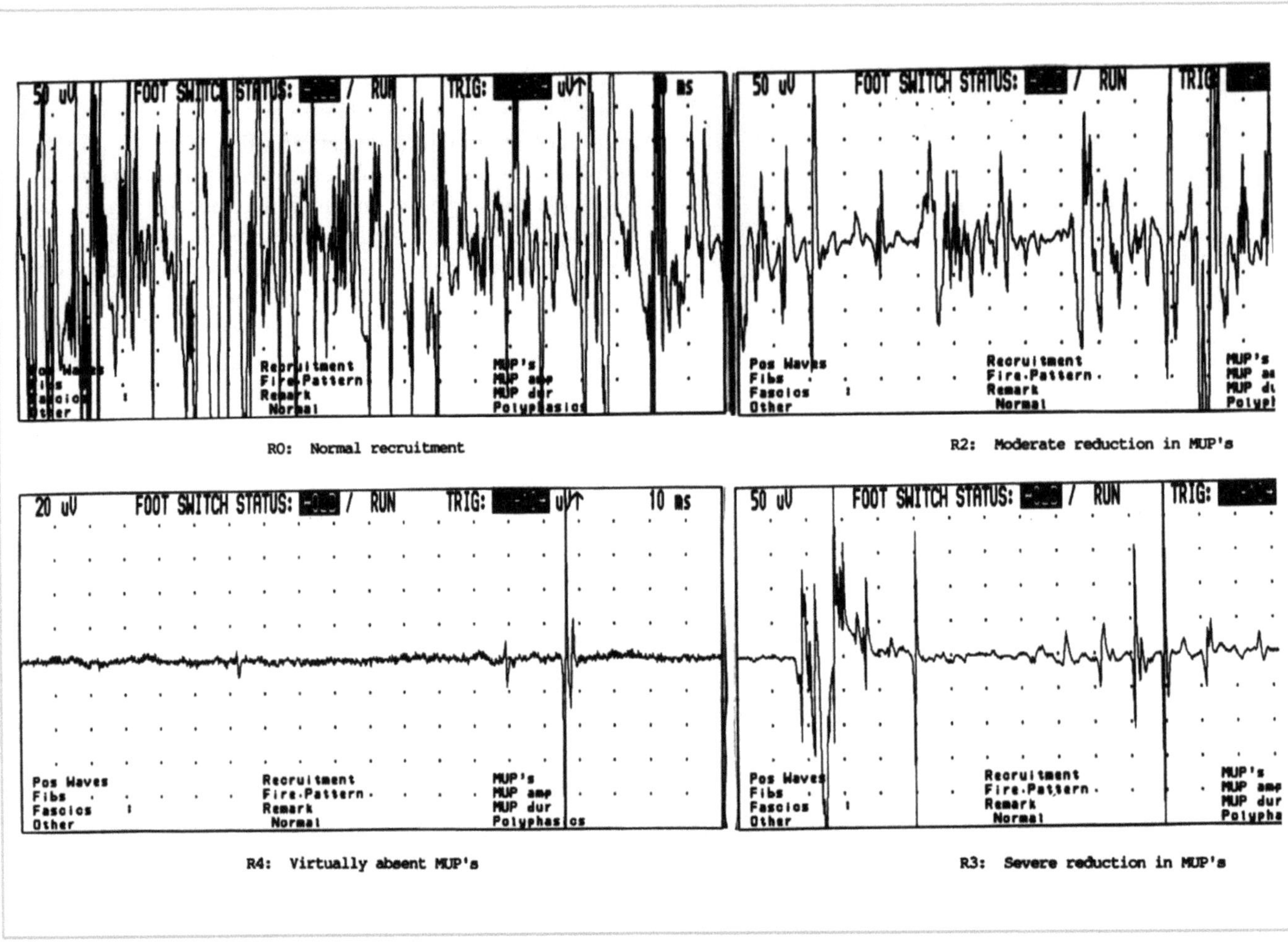

Figure 3-6.5. Motor unit potential recruitment, needle EMG of external anal sphincter muscle.

Pelvic organ prolapse potentially leads to traction injury to the pudendal nerves and branches, and to the other pelvic nerves supplying the puborectalis and pubococcygeus muscles, and denervation of these pelvic floor muscles and the periurethral and anal sphincter muscles. Pudendal nerve terminal motor latency measurements and needle EMG of these muscles may provide potentially useful information influencing surgical reparative procedures. Severely denervated parts of weakened pelvic floor muscles may not be optimal for pelvic floor repair. Neurologic disorders frequently cause bladder and/or bowel evacuatory dysfunction. Systemic neuropathies, such as diabetic neuropathy, and regional lumbar and sacral radiculopathies, such as in lumbar spine disorders, may involve the somatic and visceral (autonomic) nerve supply to the bladder detrusor muscle, rectal peristaltic muscles, and periurethral and anal sphincter muscles. Nerve conduction tests and EMG in the limbs, and autonomic studies (sympathetic skin response recording, heart rate variability during deep breathing and Valsalva maneuver) may clarify the cause of the neurogenic bladder and/or bowel syndrome in these diseases. Spinal cord disease (myelopathy), such as in multiple sclerosis, cervical or thoracic spondylosis, etc., will often cause small-capacity hyperreflexic bladder, manifesting as an overactive bladder with frequency and urgency incontinence. Pudendal nerve latencies will be normal in this setting, but lower limb somatosensory evoked potential recording may show delayed conduction along the central somatosensory afferent pathways.

References

1. Benson JT. Clinical neurophysiological techniques in incontinence. In: Ostergard DR, Bent AE, eds. Urogynecology and Urodynamics. Baltimore: Williams & Wilkins; 1996:239.
2. Snooks SJ, Swash M. Nerve stimulation techniques. In: Henry MM, Swash M, eds. Coloproctology and the Pelvic Floor. London: Butterworths; 1985:112.
3. Wexner SD, Marchetti F, Salanga VD, Corredor C, Jagelman DG. Neurophysiologic assessment of the anal sphincters. Dis Colon Rectum 1991;34:606–612.
4. Kiff ES, Swash M. Slowed conduction in the pudendal nerves in idiopathic (neurogenic) fecal incontinence. Br J Surg 1984;71:615–616.
5. Cheong DMO, Vacarro C, Salanga VD, Wexner SD, Phillips RC, Hanson MR. Electrodiagnostic evaluation of fecal incontinence. Muscle Nerve 1995;18:612–619.
6. Pfeifer J, Teoh T, Salanga VD, Agachan F, Wexner SD. Comparative study between intra-anal sponge and needle electrode for electromyographic evaluation of constipated patients. Dis Colon Rectum 1998;41:1153–1157.

3-7

Upper Gastrointestinal Evaluation Related to the Pelvic Floor

Gregory F. Bonner

Pelvic floor dysfunction is not traditionally associated with upper gastrointestinal (GI) disorders. However, most of the major pathways of the brain-gut axis, which are so critical to pelvic floor dysfunction are the same as those innervating the upper GI tract. Although sacral nerve trunks are limited to the pelvis, the vagal and splanchnic innervation is the same. Dysfunction of the enteric nervous system in the upper GI tract leads to a variety of disorders in many ways similar to those disorders more traditionally associated with pelvic floor dysfunction.

As with pelvic floor dysfunction, related upper GI disorders often manifest as disorders of motility. They can generally be categorized into esophageal motility disorders, functional dyspepsia, and disorders of biliary spasm or dyskinesia. As with other functional GI disorders, patients with these conditions often show alterations in visceral sensation such as increased sensitivity to balloon distention in various locations in the GI tract. As outlined below, there is often overlap of symptoms and presentation of the conditions.

Esophageal Motility Disorders

Esophageal motility disorders may present with a variety of symptoms including dysphagia, noncardiac chest pain, or heartburn. Dysphagia is the most serious of these symptoms and, under such circumstances, the physician's first priority is to evaluate for related diseases such as esophageal tumor or stricture, which might require urgent intervention. A barium esophagram will adequately evaluate for either tumor or stricture. Endoscopy has become more the first line intervention because it offers the added benefit of evaluating for erosive esophagitis or the precancerous condition, Barrett's esophagus. Additionally, endoscopy offers therapeutic dilatation for strictures. For patients with just pain or heartburn, an equally acceptable first diagnostic intervention is a therapeutic trial of a high-dose proton pump inhibitor such as omeprazole, 20 mg twice a day. For patients who fail to respond to this therapeutic trial, an endoscopy is indicated. Additionally, the American College of Gastroenterology recommends endoscopy to screen for Barrett's esophagus in patients with heartburn occurring three or more times weekly for 5 years or more. For patients with ongoing and troublesome symptoms, other investigational tools include an esophageal motility test and/or pH probe monitoring study. Esophageal motility testing may identify more unusual conditions, such as nutcracker esophagus or esophageal spasm, in which the patient might benefit from the addition of psychotropic agents such as trazodone. pH probe monitoring studies can differentiate if refractory heartburn symptoms are caused by excessive acid reflux as opposed to visceral hypersensitivity of the esophagus.

Dyspepsia

Dyspepsia is derived from the Greek *dys* and *peptein*, which literally interpret as bad digestion. The term dyspepsia encompasses a variety of symptoms of persistent upper abdominal pain and discomfort. The predominant symptoms may include pain, heartburn, nausea, early satiety, or postprandial fullness or bloating. This is the most common type of functional upper GI disorder, with a prevalence rate in Western countries estimated at 25%, although less than half will seek medical attention. There are subtypes of dyspepsia, which can be better approached by focusing on the predominant symptom. Ulcer-like symptoms are characterized by upper abdominal pain relieved by certain foods or antacids. Reflux-like symptoms include heartburn and acid regurgitation. Many patients with dyspepsia are thought to have a dysmotility-like disorder with less pain but more symptoms of early satiety, postprandial fullness and bloating, or nausea and vomiting. With an exhaustive evaluation, approximately 50% of patients will demonstrate some type of motility disorder such as impaired fundic relaxation, antral hypomotility, gastroparesis, small bowel dysmotility, or abnormal duodenogastric reflexes.

A traditionally defined disease with definable pathology such as an ulcer or erosive esophagitis will be found only in a minority of patients presenting with dyspepsia. The term nonulcer dyspepsia (NUD) is given to patients with no definable structural disease after a standard work-up. Patients with a histologic diagnosis of gastritis or duodenitis should be included in the category of NUD, because a clear link between the histologic abnormality and the symptoms is not established.

The aggressiveness of evaluation with endoscopy is variable depending on the circumstance. Because of concern for gastric cancer, endoscopy is recommended for all patients older than 45 years presenting with new onset dyspepsia. For younger patients, it is reasonable to delay endoscopy as long as alarm symptoms of dysphagia, weight loss, bleeding, and recurrent vomiting are not present. For younger patients, an empiric trial with an antisecretory agent is reasonable, reserving endoscopy for patients who are not helped with this therapy. Recent studies have demonstrated proton pump inhibitors such as omeprazole (Prilosec®) or lansoprazole (Prevacid®) to be more effective in treating uninvestigated dyspepsia than H_2 receptor antagonist or antacids. Other routine evaluations for older patients, and younger patients who fail an empiric trial, include routine hematologic and biochemical tests (complete blood count, liver function test, electrolytes, creatinine, and thyroid function tests). An ultrasound of the right upper quadrant would be reasonable, although it is not clear if gallstones should be considered causative of symptoms of dyspepsia. Other tests to be considered for patients with refractory and disabling symptoms include 24-hour pH testing, gastric emptying testing, and, where available, gastroduodenal motility testing.

The role of Helicobacter pylori in patients with dyspeptic symptoms remains controversial. It is reasonable to perform serologic testing and to treat for Helicobacter if positive, especially in the uninvestigated patient with dyspepsia. However, for patients with functional dyspepsia, that is, patients who have undergone an investigation that does not reveal evidence of ulcer disease, large studies have shown that treatment for H. pylori does not result in symptomatic improvement.

Sphincter of Oddi Dysfunction

Sphincter of Oddi dysfunction (SOD), also referred to as biliary dyskinesia, presents with typical symptoms of biliary colic of right upper quadrant pain with nausea and vomiting. However, this pain is not actually caused by gallstones but instead a motility disorder with spasms of the sphincter of Oddi. By definition, an ultrasound would not show gallstones. In fact, the majority of these patients have already undergone a cholecystectomy, either for gallstones or suspected microlithiasis, before the diagnosis of SOD is seriously considered. When suspected, liver function tests should be checked in the midst of an episode. Abnormal aspartate aminotransferase and alanine aminotransferase during episodes of pain, which later normalize, are highly suggestive in the appropriate clinical setting of SOD (assuming the patient has already undergone cholecystectomy). Abdominal ultrasound should be obtained, because a dilated common bile duct is also suggestive. An additional helpful noninvasive test is a hepatobiliary scan, which may show delay in emptying of the nuclear tracer into the duodenum. The most definitive diagnosis is established with biliary manometry obtained during an endoscopic retrograde cholangiopancreatography. This study is only available at certain centers and is generally performed only with hesitation, because the incidence of pancreatitis induced by this study is approximately 10%.

Summary

A variety of upper GI motility disorders have been described. These conditions are known to occur in high frequency in patients with irritable bowel syndrome and may be associated with other disorders of pelvic floor dysfunction.

Suggested Reading

1. Hunt RH. Evolving concepts in the pathophysiology of functional gastrointestinal disorder. J Clin Gastroenterol 2002;35(suppl 1): S2–S6.
2. Thumshirn M. Pathophysiology of functional dyspepsia. Gut 2002; 51(suppl 1):i63–i66.
3. Talley NJ. Dyspepsia: management guidelines for the millennium. Gut 2002;50(suppl IV):iv72–iv78.
4. Talley NJ, Janssens J, Lauritsen K, Racz I, Bolling-Sternevald E. Eradication of Helicobacter pylori in functional dyspepsia: randomized double blind placebo controlled trial with 12 months' follow up. The Optimal Regimen Cures Helicobacter Induced Dyspepsia (ORCHID) Study Group. BMJ 1999;318:833–837.

Section IV

Anatomic Correlates

Section IV

Anatomic Correlates

G. Willy Davila

A clear and concise understanding of pelvic anatomy is crucial for the management of pelvic floor dysfunction. In the past, the simplistic view that anatomic alterations were primarily responsible for symptomatic dysfunction led to surgical interventions with decreased success rates. Our current concepts intertwine the anatomic and functional aspects of pelvic floor integrity. Therefore, as we evaluate a patient with a clear anatomic defect, emphasis must be placed on what symptomatic, or asymptomatic, dysfunction is present alongside the identified anatomic defect.

Our current concept is one of viewing the pelvic floor as a horizontal unit rather than as a set of vertical compartments. This concept allows us to better understand the interrelationships that occur between organ systems at fascial, muscular, and mucosal levels. A woman with a large posterior vaginal bulge may have a rectocele, an enterocele, both, or neither. A clear knowledge of the rectovaginal septum anatomy will allow the clinician to better evaluate the patient to determine what anatomic alteration is present. Of interest is that clinicians have markedly different terminology for the same anatomic alteration. A clear example of this is the description of a posterior vaginal bulge as a rectocele by gynecologists. Colorectal surgeons tend to view rectoceles with much greater functional emphasis. Understanding anatomic alterations in the rectovaginal septum and clearly describing them is therefore of utmost importance when describing a patient's pelvic floor dysfunction, especially when multiple specialists are involved. The same concept applies to an anterior vaginal bulge. Although cystoceles are common, an anterior vaginal wall bulge can include a larger diverticulum or large inclusion cysts in the urethrovaginal septum, which can occur after a sling procedure. A careful preoperative evaluation is therefore key to determining the etiology of an anatomic alteration.

Physical anatomy must be evaluated and treated within the constraints of neuromuscular integrity. Simply replacing prolapsed organs to their anatomic position will not lead to durable results unless neuromuscular function is improved by postoperative physiotherapy. Optimal results may only be achievable if neurologic integrity is present.

Endofascial layer integrity has been given a significant amount of importance by gynecologic reconstructive surgeons. This concept has not been fully espoused by colorectal surgeons and urologists. Sharing information regarding newly understood anatomic concepts is important when planning a multidisciplinary approach to treatment of pelvic floor dysfunction, especially when surgery is involved.

In this chapter, we have separated the description of pelvic anatomy in a typical compartmentalized format. This is not to de-emphasize our view of the pelvic floor as a horizontal unit. However, to better allow for understanding by the reader, we have opted to present these data in this format. This will undoubtedly lead to some degree of repetition and some variation in the described concepts in pelvic floor anatomy. However, the reader should attempt to place the provided information into a three-dimensional unitary model with underlying common neuromuscular, mucosal, and fascial integrity variables.

4-1

Urologic Anatomic Correlates

Jonathan Jay

The structures of the urinary system as it relates to the pelvis are the ureters, the bladder, and the urethra. The goal of this chapter is to describe the location, basic function, vascular supply, nerve supply, and support of these structures within the pelvis.

Anatomy of the Pelvic Urinary System

Ureter

The ureters are paired, thick muscular tubes with a lumen of approximately 3 mm in diameter and are 24 to 30 cm in length. They originate at the renal pelvis and function to propel urine from the kidney to the bladder. In approximately 1% of the population, the ureter is duplicated. Duplications of the ureter are characterized as partial or complete. In partial duplications, the second ureter joins the first before reaching the bladder. In complete duplications, both ureters travel side by side to the bladder. In the abdomen, the ureters lie on the medial surface of the psoas major muscle, within the retroperitoneum. The right ureter lies underneath the terminal ileum, cecum, appendix, and ascending colon, and their mesenteries. The ovarian vessels cross the right ureter at its midsection. The left ureter is adherent to the underside of the mesentery of the descending and sigmoid mesocolon, and is crossed by the inferior mesenteric and ovarian vessels. The inferior mesenteric artery is near and looks similar to the left ureter; therefore, care must be taken to distinguish these structures during dissection within this area. The ureters pass into the pelvis near the bifurcation of the common iliac vessels. As they descend into the pelvis, at the level of the ischial spine, they are in close proximity to the suspensory ligament of the ovary and form the posterior limit of the ovarian fossa. This relationship is important to recognize because injury to the ureter can occur during ligation of the infundibulopelvic ligament during oophorectomy. Both ureters travel parallel and medial to the obturator fossa and the internal iliac vessels. In this region of the pelvis, they are lateral to the sacrospinous ligament and pass through the parametrium of the broad ligament to a location approximately 1.5 cm lateral to the cervix. As the ureters descend to the bladder within the broad ligament, the uterine artery crosses anteriorly (Figure 4-1.1). Within centimeters below the crossing of the ureter by the uterine artery, the inferior vesicular artery can cross the ureter anteriorly or posteriorly. The vaginal artery is posterior and medial to the ureter at this location. Here numerous arteries and veins surround both ureters and this is another common site of ureteral injury during hysterectomy. At the level of the cervix, the ureters traverse the cardinal ligament on its way to the bladder base. They travel medially on the anterior surface of the vagina for 1 to 3 cm before reaching the bladder. The ureters perforate the bladder 5 to 6 cm apart and run obliquely through the detrusor wall for 1.5 cm. The internal ureteral orifices are much closer to each other than to their external penetrations of the bladder wall. As the ureters enter the bladder, they form the trigone, which is a triangular region of the bladder base (Figure 4-1.2). The trigone is made up of a superficial and deep layer that is separate from the detrusor muscle. The superficial layer of the trigone is an extension of the inner muscular layer of the ureter. The fibers from each ureter meet to form a triangular sheet of muscle that extends from the two ureteral orifices and continues distally into the posterior aspect to the proximal urethra. The superior portion of the ureteral muscle of the trigone forms the interureteric ridge. The lateral portions of this muscle are the ureteral bars. Waldeyer's sheath is a fibromuscular sheet of tissue that originates 2 cm above the bladder and is wrapped around the ureter. This sheath extends longitudinally to the bladder neck and forms the deep portion of the trigone. The trigone sits on the muscles of the detrusor wall and anchors the ureters to the bladder. The distal intravesical portion of the ureter is submucosal and is supported by the detrusor muscle backing. As the bladder fills, urine compresses the ureter against the muscle backing, creating a flap valve, which prevents reflux of urine from the bladder into the ureter.

The arterial supply of the ureter arises from branches of various vessels as it descends into the pelvis. The abdominal sources of arterial blood supply to the ureter are the renal artery (30%), the aorta (15.4%), and the gonadal arteries (7.7%). The most significant sources of blood to the pelvic ureter are the superior vesicular arteries (12.8%), the inferior vesicular arteries (12.9%), and the internal iliac arteries (8.5%).[1] The ureteral veins drain at either end of the ureter and along its length. In the abdominal portion of the ureter, the main veins drain into the renal and gonadal vessels. Additional drainage occurs along other veins in the proximity of the ureter. The pelvic portion of the ureter drains into the plexus of veins within the broad ligament and other adjacent veins.

The nervous innervation of the abdominal ureters arises from the renal and aortic plexus. The pelvic ureter receives its nerve supply from the inferior hypogastric and pelvic plexus. These nerves contain cholinergic and adrenergic fibers that regulate ureteral peristalsis. Ureteral peristalsis is not activated by the nervous system, but is thought to modulate its actions.[4] Afferent fibers in the lamina propria relay messages of stretch, osmolarity, and pH to the brain. There are few pain receptors in the ureter itself. Most of the pain that we perceive from ureteral obstruction is secondary to distention of the renal capsule, rather than stimulation from the ureter. The pain perceived by ureteral and renal capsular distention is likely relayed through the parasympathetic nervous system and can be referred to various sites that share the nerve roots of T11-L2, such as the genitalia, groin, and upper thigh. These are common sites of referred pain during ureteral obstruction.

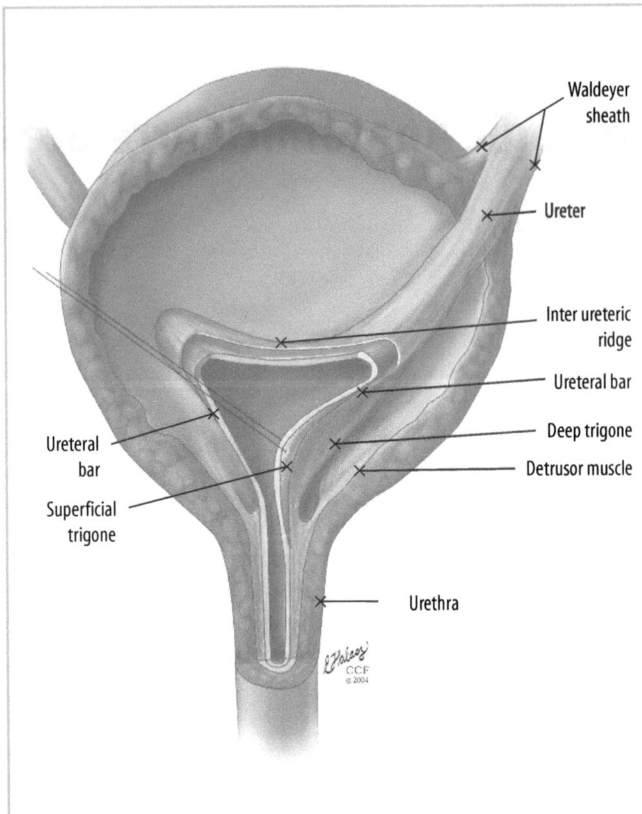

Figure 4-1.2. The trigone is a triangular region of the bladder base. It comprises a superficial and deep layer, which is derived from the ureteral musculature and Waldeyer's sheath, respectively. (Reprinted with the permission of The Cleveland Clinic Foundation.)

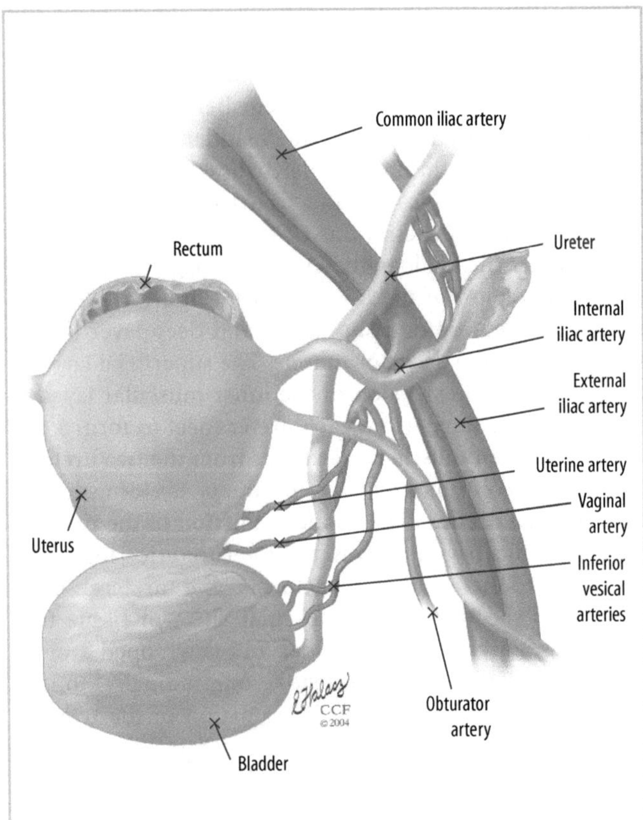

Figure 4-1.1. As the ureter descends to the bladder within the broad ligament, the uterine artery crosses it anteriorly and the vaginal artery passes underneath it. The inferior vesicular artery can cross the ureter anteriorly or posteriorly. (Reprinted with the permission of The Cleveland Clinic Foundation.)

Bladder

The urinary bladder is a hollow muscular organ, which serves as a reservoir for the storage and voluntary expulsion of urine. When filled to capacity, the bladder is spherical and holds approximately 500 mL of fluid; however, this capacity can vary based on one's size, sex, or age. The bladder lies in the anterior half of the pelvis and it is bounded anteriorly by the symphysis pubis, laterally by the pelvic sidewalls, and posteriorly by the vagina, cervix, and uterus. The urachus terminates in the umbilicus and anchors the apex of the bladder to the anterior abdominal wall via the median umbilical ligament. Fascia intimately surrounds the bladder surface. The peritoneal lining and cavity cover the superior surface. The transversalis fascia covers the anterolateral surface and the posterior surface is covered by endopelvic fascia.

As described by DeLancey,[3] the bladder, uterus, vagina, and rectum are attached to the lateral pelvic walls by a network of connective tissue, which is collectively called the endopelvic fascia. This fascia is a continuous unit that is divided into sections that have named parts. The fascia, which attaches the uterus to the lateral pelvis is called the parametria and consists of the broad, cardinal, and uterosacral ligaments. The fasciae that attach the vagina to the pelvis are collectively called the paracolpium. These fasciae contain smooth muscle, blood vessels, lymphatics, and nerves. The paracolpium, cardinal, and uterosacral ligaments are displayed in Figure 4-1.3. The endopelvic fascial support of the vagina is divided into three levels: I, II, and III [Figure 4-1.4; please see Figure 4-2.7, Chapter 4-2 (Genital Anatomic Correlates)]. The most cephalic 3 cm of the vagina is suspended by endopelvic fascia, which extends from the vagina posteriorly and superiorly over the piriformis muscle to the lateral portion of the sacrum. This constitutes level I support, and alteration of this fascia will result in vaginal apex and uterine prolapse. The mid portion of the anterior vagina is attached laterally to the arcus tendineus fasciae of the pelvic sidewall. This layer lies below the bladder body, and contributes support to the bladder and vagina within the pelvis. This portion of the endopelvic fascia is known as the pubocervical fascia. Tears or disruption of the pubocervical fascia via various mechanisms will result in a cystocele. An injury to the central portion of this fascia results in a central cystocele defect. An injury to this fascia between the vagina and the tendineus arc will result in a paravaginal cystocele defect. The posterior wall of the vagina is attached to the superior fascia of the levator ani muscles and forms the rectovaginal fascia. Injury to this portion of the endopelvic fascia will result in a rectocele. The anterior and posterior endopelvic fascia of the mid vaginal wall constitutes level II support. The region of the vagina that extends 2 to 3 cm above the hymenal ring is fused to the urethra, medial surface of the levator ani muscles, and the perineal body. At this level, there is no intervening connective tissue of the endopelvic fascia that separates the vagina from the urethra. This portion of the vagina and endopelvic fascia constitutes level III support. The levator muscles provide additional support to pelvic organs by closing the vagina and forming a shelf, which supports these organs. Strain on the pelvic organ fascial supports through gravity and Valsalva, are limited by the levator ani muscle. Alteration to the integrity or function of the pelvic floor muscle and fascial supports results in pelvic organ prolapse.

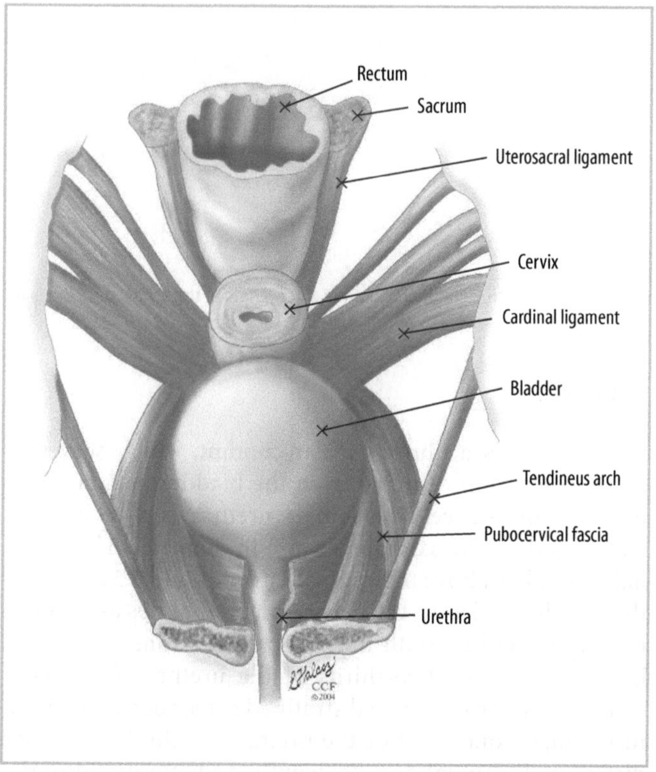

Figure 4-1.3. The bladder, uterus, vagina, and rectum are attached to the lateral pelvic walls by a network of connective tissue, which is collectively called the endopelvic fascia. This fascia is a continuous unit that is divided into sections that have named parts. (Reprinted with the permission of The Cleveland Clinic Foundation.)

On a microscopic basis, the bladder wall comprises an inner transitional cell lining, a middle muscular layer, and an outer adventitial layer. The inner transitional cell lining is covered by a glycosaminoglycan layer, which is thought to be a protective barrier from urinary irritants. The transitional cell epithelium comprises six layers of cells that rest on a basement membrane. Deep to the basement membrane is a thick fibroelastic connective tissue called the lamina propria. The lamina propria contains many blood vessels and loosely arranged smooth muscle fibers. The middle muscular layer consists of three large interlacing bundles of smooth muscle: an inner longitudinal, middle circular, and an outer longitudinal muscular layer. The outer adventitial layer consists of fat and connective tissue.

The gross evaluation of the bladder interior is done by cystoscopy, which is a common clinical tool for evaluation of intravesical pathology. On placement of the cystoscope into the urethra, the urethral mucosa is compressed and the urethra is closed. The cystoscope can be easily placed through the closed urethra. The trigone, which has been previously described, is the first structure seen on placement of the cystoscope into the bladder. The trigone is triangular in shape as a result of the internal urethral opening being equidistant to the ureteral orifices, forming an equilateral triangle. It is common to observe a fluffy white coating on the trigone surface of women, which is known as squamous metaplasia. The ureteral orifices lie approximately 3 cm apart and usually appear as small slits; however, there can be many variations to their appearance. The ureteral bars and the interureteric ridge are often used to locate the ureteral orifices; therefore, knowing the relationship of these structures is critical. The mucosa of the bladder is wrinkled or folded when the bladder is empty, and smooth when the bladder is full. This occurs because the mucosa is loosely bound to the underlying musculature on most of the detrusor surface, except for the trigone, which always appears smooth. On the surface of the bladder mucosa are numerous superficial blood vessels. An

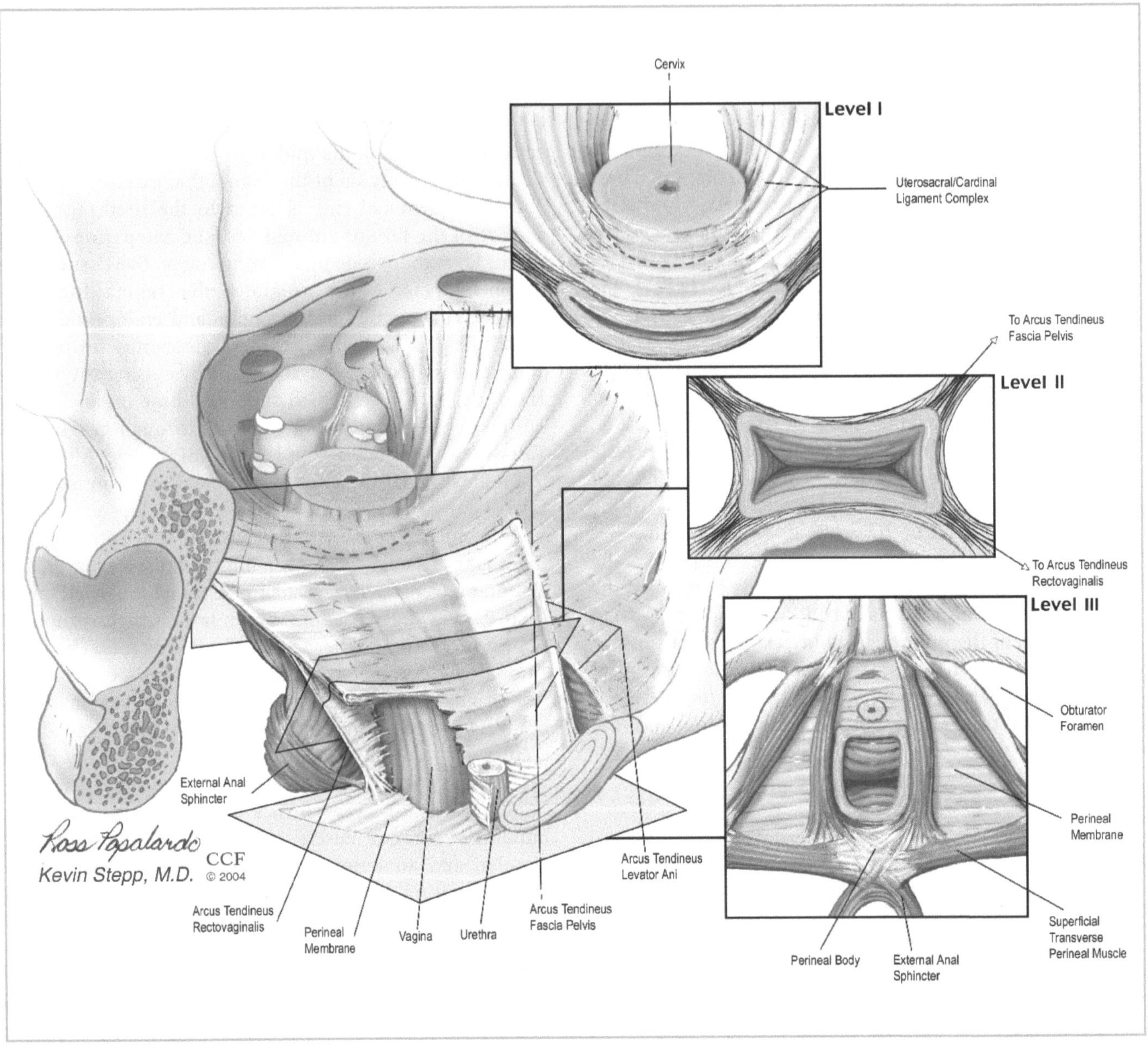

Figure 4-1.4. The three levels of pelvic support of the vagina and uterus showing the continuity of supportive structure throughout the entire length of the genital tract. (Reprinted with the permission of The Cleveland Clinic Foundation.)

air bubble is often introduced during cystoscopy and can be seen in the dome of the bladder. This bubble is used to identify the dome and allows for orientation during cystoscopy. The impression of the uterus on the anterior surface of the bladder can be appreciated in its partially filled state.

The arterial supply of the bladder arises from the superior and inferior vesical arteries, which are branches of the internal iliac vessels. The superior vesicular artery is usually a single artery, but may have 2 to 3 branches that supply the dome and posterior portions of the bladder. The venous drainage of the bladder originates from the dorsal vein of the clitoris as it bifurcates to empty into the laterally placed vaginal plexuses. This plexus of veins connects with the ovarian, uterine, and rectal plexuses to drain into the internal iliac veins.

Urethra

The urethra is a fibromuscular conduit, which serves to allow evacuation of urine from the bladder and provides urinary continence. The female urethra is 3 to 4 cm in length and is approximately 5 mm in diameter. The external orifice lies above the vaginal introitus and 2.5 cm below the glands clitoris. In the pelvis, the urethra lies anterior to the vagina and beneath the pubic bone. Transitional cells line the proximal two-thirds of the urethra. These cells change to a nonkeratinized stratified squamous epithelium in the distal one-third of the urethra. Urethral glands are distributed throughout its length and empty into the lumen. Obstruction and infection of these glands are thought to give rise to urethral diverticula. A group of these glands, known as Skene's glands, coalesce distally and

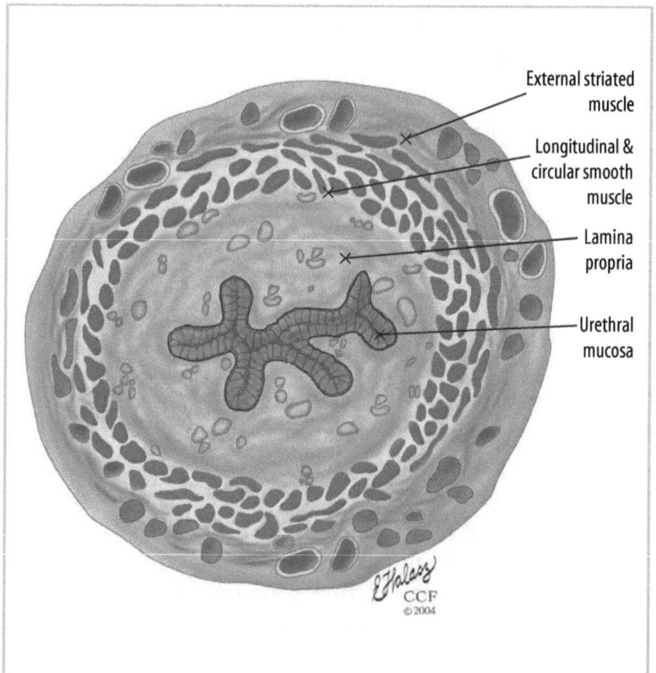

Figure 4-1.5. A transverse section through the mid portion of the female urethra. The female urethra is composed of an infolding epithelium, a fibromuscular envelope of spongy tissue, a middle smooth muscle, and an outer skeletal muscle layer. (Reprinted with the permission of The Cleveland Clinic Foundation.)

empty through two small ducts on either side of the external urethral meatus.

Histologically, the female urethra is composed of an infolding epithelium, a fibromuscular envelope of spongy tissue, a middle smooth muscle, and an outer skeletal muscle layer (Figure 4-1.5). The urethral mucosa is thrown into folds, allowing for a mucosal to mucosal coaptation. This forms a seal similar to a washer on a sink and is one of several mechanisms of continence. Under the mucosa is the lamina propria, which contains a rich vascular spongy tissue. This layer, when congested, constricts the mucosal lining enhancing mucosal coaptation. The smooth muscle of the urethra consists of longitudinal fibers, which emanate from the internal longitudinal fibers of the bladder. The outer semicircular smooth muscle fibers arise from the outer longitudinal layer of the bladder. Both sets of smooth muscle start at the bladder neck and extend to the distal urethra, where they end in a fibrous ring. The outer semicircular smooth muscle is more prominent in the mid urethra, where the smooth muscle fibers mix with the striated fibers of the external urethral sphincter. Both smooth muscle layers become sparse in the distal urethra. Constriction of these muscles extenuates the mucosal to mucosal coaptation. Collagen forms a major component, and elastin a minor component of the urethral smooth muscle layers. Collagen is thought to be a contributor to passive closure of the urethra, whereas elastic fibers are thought to prevent overdistention of the urethra.

DeLancey[4] has described the relationship of the paraurethral structures in the female. The location of these structures were described as a percentile of total urethral length, with the zero percentile defined as the internal urethral meatus and the 100th percentile as the external urethral meatus. The rationale for this scheme is that urethral length can vary, and there is no exact and reproducible length where these structures exist along the urethra from person to person. Instead, the periurethral muscles tend to lie in regions of the urethra as a percentile of its total length. The internal urethral meatus is the zero percentile. At the level of the bladder neck, the urethra passes through the wall of the bladder. The end of the bladder wall constitutes the 15th percentile of the urethra. At this level, the striated urethral sphincter muscle surrounds the urethra and extends from the 18th to the 64th percentile. At the 54th percentile, the urogenital diaphragm, the compressor urethrae and the urethrovaginal sphincter are encountered and extend to the 76th percentile (Figure 4-1.6). These periurethral muscles are continuous with the striated sphincter. The urethral compressor originates from the ischial tuberosity and extends over the urethra to the opposite tuberosity. The transverse vaginal muscle is a thin sheet, which is characterized as part of the compressor that lies below the urethra, filling the space between the urethra and the urethrovaginal sphincter. The urethrovaginal sphincter is flat and merges on the ventral side of the urethra with the urethral compressor, and extends along the sides of the urethra and vagina to enclose them in a circular manner. The pubococcygeus muscle is the most medial portion of the levator musculature and runs from the pubis to the coccyx. It is positioned on the lateral sidewall of the vagina and urethra. When this muscle contracts, it provides some urethral compression; however, it is not considered a true sphincter muscle. The point of maximal urethral pressure at rest in the supine position correlates to this area of muscle described. The maximum

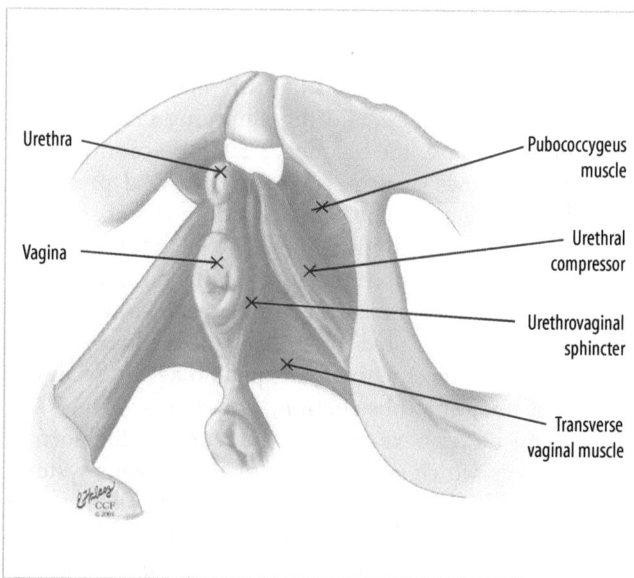

Figure 4-1.6. Periurethral muscles. (Reprinted with the permission of The Cleveland Clinic Foundation.)

voluntary increase in urethral pressure also correlates to this area. From the 79th to the 100th percentile of the urethra, the bulbocavernosus and ischiocavernosus muscles lie adjacent to, but do not connect to, the urethra. These muscles do not contribute to continence. The remaining urethra is composed of fibrous tissue and does not have a significant muscular component. This tissue is derived from the subcutaneous tissue and suspensory ligament of the clitoris. Between the 20th and 60th percentile of the urethra, the pubourethral ligament and the vaginolevator run anteriorly from the urethra and vaginal wall, respectively, to attach to the pelvic wall. The pubourethral ligament connects the urethra to the pubis and the vaginolevator connects the vagina to the pubococcygeus portion of the levator ani muscle. Fibrous tissue and muscle make up these structures. It has been implied that the support and compression of the urethra on a hammock of vaginal tissue causes compression of the urethra with Valsalva. This is thought to be one of many contributors to continence.

The urethral arterial blood supply arises from the inferior vesical arteries at the bladder neck, as well as the vaginal branches of the internal iliac vessels. The urethra's venous drainage originates from the inferior, middle, and superior vesicular veins, as well as the clitoral plexus.

The Lower Urinary Tract

The lower urinary tract (LUT) is a group of interrelated structures that are responsible for the storage and expulsion of urine. The components of the LUT consist of the bladder and the outlet. The outlet is defined as the urethra and urethral sphincters.

Lower Urinary Tract Function

As described by Wein,[5] the LUT has two phases of function, the filling/storage phase, and the emptying phase. In the filling/storage phase, the bladder is able to accommodate increasing volumes of urine at low intravesical pressures with appropriate sensation. This occurs in the absence of involuntary bladder contractions. The bladder outlet is closed and remains closed throughout this process. In the emptying phase, there is a lowering of resistance at the level of the smooth and striated sphincters. The bladder smooth muscle then contracts with an adequate magnitude, in the absence of anatomic obstruction, to evacuate urine. The bladder and outlet have reciprocal actions under normal circumstances, which are regulated by the nervous system. When the bladder is contracting, the outlet is relaxed, and when the bladder is relaxed, the outlet is contracting. This scheme describes the general process of urinary storage and micturition. Any alteration in the function of the LUT will affect one of these basic actions.

Nervous System Innervation of the Lower Urinary Tract

Three sets of peripheral nerves have a significant role in LUT function. These nerves include the sacral parasympathetics, the thoracolumbar sympathetics, and the sacral somatic nerves.

Autonomic Innervation

The parasympathetic nervous system's general role in LUT function is to facilitate emptying. The parasympathetic efferent fibers originate in the intermediolateral region of the gray matter within the sacral cord (S2-4) and emerge as preganglionic fibers from the ventral root. The pelvic nerve conveys these fibers to the LUT.

Acetylcholine is the primary neurotransmitter at the ganglia and the effector sites. The receptors of this system are muscarinic and nicotinic. There are five subtypes of known muscarinic receptors (M1–M5) and they are located on all autonomic effector cells. In the human bladder there is a predominance of M2 receptors. The M3 receptors are primarily responsible for bladder contractions. There are no known muscarinic receptors specific for the bladder. The muscarinic receptors are the targets of anticholinergics for the treatment of the overactive detrusor. The nicotinic receptors are located on autonomic ganglia and the motor end plates of skeletal muscle. Atropine competitively inhibits these muscarinic sites. High doses of nicotine inhibit nicotinic sites.

The sympathetic nervous system's general role in LUT function is to facilitate storage of urine. The efferent fibers of this system originate in the intermediolateral region of the gray matter within the thoracolumbar spinal cord (T10-L2). These nerves traverse the paravertebral ganglia and join the hypogastric plexus anterior to the aorta. This plexus divides into the right and left hypogastric nerves. These merge with the pelvic nerve to form a pelvic plexus or inferior hypogastric plexus. The primary neurotransmitters of the sympathetic nervous system are acetylcholine and norepinephrine. Acetylcholine is released at the ganglion and norepinephrine (adrenergic) at the effector sites. There are two primary types of receptors (alpha and beta) in the sympathetic nervous system and they are characterized on the basis of differential effects elicited by catecholamines. The alpha receptors are stimulated by norepinephrine and methoxamine, but not isoproterenol. When stimulated, alpha receptors elicit smooth muscle contraction. Two subtypes of alpha receptors exist, A1 and A2. These receptors are located throughout the bladder, but are predominantly in the bladder base and neck. Stimulation of alpha receptors increases outlet resistance (i.e. phenylpropanolamine). Alpha receptors are not as prominent in the female urethra and bladder neck as they are in the male. Beta receptors are stimulated the most by isoproterenol, less by epinephrine, and least by norepinephrine. When stimulated, beta receptors elicit an inhibitory effect on detrusor muscle contraction. Two subtypes of beta receptors exist, B1 and B2. Beta 1 receptors are located in the cardiovascular system. The beta 2 receptors are located throughout the bladder, but to a greater extent are within the bladder body. Beta receptor stimulation in the LUT inhibits bladder contraction and causes receptive relaxation of the detrusor to allow for increasing bladder volumes, without increasing intravesical pressure (detrusor compliance).

Somatic Innervation

The somatic nervous system originates in efferent fibers of S2–S4, which form the pudendal nerve. The pudendal nerve in turn innervates the striated sphincter and the pelvic floor. The motor ganglia are located in the anterior horn of the spinal cord and the primary neurotransmitter is acetylcholine. The receptors of the somatic system are nicotinic and their activity is blocked by curare.

Sensory Innervation

Sensory innervation of the bladder is via the pelvic and hypogastric nerves. Sensory input from the urethra is through the pudendal nerve. The tachykinins are the primary neurotransmitters and include substance P, and neurokinin A and B. These neurotransmitters relay their messages through A delta and C fibers. A delta fibers are finely myelinated, located in smooth muscle, and sense bladder fullness. The C fibers are unmyelinated, located in the mucosa and muscle, and sense nociception and overdistention of the detrusor. Up-regulation of these fibers is thought to be one of the possible etiologies of bladder pain. Other neurotransmitters of the nervous system that are thought to have a key role in LUT function are listed below.

- Detrusor contractions
  - Prostaglandins
- Detrusor relaxation
  - Adenosine 5′-triphosphate
  - Opioids
- Urethral relaxation
  - Nitric oxide
  - Opioids
- Urethral contraction
  - Serotonin
  - Epinephrine
- Supporters of micturition
  - Gamma-aminobutyric acid (inhibitory)
  - Enkephalins (inhibitory)
  - Glutamate (facilitative)
  - Dopamine (facilitative)

Nervous System Regulation of Lower Urinary Tract Function

The LUT is a dynamic and complex system. It is not a system that stores urine in a passive manner. For example, the bladder has the ability to accept increasing volumes of urine without increasing intravesical pressure (accommodation). Another example of the LUT's dynamic ability is the guarding reflex, which is an increase in urethral pressure with a cough or sneeze, a protective measure against incontinence. This complex relationship between the various components of the LUT is regulated by the central nervous system.

Storage Phase of Lower Urinary Tract Function

During filling, the normal bladder has a minimal change in intravesical pressure until capacity is reached. At low volumes, the elastic and viscoelastic properties are primarily responsible for compliance. Elasticity allows the constituents of the bladder wall to stretch without a significant increase in bladder wall tension. The viscoelasticity of the bladder causes stretch to induce an increase in tension, followed by a decay when filling stops. In the animal model, it has been shown that at a certain level of bladder distention, spinal sympathetic reflexes facilitory to bladder storage, are evoked. This allows smooth muscle relaxation of the bladder by beta receptor stimulation (accommodation). Spinal sympathetic reflexes inhibit parasympathetic activity at the level of the parasympathetic ganglia during filling. Clinically, detrusor compliance may be altered by any processes that can damage the elastic tissues (chronic cystitis, radiation, ischemia, etc.) or neurologic abnormalities, which affect smooth muscle modulation (peripheral nerve injury).

During filling, the normal outlet displays an increase in urethral pressure. This is primarily the result of striated sphincter muscle activity and to a lesser extent smooth muscle sphincter activity. Pudendal motor neurons are activated by bladder afferent input, which activates striated sphincter muscle activity.

Voiding Phase of Lower Urinary Tract Function

The cerebral cortex has facilitative and inhibitory centers of micturition, which relay signals via the pons to the bladder. The primary influence of the cortical system on the micturition reflex is inhibitory. Therefore, under normal circumstances, voiding is a reflex function under voluntary control. Sensory input from bladder wall distention is the primary stimulus for micturition. This stimulus is interpreted by the cerebral cortex, and in the appropriate social setting, there is a voluntary decrease in somatic neural discharge to the striated sphincter. There is subsequently a decrease in spinal sympathetic reflex activity and an increase in the parasympathetic neural outflow from the sacral cord, via the pelvic nerve. This results in a bladder contraction and funneling of the outlet. The pons is responsible for coordination of the micturition reflex by orchestrating the interaction of the bladder and outlet. Uncoordinated voiding activity would occur if not influenced by the pons, which regulates the sacral reflex arc, via ascending and descending spinal pathways.

References

1. McCormak LU, Anson BJ. The arterial supply of the ureter. Q Bull Northwest Univ Med School 1950;24:1.
2. Shulman CC. Innervation of the ureter. Anat Cin 1981;3:127.
3. DeLancey J. Anatomic aspects of vaginal eversion after hysterectomy. Am J Obstet Gynecol 1992;166:1717–1728.
4. DeLancey J. Correlative study of paraurethral anatomy. Am J Obstet Gynecol 1986;68:91–97.
5. Wein AJ. Neuromuscular dysfunction of the lower urinary tract and its treatment. In: Campbell MF, Walsh PC, Retik AB, eds. Campbell's Urology. 7th ed. Philadelphia: WB Saunders; 1998:953–1006.

4-2

Genital Anatomic Correlates

Kevin J. Stepp and Matthew D. Barber

The Cleveland Clinic Foundation uses a multidisciplinary approach to the patient with pelvic floor dysfunction. What follows is intended to be a clinical resource for surgical anatomy for the pelvic surgeon. This section will cover the anatomic relationships of the bones, ligaments, viscera of the pelvis, and their supportive structures as they relate to the female reproductive tract. A detailed discussion of the urinary tract and colorectal anatomy is provided elsewhere. A thorough mastery of the anatomic concepts presented here will serve as a foundation for clinical examination and surgical repair of pelvic floor dysfunction and pelvic organ prolapse.

Bones of the Pelvis

The bones of the pelvis are the rigid foundation to which all of the pelvic structures are ultimately anchored. It is important to understand and discuss the bony pelvis from the perspective of a standing woman. In the standing position, forces are dispersed to minimize the pressures on the pelvic viscera and musculature and disperse the forces to the bones that are better suited to the long-term, cumulative stress of daily life. In the upright position, the pubic rami are oriented in an almost vertical plane. Similar to the supports of an archway or bridge, the weight of the woman is transmitted along these bony supports to her femurs. Where the pubic rami articulate in the midline, they are nearly horizontal. Much of the weight of the abdominal and pelvic viscera is supported by the bony articulation inferiorly. In this way, increases in intraabdominal pressures are partially supported by the bony pelvis.

The pelvic bones are the ilium, ischium, pubic rami, sacrum, and coccyx (Figure 4-2.1). The sacrum is composed of five sacral vertebrae that are fused together. The nerve foramina are positioned anterior and laterally. Overlying the middle of the sacrum is a rich neurovascular bed. The coccyx is attached inferiorly and is the posterior border of the pelvic outlet.

Attached to the sacrum are the ilium, ischium, and pubic rami. Several landmarks are important to the pelvic surgeon. The anterior superior iliac spine is located anterior and laterally on the superior ileum. This is easily identifiable in all patients and is a clinically useful landmark. The ischium is fused to the ilium. The medial surface of the ilium has two concavities forming the lateral borders of the pelvic outlet. The superior and larger of the two is the greater sciatic notch. Inferiorly is the lesser sciatic notch. They are separated by a projection medially, called the ischial spine. The ischial spine is important clinically and anatomically because it can be palpated easily through a vaginal, rectal, or retropubic approach, and many supportive structures attach to it. The ischial spine is useful as a fixed point to describe the relative position of other anatomic structures.

The superior and inferior pubic rami are located anteriorly and articulate in the midline at the pubic symphysis. The ridge along the superior, medial surface of the superior pubic rami is called the pectineal line, or Cooper's ligament.

In the standing position, the anterior superior iliac spine and pubic symphysis are in the same vertical plane (Figure 4-2.2). This directs the pressure of the intraabdominal and pelvic contents toward the bones of the pelvis instead of the muscles and endopelvic fascia attachments of the pelvic floor. The posterior surface of the pubis symphysis is located in a plane approximately 2 to 3 cm inferior to the ischial spine. Therefore, a line drawn connecting the two structures would be almost horizontal in the standing position.

Strong ligaments hold the bones together. The ligaments of the sacroiliac joint are rarely encountered during surgery for pelvic floor dysfunction and are not addressed here. The sacrospinous ligament is a strong, easily identifiable ligament that extends from the ischial spine to the distal sacrum. The ligament fans out to attach on the S1-S4 vertebrae. This ligament divides the lateral pelvic outlet into two foramina, the greater sciatic foramen superiorly and the lesser sciatic foramen inferiorly. This is an important location for identifying the course of the pudendal nerve, artery, and vein, and will be discussed later in the chapter.

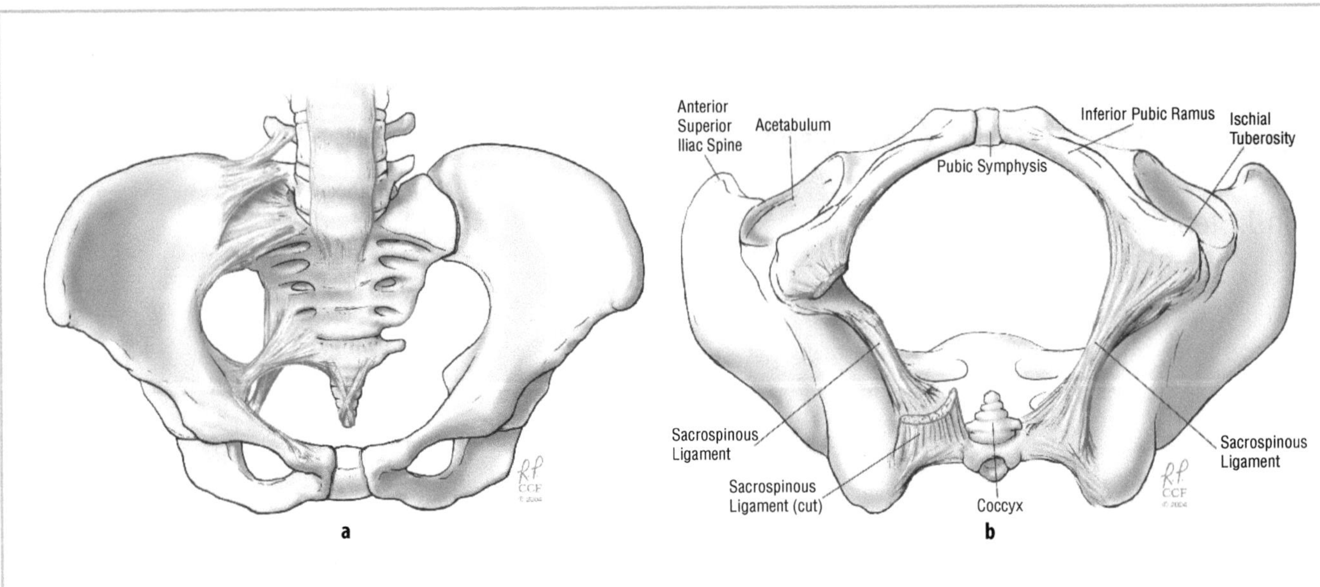

Figure 4-2.1. Bones and ligaments of the pelvis. **a**, Front view; **b**, lithotomy view. (Reprinted with the permission of The Cleveland Clinic Foundation.)

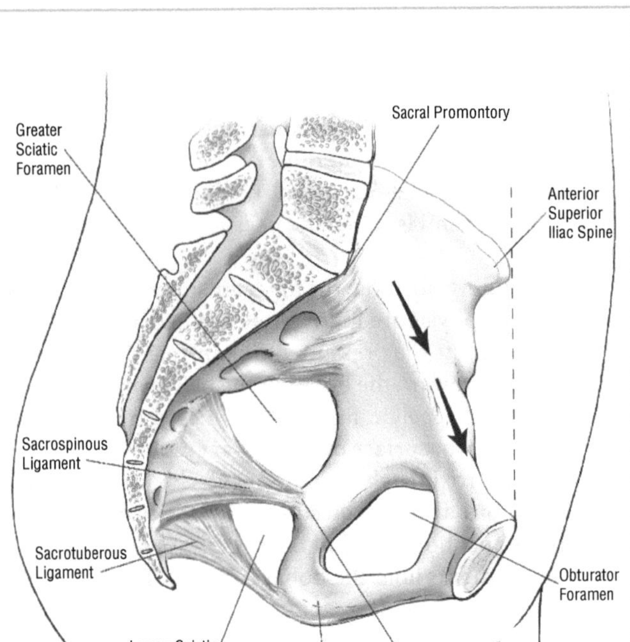

Figure 4-2.2. Bones and ligaments of the pelvis (sagittal view). The dashed line represents the vertical plane of the anterior superior iliac spine and pubic symphysis. The arrows represent the distribution of weight of the spinal column and abdominal contents and along the ilium. (Reprinted with the permission of The Cleveland Clinic Foundation.)

Muscles of the Pelvic Sidewalls and Pelvic Floor

The obturator internus and piriformis make up the pelvic sidewalls. The obturator membrane is a fibrous membrane that covers the obturator foramen. The obturator internus muscle lies on the superior (intrapelvic) side of the obturator membrane. The obturator internus origin is on the inferior margin of the superior pubic ramus and the pelvic surface of the obturator membrane. Its tendon passes through the lesser sciatic foramen to insert onto the greater trochanter of the femur to laterally rotate the thigh. The obturator internus receives its innervation from the obturator nerve originating from L5-S2. The obturator vessels and nerve pass through the anterior and lateral border of the obturator membrane to their destination in the adductor compartment of the leg.

The piriformis is part of the pelvic sidewall and is located dorsal and lateral to the coccygeus. It extends from the anterolateral sacrum to pass through the greater sciatic foramen and insert on the greater trochanter. Lying on top of the piriformis is a particularly large neurovascular plexus, the lumbosacral plexus.

There is a linear thickening of the fascial covering of the obturator internus muscle called the *arcus tendineus levator ani*. This thickened fascia forms an identifiable line from the ischial spine to the posterior surface of the ipsilateral pubic ramus. The muscles of the levator ani originate from this musculofascial attachment (Figure 4-2.3).

The skeletal muscles of the pelvic floor include the levator ani muscles, the coccygeus muscle, the external anal sphincter, the striated urethral sphincter, and the superficial perineal muscles (bulbocavernosus, ischiocavernosus, and transverse perinea). The levator ani muscle

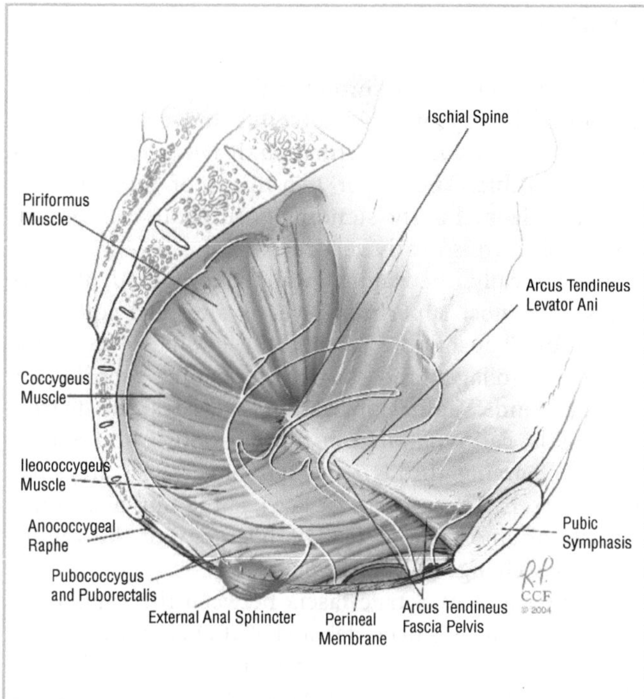

Figure 4-2.3. Muscles of the pelvis (sagittal view). (Reprinted with the permission of The Cleveland Clinic Foundation.)

complex consists of the puborectalis, pubococcygeus, and iliococcygeus muscles (Figure 4-2.4).

The puborectalis has an attachment to the posterior inferior pubic rami and arcus tendineus levator ani. Its fibers pass posteriorly forming a sling around the vagina, rectum, and perineal body to form the anorectal angle and contribute to fecal continence. Some of the fibers of the muscle may blend with the muscularis of the vagina and rectum. The pubococcygeus has a similar origin, but inserts in the midline onto the anococcygeal raphe and the anterolateral borders of the coccyx. The iliococcygeus extends along the arcus tendineus levator ani from the pubis to the ischial spine to insert in the midline onto the anococcygeal raphe.

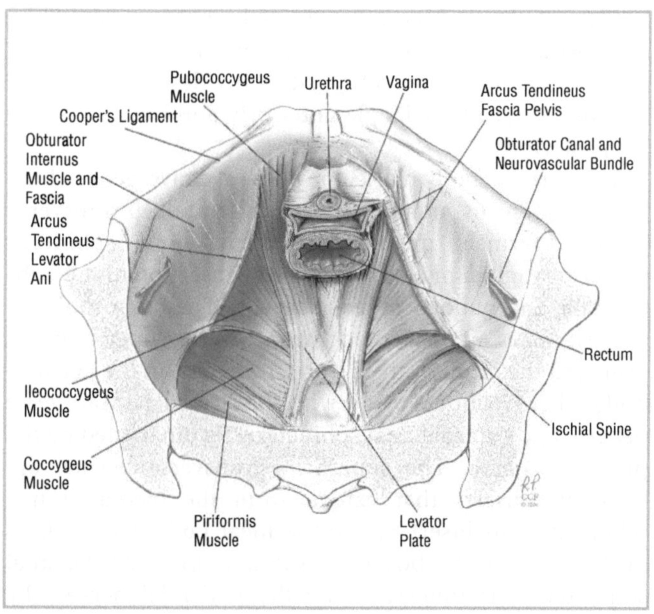

Figure 4-2.4. Muscles of the pelvis (abdominal view). (Reprinted with the permission of The Cleveland Clinic Foundation.)

The coccygeus, although not part of the levator ani, does make up the posterior part of the pelvic floor and has a role in support. Its origin is on the ischial spine and sacrospinous ligament. It inserts on the lateral lower sacrum and coccyx and overlies the sacrospinous ligament. The muscle becomes thin and fibrous with age. The coccygeus often blends with the sacrospinous ligament and, because they have the same origin and insertion, it can be difficult to distinguish the two.

The space between the levator ani musculature through which the urethra, vagina, and rectum pass is called the levator hiatus. The fusion of levator ani where they meet in the midline creates the levator plate. The levator plate forms the basis for pelvic support as will be discussed in detail later in this chapter. The levator ani may be very thin and attenuated, especially in patients with pelvic organ prolapse.

Nerves of the Pelvis

Although the muscles of the pelvic floor were initially thought to have innervation both from direct branches of the sacral nerves on the pelvic surface and via the pudendal nerve on the perineal surface, recent anatomic, neurophysiologic, and experimental evidence indicates that these standard descriptions are inaccurate and that the levator ani muscles are innervated solely by a nerve traveling on the superior (intrapelvic) surface of the muscles without contribution of the pudendal nerve[1,2] (Figure 4-2.5).

Barber et al.[1] performed systematic cadaver dissections to characterize the nerve supply to the pelvic floor muscles. The nerve supplying the coccygeus muscle and the levator ani muscles (all three) originates from S3, S4, and/or S5. The nerve exits the foramina and travels 2 to 3 cm medial to the ischial spine and arcus tendineus levator ani across the coccygeus, iliococcygeus, pubococcygeus, and puborectalis muscles. The nerve is sometimes firmly embedded in the fascia of the muscles or may be loosely attached during its course. There are small branches that penetrate the body of each muscle as the nerve traverses them. Occasionally, a separate nerve comes directly from S5 to innervate the puborectalis muscle. The piriformis receives innervation directly from sacral nerves (motor efferent) from L5-S2.

In the pelvis, the sympathetic nerves to the pelvis originate at the T5 to L2 spinal level and act to promote storage by causing relaxation of the bladder and rectum and contraction of the smooth muscle components of the urethral and anal sphincter. The parasympathetic nerve supply to the pelvic viscera originates from the second, third, and fourth sacral nerves. The parasympathetic nerves combine

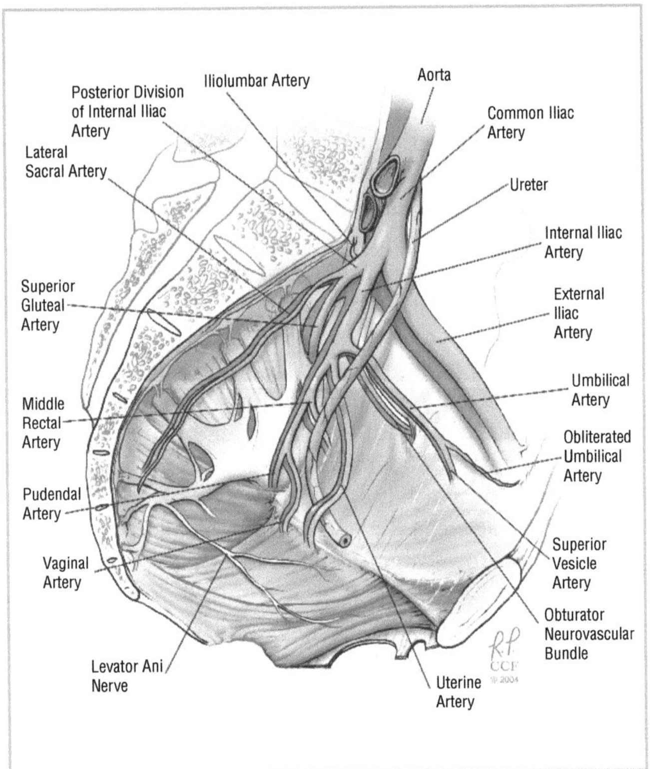

Figure 4-2.5. Nerves and vessels of the pelvis. (Reprinted with the permission of The Cleveland Clinic Foundation.)

with the hypogastric plexus and pelvic sympathetic nerves to form the pelvic nerve plexus. This plexus of nerves leaves the sacral surface to fan out on either side of the rectum approximately 3 to 4 cm superior to the pelvic floor muscles, then disperse throughout the pelvis through the endopelvic fascia.

Pelvic floor muscles have constant tone except during voiding, defecation, and during the Valsalva maneuver. This activity serves as a constant support for the pelvic viscera. The levator muscles and the skeletal components of the urethral and anal sphincters all have the ability to contract quickly at the time of an acute stress, such as a cough or sneeze, to maintain continence.

Because of the nerves' undefined course and small size, iatrogenic damage during pelvic surgery is possible. This may produce a range of effects both sensory and motor in nature. Radical hysterectomy and rectal resection are common causes of pelvic plexus injury resulting in bowel and/or bladder dysfunction.

Viscera

The organs of the female upper reproductive tract are the uterus, cervix, ovaries, and fallopian tubes. The structure and function of these organs, except as it relates to pelvic floor dysfunction, will not be covered in this chapter.

Vagina

The human vagina is a fibromuscular tube specialized for reception of the penis during coitus and delivery of the developed fetus during parturition. It needs to be mobile and distensible. The upper two-thirds of the vagina is almost horizontal in the standing position. In contrast, the lower one-third is nearly vertical.

Histologically, the vaginal wall is composed of three layers. The most superficial layer is stratified squamous epithelium. The middle layer is the lamina propria and consists of collagen and elastin. The lamina propria contains no glands. Vaginal lubrication is via a transudate from the vessels, cervix, and from the Bartholin's and Skene's glands. Coursing through the lamina propria are small blood vessels. The innermost layer is the muscularis that consists of smooth muscle. The histology of the vaginal layers may change with menopause.

The presence of a true fascia between the vagina and adjacent organs has been debated. Although at the time of surgery there appears to be an identifiable fascial plane, Weber and Walters[2] and DeLancey[3] have concluded that there is no fascia present histologically. Between adjacent organs is primarily vaginal muscularis. However, an extension of the connective tissue of the perineal body forms what some have called the rectovaginal fascia. This tissue extends 2 to 3 cm cephalad from the hymenal ring along the posterior vaginal wall.

Perineum

The perineum is divided into two compartments: superficial and deep. These are separated by a fibrous connective tissue layer called the perineal membrane. The borders of the perineum are the ischiopubic rami, ischial tuberosities, sacrotuberous ligaments, and coccyx. A line connecting the ischial tuberosities divides the perineum into the urogenital triangle anteriorly and the anal triangle posteriorly.

The perineal body marks the point of convergence of the bulbospongiosus muscles, superficial and deep transverse perinei, perineal membrane, external anal sphincter, posterior vaginal muscularis, and the insertion of the puborectalis and pubococcygeus muscles. The bulbospongiosus originates on the inferior surface of the superior pubic rami and the crura of the clitoris. It inserts on the perineal body, where its fibers blend with the superficial transverse perinei, and external anal sphincter. It is innervated by the pudendal nerve. The superficial transverse perinei are bilateral muscles that extend from the medial ischial tuberosities to insert on the perineal body. Some fibers blend with the bulbospongiosus and the external anal sphincter. It is innervated by the pudendal nerve. The ischiocavernosus originate from the medial ischial tuberosities and ischiopubic rami. They insert on the infe-

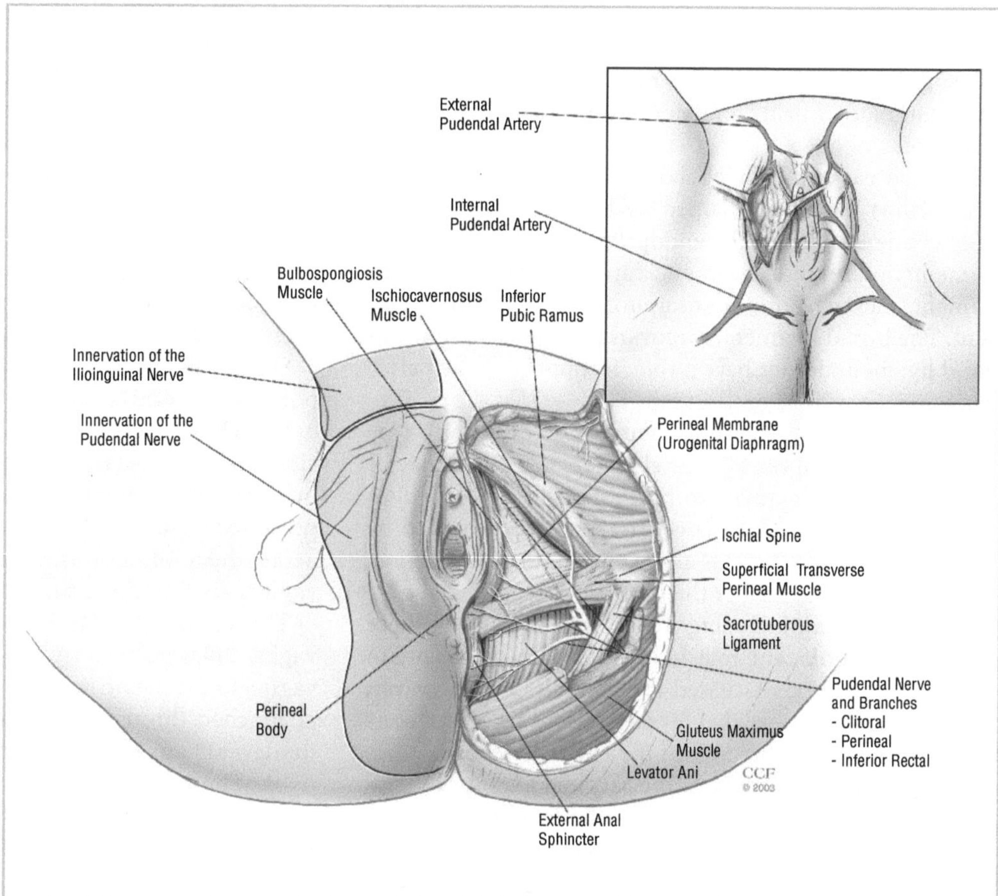

Figure 4-2.6. Perineal anatomy with Martius flap inset. The distribution of the pudendal and ilioinguinal innervation is shown. The inset shows the dual vascular supply to the labial fat pad used during a Martius flap procedure. (Reprinted with the permission of The Cleveland Clinic Foundation.)

rior aspect of the pubic angle and are innervated by the pudendal nerve.

The deep perineal compartment is composed of the deep transverse perineal muscles, portions of the external urethral sphincter muscles (compressor urethrae and urethrovaginal sphincter), portions of the anal sphincter, and the vaginal musculofascial attachments.

The neurovascular anatomy of the perineum is illustrated in Figure 4-2.6. The motor and sensory innervation of the perineum is via the pudendal nerve. The pudendal nerve originates from S2-S4 and exits the pelvis through the greater sciatic foramen, hooks around the ischial spine, then reenters the pelvis through the lesser sciatic foramen. It then travels along the medial surface of the obturator internus, through the ischiorectal fossa in a thickening of fascia called Alcock's canal. It emerges posterior and medial to the ischial tuberosity where it pierces the perineal membrane and divides into three branches to supply the perineum: clitoral, perineal, and inferior rectal (inferior hemorrhoidal). Damage to the pudendal nerve (i.e., birth trauma) can result in denervation of the periurethral muscles involved in reflex contraction during increased intraabdominal pressure resulting in stress urinary incontinence.

The blood supply to the perineum is from the pudendal artery, which travels with the pudendal nerve to exit the pelvis. Similar to the nerve, there are three main branches with rich collateral anastomosis: clitoral, perineal, and inferior rectal. It is this rich collateral anastomosis that allows a Martius flap to be utilized by pelvic surgeons. A Martius flap receives its rich blood supply anteriorly and posteriorly from branches of the external and internal pudendal arteries, respectively (Figure 4-2.6, inset). Details of the anus, urethral sphincter, and external anal sphincter, and their continence mechanisms are discussed in Chapters 4-1 and 4-3.

Mechanisms of Support

The normal axis of the pelvic organs in the standing woman places the vagina and rectum directly over the levator plate. The levator plate and muscles of the pelvic floor therefore support the pelvic organs. The remainder of this section will describe the structures and attachments that keep the pelvic organs in the proper orientation so that they may be supported by the pelvic floor musculature.

The endopelvic fascia is the loose connective tissue network appearance of the retroperitoneum that envelops all of the organs of the pelvis and connects them loosely to the supportive musculature and bones of the pelvis. The term endopelvic fascia is used here to describe the tissues located between the surfaces of the peritoneum, muscles, and pelvic organs. Histologically, it is composed of collagen, elastin, adipose tissue, nerves, vessels, lymph channels, and smooth muscle. Its properties provide stabilization and support, yet allow for the mobility of the viscera to permit storage of urine and stool, coitus, parturition, and defecation.

Several areas of the endopelvic fascia (and its associated peritoneum) have been named by anatomists. These are really condensations of the endopelvic fascia and not true ligaments: uterosacral ligament, cardinal ligament, broad ligament, mesovarium, mesosalpinx, and the round ligament. The broad ligament, mesovarium, mesosalpinx, and round ligament do not have a role in support of the pelvic organs.

DeLancey[5] described vaginal support in three levels (see Figure 4-1.4 in Chapter 4-1). Level I refers to the uterosacral ligament/cardinal ligament complex and is the most cephalad supporting structures. Level II support is provided by the paravaginal attachments along the length of the vagina. Level III support describes the most inferior or distal portions of the vagina including the perineum. Each of these areas has a significant role in maintaining pelvic organ support and will be discussed individually. It is, however, important to remember that levels I, II, and III are all connected through continuation of the endopelvic fascia.

Comprising level I support, the cardinal and uterosacral ligaments attach to the cervix from the lateral and posterior sides, respectively, with fibers intermingling. The cardinal ligament blends with the uterosacral ligament and they are difficult, if not impossible, to precisely delineate from one another. Fibers traveling predominately laterally make up the cardinal ligament, whereas fibers going to the sacrum make up the uterosacral ligament. These fibers form a three-dimensional complex attaching the upper vagina, cervix, and lower uterine segment to the sacrum and lateral pelvic sidewalls at the piriformis, coccygeus, and the levator ani and perhaps the obturator internus fascia overlying the ischial spine. Together, the uterosacral/cardinal ligament complex supports the cervix and upper vagina to maintain vaginal length and keep the vaginal axis nearly horizontal so that it rests on the rectum and can be supported by the levator plate. This keeps the cervix just superior to the level of the ischial spine.

Contiguous with the uterosacral/cardinal ligament complex at the location of the ischial spine is level II support – the paravaginal attachments. These are the connections of the lateral vagina and endopelvic fascia to the *arcus tendineus fascia pelvis* anteriorly and the *arcus tendineus rectovaginalis* posteriorly – level II support functions to keep the vagina midline directly over the rectum.

The arcus tendineus fascia pelvis is similar in composition to the arcus tendineus levator ani. It, however, arises from the levator ani fascia rather than that of the obturator internus. Similar to the arcus tendineus levator ani, it originates on the ischial spine; however, as it approaches the pubic symphysis, the arcus tendineus fascia pelvis travels medially and inferiorly to the arcus tendineus levator ani, inserting on the inferior aspect of the superior pubic rami over the origin of the puborectalis muscle. The arcus tendineus fascia pelvis or "white line" is a thickened condensation of the parietal fascia into which the paravaginal endopelvic fascia connects, supporting and creating the anterior lateral vaginal sulci. Furthermore, the axis of both of the arcus tendineus levator ani and the arcus tendineus fascia pelvis are nearly horizontal in the standing woman, creating the normal axis of the vagina. Anteriorly, the endopelvic fascia blends with the vaginal muscularis and is continuous with the supportive structures of the urethra (see Chapter 4-1).

Similar to the anterior paravaginal supports, there are posterior lateral supports as well (Figure 4-1.4 Figure 4-2.7). These fibers blend with the vaginal muscularis anteriorly, rectal muscularis posteriorly, and the perineal body inferiorly. The lateral endopelvic fascia attachments of the posterior vaginal wall do not have significant connections across the midline. Rather, they anchor the posterior lateral vaginal sulci to the ipsilateral levator ani.[4]

The endopelvic fascia extends from posterior lateral vagina sulci posteriorly around the rectum to attach the vagina to the pelvic floor. The posterior vaginal muscularis is attached through this endopelvic fascia to the fascia of the levator ani laterally at the arcus tendineus rectovaginalis.[6] The arcus tendineus rectovaginalis represents a condensation of the parietal fascia of the levator ani coursing from the perineal body inferiorly, along the levator ani laterally, where it intersects the midpoint of the arcus tendineus fascia pelvis. The arcus tendineus rectovaginalis is approximately 4 cm in length. The connection to the arcus tendineus rectovaginalis creates the change in axis toward vertical of the distal vagina.

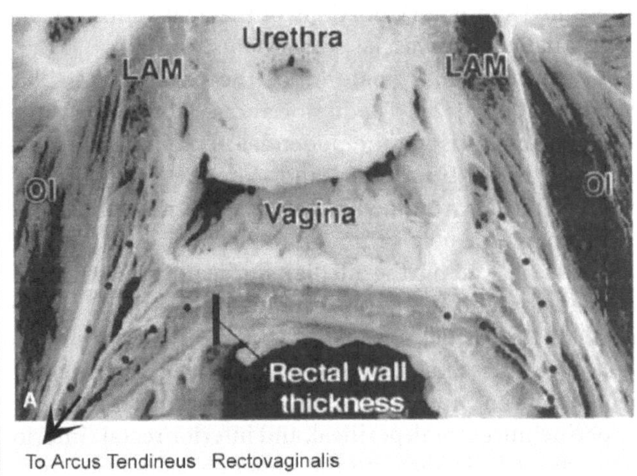

Figure 4-2.7. Photomicrograph of posterior wall attachments. Note the fibers of the endopelvic fascia that are attached (outlined by dots) to the lateral sulcus of the posterior vaginal wall. OI, obturator internus muscle; LAM, levator ani muscle. (Reprinted from American Journal of Obstetrics and Gynecology, Vol 180, JOL DeLancey, Structural anatomy of the posterior pelvic compartment as it relates to rectocele, p 815–823, © 1999 Mosby, with permission from Elsevier)

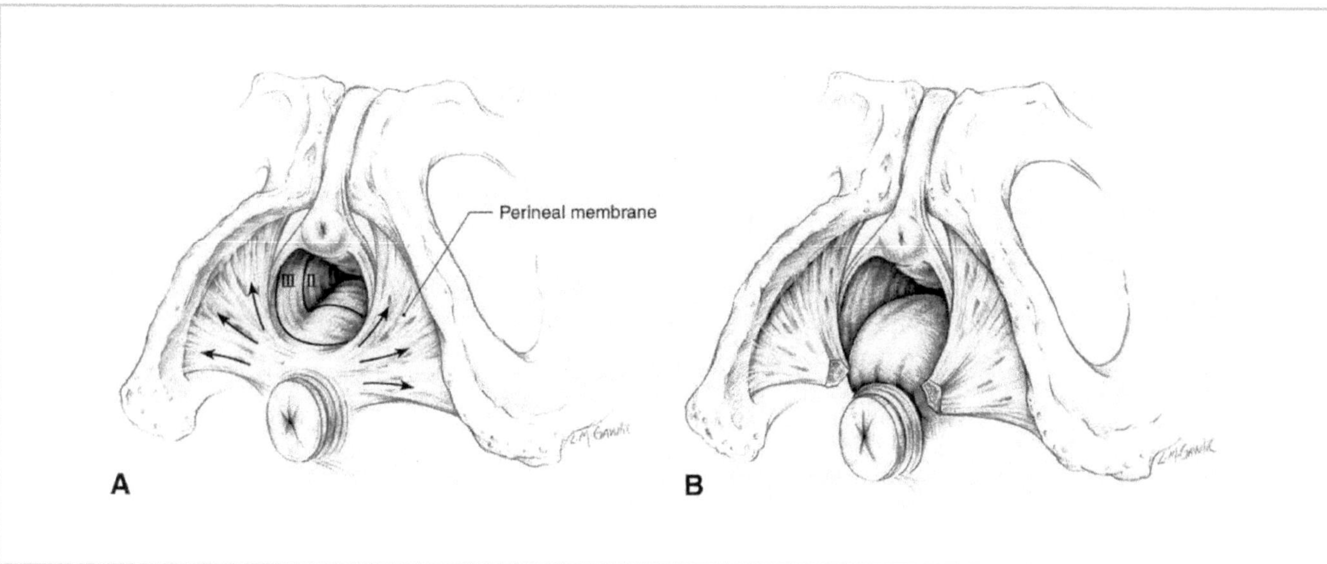

Figure 4-2.8. Perineal attachments of the perineal membrane to the inferior pubic rami and direction of tension on fibers uniting through the perineal body (arrows). (Reprinted from American Journal of Obstetrics and Gynecology, Vol 180, JOL Delancey, © Structural anatomy of the posterior pelvic compartment as it relates to rectocele, p 815–823 1999 Mosby, with permission from Elsevier)

Level III support is provided by the perineal body, perineal membrane, superficial and deep perineal muscles, and endopelvic fascia. These structures support and maintain the normal anatomic position of the distal one-third of the vagina. The condensation of connective tissue at the point of convergence of the level III structures, distal rectum, levator ani, and distal level II attachments forms the perineal body. The perineal body is critical for support of the lower part of the vagina and proper function of the anal canal. The perineal membrane anchors the perineal body and distal vagina laterally and anteriorly to the ischiopubic rami (Figure 4-2.8). According to DeLancey,[4] "When the distal rectum is subjected to increased force directed caudally, the fibers of the perineal membrane become tight and resist further displacement." Separation of the perineal body from the perineal membrane results in perineal decent and can contribute to defecatory dysfunction.

The three levels of support are connected and interdependent. Level III structures are connected to the endopelvic fascia that surrounds the vagina and rectum and are therefore continuous with level II support. Level II support is connected to level I support through the confluence of the lateral endopelvic fascia attachments and the uterosacral ligament/cardinal ligament complex. Adequate support at all levels maintains the pelvic organs in their normal anatomic positions.

When the vagina, bladder, and rectum are kept in the horizontal plane over the levator plate and pelvic floor muscles, intraabdominal and gravitational forces are applied perpendicular to the vagina and pelvic floor while the pelvic floor musculature counters those forces with its constant tone. It is this horizontal position and support by the levator ani that maintain pelvic organ support. With proper tone of the pelvic floor muscles (levator ani), the stress is relieved from the lateral paravaginal attachments. Furthermore, in times of acute stress, such as a cough or sneeze, there is a reflex contraction of the pelvic floor musculature countering and further stabilizing the viscera. The genital hiatus responds by narrowing to maintain level III support. With pelvic floor weakness, such as with neuropathic injury or mechanical muscular damage, the endopelvic fascia then becomes the primary mechanism of support. Over time, this stress can overcome the endopelvic fascial attachments and result in loss of the normal anatomic position through breaks, stretching, or attenuation of endopelvic fascia supports. This can result in changes in the vector forces applied to the viscera and may lead to pelvic organ prolapse and/or dysfunction. Recreation of these supportive connections and proper position of the organs while maintaining adequate vaginal length to keep the vaginal apex in a natural position should be the goal of pelvic reconstructive surgery.

Surgical Correlates

Iliococcygeus and Sacrospinous Vaginal Vault Suspension

Figure 4-2.9 illustrates the proper suture placement for vault suspension to the iliococcygeus muscle and to the sacrospinous ligament. Note the relationship to surrounding structures, notably the pudendal vessels and nerve. During an iliococcygeus vaginal vault suspension, sutures are placed 1 to 2 cm medial and 1 cm inferior to the ischial spine. To suspend the vaginal vault to the sacrospinous ligament, sutures are placed 1 cm medial and 1 cm cephalad to the ischial spine to avoid damage to the underlying vessels and nerve.

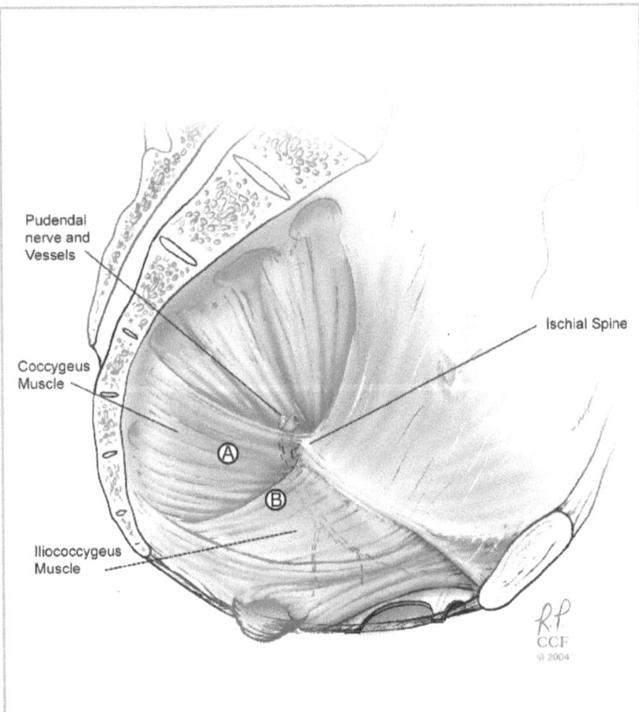

Figure 4-2.9. Surgical correlates – iliococcygeus and sacrospinous vaginal vault suspension. A, Location of suture placement for sacrospinous ligament fixation. B, Location of suture placement for iliococcygeus vaginal vault suspension. (Reprinted with the permission of The Cleveland Clinic Foundation.)

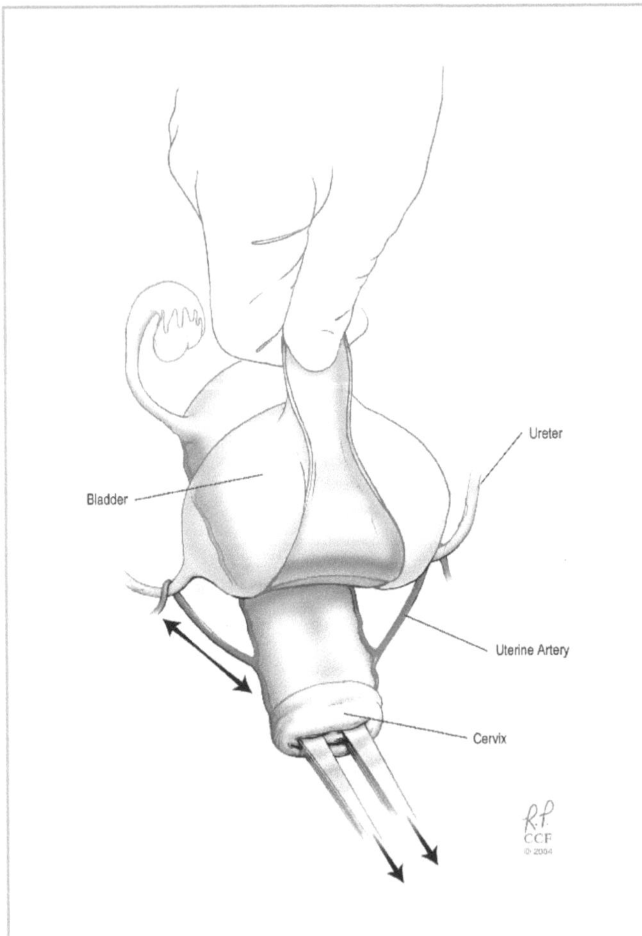

Figure 4-2.10. Surgical correlates – ureter anatomy in relation to the uterus and vaginal apex at total vaginal hysterectomy. Upward traction on the bladder during vaginal hysterectomy increases the distance of the ureter away from the area of clamp placement. (Reprinted with the permission of The Cleveland Clinic Foundation.)

Ureter Anatomy in Relation to the Uterus and Vaginal Apex at Total Vaginal Hysterectomy and Anterior Repair

In 1962, Hofmeister and Wolfgram[7] showed that while performing a vaginal hysterectomy the ureter is 1.5 to 2.1 cm from the cervix and 1.0 cm from the infundibular ligaments. During anterior colporrhaphy, sutures can be as close as 0.9 cm from the ureter. Figure 4-2.10 illustrates a simple maneuver to increase the safe distance between the cervix and the ureter by 1 cm during vaginal hysterectomy by using a right-angled retractor placed in the vesicovaginal septum.

Anatomy Surrounding the Uterosacral Ligament

Knowledge of the anatomy surrounding the uterosacral ligament is paramount to safety when performing vault suspension procedures involving the ligament. Figure 4-2.11 illustrates the close proximity of the ureter and vasculature to the uterosacral ligament. The ureter is closest to the uterosacral ligament at its distal end – approximately 1 cm. The ureter diverges laterally as the uterosacral ligament is traced toward the sacrum. At the level of the spine, the

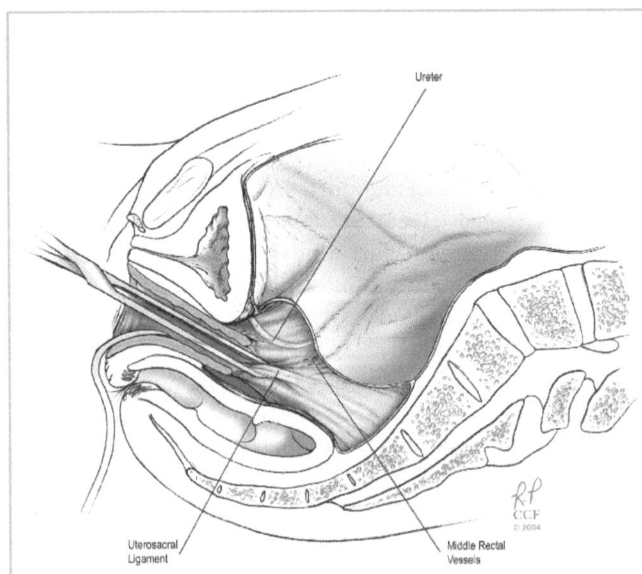

Figure 4-2.11. Surgical correlates – anatomy surrounding the uterosacral ligament when an Allis clamp is applied at the time of vaginal vault suspension. (Reprinted with the permission of The Cleveland Clinic Foundation.)

ureter is approximately 2.3 cm lateral to the uterosacral ligament.[8] Note the location of the middle rectal and superior gluteal vessels beneath the uterosacral ligament. When using the uterosacral ligament for suspension procedures, the surgeon should understand the course and proximity of these adjacent structures.

References

1. Barber MD, Bremer RE, Thor KB, Dolber PC, Kuehl TJ, Coates KW. Innervation of the female levator ani muscles. Am J Obstet Gynecol 2002;187:64–71.
2. Pierce LM, Reyes M, Thor KB, et al. Innervation of the levator ani muscles in the female squirrel monkey. Am J Obstet Gynecol 2003; 188:1141–1147.
3. Weber AM, Walters MD. Anterior vaginal prolapse: review of anatomy and techniques of surgical repair. Obstet Gynecol 1997;89:331–338.
4. DeLancey JOL. Structural anatomy of the posterior pelvic compartment as it relates to rectocele. Am J Obstet Gynecol 1999;180: 815–823.
5. DeLancey JOL. Anatomic aspects of vaginal eversion after hysterectomy. Am J Obstet Gynecol 1992;166:17–28.
6. Leffler KS, Thompson JR, Cundiff GW, Buller JL, Burrows LJ, Schon Ybarra MA. Attachment of the rectovaginal septum to the pelvic sidewall. Am J Obstet Gynecol 2001;185:41–43.
7. Hofmeister FJ, Wolfgram RC. Methods of demonstrating measurement relationships between vaginal hysterectomy ligatures and the ureters. Am J Obstet Gynecol 1962;83:938–948.
8. Buller JL, Thompson JR, Cundiff GW, et al. Uterosacral ligament: description of anatomic relationships to optimize surgical safety. Obstet Gynecol 2001;97:873–879.

4-3

Colorectal Anatomic Correlates

James Doty and Eric G. Weiss

An understanding of the anorectal anatomy and its relationship to the pelvic floor is essential to understanding the pathophysiology of pelvic floor dysfunction and thus how to evaluate and manage its disorders.

The pelvic floor is composed of the pelvic floor musculature, the fascia of the pelvic floor, the associated viscera that pass through, and the blood vessels and nerves that supply these structures. The major components of the pelvic floor from cephalic to caudal are the peritoneum, viscera, endopelvic fascia, levator ani muscles, and external genital muscles.

Support of the Pelvic Floor

The pelvic organs when removed from the body are a limp and formless mass. Their shape and position in vivo are dependent on their various attachments to the bony skeleton through the pelvic muscles and connective tissue. These tissues can have an active or passive role in pelvic visceral support. The passive support structures are the sacrum, coccyx, pubic rami, parietal fascia, endopelvic fascia, and levator tendons. The primary active support structures are the levator ani muscles.

Fascia and Ligaments of the Pelvic Floor

The walls and floor of the pelvis are lined by the parietal endopelvic fascia, which continues on the internal organs as visceral fascia and serves to attach the pelvic organs to the pelvic walls (Figure 4-3.1). Unlike fascia in the abdominal wall, which contains regularly arranged collagen bundles, this fascia has variable amounts of collagen, elastin, and fibrovascular elements. Much of the strength of this endopelvic fascia is derived from the walls of arteries and veins that run within it.

In the female, on each side of the pelvis, the endopelvic fascia connects the cervix and vagina to the pelvic wall. The attachment forms a broad sheet, laterally extending from the cephalad parametrium, which attaches the uterus to the sidewall, to the inferior paracolpium, which attaches the vagina to the side wall at the level of the levators. The cephalic paracolpium is lengthy and attaches the vagina to the pelvic walls. More caudally the attachment is more direct. It is this attachment that stretches the vagina between the rectum and the bladder. Support of the bladder is dependent on the attachment of the bladder to the vagina posteriorly and the support of the vagina by the more caudal paracolpium. Similarly, the posterior vaginal wall and rectovaginal fascia form a barrier to the anterior bulging of the rectum and thus prevent formation of a rectocele. In the most distal vagina, the vaginal wall is attached directly to surrounding structures without a paracolpium. Anteriorly, it is fused to the urethra, posteriorly with the perineal body and laterally with the levator ani muscles. Damage to the upper supports of the vagina results in vaginal and uterine prolapse whereas damage to the lower supports results in a cystocele and/or rectocele formation.

Posterior to the rectum is the mesorectum, which contains both blood vessels and lymphatics that supply and drain the rectum. This is loosely bound down the front of the sacrum and coccyx by connective tissue known as the fascia propria. The lateral ligaments, which attach the rectum to the pelvic walls, are condensations of the fascia propria and contain loose areolar tissue, nerves, and small blood vessels. Thus, the mesorectum can be mobilized by dissection in the "mesorectal plane" leaving the mesorectum invested in this thin layer of fascia. The sacrum and coccyx are also covered in a thicker fascia, which extends downward and forward, just superficial to the anococcygeal ligament known as Waldeyer's fascia. Anteriorly the rectum is covered with a layer of visceral fascia that extends from the anterior peritoneal reflection to the urogenital diaphragm. This is Denonvilliers fascia and lies between the rectum and vagina (or prostate in men). Nerves important to bladder control and male sexual function pass through this fascia. The hiatal ligament, originat-

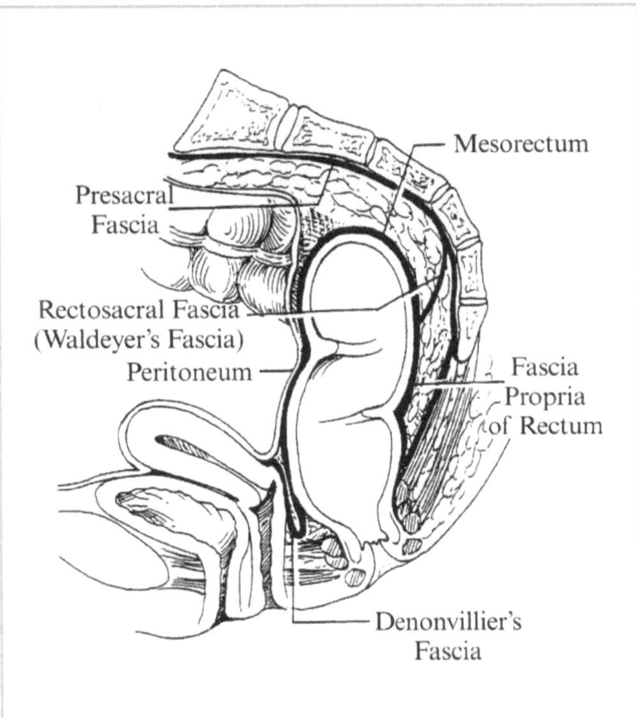

Figure 4-3.1. Fascial layers of the female posterior pelvic floor. (Reprinted from Fundamentals of Anorectal Surgery, 2nd ed, DE Beck, SD Wexner, page 3, © 1996, with permission from Elsevier)

ing from the pelvic fascia, surrounds the rectum and vagina and maintains their patency during levator contraction.

In addition to the bony support of the pelvis, there are two true tendons. The arcus tendineus fascia pelvis (ATFP) and arcus tendineus levator ani (ATLA). These are dense aggregations of connective tissue, predominantly collagen, that provide lateral passive pelvic support. These tendons are condensations of the obturator and levator ani fascia. The ATLA inserts anteriorly at the pubic rami and posteriorly at the ischial spine. The ATFP lies medial to the ATLA at the anterior insertion of the pubic rami and inserts posteriorly at the ischial spine. These tendons provide anchoring sites for the levators and the vagina and thus are key to the support of the pelvic floor.

Muscles of the Pelvic Floor and Perineum

The pelvic diaphragm, composed of the levator muscles and their fascia, form a muscular sheet through which the pelvic visceral structures (lower rectum and vagina) pass. It functions to support the pelvic viscera and helps to maintain urinary and fecal continence.

Levator Ani

The levator ani muscular sling is composed of three muscles: the pubococcygeus, the iliococcygeus, and the ischiococcygeus (see Figure 4-2.4, Chapter 4-2). The ilio- and ischiococcygeus originate from the ischial spine and posterior obturator fascia and insert into the anococcygeal raphe, the coccyx, and the sacrum, forming a shelf on which the pelvic organs may lie.

The pubococcygeus arises from the posterior pubis and anterior obturator and inserts into the anococcygeal raphe, the sacrum, and coccyx. Various muscle subdivisions have been assigned to the medial portion of the pubococcygeus depending on its attachments. These include puborectalis, pubovaginalis, and pubourethralis.

The puborectalis originates from the pubis and inserts into the anococcygeal raphe. It is the medial and inferior portion of the pubococcygeus. The puborectalis is a U-shaped muscle that originates from the pubic bones and passes behind the rectum forming a sling. The puborectalis passes beside the vagina to which it is attached laterally (here named the pubovaginalis) and then passes posterior to the anorectal junction. It provides support for the rectum and indirect support for the vagina, bladder, and urethra by drawing these structures anterior toward the pubic bone. Indirect elevation of the anterior vaginal wall and urethrovesical neck is provided by the bulk of the puborectalis muscle as it draws the rectum and posterior vaginal wall forward with contraction. The tonic contraction of the puborectalis closes the urogenital hiatus, contributes to the posterior curve of the vagina, and reduces pressure on the pelvic outlet. When its tone is lax, the urogenital hiatus opens, the anorectal angle becomes obtuse, and the levator plate sags.

The levator muscles maintain constant tone and, provided they are functioning, the supportive ligaments and fascia are under no tension. When the pelvic floor muscles relax or are damaged, the intraabdominal pressures are applied to the pelvic organs and ligaments. The ligaments function well for short periods under this stress but will stretch and weaken over time, eventually leading to organ prolapse and problems with incontinence.

Rectal and Anal Muscles

The rectal muscles, from mucosa to serosa are the muscularis mucosae, an inner circular layer followed by an outer longitudinal layer (Figure 4-3.2). The inner muscular layer forms the rectal valves and transitions into the internal anal sphincter (IAS). The outer longitudinal layer extends from the sigmoid colon where it envelops the circumference but is thickest at the taenia coli. This muscle splays and becomes confluent at the rectosigmoid junction descending down the rectum to the anorectal junction. Fibers from this muscle descend into the intersphincteric groove where they splay out and may cross both the IAS and external anal sphincter (EAS) and ultimately insert on the perineal and perianal skin. Some of the fibers above the anorectal junction insert into the perineal body and the coccyx.

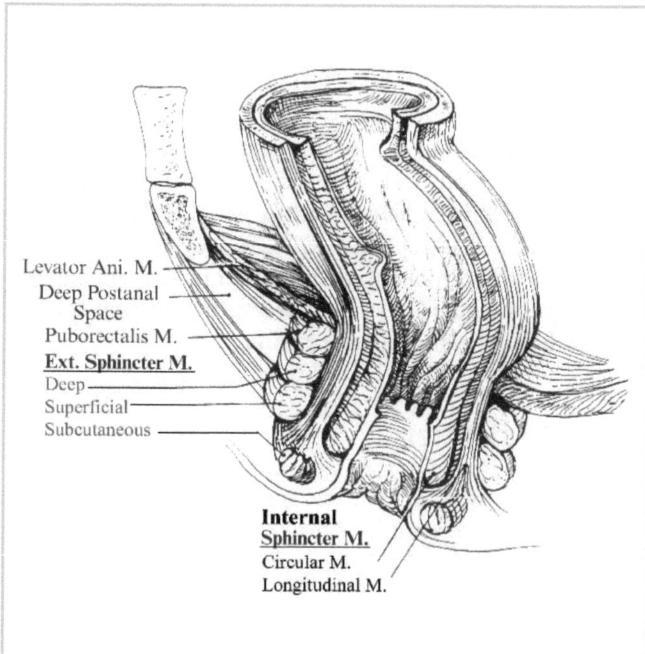

Figure 4-3.2. Muscular anatomy of the anus. (Reprinted from Fundamentals of Anorectal Surgery, 2nd ed, DE Beck, SD Wexner, page 7, © 1996, with permission from Elsevier)

The EAS is a cylindrical-shaped voluntary skeletal muscle that lies outside the IAS. Proximally it abuts the puborectalis at the anorectal ring and extends distally beyond the level of the ZAS. The muscle is attached posteriorly to the anococcygeal ligament and the perineal body anteriorly.

Perineal Muscles

The anal canal is separated from the urogenital organs by the perineum, which contains the perineal body (see Figure 4-2.6, Chapter 4-2). The perineal body, from superficial to deep, contains the superficial and deep transverse perineii muscles and the anterior extension of the external sphincter, which inserts upon the bulbocavernosus. Below the pelvic diaphragm is the triangular urogenital diaphragm. It lies at the level of the hymenal ring and attaches to the urethra, vagina, and perineal body and to the ischiopubic rami. Just above the urogenital diaphragm membrane are compressor muscles for the urethra and vagina.

Pelvic Viscera

The Rectum

The rectum is identified at the level of the sacral promontory, distinguished by loss of complete peritoneal covering, absence of appendiceal epiploicae, absence of a true mesocolon, and divergence of the three taenia coli. The rectum is approximately 15 to 20 cm in length and ends distally at the anal canal. The upper third of the rectum is peritonealized anteriorly and laterally, the middle third only anteriorly, and the lower third is retroperitoneal. The distal third of the rectum is related anteriorly to the vagina and uterus (prostate and seminal vesicles in men) and forms the rectouterine (or rectovesical) pouch or pouch of Douglas. The anterior peritoneal reflection lies approximately 5 to 8 cm from the perineal skin. The rectum has three folds, two on the left at 7 to 8 cm and 12 and 13 cm, and one on the right at 9 to 11 cm. The middle valve in the rectum corresponds to the level of the anterior peritoneal reflection. There are also three lateral curves. The upper and lower curves are convex to the left, and the middle is convex to the right. The rectum sits in a hollow anterior to the sacrum passing downward and posteriorly and then down and anteriorly to become the anal canal at the level of the pelvic floor. The anorectal ring (palpated as the puborectalis) is at the junction of the IAS and the levators.

Anal Canal

The anal canal is approximately 4 cm in length extending from the top of the EAS (or anorectal ring) to the anal verge. This definition is clinical. Histologically, the anal canal mucosa extends from the anal verge to approximately 1 cm above the dentate line. The anal canal, similar to its sphincters, is related anteriorly to the perineal body and the lower posterior part of the vagina. Posterior to the anal canal is the presacral fascia, the anococcygeal ligament, the anococcygeal raphe (an extension of the iliococcygeus), and the posterior extension of the puborectalis and external sphincter musculature, which inserts into the coccyx.

Skin from the buttock is continuous with the anal margin and continues to the lower border of the IAS. This epithelium is keratinized stratified squamous with hair follicles, sweat glands, and sebaceous glands. Proximal to the level of the dentate line, the epithelium is nonkeratinized squamous with no dermal appendages. There is a transition zone where squamous and columnar epithelium are mixed and then the columnar epithelium of the rectum predominates. Vertical mucosal folds known as anal columns are found at the upper anal canal just above the dentate line. Anal valves connect these folds at the inferior margins. Above each valve is the anal pit or sinus, which drains on average eight anal glands.

Blood Supply

Arterial Supply

The superior and inferior hemorrhoidal arteries supply the rectum and anal canal (Figure 4-3.3). The superior hemorrhoidal artery is a continuation of the inferior

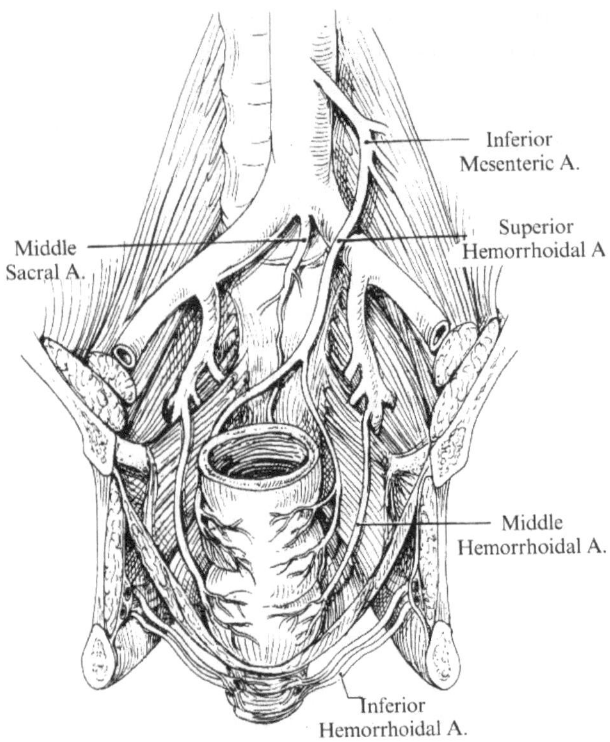

Figure 4-3.3. Anorectal vascular anatomy. (Reprinted from Fundamentals of Anorectal Surgery, 2nd ed, DE Beck, SD Wexner, page 6, © 1996, with permission from Elsevier)

mesenteric artery. The inferior hemorrhoidal artery is a branch of the pudendal artery, which in turn is a branch from the internal iliac artery. The inferior hemorrhoidal artery, after branching off the pudendal, crosses the ischiorectal fossa and goes through the EAS to enter the submucosa of the anal canal and ascend in this plane. The middle hemorrhoidal artery may or may not be present depending on the size of the superior hemorrhoidal artery. It originates from the internal iliac or proximal pudendal and goes anterolateral at the level of the levators. It can be injured at this level when dissecting the upper vagina from the rectum.

Venous Drainage

The three venous plexuses drain through their respective veins. The inferior hemorrhoid plexus drains via the inferior rectal (or hemorrhoidal) vein into the pudendal vein and then into the internal iliac vein. The internal hemorrhoid plexus above the dentate line drains into the middle rectal vein. The perirectal plexus drains into the single superior rectal vein.

Innervation

Levator Ani

Motor innervation of the levator muscles is from the 2nd, 3rd, and 4th sacral roots from above the muscle and the pudendal nerve from below. Controversial in the literature, the puborectalis is reported to be innervated by the 4th sacral root, the pudendal, or both. Some authors suggest that the puborectalis belongs more to the external sphincter than to the levator muscle group.

External Sphincter

The EAS is a voluntary muscle innervated by the inferior rectal branch of the pudendal nerve. The pudendal nerve arises from sacral roots 2–4, leaves the pelvis through the greater sciatic foramen, crosses the ischial spine, and continues in the pudendal canal. This nerve innervates the EAS, as well as the penis and clitoris.

Internal Sphincter

Sympathetic innervation is derived from the hypogastric and pelvic plexus. Parasympathetic innervation is from sacral roots 1–3 via the pelvic plexus. Sympathetic tone is believed to be excitatory, but it is unclear if parasympathetic is inhibitory or excitatory.

Anal Canal

The anal canal's sensation to touch, sharp, and temperature is present from the anal verge to about 1 cm above the dentate line. This sensation is mediated by the inferior rectal branch of the pudendal nerve, which is derived from sacral roots 2–4. This branch of the pudendal and the perineal nerve are the two pudendal branches that arise within Alcock's canal in the ischiorectal fossa.

Rectum

The rectum is only sensitive to distension. This sensation is thought to be a combination of "sensors" in the rectal wall as well as surrounding structures such as the pelvic floor muscles and pelvic fascia. Rectal sensation is parasympathetic through the pelvic plexus sacral roots 2–4.

Sympathetic and Parasympathetic Innervation to the Pelvis

Sympathetic innervation originates from lumbar roots L1 to L3. These fibers pass through the sympathetic ganglion and leave as lumbar sympathetic nerves that join the preaortic plexus. This extends caudally as the mesenteric plexus and then the superior hypogastric plexus. This plexus is formed both by the lumbar sympathetic nerves through the plexus and separately by lumbar splanchnic nerves (from the sympathetic chains). The plexus then divides into the two hypogastric nerves descending along the iliacs until joined by the parasympathetic nervi erigentes to form the pelvic or inferior hypogastric plexus.

Parasympathetic nerve fibers arise from the sacral nerve roots and emerge as pelvic splanchnic nerve or nervi erigentes, which join the sympathetic nerve to form a pair of inferior hypogastric (or pelvic) plexuses on each of the pelvic side walls. Fibers from this pelvic plexus (sympathetic and parasympathetic) then branch out and innervate the bladder, ureters, corpora cavernosa, and rectum. Somatic fibers also pass through the pelvic plexus to innervate the levators and ureteral striated muscles. These pelvic plexuses run anterolateral to the rectum necessitating close rectal dissection to avoid nerve injury.

Anorectal Spaces

Knowledge of the anorectal spaces is important in the understanding of anorectal and pelvic floor disorders and to avoid injury to important structures during surgical therapy.

The *perianal space* is that area surrounding the anal canal in the immediate area of the anal verge. Laterally it is continuous with the fat of the buttocks. Cephalically, it is continuous with the intersphincteric space and it contains the distal EAS, branches of the inferior rectal vessels, nerves, and lymphatics. The external hemorrhoid plexus lies in the perianal space and communicates with the internal hemorrhoid plexus at the dentate line.

The *ischiorectal space* is lateral to the perianal space. Its apex is formed by the joining of the levator and obturator muscles while the inferior boundary is the skin of the perineum. Anteriorly, it is bound by the transverse perineal muscles and posteriorly by the gluteal skin. This space contains the pudendal vessels and pudendal nerve in Alcock's canal. Alcock's canal, or the pudendal canal, is on the medial side of the obturator internus in the ischiorectal fossa.

The *intersphincteric space* is between the anal sphincters and communicates with the perianal space. Most of the anal glands are in the intersphincteric space.

The *superficial postanal space* connects the perianal spaces with each other below the level of the anococcygeal ligament. The deep postanal space, the right and left ischioanal/rectal spaces communicate with this space posteriorly deep to the anococcygeal ligament. Infection in this space leads to a horseshoe abscess.

References

1. Aigner F, Zbar AP, Ludwikowski B, Kreczy A, Kovacs P, Fritsch H. The rectogenital septum: morphology, function, and clinical relevance. Dis Colon Rectum 2004;47(2):131–140.
2. Beck DE, Wexner SD, eds. Fundamentals of Anorectal Surgery. 2nd ed. London: WB Saunders; 1998.
3. Corman ML, ed. Colon and Rectal Surgery. 4th ed. Philadelphia: Lippincott-Raven; 1998.
4. Gordon PH, Nivatvongs S, eds. Principles and Practice of Surgery for the Colon, Rectum and Anus. St. Louis: Quality Medical-Publishing; 1992.
5. Pemberton JH, Swash M, Henry MM, eds. The Pelvic Floor: Its Function and Disorders. London: WB Saunders; 2002.
6. Strohbehn K. Urogynecology and pelvic floor dysfunction. Obstet Gynecol Clin 1998;25(4):684.

Section V

Sexual Function

5-1

Female Sexual Dysfunction

Lawrence S. Hakim and Giovanna M. DaSilva

Sexual dysfunction is a couple's disease, and is common in both women and men. With the introduction of effective oral therapies for the treatment of erectile dysfunction (ED) in men, increased media attention has brought "female sexual dysfunction" (FSD) into the limelight. Epidemiologic studies have demonstrated that whereas more than 50% of men aged 40 to 70 years have sexual dysfunction, this problem may in fact be even more prevalent in women. Whereas psychological and interpersonal relationship issues had been thought to be the primary causative factors of FSD, we are now learning that, similar to sexual dysfunction in men, FSD can often be attributed, at least in part, to pathophysiologic changes, such as chronic disease states, aging, and medications, as well as other physical factors, such as pelvic and colorectal surgeries. In addition, FSD has been shown to respond to treatment of the underlying condition and to other specific therapies. In this chapter, we will look at the various classifications, causative factors, diagnostic evaluations, and therapeutic alternatives for women with sexual dysfunction, with a special emphasis on the impact of pelvic floor disorders and colorectal surgery (CRS) on FSD.

Epidemiology

Sexual dysfunction is a disease affecting men and women of all ages.[1] The Massachusetts Male Aging Study,[2] one of the largest epidemiologic studies, revealed the high prevalence of ED. In fact, it was shown that 52% of men between the ages of 40 and 70 had ED. Subsequent epidemiologic studies in women suggested that FSD was also an extremely prevalent problem. A report in the *Journal of the American Medical Association*[3] suggested that 43% of women, aged 18 to 59, had FSD. In fact, the prevalence of FSD in this study was greater than that of ED.[3] Of note, FSD is more prevalent in patients with a history of sexual abuse and is often reported in women with a history of sexual coercion. In a study reported at the Annual Meeting of the American Urologic Association in 2000, Nehra and associates looked at the prevalence of FSD in the partners of men with ED. One hundred fifty women were evaluated, aged 25 to 82 years. The study revealed that 56% of these women had FSD, and demonstrated specific vascular risk factors, including cigarette smoking, hypercholesterolemia, hypertension, and prior pelvic surgery. Based on population surveys, this would suggest that more than 30 million women in the United States might have FSD. Unfortunately, despite this high prevalence, it has been estimated that less than 5% of women with FSD are being treated.

Several explanations for the phenomenon of FSD have been proposed. These include embarrassment on the part of the patient to discuss personal sexual matters or a feeling that her concerns would be "brushed aside," embarrassment or lack of time on the part of the physician (man or woman), and lack of education regarding the prevalence, significance, and treatment options available for FSD.

The question arises as to why the identification and treatment of FSD is important. Despite the belief of many physicians that the treatment of sexual dysfunction is not a "medical" priority, it is crucial to remember that normal sexual function is an important part of the essential intimacy between a woman and man ("the COUPLE"). Furthermore, sexual dysfunction often leads to loss of self-esteem, depression, and alienation from one's partner. In addition, similar to its counterpart in the male (ED), FSD is a spectrum of disease and may be an early warning sign of significant unrecognized systemic vascular disease indicating an increased risk for heart attack or stroke.

Sexual Dysfunction and Aging

It has been suggested that in both women and men, sexual dysfunction is age-related and progressive. The Massachusetts Male Aging Study[2] demonstrated that the incidence of ED clearly increases as men enter the sixth, seventh, and eighth decades of life. This increased prevalence with aging is less clear in women. Unlike ED, we can see a significant prevalence of FSD in women of all ages, such as young mothers after childbirth, premenopausal smokers, as well as peri- and postmenopausal women.

There are, however, specific factors related to aging that are often associated with FSD and need to be addressed. As women age, there is a progressive decrease in overall physiologic function. In addition, as estrogen levels diminish, genital and vaginal atrophy may be seen. This can be associated with decreased vaginal lubrication and pain during intercourse, or dyspareunia.

Female sexual dysfunction can also be the result of pathophysiologic changes resulting from chronic disease processes, such as atherosclerosis, cardiovascular disease, lipid disorders, and diabetes. Psychological issues, such as depression, are often seen in association with women with sexual dysfunction, and may be a causative factor. In addition, the loss of a lifelong partner, or other partner-specific issues, including ED and chronic disease, are often contributory.

Sexual Dysfunction and Incontinence

The association of FSD and urinary incontinence has also been shown.[4] In fact, it has been estimated that in up to 43% of women with urinary incontinence, evidence of sexual dysfunction is also present. Considering that less than 30% of physicians ask their patients with urinary incontinence about their sexual function, this may be an underestimate.

Diagnostic Classifications

In 2000, a new classification of FSD was developed by a consensus panel of experts in the field of sexual medicine[5] (Table 5-1.1). The categories include sexual desire disorders, sexual arousal disorders, orgasmic disorders, and sexual pain disorders.

The definitions of the new classification system took into account that there should be some degree of "personal distress" caused by the disorder for it to be a problem. Additionally, the presence of sexual dysfunction in the male partner or any impact of the FSD on the partner was not part of the classification or definition. This system can be used whether FSD results from medical or psychosocial factors. In addition, each of these categories was independent of the other, and overlap between symptoms could often be seen. For example, women may complain of both poor arousal and pain with intercourse. Additionally, a woman with FSD may experience loss of libido, but arousal and lubrication can be normal.

Table 5-1.1. Classification of female sexual dysfunction*

I – Sexual desire disorders
II – Sexual arousal disorders
III – Orgasmic disorders
IV – Sexual pain disorders

*Overlap between symptoms is often seen.

Hypoactive Sexual Desire Disorder

Hypoactive sexual desire disorder is the persistent or recurrent lack of sexual thoughts and/or receptivity to sexual activity, which causes personal distress. Hypoactive sexual desire may be associated with psychological or emotional disorders, as well as physiologic factors, such as androgen insufficiency. Sexual aversion disorder is a subcategory of hypoactive sexual desire.

Sexual Arousal Disorder

Sexual arousal disorder is the persistent or recurrent inability to attain or maintain sexual excitement, which causes personal distress. This disorder includes poor vaginal lubrication, decreased genital sensation, and poor vaginal smooth muscle relaxation. Arousal disorders are primarily physiologic in nature and can often result from pelvic and colorectal surgery and other pelvic disorders, various medications, atherosclerosis, cigarette smoking, and vascular disease. This disorder most closely parallels ED in the male. In fact, a condition of *hyperactive sexual arousal disorder*, analogous to the priapism state in the male, has also been described.

Orgasmic Disorder

Orgasmic disorder is the persistent or recurrent loss of the ability to achieve orgasm with sufficient sexual stimulation and arousal, and which causes personal distress. Orgasmic failure may be associated with changes after nerve injury from colorectal and pelvic surgery, spinal cord injury, androgen insufficiency, as well as psychosexual factors.

Sexual Pain Disorder

Sexual pain disorder is the persistence or recurrence of genital pain associated with sexual stimulation and intercourse, which causes personal stress. Dyspareunia, pain upon intromission, and vaginismus, or the reflexive "closing" of the vaginal introitus, are types of sexual pain disorders. Pelvic trauma, such as seen with childbirth injuries and CRS, as well as psychological trauma, may be associated with this disorder.

Table 5-1.2. Female sexual response cycle

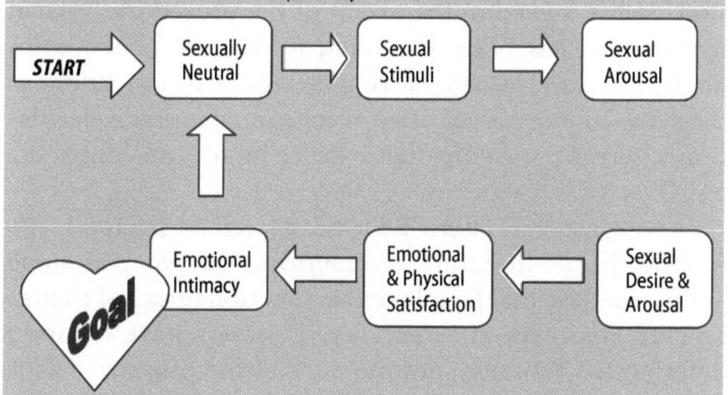

Source: Adapted from Basson.[8]

The Female Sexual Response

The female sexual response, as described in the mid 1960s by Masters and Johnson,[6] begins with *excitement*, leading to *plateau*, *orgasm*, and finally, *resolution*. In 2000, Dr. Rosemary Basson,[7] a pioneering researcher in the field of FSD, proposed a new, nonlinear female sexual response cycle. Dr. Basson suggested that the sexual response is driven by the desire to enhance intimacy, and begins with a state of *sexual neutrality*[8] (Table 5-1.2). As the woman seeks a sexual stimulus and responds to it, she becomes sexually aroused. Arousal leads to desire, thus stimulating a woman's willingness to receive or provide additional stimuli. Emotional and physical satisfaction are gained by an increase in sexual desire and arousal. Emotional intimacy is then ultimately achieved. Various biological and psychological factors can negatively affect this cycle, thus leading to FSD.

Physiology of Female Sexual Function

The underlying physiologic processes in both normal female sexual function and FSD are not yet well understood. Normal female sexual function is based on an interaction between intact anatomic, vascular, and neurologic factors. Sexual arousal is marked by physiologic changes secondary to increased genital blood flow, which leads to vaginal congestion and lubrication, facilitating intercourse. The normal vascular response is a result of cavernosal and arteriole smooth muscle relaxation via the androgen-dependent nitric oxide synthase system during sexual stimulation and arousal. This produces a vascular engorgement of the vestibule and clitoris. Vaginal lubrication is a transudate of serum that results from this normal increase of pelvic blood flow with arousal. Patients with arousal disorders may complain of decreased vaginal lubrication and dyspareunia.

Normal sensation is also critical in allowing sexual arousal, and the ability to achieve orgasm requires an intact sympathetic outflow tract. In women with diabetes or spinal cord injuries, the neuropathy may result in diminished levels of sexual functioning and orgasmic disorders. Similarly, after pelvic floor surgery, orgasmic function can be potentially affected by changes in the pelvic floor muscles because they modulate vaginal receptivity as well as the motor responses during orgasm.

Causes of Female Sexual Dysfunction

Similar to its counterpart in the male, there are numerous causes of FSD (Table 5-1.3). In fact, any factors influencing normal physiology, or any step of the sexual response in women, may contribute to FSD. The etiologies can be grouped as being either psychogenic or organic in nature, although overlap often exists.

Psychogenic FSD may be noted as either a result of, or a causative factor for, depression. In fact, it behooves the physician to rule out a primary diagnosis of depression in any women presenting with symptoms of FSD, especially poor libido. Likewise, the temptation to assign a diagnosis of depression to a woman without at least screening for the presence of sexual dysfunction should be avoided. Other facets of psychogenic FSD may include interpersonal relationship issues, abuse, performance anxiety, and psychological distress. These psychological factors often accompany organic factors.

As in men, many women with sexual dysfunction will have an underlying or contributing organic etiology of their sexual dysfunction. Any factor, disease state, or medication affecting the vascular, hormonal, or neurologic

Table 5-1.3. Risk factors for female sexual dysfunction

Atherosclerosis	Radical pelvic surgery
Smoking	Colorectal surgery
Diabetes	Hysterectomy
Dyslipidemia	Neuropathies
Peripheral vascular disease	Genital atrophy
Hypertension	Dermatitis
Pelvic or perineal trauma	Clitoral adhesions
Arterial compression from cycling	Bartholin's gland cysts
Endocrinopathies	Episiotomy scars
Postmenopausal changes	Vestibulitis
Hypothyroidism, hyperthyroidism	Vulvar cancer
Low estrogen levels	Lichen sclerosis
Pituitary tumor, hyperprolactinemia	Liver and/or renal failure
Androgen insufficiency syndrome	Sexual abuse
Pudendal nerve injury	Psychological factors
Stroke, Alzheimer's, Parkinson's disease	Life stressors
Endocrinopathies	Interpersonal relationship issues
Pelvic/spinal cord injury	Medications
Multiple sclerosis	

Table 5-1.4. Medications associated with female sexual dysfunction

Antidepressants	Beta-blockers
SSRI agents	Antihistamines
Antihypertensives	Sympathomimetic amines
Narcotics	Anticonvulsants
Diuretics	Metronidazole
Antiandrogens	Metoclopramide
H2 blockers	Marijuana
Anticholinergics	Calcium channel blockers
Psychotropics	Antiandrogens
Cocaine	Alkylating agents
Alcohol	Oral contraceptives
Lipid-lowering drugs	Sedatives
Nonsteroidal antiinflammatory drugs	Tamoxifen
Oncologic agents	Gonadotropin-releasing hormone agonists
Anxiolytics	Analgesics

components of the sexual response can cause FSD. Vascular causes of FSD include atherosclerosis, smoking, diabetes, dyslipidemia, peripheral vascular disease, hypertension, renal failure, and perineal or pelvic trauma, such as that seen with bicycle riding. Neurologic diseases associated with FSD may include stroke, multiple sclerosis, spinal cord injury, Alzheimer's disease, Parkinson's disease, pudendal nerve injury, and iatrogenic surgical changes during radical pelvic surgery, CRS, or hysterectomy. Hormonal or endocrinologic causes of FSD include hypogonadism, androgen insufficiency, menopausal changes and genital atrophy, hypothyroidism, hyperthyroidism, diabetes, low estrogen levels, pituitary tumor, and hyperprolactinemia. Various specific gynecologic causes of FSD have been identified. These include genital atrophy, vulvar dystrophy and cancer, dermatitis, clitoral phimosis and adhesions, Bartholin's duct cysts, episiotomy scars, vestibulitis, vulvar cancer, and lichen sclerosis.

Numerous medications may contribute to sexual dysfunction in women (Table 5-1.4). These include antihistamines, sympathetics, anticonvulsants, metronidazole, antihypertensive agents, diuretics, beta-blocking agents, calcium channel blockers, antiandrogens, cancer drugs, anxiolytics, oral contraceptive agents, antidepressants, psychotropic agents, narcotics, alcohol, lipid-lowering agents, anti-estrogens, and gonadotropin-releasing hormone agents, to name a few.

Female Sexual Dysfunction and Colorectal Surgery

The occurrence of sexual dysfunction after CRS is a well-recognized phenomenon. Prior studies in men have demonstrated that the incidence of sexual dysfunction is between 20% and 34% in patients undergoing rectal excision, and is related to the extent of the procedure.[9] Sexual dysfunction has been attributed to damage of the autonomic nerves, especially during pelvic dissection. Injury to the sympathetic nerves often results in retrograde ejaculation whereas parasympathetic nerve injury can contribute to ED.

In contrast, FSD after CRS is not as well understood. The findings in the literature are controversial; whereas some authors believe that female sexual function does not change or even improves after surgery,[10,11] others have reported deteriorated function, despite cure of the disease.[6,12] This discrepancy in results may be attributed to the fact that most studies were retrospective, included a small number of female patients, lacked a baseline functional status, and did not use a validated sexual inventory. Therefore, the available literature data warrant cautious evaluation. Table 5-1.5 summarizes the results of 14 studies in women who underwent major colorectal procedures.

After CRS, sexual dysfunction in women may be attributed to disorders of desire, arousal, orgasm, and/or pain. Sexual desire has been shown to maintain or improve in 76% to 80% of women after CRS and is often accompanied by increased frequency in sexual activity and sexual satisfaction.[13,14] Such an improvement may be expected, because general health is likely to improve after surgery. However, previous studies have demonstrated that up to 78% of women complain of decreased or complete loss of libido after surgery.[15] Factors such as impaired body image, concerns of partner negative reaction, fear of stool leakage, and use of a stoma are known to adversely impact sexual function, and may account for these findings. Among these factors, the negative impact of a stoma has been specifically addressed by several studies. Gloeckner and Starling[16] conducted an interview with 40 subjects with a permanent stoma (24 men, 16 women) and found that 60% of the patients had impaired sexuality after surgery. One year after surgery, 67.5% of the patients reported improvement on sexual attractiveness, and this significantly correlated with gender (female vs. male), type of disease (inflammatory bowel disease vs. malignancy), type of stoma (ileostomy vs. other stoma types), duration of disease before surgery (>10 years vs. <10 years), and the ability to manage the stoma (no problems to manage vs. problems to manage). In a study recently published by Engel et al.,[17] the long-term quality of life of 329 patients with rectal cancer was prospectively assessed. Overall, patients who underwent an anterior resection had better quality-of-life scores than those who had an abdominoperineal resection with a permanent stoma; the effects of the abdominoperineal resection on the patients' quality of life did not improve over time in this series. Patients with stoma had significantly worse quality-of-life scores than patients without stoma; however, quality of life greatly improved for patients whose stoma was reversed. In their study, Sjögren and Poppen[14] also demonstrated significant improvement in patients' sexuality after stoma closure.

Table 5-1.5. Female sexual dysfunction after major colorectal surgery

Author, year	Surgery	*n*	Decreased Desire	Decreased Lubrication	Altered Orgasm	Dyspareunia	Decreased Satisfaction
Williams and Slack,[22] 1980*	APR	3	—	—		1 (33)	—
Deixonne, 1983[36]	APR	26	12 (46)	—	11 (42)	8 (31)	—
Hjortrup, 1984*[37]	LAR	20	—	—	1 (5)	2 (10)	—
Fegiz et al.,[21] 1986	APR	15	—	—	(70)	(65)	—
	LAR – stapled	9			(24)	(65)	
	LAR – manual	17			(44)	(44)	
Metcalf et al.,[11] 1986	RPC	50	—	—	2	(38)	—
	Kock pouch	50			2	(48)	
Cirino et al.,[15] 1987	APR	18	15 (77)	—	1 (5)	6 (33)	8 (44)
Wikland et al.,[19] 1990	Pouch	71	—	—	—	27	—
Cunsolo, 1990[38]	APR	8	2 (25)	—	2 (25)	4 (50)	
Öresland et al.,[13] 1994	J pouch	20	5 (25)	—	1 (5)	5 (25)	—
Sjögren and Poppen,[14] 1995	S pouch	30	6 (20)	7 (23)	10 (33)	5 (16)	—
Damgard et al.,[10] 1995	J pouch	23	—	—	0	0	—
Bambrick et al.,[12] 1996	J pouch	92	16 (18)	22 (26)	13 (15)	22 (26)	21 (26)
Tiainen et al.,[18] 1999	J pouch	95	—	—	2 (5)	9 (22)	4 (9)

* Prospective studies. APR, abdominoperineal resection; LAR, low anterior resection.

Although patients with a J pouch do not have to deal with the inconvenience of a stoma, 3% to 43% of women fear stool leakage from the pouch, causing a hindrance of sexual activity.[18] In addition, special precautions such as emptying the pouch before intercourse are often necessary in these patients.

Sjögren and Poppen[14] evaluated 30 women who underwent total proctocolectomy with S pouch and found that 13.3% of the patients had dyspareunia and 10% complained of vaginal dryness, which could not be explained by estrogen deficiency. In a larger series including 92 patients who underwent restorative total proctocolectomy, Bambrick et al.[12] found that 22 patients (26.8%) had dyspareunia, of whom 30.5% experienced decreased sexual pleasure, and 22 patients (26.5%) had vaginal dryness, 7% of whom experienced this 75% or more of the time.

Few factors have been postulated to cause dyspareunia in women after rectal excision. It has been suggested that neurovascular damage during dissection interrupts the physiologic response of arousal, leading to decreased lubrication and dyspareunia. In patients with rectal cancer, radiation therapy can cause damage to the small vessels and induce fibrosis with loss of elasticity of the vaginal canal, resulting in arousal impairment.

Anatomic changes after rectal excision and pelvic floor closure have also been postulated to cause sexual discomfort or dyspareunia. Metcalf et al.[11] compared sexual dysfunction between 50 patients who had a Kock pouch (continent stoma) and 50 patients who had an ileoanal anastomosis. Although the overall rate of dyspareunia decreased in both groups after surgery, patients with a Kock pouch had significantly more persistent dyspareunia and excessive positional vaginal discharge, than patients who underwent ileoanal anastomosis.

Wikland et al.[19] evaluated the anatomic changes of the genital tract in women after proctocolectomy. The authors performed a gynecologic examination in 71 women and compared physical findings to the gynecologic complaints. Of the 71 women, 35 (49%) had a distressing vaginal discharge after proctocolectomy, compared with 6 (9%) before surgery. Dyspareunia was reported by 8 women (12%) before surgery and 18 (27%) after surgery. In 44 patients (61%), gynecologic examination revealed caudal fixation and dilatation of the posterior vaginal fornix. Thirty (68%) of these 44 women had heavy vaginal secretion (upper thirds of the vagina filled with fluid, cervix identified after washout) associated with the anatomic changes. In a subsequent study, some of these women underwent vaginography, which confirmed anatomic changes assessed by physical examination.[20] Öresland and associates[13] from the same institution demonstrated that such abnormalities are not present when a bowel reconstruction is performed. Twenty-one women who underwent restorative proctocolectomy underwent gynecologic examination and vaginography after surgery. Five experienced occasional dyspareunia and one complained of vaginal discharge. Gynecologic examination demonstrated unaltered vaginal position and angulation in all patients, which was also proven by the vaginography. Corroborating these findings, Sjögren and Poppen[14] found normal vaginal anatomy in 29 (96.6%) of 30 patients who underwent proctocolectomy with pouch formation. These data suggest that the interposed pelvic pouch prevents its dorsal displacement in contrast to proctocolectomy and permanent stoma.

Orgasmic function can potentially be affected by changes in the pelvic floor muscles after radical pelvic and colorectal surgery. In a study performed by Fegiz and associates,[21] orgasmic function was more affected in patients after abdominoperineal resection (70%) than patients who underwent colorectal hand-sewn (65%) or stapled anastomosis (44%). The majority of the studies, however, have shown that most women maintain the ability to achieve orgasm after surgery.[22] This may be explained by the fact that the pudendal nerve supply to the pelvic floor muscles, which is considered essential for female orgasmic response, remains intact.

At the Cleveland Clinic Florida, a recent study looked at the incidence of changes in sexual function in women who underwent CRS as a result of rectal cancer and benign diseases.[23] A survey was sent to 225 patients, and 74 (32.8%) returned the questionnaire; 31 had rectal cancer and 43 had benign diseases. The mean age was 54 years in the patients with rectal cancer and 36 years in the group with benign disease ($P < 0.01$). The mean time of follow-up postsurgery was 41.3 (18–93) months. Forty-five patients (60.8%) were married, and 48 (74.9%) continued their education beyond high school. Fifty women (67.5%) indicated being sexually active at the time of surgery. Of these, 32 patients (64%) reported worse sexual function after surgery, and 16 (55%) sought help. Fourteen patients (43.7%) expressed an interest in receiving treatment for FSD, which was more likely in women younger than 60 years of age ($P = 0.012$). Sixteen patients (32%) reported no difference, and only 1 (2%) indicated improved sexual function after surgery. Interestingly, 65 patients (87.8%) indicated that their physician never discussed the risk of postsurgical sexual dysfunction before surgery. However, only 24 patients (typically younger women) reported that they would have wanted to discuss such issues preoperatively ($P = 0.001$). There was no statistically significant association between the desire to discuss sexual matters and level of education, marital status, diagnosis, or the presence of a stoma. This study suggests that CRS may significantly and adversely impact a woman's sexual function and her sexuality, and, therefore, all women should be preoperatively counseled regarding the potential for sexual dysfunction after surgery.

Clinical Evaluation

The clinical evaluation of FSD can be approached similar to any other disease process, and typically begins with a simple "step-care approach." The first step is the identification of the problem by the woman and her partner. The physician or health care professional is then notified, and a thorough evaluation is performed. Identification of reversible causes and incorporation of various lifestyle changes (e.g., stop smoking) are often the first steps to maximizing and improving sexual function. Medical and surgical therapies are used when appropriate; however, alternative and complementary therapies certainly may have a role in successful therapy. The importance of communication among the woman, her partner, and the physician cannot be overemphasized, to help the patient achieve an improved sexual function and youthful sexual rejuvenation. These steps are best summarized by the term I.N.T.I.M.A.C.Y.™, as illustrated in Table 5-1.6.

As noted above, communication is critical, although the majority of patients and physicians alike tend to avoid discussing this topic in the general health care arena. However, evidence is now clear that sexual dysfunction is often the earliest sign of other underlying disease, such as cardiovascular disease, diabetes, and hyperlipidemia. In fact, it has been suggested that by identifying sexual dysfunction early and treating the underlying causes, serious cardiovascular events, such as heart attack and stroke may be avoided. Certainly, it can significantly improve the woman's quality of life and improve intimacy in the couple.

The first step in the diagnostic work-up is identification of the problem. Unfortunately, few women volunteer any history of FSD, and therefore information should be actively elicited as part of the routine medical history. A simple approach for a clinician could be, "Many woman report that once they reach menopause, they experience changes in sexual function.... Have you noticed any changes in your sexual function or desire?" The idea is to offer an open-ended question and opportunity for the patient to realize 1) she is not alone, 2) you are interested in helping her, and 3) successful treatments are available. Various questions regarding the length of time since the onset of sexual dysfunction, libido issues, ability to become sexually aroused, poor lubrication, ability to achieve orgasm, and dyspareunia should be asked to further classify the sexual dysfunction (e.g., hypoactive sexual desire disorder). Any history of abuse should be elicited. The history should also include any cardiovascular, neurologic and hormonal risk factors, medications (both prescribed and over-the-counter), illicit drug use, smoking history, and partner issues including ED.

Validated questionnaires can be extremely effective in helping to identify the presence of FSD and define the type of sexual dysfunction in a particular patient.[24] The clinical

Table 5-1.6. Management of female sexual dysfunction: I.N.T.I.M.A.C.Y™

I.	Identification
N.	Notification
T.	Thorough evaluation
I.	Identify reversible causes/incorporate lifestyle changes
M.	Medical and surgical options
A.	Alternative therapies
C.	Communication
Y.	Youthful rejuvenation

assessment of FSD may include the use of tools to monitor treatment efficacy. The Female Sexual Function Index questionnaire is one such validated tool that assesses specific domains of sexual function, including desire, lubrication, dyspareunia, and orgasm.[25,26] Each question is ranked from 0 to 5. The questionnaire can also be used to objectively measure the response to therapy over time.

Laboratory evaluation is an important part of the overall diagnostic evaluation of women with sexual dysfunction. Laboratory tests include a complete lipid profile, thyroid function tests, and fasting glucose levels; a complete endocrine evaluation in appropriately selected patients is typically performed. The hormonal evaluation should include serum estradiol levels, total and free testosterone levels, dehydroepiandrosterone (DHEA-S and DHEA), serum follicle-stimulating hormone, and leuteinizing hormone and serum prolactin levels.

Various specialized diagnostic tests may be performed in selected patients and centers, including measurement of vaginal pH, duplex Doppler measurement of genital blood flow (pre- and poststimulation), neurologic and temperature testing, vaginal wall compliance, and genital vibratory sensation thresholds or biothesiometry.

A directed physical examination should be part of the diagnostic work-up, including a pelvic examination to search for evidence of clitoral phimosis, genital atrophy, a narrowed introitus, and point tenderness. Care should be taken to assure that a recent breast examination and Pap smear have been performed and the results obtained. Further examination and testing (e.g., neurologic tests) are performed on an individual basis.

Regardless of the underlying etiology of the sexual dysfunction, consideration for evaluation by a licensed sex therapist should be entertained. It is important to remember that even with a clear physical etiology, the couple's problem may be multifactorial, and psychosexual and relationship issues should be addressed early. In addition, evaluation of the partner by the appropriate specialist is critical when addressing this couple's disease.

Antidepressants and Female Sexual Dysfunction

Although numerous medications can have an adverse impact on sexual function, few classes of medications have been demonstrated to have more impact on sexual function than antidepressants. Although many of the various classes of antidepressant agents can affect sexual function, the selective serotonin reuptake inhibitors (SSRIs), such as paroxetine, sertraline, or fluoxetine, have been shown to adversely affect libido, arousal, as well as the ability to reach orgasm.[27] Under a physician's care, options may include switching agents, dose reduction, or drug holiday. Bupropion (Wellbutrin) may be an effective alternative.[28] In addition, studies have suggested that the use of sildenafil with selective SSRI medications may effectively combat the FSD associated with the SSRI. It is also important to recognize that FSD may present with many of the same symptoms as depression, such as poor energy and lack of interest. We have seen numerous patients who were misdiagnosed as having depression, and placed on SSRI drugs (making their FSD worse!), when in fact the primary problem was FSD. Successful management of the underlying cause of the FSD (e.g., hormone replacement therapy for androgen insufficiency syndrome) has allowed many women to stop using antidepressant medications and regain a normal sex life.

Treatment

Once a complete diagnostic evaluation has been performed, the patient's sexual dysfunction can be accurately classified. One should note that there is often overlap of symptoms, such as hypoactive sexual desire and arousal insufficiency. It is also important to educate the patient and partner about normal physiologic response and sexual function, especially as it pertains to such issues as aging and childbirth. It is important to identify to the patient any obvious reversible causes of FSD or potential lifestyle changes that may be beneficial to her sexual function, such as stopping smoking, eating a healthy diet, regular exercise, improving partner communication, and stress reduction, because clearly there is a correlation between the patient's general health and her sexual function.

Various medication schedules may need to be addressed and prescriptions altered to maximize sexual wellness. This is best done in collaboration with the woman's primary care physician.

Especially in postmenopausal women, vaginal dryness may be the result of hormonal and vascular changes. Water-soluble lubricants can be very effective in alleviating any symptoms of dyspareunia and discomfort. In addition, topical estrogen replacement therapy (Estrace Cream®, Vagifem Suppositories®, Estring®) may be helpful in treating vaginal atrophy, decreasing coital pain, and improving clitoral sensitivity. For women with androgen insufficiency, demonstrated by diminished androgen levels, including low free serum testosterone levels and DHEA, androgen replacement therapy may be beneficial.[29] The goal of treatment is to restore the levels to "normal" (mid to upper normal) and correlate with sexual dysfunction, not to strive for supraphysiologic levels, which may be unsafe and lead to significant side effects (i.e., acne, facial hair). Patient education and care to avoid any possibility of pregnancy when taking hormone replacement therapy is paramount. Close and regular monitoring of hormone levels is important in the close management of these women, who will often report significant improvement in arousal, lubrication, and libido after 8 to 12 weeks of therapy. In fact, many will report even sooner that they are now "dreaming" or fantasizing about sex, often for the first time in years.

Testosterone replacement in women has associated risks. Oral administration of testosterone may be associated with hepatic disease and should be avoided. Certain androgen therapies may adversely affect lipid profiles and may cause polycythemia, alopecia, acne, and female hirsutism. Hence, lipid profiles should be measured by the primary care physician every 3 months in androgen-deficient women with secondary FSD taking androgen replacement therapy. After 6 to 12 months, if the medication is ineffective, it should be discontinued. Purely empiric treatment with hormones has not been shown to be effective; therefore, it is inappropriate to use androgen therapy in the absence of an FSD with androgen deficiency that is supported by laboratory data. The benefits of transdermal testosterone replacement therapy for postmenopausal women were suggested by a recent study of women with FSD who had undergone bilateral oophorectomy and hysterectomy.[30] The authors concluded that testosterone replacement may be helpful in this subset of patients, as measured by such endpoints as frequency of sexual activity, receptivity, and orgasm.

Dehydroepiandrosterone has been demonstrated by several investigators to have a positive impact on sexual dysfunction in women.[31] Improved libido, arousal, and orgasmic ability have been reported in women receiving up to 50 mg daily of DHEA. In a recent study presented at the European Society of Sexual Medicine, Alici and colleagues evaluated 55 women with sexual dysfunction and low androgen (DHEA) levels, ranging from age 23 to 54 (mean 43) years. All of their partners had normal erectile function. The women were treated with DHEA 50 mg/d for 3 months. All of the women demonstrated increased DHEA levels, and more than 95% demonstrated improvement in Female Sexual Function Index testing in areas of desire (libido), arousal, and orgasm. The investigators concluded that DHEA is effective in improving sexual function in women with low DHEA levels and FSD.

In women receiving DHEA therapy, care should be taken to monitor androgen levels on a regular basis and to adjust the dose accordingly. Side effects of acne and hirsutism were relatively rare in these patients, but responded to dose adjustments. In October 2000, the American College of Obstetricians and Gynecologists (ACOG) issued a committee opinion cautioning physicians considering using androgen therapy, including testosterone and DHEA supplements, to treat low sex drive in women.[32]

There seems to be a significant role for various other herbal preparations in the treatment of FSD. Nonprescription topical solutions, such as Viacreme® or Viagel®, have been purported to increase genital sensitivity and assist in achieving orgasm. Viacreme® is an amino acid-based (L-arginine) solution that contains menthol. L-Arginine is involved in nitric oxide synthesis, which is responsible for vascular and nonvascular smooth muscle relaxation. When applied to the clitoris, Viacreme® may increase blood flow by dilating clitoral blood vessels. While these and other agents await well-designed placebo-controlled trials to prove their efficacy, safety should remain of primary concern for any patient using these or other non-FDA-approved therapies.

A better understanding of the physiology of the sexual arousal response in women, including the role of an androgen-dependent, nitric oxide-mediated stimulation of genital and cavernosal smooth muscle, leading to increased clitoral blood flow and genital engorgement, has led to the use of various phosphodiesterase type 5 inhibitors, such as sildenafil for the treatment of female sexual arousal insufficiency. A recent study by Berman and associates[33] demonstrated that in women with female sexual arousal insufficiency and androgen insufficiency, once the androgen levels were restored with hormone replacement therapy, sildenafil was not only safe, but was quite effective in improving the parameters of sexual function, especially arousal. Other investigators have demonstrated that, in selected subpopulations such as premenopausal women with sexual arousal disorder, or those with spinal cord injuries, sildenafil may have a much higher efficacy.[34]

Further trials are currently underway to evaluate the safety and efficacy of topical prostaglandin E_1 to augment genital blood flow in women with FSD. In addition, various new topical testosterone preparations, intranasal therapies, centrally acting oral agents (apomorphine), and phosphodiesterase inhibitors are being evaluated in women with sexual dysfunction.

Eros

Currently, the only FDA-approved device for the treatment of FSD is the EROS-CTD clitoral therapy device). This small, hand-held, battery-operated unit has a soft, cup-like attachment that, when placed over the clitoris, allows the woman to apply a gentle vacuum to the region, which directly stimulates the clitoral area and causes increased cavernosal blood flow and engorgement. The device functions similar to vacuum erection devices used by men with ED. Improvements in genital sensation, arousal, orgasmic ability, and vaginal lubrication, all leading to enhanced sexual satisfaction have been reported.[35]

Summary

Female sexual dysfunction is an important clinical condition affecting millions of women and couples each year. Because the underlying cause may vary, from vascular, to neurologic, to hormonal, surgical, and/or psychological, it is important to perform a thorough evaluation in all women with FSD. The first step for the physician is to *ask the question*. It is important that physicians treat sexual dysfunction like any other disease process and remember that FSD can occur at any age and may be the first sign of other underlying disease. It is also paramount to understand that no single treatment option is right for everyone. At Cleveland Clinic Florida, we have found that a multidisciplinary approach is extremely effective in dealing with the various issues facing the woman or couple with sexual dysfunction. This typically includes the urologist, gynecologist, primary care physician, sex therapist, and various other specialists and health care providers. By talking to all patients about their sexual health, by encouraging communication with the woman and her partner, and by educating women and their partners regarding successful treatment alternatives, physicians can go a long way to improving women's sexual health and the overall quality of life for the couple.

References

1. Hakim LS, Platt D. The Couple's Disease: Finding a Cure for Your Lost Love Life. Grand Rapids, MI: DHP Publishers; 2002.
2. Feldman HA, Goldstein I, Hatzichristou DG, Krane RJ, McKinlay JB. Impotence and its medical and psychosocial correlates: results of the Massachusetts Male Aging Study. J Urol 1994;151(1):54–61.
3. Laumann EO, Paik A, Rosen RC. Sexual dysfunction in the United States: prevalence and predictors. JAMA 1999;281:537–544.
4. Salonia A, Zanni G, Nappi RE, et al. Sexual dysfunction is common in women with lower urinary tract symptoms and urinary incontinence: results of a cross-sectional study. Eur Urol 2004;45(5): 642–648.
5. Basson R, Berman J, Burnett A, et al. Report of the international consensus development conference on female sexual dysfunction: definitions and classification. J Urol 2000;163(3):888–893.
6. Masters EH, Johnson VE. Human Sexual Response. Boston: Little Brown; 1966.
7. Basson R. The female sexual response: a different model. J Sex Marital Ther 2000;26:51–65.
8. Basson R. Human sex-response cycles. J Sex Marital Ther 2001;27: 33–43.
9. Havenga K, Maas CP, DeRuiter MC, Welvaart K, Trimbos JB. Avoiding long-term disturbance to bladder and sexual function in pelvic surgery, particularly with rectal cancer. Semin Surg Oncol 2000; 18:235–243.
10. Damgard B, Wettergren A, Kirkegaard P. Social and sexual function following ileal pouch anal anastomosis. Dis Colon Rectum 1995;38: 286–289.
11. Metcalf AM, Dozois RR, Kelly KA. Sexual function in women after proctocolectomy. Ann Surg 1986;204:624–627.
12. Bambrick M, Fazio VW, Hull TL, Pucel G. Sexual function following restorative proctocolectomy in women. Dis Colon Rectum 1996;39: 610–614.
13. Öresland T, Palmblad S, Ellström M, Berndtsson I, Crona N, Hulten L. Gynaecological and sexual function related to anatomical changes in the female pelvis after restorative proctocolectomy. Int J Colorectal Dis 1994;(9):77–81.
14. Sjögren B, Poppen B. Sexual life in women after colectomy-proctomucosectomy with S-pouch. Acta Obstet Gynecol Scand 1995;74: 51–55.
15. Cirino E, Pepe G, Pepe F, Panella M, Rizza G, Cali V. Sexual complications after abdominoperineal resection. Ital J Surg Sci 1987;17: 315–318.
16. Gloeckner MR, Starling JR. Providing sexual function information to ostomy patients. Dis Colon Rectum 1982;25:575–579.
17. Engel J, Kerr J, Schlesinger-Raab A, Eckel R, Sauer H, Holzel D. Quality of life in rectal cancer patients: a four-year prospective study. Ann Surg 2003;238(2):203–213.
18. Tiainen J, Matikainen M, Hiltunen KM. Ileal J-pouch-anal anastomosis, sexual dysfunction, and fertility. Scand J Gastroenterol 1999; 34:185–188.
19. Wikland M, Jansson I, Asztély M, et al. Gynecological problems related to anatomical changes after conventional proctocolectomy and ileostomy. Int J Colorectal Dis 1990;5:49–52.
20. Asztély M, Palmblad S, Wikland M, Hulten L. Radiological study of changes in the pelvis in women following proctocolectomy. Int J Colorectal Dis 1991;6:103–107.
21. Fegiz G, Trent A, Bezzi M, et al. Sexual and bladder dysfunction following surgery for rectal carcinoma. Ital J Surg Sci 1986;16:103–109.
22. Williams JT, Slack WW. A prospective study of sexual function after major colorectal surgery. Br J Surg 1980;67:772–774.
23. da Silva GM, Gurland B, Coulqhoun P, et al. Sexual function in female colorectal patients: frequently affected, frequently of interest and frequently ignored. Dis Colon Rectum 2002;45:A14.
24. Derogatis L, Rust J, Golombok S, et al. Validation of the profile of female sexual function (PFSF) in surgically and naturally menopausal women. J Sex Marital Ther 2004;30:25–36.
25. Kaplan SA, Reis RB, Kohn IJ, et al. Safety and efficacy of sildenafil in postmenopausal women with sexual dysfunction. Urology 1999;53: 481–486.
26. Rosen RC. The female sexual function index (FSFI): a multidimensional self-report instrument for the assessment of female sexual function. J Sex Marital Ther 2000;26:191–208.
27. Kanaly KA, Berman JR. Sexual side effects of SSRI medications: potential treatment strategies for SSRI-induced female sexual dysfunction. Curr Womens Health Rep 2002;2(6):409–416.
28. Clayton AH, Warnock JK, Kornstein SG, Pinkerton R, Sheldon-Keller A, McGarvey EL. A placebo-controlled trial of bupropion SR as an antidote for selective serotonin reuptake inhibitor-induced sexual dysfunction. J Clin Psychiatry 2004;65(1):62–67.
29. Bachmann G, Bancroft J, Braunstein G, et al. Female androgen insufficiency: the Princeton consensus statement on definition, classification and assessment. Fertil Steril 2002;77(4):660–665.
30. Shifren JL, Braunstein GD, Simon JA, et al. Transdermal testosterone treatment in women with impaired sexual function after oophorectomy. N Engl J Med 2000;343(10):682–688.
31. Shifren JL. The role of androgens in female sexual dysfunction. Mayo Clin Proc 2004;79(4 suppl):S19–S24.
32. American College of Obstetricians and Gynecologists. Androgen replacement no panacea for women's libido. Press release. October 31, 2000.
33. Berman JR, Berman LA, Toler SM, Gill J, Haughie S, Sildenafil Study Group. Safety and efficacy of sildenafil citrate for the treatment of female sexual arousal disorder: a double-blind, placebo controlled study. J Urol 2003;170(6 Pt 1):2333–2338.
34. Sipski ML, Rosen RC, Alexander CJ, Hamer RM. Sildenafil effects on sexual and cardiovascular responses in women with spinal cord injury. Urology 2000;55(6):812–815.
35. Billups KL, Berman L, Berman J, Metz ME, Glennon ME, Goldstein I. A new non-pharmacological vacuum therapy for female sexual dysfunction. J Sex Marital Ther 2001;27:435–441.
36. Deixonne B, Baumel H, Domergue J. Sexual disorders following abdominoperineal resection of the rectum. Sem Hop 1983 10;59(10):677–682.
37. Hjortrup A, Kirkegaard P, Friis J, Sanders S, Andersen F. Sexual dysfunction after low anterior resection for midrectal cancer. acta chir Scand 1984;150(8):687–688.
38. Cunsolo A, Bragaglia RB, Manara G, Poggioli G, Gozzetti G. Urogenital dysfunction after abdominoperineal resection for carcinoma of the rectum. Dis Colon Rectum 1990;33(11):918–922.

Section VI

Incontinence Therapy

Section VI

Incontinence Therapy

Steven D. Wexner

In perhaps no area of colorectal management has therapy been as challenging as for the satisfactory treatment of fecal incontinence. Numerous surgical and nonsurgical therapies exist, none of which is a panacea. In fact, the veritable plethora of options for the treatment of both urinary and fecal incontinence is attestation to the lack of a uniformly satisfactory "cure." Certainly milder cases of fecal incontinence may be amenable to noninvasive interventions, such as dietary modifications, pelvic floor exercises, and pharmacologic agents. Should these therapies fail to afford the patient satisfactory improvement, the next level of interventions includes injection therapy. Within this realm are bulking agents, as used for urinary incontinence, and more recently, include carbon beads, silicone, and autologous fat for fecal incontinence. If bulking fails to give satisfactory results, then either direct repairs or plications of the native muscle may in some circumstances be satisfactory. Certainly for fecal incontinence, however, the quite acceptable short-term results have been very disappointing during long-term follow-up. Specifically, the initially reported success rates ranging from 70% to 90% in short-term assessment have recently been reported to decrease to 50% or less during longer-term evaluation.

The lack of adequate first (medical) and second (injection therapies) line therapies and the lack of durability of third line therapies (direct muscular repair) have led to a quest for more intensive interventions. For urinary incontinence, laparoscopic and retropubic procedures, as well as sacral nerve stimulation, have been used with increasing frequency. The artificial urinary sphincter also has a role. Many of these therapies, after a successful use for urinary incontinence, have been modified and adapted for the treatment of fecal incontinence. Specifically, increasing experience with sacral nerve stimulation and artificial bowel sphincter has led to wider choices for the patient with fecal incontinence. The 11 chapters within this section on "Incontinence Therapy" highlight some of the difficulties in managing incontinence and offer glimpses into the future, which hopefully holds significant and much-needed improvements for patients with these problems.

6-1

Device Therapy for Stress Incontinence

Jennifer T. Pollak and G. Willy Davila

The majority of patients with stress urinary incontinence are successfully managed with surgery; however, not all patients desire surgery, nor are all patients operative candidates. Nonsurgical treatment options include pelvic floor muscle exercises, biofeedback, electrical stimulation, and pharmacologic agents. Anti-incontinence devices serve as another alternative to treat these patients, and include both vaginal and urethral prosthesis (Table 6-1.1; Figure 6-1.1).

These devices have several advantages. Patients who are awaiting surgery or who are not optimal surgical candidates may successfully use these devices. This may be an ideal treatment for women who have only exercise-induced stress incontinence. They may be inserted and removed by the patients themselves. When cared for properly, these devices have minimal complications.

The devices, however, have several disadvantages. They require motivated patients. They should not be used with urinary tract infections, and they may predispose women to infection. Self-removal requires good manual dexterity. Patients at risk for endocarditis should not use these devices.[1] Women may also use them temporarily until they complete childbearing.

Vaginal Devices

Pessaries have long been used to treat stress urinary incontinence. Standard-shaped pessaries, or those that have been modified with a ball or a protuberance, may successfully treat incontinence. These pessaries stabilize the bladder neck and increase urethral resistance.

Vierhout and Lose[2] found a 63% subjective cure or improvement rate with pessary use in their review of the literature. A 36% to 66% rate of complete continence has been reported when pessaries were used to treat exercise-induced stress incontinence.[3,4] However, a long-term assessment of the incontinence ring pessary demonstrated only a 16% success rate for the treatment of mixed and stress incontinence.[5] Complications from pessary use are uncommon and include vaginal abrasions, ulcerations, urinary tract infections, and vaginal infection. These problems are rare if the pessary is cared for properly, removed and cleaned at least every 3 months, and if the patients concomitantly use estrogen locally in the vagina. We have not found any particular type of pessary to be superior in treating stress urinary incontinence. If the Smith-Hodge remains in place, it has theoretical advantages in providing bladder neck support.

Other vaginal devices that have been used to treat stress incontinence include the contraceptive diaphragm, tampons, and the bladder neck support prothesis (Introl; UroMed, Needham, MA). Contraceptive diaphragms have a reported success rate of 46% to 83%.[2] Tampons have been successful in treating incontinence. In fact, standard super tampons were shown to be more efficacious than pessaries in treating exercise-induced incontinence.[4] Special types of tampons have been developed to specifically treat incontinence. The Conveen Continence Guard (Coloplast a/s, Humlebaek, Denmark) is a hydrophilic polyurethane tampon that has two wings to support the bladder neck. It must be soaked in water, and then attached to an applicator before insertion. It has a 53% to 73% success rate.[2] A newer device, the Contrelle Continence Tampon (Coloplast a/s) is made of hydrophobic polyurethane and has its own applicator that inserts similarly to a standard tampon. In a study comparing these two devices, both devices were found to reduce the amount of leakage significantly, but the Contrelle reduced urine loss significantly more than the Conveen.[6] These tampons can be left in the vagina for up to 16 hours, and do not need to be removed in order to void. They should, however, be replaced monthly.

The Introl prosthesis (UroMed) is a vaginal device with two prongs that sit against the anterior vaginal wall at the level of the bladder neck. It has been suggested that the device may mimic the results of a retropubic urethropexy. Success rates are reported between 50% and 94%.[2]

Table 6-1.1. Devices for the treatment of stress urinary incontinence

Vaginal Devices
Standard pessaries
• Ring with support
• Hodge
Anti-incontinence pessaries
• Incontinence ring
• Incontinence dish
• Incontinence dish with support
Contraceptive diaphragm
Tampon
• Standard tampon
• Contrelle Continence Tampon
• Conveen Continence Guard
Bladder neck support prothesis (Introl)
Urethral Devices
Urethral cap (FemAssist)
Urethral patch
Urethral insert
• FemSoft
• Reliance Urinary Control Insert
• Viva urethral plug

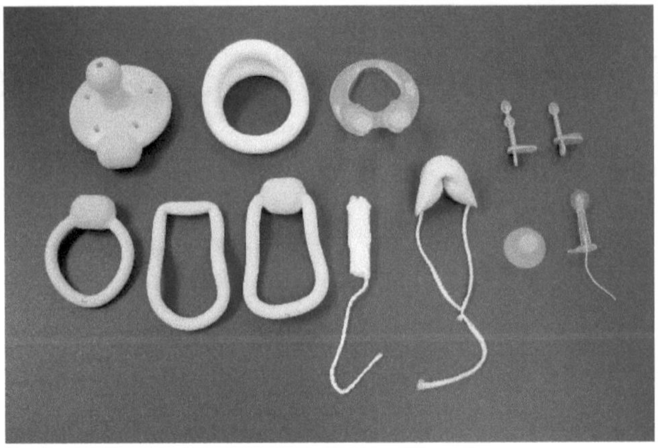

Figure 6-1.1. Devices for stress urinary incontinence include intravaginal pessaries, intravaginal disposable devices, urethral meatal barriers, and intraurethral plugs.

Urethral Devices

Urethral devices may be worn external to the urethral meatus or inserted into the urethra. These devices are contraindicated in patients with bladder or urethral cancer, urinary tract infection, interstitial cystitis, neurogenic bladder, and in those at risk for endocarditis. Patients with limited manual dexterity may have difficulty using them.

The FemAssist device (Insight Medical, Bolton, MA) is a dome-shaped urethral cap made of silicon. It is placed over the urethral meatus with mild suction. This creates a watertight seal between the device and the urethra. The device may be worn for up to 4 hours. When the patient wishes to void, she releases the vacuum seal. It may be washed and worn for up to a week. Trincello et al.[7] reported a significant reduction in the volume of urine loss, and a 56% satisfaction rate. Rabin[8] did not find significant urinary tract infection rates or bacteriuria, and noted that the patients were dry 82% of the time.

The urethral patch (UroMed) is another external urethral barrier. It is a single-use foam pad that is coated with an adhesive and fits over the urethral meatus, providing opposing pressure and a barrier to leakage. The patch is removed for urination. Women who used this device reported a significant decrease in urinary leakage and an improvement in their quality of life.[9]

Other devices are inserted into the urethra. The FemSoft (Rochester Medical Corp., Stewartville, MN) insert is a silicone catheter encased in an oil-filled sleeve that has a balloon on its tip. When the FemSoft is placed into the urethra, it conforms to the urethra to decrease incontinence. The device is a sterile, single-use product that is removed and discarded when the patient voids. Women using the device had significantly less urinary leakage, and the majority of them were satisfied with the product. There was, however, a 25% rate of urinary tract infections.[1]

The Reliance Urinary Control Insert (UroMed), or urethral plug, is a single-use, balloon-tipped cylinder that is inserted into the urethra. The air-filled balloon holds the device in place and prevents incontinence. In order to void, a string is pulled to deflate the balloon. This insert is also a single-use, disposable item. Patients who used this device had an 89% subjective success rate, 31% incidence of infection, and a 21% incidence of gross hematuria.[10] Another urethral insert, the Viva urethral plug, had a 33% to 43% success rate. This device was also associated with urinary tract infection and was hindered by plug migration to the bladder.[2]

Conclusion

Although at first thought, vaginal and urethral devices may seem to be ideal means to manage mild to moderate stress incontinence, in practice, their acceptance has been low. Nonsurgical therapy requires motivation, patience, and long-term commitment. In addition, the vaginal and urethral devices cure far less women than surgery. However, they may increase the quality of life for patients who are medically unfit for surgery or who do not desire an operation. The "nuisance factor" associated with device care is the major barrier to greater acceptance. The devices may be ideal for women with only exercise-induced incontinence or for those who need a temporary treatment until they are able to undergo surgery. Surgery and pelvic floor rehabilitation remain the most common and popular treatments for stress urinary incontinence.

References

1. Elliott DS, Boone TB. Urethral devices for managing stress urinary incontinence. J Endourol 2000;14(1):79–83.
2. Vierhout ME, Lose G. Preventive vaginal and intra-urethral devices in the treatment of female urinary stress incontinence. Curr Opin Obstet Gynecol 1997;9(5):325–328.
3. Glavind K. Use of a vaginal sponge during aerobic exercises in patients with stress urinary incontinence. Int Urogynecol J 1997;8: 351–353.
4. Nygaard I. Prevention of exercise incontinence with mechanical devices. J Reprod Med 1995;40:89–94.
5. Robert M, Mainprize TC. Long-term assessment of the incontinence ring pessary for the treatment of stress incontinence. Int Urogynecol J Pelvic Floor Dysfunct 2002;13:326–329.
6. Thyssen H, Bidmead J, Lose G, Moller Bek K, Dwyer P, Cardozo L. A new intravaginal device for stress incontinence in women. BJU Int 2001;88(9):889–892.
7. Trincello D, Bolderson J, Richmond D. Preliminary experience with a urinary control device in the management of women with genuine stress incontinence. Br J Urol 1997;80:752–756.
8. Rabin JM. Clinical use of the FemAssist device in female urinary incontinence. J Med Syst 1998;22:257–271.
9. Brubaker L, Harris T, Gleason D, Newman D, North B. The external urethral barrier for stress incontinence: a multicenter trial of safety and efficacy. Miniguard Investigators Group. Obstet Gynecol 1999; 93(6):932–937.
10. Staskin D, Bavendam T, Miller J, et al. Effectiveness of a urinary control insert in the management of stress urinary incontinence: early results of a multicenter study. Urology 1996;47:629–636.

6-2

Medications for Stress Urinary Incontinence

G. Willy Davila

Pharmacologic therapy of stress urinary incontinence (SUI) is directed toward enhancing urethral sphincteric function, especially during times of increased intraabdominal pressure such as coughing or lifting. Appropriate function of the intrinsic and extrinsic urethral sphincteric musculature, along with appropriate urethral support, as well as integrity of the urethral mucosa and submucosal vascular plexus are the key components of the delicate female continence mechanism. The extrinsic muscular support of the urethra is dependent on intact pelvic floor musculature and its innervation. Physiotherapeutic approaches to SUI are discussed in Chapter 12-2. The urethral mucosal and vascular factors are dependent on estrogen availability. In menopausal women, there is a prompt reduction in blood flow to the urethral submucosa and thinning of the urethral mucosa. Usage of local estrogen cream is discussed in Chapter 11-1.

Alpha 1A/1L adrenoceptors are present at the level of the urethra and bladder neck smooth muscle. They are amenable to pharmacologic manipulation in order to enhance urethral sphincteric tone and contractility. Alpha-adrenergic stimulants have therefore been used for many years in the treatment of SUI. These include phenylpropanolamine, pseudoephedrine, ephedrine, and phenylephrine. These medications result in enhanced urethral sphincteric tone attributed to a direct effect on the alpha-adrenergic receptors in the urethral smooth muscle. In the United States, phenylpropanolamine was removed from the market in the year 2000, because of the reported occurrence of subarachnoid and intracerebral hemorrhage in adults taking this medication. It was found that in women who were admitted to the hospital with intracerebral bleeding, the odds ratio was 16.58 for usage of phenylpropanolamine. Before its removal from the market, phenylpropanolamine was widely used for mild SUI symptoms. It was used either as a single-dose extended-release formulation, or three-times-daily 25-mg tablets. The improvement in SUI was mild, but predictable. Previous urodynamic studies have demonstrated an increase in urethral closure pressure with alpha-adrenergic agents. Importantly, the combination of an alpha-adrenergic oral agent along with local estrogen replacement therapy resulted in a synergistic improvement in SUI symptoms.[1] Thus, we recommended the usage of local estrogen cream along with an oral alpha-adrenergic agent before their removal from the market.

Imipramine (Tofranil) is a drug with combined properties as an anticholinergic, alpha-adrenergic, and mild anesthetic agent. It was primarily used as a tricyclic antidepressant and anxiolytic. Its usage for depression has been limited because of the recent availability of serotonin reuptake inhibitors. Because of its combined properties, imipramine can be very useful in mixed urinary incontinence, as well as in SUI. It can be used as a once-a-day 25- to 50-mg tablet, or can be given in multiple daily dosages. Care must be taken in the elderly because of the possibility of cardiovascular side effects. Various studies have demonstrated an improvement in urethral closure as well as SUI symptoms with imipramine. Cure rates are variable but have been shown to be as high as 35%, especially in those patients who demonstrated an increase in urethral closure pressure during therapy.[2]

Most recently, duloxetine has been studied widely for the treatment of stress as well as mixed urinary incontinence. Although it is not yet approved by the Food and Drug Administration for incontinence, it is available for clinical use for depression, and multiple studies have been completed and data have been reported. Duloxetine's effect in the central nervous system at the level of Onuf's nucleus leads to a stimulation of the external urethral sphincter. Because of this stimulation, reduction in SUI of 50% to 65% has been reported with an approximate 20% cure of SUI symptoms.[3] It is also being studied in the treatment of mixed urinary incontinence. Other agents are also being studied with alpha-adrenergic properties. Once duloxetine and other agents are available, they may provide an effective means for addressing SUI pharmacologically.[4] As with any nonsurgical treatment modality for urinary incontinence, a multimodality approach, including behavioral modifications and pelvic floor exercises, is likely to be most beneficial for patients.

References

1. Hilton P, Tweddell AL, Mayne C. Oral and intravaginal estrogens alone and in combination with alpha-adrenergic stimulation in genuine stress incontinence. Int Urogynecol J 1990;1:80–86.
2. Lin H-H, Sheu B-C, Lo M-C, Huang S-C. Comparison of treatment of outcomes of imipramine for female genuine stress incontinence. Br J Obstet Gynaecol 1999;106:1089–1092.
3. Dmochowski RR, Miklos JR, Norton PA, et al. Duloxetine versus placebo for the treatment of North American women with stress urinary incontinence. J Urol 2003;170:1259–1263.
4. Norton PA, Zinner NR, Yalcin I, et al. Duloxetine versus placebo in the treatment of stress urinary incontinence. Am J Obstet Gynecol 2002; 187:40–48.

6-3

Sling/Tension-Free Vaginal Tape

Gamal M. Ghoniem and Usama M. Khater

Over the past decade, suburethral slings have emerged as the procedure of choice for the surgical correction of most types of female stress urinary incontinence (SUI). In 1997, the Female Stress Urinary Incontinence Clinical Guidelines Panel of the American Urological Association analyzed the literature regarding surgical procedures for treating SUI and established that retropubic suspension and sling procedures were the most durable procedures with the longest outcome for dryness.[1] In recent years, slings have become more common, because needle suspension procedures have failed the test of time, and suburethral sling procedures have proved to be less morbid than retropubic suspensions.

Pubovaginal Sling

In 1907, Von Girodano introduced the sling concept for treatment of urinary incontinence when he wrapped a gracilis graft around the urethra. However, credit for the first pubovaginal sling went to Goebell in 1910 when he rotated the pyramidalis muscles beneath the urethra and joined them in the midline. In 1914, Frangenheim used rectus abdominis muscle and fascia for slings. Stoeckel argued that the material used for the slings was not important in the outcome, and the success depends on a high urethral position and attachment of the sling to the abdominal muscles. Price described the first fascial sling in 1933. Millin used strips of rectus fascia, looped them under the urethra, and tied them over the top of the urethra. In 1942, Aldridge used fascial slings in conjunction with vaginal plastic operations. He mobilized strips of abdominal fascia, leaving the edges attached to the recti muscles medially, and tunneled the strips through the recti 4 cm above the pubis. The two ends were sutured together through a vertical vaginal wall incision. The Aldridge procedure dominated sling surgery for many decades. Green and Robertson in 1962 and 1970, respectively, described incontinence cases that were well supported but leaked with increases in abdominal pressure. McGuire also reported a high failure rate for anterior colporrhaphy and retropubic suspension in women who had type III SUI. When these patients were treated with a pubovaginal sling, they had a 91% cure rate at a mean of 2.3 years.[2] The sling was reintroduced to urologists and, since then, sling material, size, suture, and point of fixation have undergone numerous revisions. The sling may be placed abdominally or transvaginally. The common goal is to restore sufficient urethral outlet resistance while avoiding urethral obstruction and allowing spontaneous micturition. The mechanism by which the sling accomplishes these goals remains controversial. Ghoniem and Shaaban[3] performed videourodynamics on 35 women undergoing a pubovaginal sling for intrinsic sphincter deficiency. Fluoroscopic monitoring during straining showed upward and outward movement of the sling, and downward and inward movement of the bladder base. The opposing forces closed the bladder neck, preventing leakage of contrast material and restoring outlet resistance. The sling restores anatomic position of the urethra. Thus, it is effective for treating intrinsic sphincter deficiency and urethral hypermobility. A variety of materials have been used successfully as suburethral slings. Slings can be made from autologous materials, allografts, or synthetic materials.

Autologous Slings

Until the 1950s, only autologous tissue such as rectus fascia or fascia lata was available for transposition beneath the urethra. Autologous materials are generally associated with less urethral erosion and are easily shaped. The disadvantages of using such materials include increased operating time, morbidity, and discomfort caused by harvesting of the graft. Cosmesis may also be an issue for many patients. Furthermore, the patient's own tissue may have insufficient strength to withstand substantial forces as a consequence of aging and hypoestrogenicity.

Rectus Fascia

Rectus fascia is one of the most popular autologous materials used for a sling procedure. It is easily harvested, even in patients with multiple abdominal operations. It is durable and rarely causes urethral erosion. Histologic changes after placement of a rectus fascia sling showed extensive remodeling with increased fibroblasts and connective tissue on biopsy specimens.[4] The disadvantage of using rectus fascia is that the fascia may be scarred and thickened because of prior operations.

Fascia Lata

The advantages of fascia lata include the easily obtainable long graft that is unscarred and of uniform thickness. Because the graft is long, it is easier to achieve adequate tension on the sling. This tension is obtained by tacking the fascia to itself, which achieves uniform pressure distribution over the urethra. Because there is only a small incision, recovery time is less, and there is no risk for abdominal hernia. The disadvantage is that additional operative time is required and there is pain at the harvesting site.

Vaginal Wall Sling

The anterior vaginal wall sling was introduced by Raz et al.[5] in 1989. In this procedure, the sling is constructed from the anterior vaginal wall and underlying fascia to compress and support the urethra. This technique was modified in 1998 by Vasavada et al.,[6] as they anchored the vaginal wall patch to the pubic tubercle with two sutures and preserved the endopelvic fascia. These modifications were believed necessary to reduce the rate of suture pull-through from the vaginal side and to reduce the rate of prolapse caused by breakdown of the endopelvic fascia.

Allograft Slings

Allografts are tissues harvested from a human donor, which are transplanted into human recipients. Allograft materials were introduced in an attempt to decrease morbidity associated with harvesting autologous slings. Human cadaveric fascia and cadaveric dermis are now the two most often used allograft materials. Organic materials seem to be a good choice for the sling because of lower risk of erosion. The allograft is obtained from a licensed tissue bank regulated by the American Association of Tissue Banks (AATB). A multistep sterilization process is conducted to eliminate risk of disease transmission. A graft's processing may affect its quality. Failure caused by graft degradation is more common with fresh frozen irradiated grafts than with chemically processed grafts. The latter retain their quality and architecture more readily.

Donor and recipient matching is not required, although presence of DNA material has been reported in some commercially available cadaveric allografts. The clinical significance of this DNA material is undetermined. Although the risk of disease transmission is remote, it is imperative that it be explained to patients. Several studies have attempted to quantify this risk. Only one case of transmission of the human immunodeficiency virus (HIV) has been reported from tissue (not fascia) transplantation since screening for HIV and other viral pathogens was initiated. If guidelines for tissue banking set forth by the Food and Drug Administration are followed, risk of acquiring tissue from a donor with HIV is estimated at 1 of 1 667 600. By comparison, risk of HIV infection from a blood transfusion is much higher than from tissue transplant.[7]

Another potential risk of cadaveric fascia transplantation is transmission of prion diseases. Prions are protein molecules associated with transmission of spongiform encephalopathy. The most common form of this encephalopathy is Creutzfeldt-Jacob disease. A literature review did not document spongiform encephalopathy associated with cadaveric fascia lata or dermis.[8]

Allograft fascia lata slings have been proven as efficacious as autologous fascia in short-term follow-up studies.[9] Long-term follow-up and evaluation are required. Processed dermal matrix is a human cadaveric allograft available under the brand names DuraDerm™ (C.R. Bard Inc., Murray Hill, NJ) and Urogen™ (AMS, Minnetonka, MN). The acellular collagen matrix contains proteins, proteoglycans, and other ground substances, and serves as structural support for effective ingrowths of the host's collagen. Being acellular serves to eliminate immune response, and tissue rejection. Repliform™ (LifeCell Corp., Branchburg, NJ) is an acellular human dermal matrix allograft that is available for tissue reinforcement in pelvic floor and incontinence surgery.

Xenograft Slings

Xenografts, or tissues derived from nonhuman donors, have been used as slings. Pelvicol™ (C.R. Bard Inc.) is a flat sheet of fibrous acellular porcine dermal collagen. It is claimed to have a composition that minimizes the risk of urethral erosions. Crosslinking with hexamethylene diisocyanate (HMDI) is believed to resist breakdown by collagenase.

DermMatrix™ (Advanced UroScience), a new biocompatible, acellular matrix, is a heterologous material made of porcine corium. It is claimed to have high omni direction tensile strength, which lessens the incidence of suture pull-through. It is non-antigenic and is adapted by the host tissue by ingrowths.

Stratasis™ (Cook Urological, Bloomington, IN) is a sub-urethral sling material constructed from porcine small intestinal submucosa. It is an acellular collagen matrix, which therefore theoretically, is nonimmunogenic. It consists predominantly of type I collagen. It is not crosslinked, and therefore does not lead to encapsulation. It is biodegradable and resorbable. It is gradually remodeled and disappears, leaving behind normal tissue.

Fortaflex™ (Organogenesis, Canton, MA) consists of acellular collagen matrix material derived from porcine small intestine submucosa. The intestinal submucosa had a stretch capacity not seen with cadaveric fascia.

Tissue-Guard™, or Veritas® (Synovis Surgical Innovations, St. Paul, MN), is biological material derived from bovine pericardium. It is crosslinked with glutaraldehyde, which preserves its natural collagen structure. High content of glutaraldehyde crosslinking is associated with a high incidence of calcification. Their processing reduces this to less than 6 parts per million.

Clinical data supporting long-term safety and efficacy of these new allografts are not available and, thus, their use is endorsed with caution.

Synthetic Slings

There are several obvious advantages of using synthetic materials for slings. The supply of artificial material is unlimited in quantity, consistent in quality, and can be fashioned into any size or shape. The use of synthetic materials obviates the need for harvesting tissue from a second operative site and minimizes dissection when compared with biomaterials; synthetic materials are more uniform, more consistent, and more durable. In addition, artificial graft materials are sterile and free of biomaterials. The ideal synthetic sling material should be of consistent strength and quality, nonabsorbable, mildly elastic, and easily available in adequate sizes. It should also be at minimal risk of evoking a foreign body inflammatory response and erosion into surrounding structures. No such material exists at present. The drawbacks of synthetic graft materials include graft infection, erosion into the urethra, or extrusion into the vagina.

Short-term objective cure rates range between 73% and 93%.[17] Intermediate and long-term results suggest that the continence rate at 10 years is similar to that at 1 year.[17,18]

The first slings were made of multifilament polyethylene (Mersilene™; Ethicon, Somerville, NJ) to support the urethra and bladder neck. These were replaced by monofilament polypropylene grafts such as Marlex™ (C.R. Bard, Cranston, RI) and Prolene (Ethicon), which were designed to overcome urethral fixation and scarring. The interwoven mesh acts as scaffolding into which the patient's endogenous tissue can grow and galvanize the repair. Consequently, the larger pore size of polypropylene grafts leads to lesser scarring than the polyethylene. Many of the same complications occurring with polyethylene were seen with polypropylene, including urethral erosion, chronic retention, and postoperative urgency and frequency. Gore-Tex (WL Gore and Associates, Flagstaff, AZ) is another synthetic sling, which is made of expanded and reinforced polytetrafluorethylene. Problematic graft infection, rejection, and erosion resulted from its use. A Silicon sling was introduced in 1985 and thought to be superior to Marlex™ or Mersilene™ because of its smooth surface, which would promote formation of a fibrous sheath. Although initial cure rates were promising, long-term results were disappointing. Silicone slings have a high rate of vaginal erosions and sinus formation.[10] Additional synthetic sling materials have been used and have also met with high complication rates. ProteGen™ (Boston Scientific, Natick, MA) is a woven polyester sling treated with bovine collagen, and was associated with unacceptable rates of vaginal and urethral erosions.

In 1993, Ulmsten and Petros proposed the Integral Theory, postulating that stress and urge are defective flow symptoms that may arise from laxity in the vagina or its supporting ligaments. Using this theory and the hammock hypothesis previously proposed by DeLancey and Richardson,[11] Ulmsten et al. postulated that re-creation of the pubourethral ligament and support of the suburethral vagina are essential in treating SUI. This finding led to development of the original tension-free urethropexy, the intravaginal slingplasty, and later the transvaginal tape. Tension-free vaginal tape (TVT) (Ethicon, Inc.) is another type of synthetic suburethral sling, which was introduced by Ulmsten and has achieved popularity. Since its introduction in the treatment of SUI, TVT has gained wide acceptance because of its simplicity, ability to be performed under local anesthesia, decreased operative time, decreased recovery, and good outcomes. Ulmsten et al.[12] described an 86% cure rate with significant improvement in another 11% in a 3-year follow-up.

The TVT sling is placed at the mid urethra. The rationale behind transvaginal tape is that a tissue reaction to the polypropylene mesh tape produces a controlled longitudinal deposition of collagen along the length of the tape, forming a collagen scar that stimulates the urethral support mechanism of the pubourethral ligaments. This scar secures proper fixation of the mid urethra to the pubic bone and simultaneously reinforces the suburethral vaginal hammock and its connection to the pubococcygeus muscles. Bladder perforation is the most common complication of this procedure. In an effort to reduce this complication, downward needle-carrier placement through the retropubic space, such as the SPARC™ (American Medical Systems, Minneapolis, MN) female sling system was introduced in 2001. The manufacturer of the traditional TVT also recently introduced suprapubic needle passers. However, because these slings still require passage through the retropubic space, their use may lead to vascular, bowel, and bladder injury, and cystoscopy is still required. In France in 2001, Delorme introduced a new approach for placement of sling through the obturator foramen. The aim of this approach is to substitute the lost hammock action and support to the mid urethra as described by DeLancey[11] and provide a backboard support to the urethra without penetration of the retropubic space. It is a perineal approach compared with the retropubic approach, avoiding the pelvic cavity. It is ideal for obese patients or those with previous retropubic surgery. The tape trajectory runs deeper and parallel to the deep perineal membrane, and below the pubocervical fascia and levator ani muscles, providing support and minimal urethral compression, and is less obstructive than retropubic slings. Delmas et al.[13] studied the tape trajectory in 10 fresh female cadavers, after passing it in a fashion as performed clinically. The tape is located at the medial and inferior border of the obturator foramen and away from the lateral and superior femoral vessels. It is 1 cm away from the insertion of adductor longus and gracilis muscles. The obturator vessels and nerves are situated far on the opposite side of the obturator foramen, and directed toward the thigh. At the perineum, the tape is in a deeper plane to the urogenital triangle with its pudendal nerve branches, including the branches to the clitoris. These branches are protected by the ischiopubic ramus. Laterally, the tape passes under the levator ani muscles and endopelvic fascia, but over or below the arcus tendinous fascia pelvis and 4 cm away from the obturator neurovascular bundle. The accessory pudendal artery (3% of population) runs above the levator before it exits under the symphysis pubis.

The transobturator route transverses muscles and fascia, with no major vessels, nerves, or viscera along the needle route. At Cleveland Clinic Florida, we use the transobturator approach for placement of ObTape™ (Mentor, Santa Barbara, CA) and Monarc™ (AMS) slings. The senior author uses this approach with Ghoniem pinch maneuver. In this technique, the patient is placed in the lithotomy position, under local, general, or regional anesthesia. The anterior vaginal wall is infiltrated with 1:1 mix of 0.5% bupivacaine hydrochloride and 1% lidocaine hydrochloride with diluted epinephrine (1:100 000) to facilitate dissection and hemostasis. The incision sites extending lateral to the urethra and up to the vaginal sulcus are infiltrated. The incision is made at the level of the mid urethra enough to introduce the index finger. The dissection is performed to the lateral vaginal sulcus until the inner aspect of the ischiopubic ramus is felt with the index finger. The ischiopubic ramus is felt, and the inner edge of the obturator foramen is pinched between the thumb and the index finger "pinch maneuver" (Figure 6-3.1). Once a good pinch is achieved, all the following steps are performed in this position.

The skin is punctured with a 15-blade knife 1 cm above the external meatus level and at the level of the clitoris. The puncture site is directly above the thumb and at the inner edge of the ischiopubic ramus. The passer is inserted in two movements/steps guided by the index finger (Figure 6-3.2).

The first step is up and down penetration of the obturator membrane with a distinct "give away" feeling. Once the tip of the passer is felt with the index finger, the passer is rotated from a lateral to medial direction and guided to the vaginal incision. It is kept on the ventral surface of the finger to protect the urethra at all times. The end of the tape is threaded into the eye of the passer and is pulled through to bring the wider portion of the tape for a better hold (Figure 6-3.3).

The passer is then pulled in a reverse movement, bringing the end of the tape through the punctured skin. The same steps are repeated on the contralateral side. The ends of the tape are pulled and the midportion of the tape (narrow) is kept under the urethra. Appropriate tension is achieved when a space between the urethra and tape is

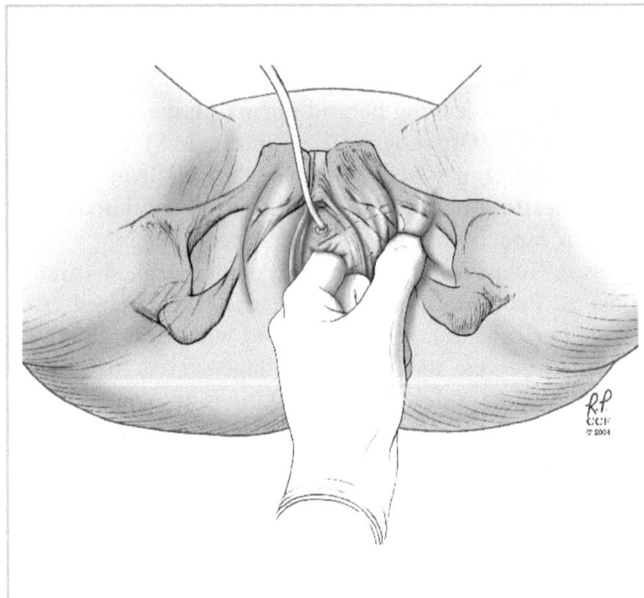

Figure 6-3.1. The inner edge of the ischiopubic ramus is pinched between the index finger (inside the incision) and thumb on the skin. (Reprinted with the permission of The Cleveland Clinic Foundation.)

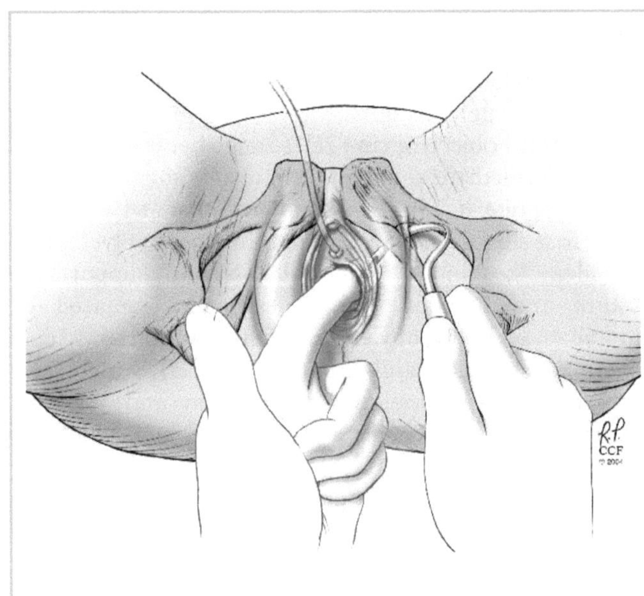

Figure 6-3.2. The ventral aspect of the index finger should guide the tip of the passer from the lateral position through the vaginal incision. (Reprinted with the permission of The Cleveland Clinic Foundation.)

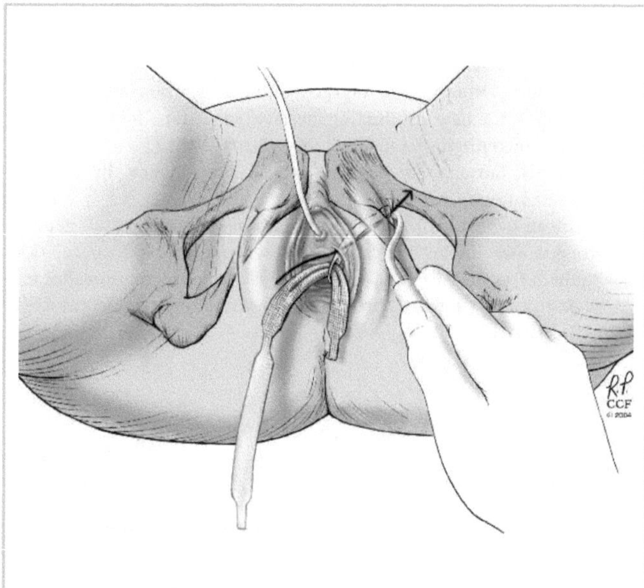

Figure 6-3.3. The narrow end of the tape is threaded into the eye of the passer and pulled until the wider portion of the tape is secured into the eye of the passer. (Reprinted with the permission of The Cleveland Clinic Foundation.)

ensured using a clamp. The ObTape™ is nonelastic and does not recoil. The ends of the tape are cut, by pressing on the skin, avoiding tension on the tape. Skin adhesive is applied to the punctured sites, and the vaginal incision is closed with 2-0 polyglactin 910 suture.

In case of a cystocele repair, the transobturator tape (TOT) or sling can be placed through the same incision. It is preferable to fix the edge of the tape on each side of the urethra to the pubocervical fascia to prevent early movement of the tape.

Transobturator tape is effective for treatment of stress urinary incontinence.[14] Delorme[15] reported his satisfactory initial feasibility and safety study on 40 patients with a follow-up of 3 to 12 months, and more recently on 32 patients with a minimum of 1-year follow-up (mean 17 months).[16] He reported cure in most of his patients with failure in only one patient.

We obtained promising initial results when we evaluated 27 women with SUI who underwent transobturator ObTape™ at our institution. Eight of the patients underwent TOT alone and 19 had TOT with other pelvic reconstructive procedures. There were no urethral or bladder injuries. Vaginal erosion was seen in one patient, and it was treated by local excision and closure. No bowel or vascular injuries occurred. Incontinence was cured or improved in more than 80% of patients.

Complications of Slings

The risk of *sling erosion* is high among patients who underwent synthetic sling placement compared with nonsynthetic material grafts. The Female Stress Urinary Incontinence Panel of the American Urological Association reported 12 vaginal or urethral erosions in 1715 patients who underwent autologous pubovaginal sling placement (0.007%) compared with 41 erosions in 1515 patients treated with various synthetic slings (0.027%).[1] The mechanism of erosion is thought to be secondary to delayed infection of synthetic materials resulting in vaginal erosion, and excessive tension applied to the sling or unrecognized urethral injury at the time of operation resulting in urethral erosion. The symptoms of erosion may include vaginal and urethral pain, irritative voiding symptoms, vaginal discharge, or bleeding. Erosions can be detected by physical examination and urethroscopy.

Overall, new-onset urge incontinence has been reported in as many as 20% of patients after sling operations.[17]

Urinary retention, or an inability to void after a pubovaginal sling procedure, is common especially during the initial postoperative period. This complication may occur because of impaired detrusor contractility, bladder outlet obstruction, or both. To reduce the risk of this complication, prior identification of poor detrusor contractility is required, and the patient should be advised of the risk of postoperative voiding difficulties. Placement of slings without tension as in the TVT procedure has led to a substantial decrease in obstructive complications. Further reduction of this complication has been achieved by adopting the transobturator technique (TOT).

Recurrent stress incontinence, urinary tract infections, incisional hernias, and pelvic hematoma may also develop as complications of sling procedures. Bladder perforation, blood vessel, or nerve injury can occur during placement of slings. Vaginal mucosal slings are associated with the risk of epithelial inclusion cyst formation.[19]

References

1. Leach GE, Dmochowski RR, Appell RA, et al. Female Stress Urinary Incontinence Clinical Guidelines Panel summary report on surgical management of female stress urinary incontinence. The American Urological Association. J Urol 1997;158:875–880.
2. Niknejad K, Plzak LS III, Staskin DR, Loughlin KR. Autologous and synthetic urethral slings for female incontinence. Urol Clin North Am 2002;29:597–611.
3. Ghoniem G, Shaaban A. How does the suburethral sling work? J Urol 1994;151(suppl):514A. Abstract 1146.
4. FitzGerald MP, Mollenhauer J, Brubaker L. The fate of rectus fascia suburethral slings. Am J Obstet Gynecol 2000;183:964–966.
5. Raz S, Stothers L, Young G, et al. Vaginal wall sling for anatomical incontinence and intrinsic sphincter dysfunction: efficacy and outcome analysis. J Urol 1996;156:166–170.
6. Vasavada S, Rackley R, Appel RA. In situ vaginal wall sling formation with preservation of endopelvic fascia for treatment of stress urinary incontinence. Int Urogynecol J Pelvic Floor Dysfunct 1998; 9:379–384.
7. Choe J, Bell T. Genetic material is present in cadaveric dermis and cadaveric fascia lata. J Urol 2001;166:122–124.
8. Cashman NR. A prion primer. CMAJ 1997;157:1381–1385.
9. Brown SL, Govier FE. Cadaveric versus autologous fascia lata for pubovaginal sling: surgical outcome and patient satisfaction. J Urol 2000;164:1633–1637.

10. Duckett JR, Constantine G. Complications of silicon sling insertion for stress urinary incontinence. J Urol 2000;163(6):1835–1837.
11. DeLancey JOL, Richardson AC. Anatomy of genital support. Clin Obstet Gynecol 1993;175:311–319.
12. Ulmsten U, Johnson P, Rezapour M. A three-year follow up for tension free vaginal tape for surgical treatment of female stress urinary incontinence. Br J Obstet Gynaecol 1999;106:345–350.
13. Delmas V, Hermieu JF, Dompeyre P, et al. The trans obturator tape (T.O.T.™): anatomical dangers. Abstract 777. XVIIIth Congress of the European Association of Urology, Madrid; 2003.
14. Delorme E, Droupy S, de Tayrac R, Delmas V. Transobturator tape (Uratape): a new minimally-invasive procedure to treat female urinary incontinence. Eur Urol 2004;45(2):203–207.
15. Delorme E. Transobturator urethral suspension: mini-invasive procedure in the treatment of stress urinary incontinence in women. Prog Urol 2001;11:1306–1313.
16. Delorme E, Droupy S, de Tayrac R, Delmas V. Transobturator tape (Uratape®): a new minimally invasive procedure to treat female urinary incontinence. Eur Urol 2004;45:203–207.
17. Jarvis GJ. Surgery for genuine stress incontinence. Br J Obstet Gynaecol 1994;101:371–374.
18. Bidmead J, Cardozo L. Sling techniques in the treatment of genuine stress urinary incontinence. Br J Obstet Gynaecol 2000;107:147–156.
19. Baldwin DD, Hadey HR. Epithelial inclusion cyst formation after free vaginal wall swing sling procedure for stress urinary incontinence. J Urol 1997;157:952.

6-4

Bulking Agents

Raymond R. Rackley and Ahmed Elazab

For the past several decades, injectable bulking agents have been used for a growing number of people with varying degrees of stress urinary incontinence (SUI). The goal of this therapy for incontinence is to create coaptation and a better mucosal seal for the incompetent urethra. These agents improve intrinsic sphincter function as evidenced by an increase in posttreatment abdominal leak pressure measurements.

Although first introduced in 1938, this conceptual form of therapy has received renewed interest since 1988, with the introductory use of bovine glutaraldehyde crosslinked (GAX) collagen for the treatment of intrinsic sphincteric deficiency (ISD). In the United States (US), most physicians' experiences with injectable bulking agents have been limited to GAX collagen (Contigen; Bard Urological, Covington, GA), autologous fat, and carbon-coated beads (Durasphere; Boston Scientific Corp., Boston, MA) whereas non-US physicians have used such agents as silicon, polytetrafluoroethylene paste, calcium hydroxylapatite, hyaluronic acid, and injectable microballoons. New and revised bulking agents are currently under investigation in the US, as properties of the ideal bulking agent make this form of therapy a highly competitive field for innovations in technology.

Contemporary application of bulking agents has become increasingly popular by both patients and physicians, because the therapy entails a minimally invasive procedure performed under a local anesthesia in an outpatient setting. Although this choice of therapy is largely patient driven, selection remains important to the ultimate success of injection therapy, because the ideal candidates are usually those who have SUI with good anatomic support and healthy tissue.

Despite a tremendous growth in contemporary interest in injectable bulking agents, there have been only a few prospective randomized trials comparing different bulking agents or even injectable therapy to other forms of treatment for SUI. This chapter will summarize the properties, technical approaches, published results, and complications of the various agents available in the US, as well as explore some of the controversies surrounding these issues that have promoted or limited the application of this therapy.

Bulking Agents

The ideal bulking agent must be hypoallergenic, biocompatible, and nonimmunogenic. Additional considerations for selecting a particular agent include cost, implant handling and storage, special equipment for injection, technical ease of application, local and distal migration of the material, and tissue reaction such as inflammation and wound healing. The bulking agents currently approved in the US by the Food and Drug Administration (FDA), include GAX collagen, autologous fat, and carbon-coated beads and are reviewed below.

Contigen

Contigen is a GAX collagen, a highly purified suspension of bovine collagen in approximately 60% vol/vol phosphate-buffered saline containing approximately 95% type I collagen and approximately 1% to 5% type III collagen. The GAX makes collagen less susceptible to fibroblast-secreted collagenase that may result in resorption of the implant and loss of long-term efficacy. Although the resorption of GAX collagen is reduced, this implant is clinically characterized as having resorptive properties resulting in the need for repeated injections (top-off injections once or twice a year) to maintain its long-term efficacy. Persistence of the implant itself has been demonstrated with magnetic resonance imaging of the urethra at intervals of up to 22 months after injection, but the measured volume was less than that injected. Because 2% to 5% of patients are sensitized to collagen through dietary exposure, all patients must undergo a skin test 30 days before treatment; the implant itself at the submucosal location elicits a minimal inflammatory response without foreign body reaction. The properties of this bulking agent allow versatility in its technical application (periurethral, submucosal, or transurethral approach) depending on physician preferences.

As with most anti-incontinence procedures, it is difficult to analyze the outcome of Contigen injection therapy for urinary incontinence because of varying definitions of reported "cure." Early results have reported subjective cure rates of 85% to 95%. In a multicenter trial reported by O'Connell et al.[1] in 1995 on periurethral Contigen injection for 160 females with ISD, 93.8% had significant improvement and 78.1% were completely dry after the injection. That same year, Monga et al.[2] reported on 60 women with genuine stress incontinence using periurethral collagen injections. Subjective success rates were 86% at 3 months, 77% at 12 months, and 68% at 24 months. The long-term (>2 years) success rates are not as favorable (26%–65%), but some long-term cures are noted in this and other trials.

Despite the widespread and long-term application of this implant for the treatment of SUI, treatment-related morbidity has been minimal. Reports of transient urinary retention ranges from 1% to 21% and is managed with intermittent catheterization or short-term catheter use. Urinary tract infection occurs in 1% to 25% of cases, and self-limited hematuria may occur in 2% of cases. Other reported but rare complications include periurethral abscess formation, de novo instability, and a reaction at the previously negative skin test site after a Contigen implant treatment.

When initially introduced for the treatment of ISD in patients with nonmobile urethras, the use of the Contigen implant as well as other bulking agents for patients with hypermobility was somewhat controversial. There have been numerous and well-characterized reports by several centers showing no significant difference in outcome in SUI patients with or without urethral hypermobility. This finding has continued to serve as a strong argument for the use of injectable bulking agents as a first line, minimally invasive therapy for the treatment of SUI, even before the use of oral medications with their requirement and cost of daily administration and attendant side-effect profiles. Continued outcome studies with defined case selection criteria and improved implant material with outpatient technical applications under local anesthesia will continue to define the proper place for this form of SUI treatment, especially in choosing more invasive treatments. In 1997, Berman and Kreder[3] analyzed the cost effectiveness of collagen compared with a sling procedure for type III incontinence, and concluded that surgery was more cost effective than collagen.

Autologous Fat

Whereas autologous fat is one of the bulking agents that is approved by the FDA, several other autologous materials are under consideration, such as autologous chondrocytes and bladder muscle cells. Periurethral autologous fat has been used as an inexpensive, biocompatible, and readily available periurethral bulking agent. The main disadvantages to using fat are the variability of reabsorption (50%–90%) attributed to lack of graft survival and the degree of eventual connective tissue replacement. As in all autologous grafts, survival depends on minimal handling that is afforded through the use of low suction pressure during liposuction and the use of large-bore needles. The procedure involves harvesting abdominal wall fat by liposuction usually under general anesthesia. The injection is usually performed via the periurethral/submucosal approach with a 16- or 18-gauge needle, to reduce the sheering forces generated through the longer needle of the transurethral approach.

Short-term success rates of autologous fat injection for ISD have ranged from 56% to 66%. Long-term survival rates of fat grafts of 10% to 50% have been reported. In 2001, Lee et al.[4] reported a randomized, double-blind, placebo-controlled trial comparing autologous fat for periurethral injection versus saline. They reported a cure/improvement rate at 3-month follow-up of 22%. Interestingly, the cure/improvement rate for the placebo group was 20.7%. Haab et al.[5] reported in a nonrandomized, prospective comparison of periurethral fat versus collagen injections. At 7-month follow-up, 13% of the fat group and 24% of the collagen group were cured. They also reported that the mean percentage of subjective improvement was significantly greater in the collagen group than in the fat group. The failure rate was significantly greater in the fat group compared with the collagen group.

Complications related to periurethral fat injections have been rare and include voiding difficulties (7.6%), de novo instability (3.8%), rectus muscle hematoma formation (3.4%), and urinary tract infection (3.4%–4.7%). Complications from the donor graft site harvest, such as abdominal wall pain, hematoma, and infection are expected depending on the size of graft harvest. Additional complications of urethral pseudolipoma and fat embolism, one of which was fatal, have been reported.

Durasphere

The newest FDA-approved bulking agent is Durasphere. It consists of pyrolytic carbon-coated zirconium oxide beads suspended in an approximately 50% vol/vol water-based carrier gel containing glucan. The targeted radio-opaque bead size originally ranged from 250 to 300 microns in diameter, but now consists of non–radio-opaque beads measuring between 100 to 200 microns. This size reduction was made to facilitate the technical aspects of injection without risking particle migration associated with agents less than the known 80-micron threshold. This synthetic bulking agent is permanent and nonreactive. The potential advantages over other injectable bulking agents include nonimmunogenicity (eliminating the need for skin testing before implantation) and a bead diameter size of more than 80 microns (decreasing the risk of potential macrophage phagocytosis and migration).

The short-term results of Durasphere seem to be comparable to those of other established bulking agents. Lightner et al.[6] reported prospective results in 355 women randomized to receive Durasphere or Contigen for the treatment of ISD. At 12 months after injection, 80.3% of patients in the Durasphere group had improvement in continence grade versus 69.1% in the bovine collagen group. Improvement in urine loss, as measured by reduction in pad weight, was 27.9 g for the Durasphere group versus 26.4 g for the Contigen group. Mean total volume injected was 7.55 mL for Durasphere versus 9.58 mL for collagen. An increase in durability was not proven, because Durasphere required a mean of 1.69 injections versus 1.55 for collagen.

Complications associated with Durasphere are similar to those associated with Contigen. However, the incidence of self-limited acute urinary retention and urgency are higher with Durasphere (24.7% and 16.9%, respectively) than with Contigen (11.9% and 3.4%, respectively).

Although Durasphere is reported as being nonmigratory, some controversy regarding this issue was generated early on in the clinical use of this agent. In 2001, Pannek et al.[7] described 20 patients with ISD who underwent transurethral submucosal injection of Durasphere. They reported six cases with evidence of distant migration of the Durasphere particles on X-ray. The authors could not explain the reason for bead migration or its clinical implications, yet it seemed that all these patients were males injected under a general anesthetic using large amounts of material in excess of typical volumes used for the treatment of female SUI. This invariably led to venous injection and extravasation of the material.

Evaluation and Patient Selection

As with any patient undergoing evaluation for urinary incontinence, the patient's general medical history should be reviewed, especially the incontinence history, previous incontinence therapies, and current incontinence symptoms. In addition to a pelvic examination, urinalysis and/or urine culture should be used to check for infection to ensure that an injectable agent will not be implanted in the setting of an active infection. Urodynamics or a supine stress test may be used to confirm the diagnosis of SUI and significant ISD. Cystoscopy may be performed before and certainly at the time of the injection procedure to inspect the urethra for pathologic conditions such as stricture, urethral diverticulum, or poor mucosal closure and tissue integrity.

Patients with SUI and normal detrusor function are ideal candidates for injectable bulking agents. The presence of significant ISD was originally the primary indication for the use of injectable bulking agents in patients with SUI. However, varying degrees of ISD do coexist or are transiently introduced with hypermobility and, thus, injectable bulking agents have been successfully used in patients with hypermobility to improve the ISD component of their incontinence. Furthermore, women who would require regional or general anesthesia for their SUI treatment, but are poor operative risks (i.e., use of hypocoagulable medications), should be offered injectable bulking agents under a local anesthetic as a viable option for treating their condition.

Detrusor overactivity may be present or unrecognized in many patients with confirmed SUI. Although injectable bulking agents are indicated primarily for SUI, many patients with mixed signs and symptoms of SUI and overactive bladder notice improvement in their overactive bladder condition as a result of improving the SUI condition. These findings are tempered by reports of de novo or worsening overactive bladder conditions in studies using injectable bulking agents. However, these cases typically represent under-recognized and previously unmanaged coexisting overactive bladder conditions.

When using Contigen, a sensitivity skin test should be administered 30 days before the treatment to detect hypersensitivity to bovine collagen. Testing involves injecting 0.1 mL of bovine collagen intradermally in the inner aspect of the forearm. The patient is instructed to report any redness or swelling at the test site for 30 days.

A relative contraindication to injectable therapy is severe vaginal tissue atrophy as evident by mucosal prolapse and urethral caruncles. It is generally recommended that hormonal deficiency resulting in vaginal atrophy be treated with topical estrogens for several months before injection therapy to improve urethral compliance, tissue manipulation, and acceptance of the bulking agent, as well as the ability of the periurethral tissue to retain the injectable bulking agent long-term. Absolute contraindications to therapy with injectable bulking agents include significant urethral pathology such as a urethral diverticulum, an untreated urinary tract infection, and known hypersensitivity to injectable bulking agents, such as Contigen.

Injection Techniques

Transurethral Approach

A transurethral approach for a bulking agent injection is usually performed in an office or outpatient setting with local urethral anesthesia. The patient is placed in the dorsal lithotomy position, and the surgical field is prepared and draped in a sterile manner. Local anesthesia consists of 2% lidocaine jelly placed intraurethrally and left in place for approximately 5 to 10 minutes. With the cystoscope inserted into the mid urethra, the tip of the injection needle is inserted underneath the mucosa and the bulking agent is deposited precisely into the sub-lamina propria tissue plane until complete coaptation of the urethral mucosa is visualized (Figure 6-4.1). To assess for potential cure in patients given local anesthesia, the surgeon performs a

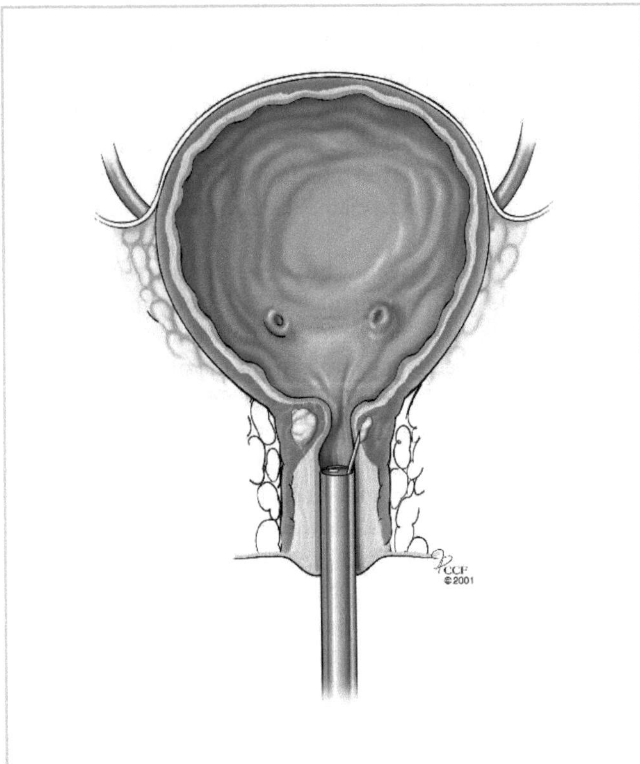

Figure 6-4.1. Transurethral approach for bulking agents. (Reprinted with the permission of The Cleveland Clinic Foundation.)

cough stress test before the injection and then may use this finding as a baseline guide for determining if further implant material is needed during the injection procedure.

Periurethral Approach

The bent-needle technique for a periurethral approach for injectable bulking agents is a simplified, reproducible technique with a one-case learning curve. In comparison to traditional transurethral and periurethral approaches using long, straight needles for various implant materials, the bent-needle technique produces superior results in terms of needle localization and tissue acceptance of bulking under lower pressure without the need of regional or general anesthesia. With cystoscopic confirmation of localizing the needle tip, there is less potential for injecting into vascular spaces or into the deeper muscle layers of the urethra. The bent-needle technique of a periurethral approach produces a longer tract of tissue resistance to the implant material exuding from the injection site that is typical of transurethral procedures.

Preparation of the patient and anesthesia used are the same as for the transurethral route. Additionally, 2 mL of 1% plain lidocaine solution may be injected with a 1.5-inch 30-gauge needle at the 3- and 9-o'clock periurethral positions.

As depicted in Figure 6-4.2, we have opted to use a preformed, approximately 15-degree bent needle that does not compromise the inner luminal radius of the needle, which would potentially increase the resistance to injection of the implant material. Because of the shorter working length and variation of gauge selection available with needles that may reduce resistance to implant flow characteristics, the bent-needle technique for a periurethral approach can be used for all implant materials available for injectable bulking agents and is generally thought to be easier to perform and more efficacious than the original periurethral and transurethral approaches. The smaller needle gauges selected for a periurethral approach of this technique allow the injection of bulking agents for patients taking medications, producing a hypocoagulable condition without the need for interrupting their therapy. Using a transurethral approach with larger-gauge needles not only increases the potential for persistent bleeding and obscures direct observation of the resolution in the postprocedural time period, but the shorter needle track also lessens the compressive hemostatic influences provided from a longer needle track of the periurethral approach.

Depending on the material chosen for injection, the bent-needle technique requires the selection of a preformed needle (22 gauge for Contigen or a tapered-hub 19–21 gauge for Durasphere) bent at the midpoint to 15 degrees toward the luminal side of the needle. At the 3- or 9-o'clock position of the urethral meatus, the point of the

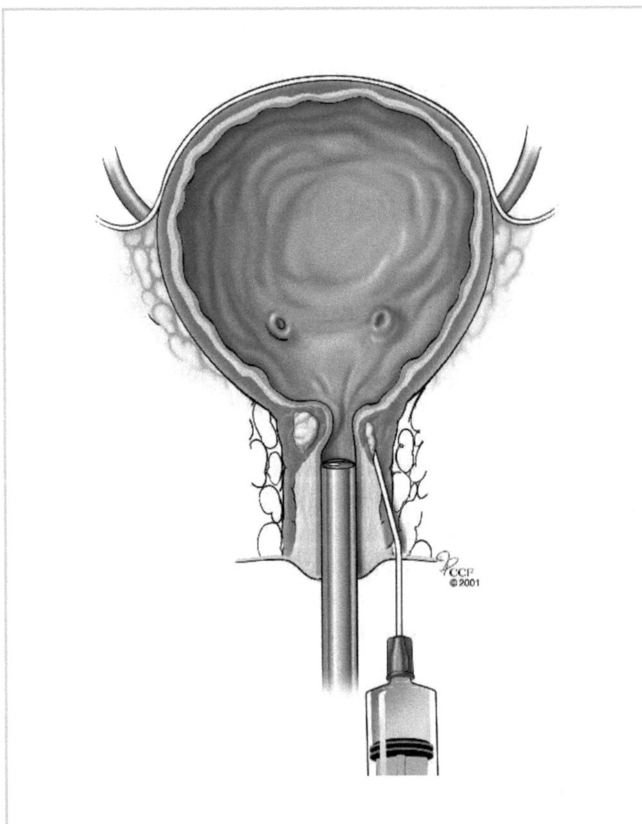

Figure 6-4.2. The bent-needle technique for the submucosal/periurethral approach. (Reprinted with the permission of The Cleveland Clinic Foundation.)

needle is placed at the urethral/vaginal wall crease. With the needle positioned with the bend toward the urethral lumen, the needle is advanced to the proximal urethra and bladder neck. The length and the bend of the needle allow the needle to transverse the urethral musculature and to arrive at the distensible plane between the lamina propria and muscularis muscle at the proximal urethra and bladder neck area. Using a bent needle instead of the traditional straight needle allows for a consistent, reproducible placement of the needle tip within the proper plane of the proximal urethra with minimal manipulation. Depending on the material chosen for injection, hydrodissection of the distensible pocket for implant injection between the lamina propria and the muscularis muscle may be achieved first by injecting 0.5 mL of saline or 1% lidocaine before introducing the injectable implant material.

To assess for immediate cure in cases performed under local anesthesia, the surgeon may perform a cough/stress test on the patient. If stress incontinence persists, additional material should be injected during the same session. The bent needle may also be inserted on the opposite side at the 9-o'clock position and the bulking agent injected with the same technique if further coaptation is required based on clinical judgment or using the cough/stress test maneuver as a guide for achieving a satisfactory treatment endpoint.

The merits reported by physicians for adopting the bent-needle technique for a periurethral approach include the following highlights. First, the technique is reproducible and has a one-case learning curve compared with other methods for this therapy. Second, superior needle-tip localization at the level of the proximal urethra is achieved compared with traditional periurethral and transurethral techniques. Third, cystoscopic confirmation of the needle tip, as well as lateral to medial movement of the needle prevents vascular or deep urethral muscle injection of the bulking agent. Fourth, the formation of a longer periurethral needle track for the injection procedure prevents costly implant extrusion, decompression of the bulking achieved, and better management of patients with a hypocoagulable condition. Fifth, less implant material is needed for the injection procedure compared with other less consistent techniques. Finally, the technique is widely applicable for any implant material chosen without new techniques to be learned by the physician and supporting staff for every implant introduced into the market.

Postprocedural Care

After the procedure, the patient is instructed not to leave the outpatient care area without confirming her ability to void after the treatment. Transient urinary retention is a rare complication (<1%) that may sometimes be associated with a temporary sensory nerve deficit from the amount of periurethral lidocaine injected at the start of the procedure rather than actual urethral obstruction from the injection medium and bulking agent. Should the patient be unable to void despite bladder fullness, she is catheterized once in the office and instructed on self-catheterization with a 12-French catheter if needed until near-normal voiding resumes (usually within 24 hours). The use of indwelling urethral catheters should be avoided to prevent molding and displacement of the bulking agent. A one-time prophylactic oral antibiotic is administered at the conclusion of the procedure. Routine over-the-counter oral analgesics are suggested for any rare postoperative urethral discomfort. Although restriction of physical activity or intercourse after the procedure has never been shown to influence the outcome of this therapy, the use of common sense surrounding these issues is encouraged for the first several days after the implant procedure.

The patient is informed that additional injections may be necessary to establish complete urinary continence because of resorption or displacement of the injection medium, injectable bulking agent, or local anesthetic agent. Additional factors include the inability to have achieved complete tissue coaptation or, in some instances, the injectable material leaks or displaces from its implanted site. Repeat injections are usually scheduled with a 4- to 6-week waiting period between sessions to allow for local tissue healing, and depending on the surgeon's preference.

Summary

The aim of injectable bulking agent therapy for SUI is to create coaptation and a mucosal seal for the incompetent urethra. Injectable bulking procedures have the advantages of simplicity, repeatability, and minimal risk of complications. Performed in an outpatient setting under local anesthesia, injection therapy is suitable for first line therapy for SUI, especially in people who are not candidates for surgery requiring regional or general anesthesia, or who cannot discontinue their anticoagulation therapies. Whereas the best candidates for injection therapy were originally thought to be only those with pure ISD and good anatomic support, the indications for its use has been supported for all females with SUI.

Results of the injectable therapy are variable and dependent on the material and route of administration, the number of injections used per session, and the number of sessions needed to achieve a continent status. The definition of successful patient outcomes is not standardized for injectable therapy and the minimum follow-up period reported varies. Despite these issues, the contemporary clinical experience warrants continued enthusiasm for injectable bulking agent therapy as an option for treating female SUI. Clinical research optimizing the reporting of standardized outcome measures, as well as defining the natural history of the condition, will help to clarify the role of this therapy among all the options for the treatment of SUI.

Overall, injectable therapy complication rates are minimal and the severity is usually minor. It is our belief that complications such as extravasation of the implant material are attributable to the injectable material placed using high injectable pressure systems or in cases in which the patient is given a regional or general anesthetic that blocks the ability to report discomfort when the material is implanted too deep into the muscularis wall of the urethra. For these same reasons, prolonged urinary retention may develop when too much material is injected. Thus, ideal techniques, such as the bent-needle approach to periurethral injections, will minimize the risk of the above complications. Furthermore, costs and rare complications caused by regional or general anesthesia for a procedure that can be performed under a local anesthetic support the enthusiasm for adopting the use of techniques that best achieve this goal. As with many minimally invasive therapies for the treatment of SUI, complications seem to be unique to the material used and the technique of application as listed below.

The one major drawback to contemporary injectable therapy is that it may not produce a complete or long-term permanent treatment solution for the SUI condition because we have little knowledge of its various etiologies and respective natural histories. Conversely, one major attraction of this therapy is its perceived advantage in the SUI patient with borderline bladder function such as hypocontractility, requiring a near-reversible solution to their management of complex voiding dysfunction. Frequently, reinjections may be necessary if this option is selected for the management of SUI, and the quest for innovations in injectable therapy continues to support the clinical application and research for discovering potential permanent treatment options for SUI.

References

1. O'Connell HE, McGuire EJ, Aboseif S, Usui A. Transurethral collagen therapy in women. J Urol 1995;154:1463–1465.
2. Monga AK, Robinson D, Stanton SL. Periurethral collagen injections for genuine stress incontinence: a 2-year follow-up. Br J Urol 1995; 76(2):156–160.
3. Berman CJ, Kreder KJ. Comparative cost analysis of collagen injection and fascia lata sling cystourethropexy for the treatment of type III incontinence in women. J Urol 1997;157:122–124.
4. Lee PE, Kung RC, Drutz HP. Periurethral autologous fat injection as treatment for female stress urinary incontinence: a randomized double-blind controlled trial. J Urol 2001;165(1):153–158.
5. Haab F, Zimmern PE, Leach GE. Urinary stress incontinence due to intrinsic sphincteric deficiency: experience with fat and collagen periurethral injections. J Urol 1997;157:1283–1286.
6. Lightner D, Calvosa C, Andersen R, et al. A new injectable agent for treatment of stress urinary incontinence: results of a multicenter, randomized, double-blind study of Durasphere. Urology 2001;58: 12–15.
7. Pannek J, Brands FH, Senge T. Particle migration after transurethral injection of carbon coated beads for stress urinary incontinence. J Urol 2001;166:1350–1353.

6-5

Retropubic Therapy for Stress Incontinence

Mark D. Walters

Since 1949, when Marshall et al.[1] first described retropubic urethrovesical suspension for the treatment of stress urinary incontinence, and since Burch's landmark article in 1961,[2] retropubic procedures have emerged as consistently curative. Although numerous terminologies and variations of retropubic repairs have been described, the basic goal remains the same: to suspend and stabilize the anterior vaginal wall, and thus the bladder neck and proximal urethra, in a retropubic position. This prevents their descent and allows for urethral compression against a stable suburethral layer. We select a retropubic approach (versus a vaginal approach) depending on many factors, including the need for laparotomy for other pelvic disease, the amount of pelvic organ relaxation, and whether a vaginal or abdominal procedure will be used to suspend the vagina. Additionally, the age and health status of the patient, and the preferences of the patient and surgeon are also determining factors. We generally do not perform a retropubic procedure for intrinsic sphincter deficiency.

Few data differentiate one retropubic procedure from another, although all have advantages and disadvantages. The three most studied and popular retropubic procedures are the Burch colposuspension, the Marshall-Marchetti-Krantz (MMK) procedure, and the paravaginal defect repair. At the Cleveland Clinic, we prefer the Burch colposuspension for urodynamic stress incontinence with bladder neck hypermobility and adequate resting urethral sphincter function, and we combine it with a paravaginal defect repair when the patient has stage II anterior vaginal prolapse or when a concurrent sacral colpopexy is to be done. We do both open and laparoscopic retropubic repairs; laparoscopic techniques will be discussed elsewhere. We no longer perform MMK procedures, so this operation will not be described. The surgical techniques described herein are contemporary modifications of the original operations: Tanagho in 1976[3] described the modified Burch colposuspension; the paravaginal defect repair has been described by Richardson et al.[4] and Shull and Baden.[5] Although less critically studied, the paravaginal defect repair is regionally popular and widely performed in the United States.

This chapter describes only retropubic suspension procedures that utilize an abdominal wall incision for direct access into the space of Retzius. The use of laparoscopy and mini-incision laparotomy to enter the retropubic space and perform these and similar procedures is expanding, both in terms of clinical experience and in research. A thorough critique of the use of operative laparoscopy for urinary incontinence and prolapse is described elsewhere.

Indications for Retropubic Procedures

Retropubic urethrovesical suspension procedures are indicated for women with the diagnosis of urodynamic stress incontinence and a hypermobile proximal urethra and bladder neck. These procedures yield the best results when the urethral sphincter is capable of maintaining a watertight seal at rest, but cannot withstand the unequal transmission of abdominal pressure to the proximal urethra, relative to the bladder, with straining. Although retropubic procedures can be used for intrinsic sphincter deficiency with urethral hypermobility, sling operations probably yield better long-term results.

Women with stress incontinence generally should have a trial of conservative therapy before corrective surgery is offered. Conservative treatments come in the form of pelvic muscle exercises, bladder retraining, pharmacologic therapy, functional electrical stimulation, and mechanical devices, such as pessaries. Eligible postmenopausal patients with atrophic urogenital changes should be prescribed vaginal estrogen before surgery is considered.

Surgical Techniques

Operative Setup and General Entry into the Retropubic Space

The patient is supine, with the legs supported in a slightly abducted position, allowing the surgeon to operate with one hand in the vagina and the other in the retropubic space. The vagina, perineum, and abdomen are sterilely prepped and draped in a sterite manner that permits easy access to the lower abdomen and vagina. A three-way 18-French Foley catheter with a 20-mL balloon is inserted into the bladder and kept in the sterile field. The drainage port of the catheter is left to gravity drainage and the irrigation port is connected to sterile water with or without blue dye. One perioperative intravenous dose of an appropriate antibiotic should be given as prophylaxis against infection.

A Pfannenstiel or Cherney incision is made. During intraperitoneal surgery, the peritoneum is opened, the surgery is completed, and the cul-de-sac is obliterated, if necessary. The retropubic space is then exposed. Staying close to the back of the pubic bone, the surgeon's hand is introduced into the retropubic space and the bladder and urethra are gently moved downward. Sharp dissection is not usually necessary in primary cases. To aide visualization of the bladder, 100 mL of sterile water with methylene blue or indigo carmine dye may be instilled into the bladder after the catheter drainage port is clamped.

If previous retropubic or needle suspension procedures have been performed, dense adhesions from the anterior bladder wall and urethra to the symphysis pubis are frequently present. These adhesions should be dissected sharply from the pubic bone until the anterior bladder wall, urethra, and vagina are free of adhesions and are mobile. If identification of the urethra or lower border of the bladder is difficult, one may perform a cystotomy, which, with a finger inside the bladder, helps to define the bladder's lower limits for easier dissection, mobilization, and elevation.

Burch Colposuspension

After the retropubic space is entered, the urethra and anterior vaginal wall are depressed downward. No dissection should be performed in the midline over the urethra or at the urethrovesical junction, thus protecting the delicate musculature of the urethra from surgical trauma. Attention is directed to the tissue on either side of the urethra. The surgeon's nondominant hand is placed in the vagina, palm facing upward, with the index and middle fingers on each side of the proximal urethra. Most of the overlying fat should be cleared away, using a swab mounted on a curved forceps. This dissection is accomplished with forceful elevation of the surgeon's vaginal finger until glistening white periurethral fascia and vaginal wall are seen. This area is extremely vascular, with a rich, thin-walled venous plexus that should be avoided, if possible. The position of the urethra and the lower edge of the bladder is determined by palpating the Foley balloon and by partially distending the bladder to define the rounded lower margin of the bladder as it meets the anterior vaginal wall.

Once dissection lateral to the urethra is completed and vaginal mobility is judged to be adequate by using the vaginal fingers to lift the anterior vaginal wall upward and forward, sutures are placed. Number 0 or 1 delayed absorbable or nonabsorbable sutures are placed as far laterally in the anterior vaginal wall as is technically possible. We apply two sutures of No. 0 braided polyester on a SH needle (Ethibond; Ethicon, Inc., Somerville, NJ) bilaterally, using double bites for each suture. The distal suture is placed approximately 2 cm lateral to the proximal third of the urethra. The proximal suture is placed approximately 2 cm lateral to the bladder wall at, or slightly proximal to, the level of the urethrovesical junction. In placing the sutures, one should take a full thickness of vaginal wall, excluding the epithelium, with the needle parallel to the urethra (Figure 6-5.1). This maneuver is best accomplished by suturing over the surgeon's vaginal finger at the appropriate selected sites. On each side, after the two sutures are placed, they are passed through the pectineal (Cooper's) ligament, so that all four suture ends exit above the ligament (Figure 6-5.1). Before tying the sutures, a 1 × 4 cm strip of Gelfoam may be placed between the vagina and obturator fascia below Cooper's ligament to aid adherence and hemostasis.

As noted previously, this area is extremely vascular, and visible vessels should be avoided if at all possible. When excessive bleeding occurs, it can be controlled by direct pressure, sutures, or vascular clips. Less-severe bleeding usually stops once the fixation sutures are tied.

After all four sutures are placed in the vagina and through the Cooper's ligaments, the assistant ties first the distal sutures and then the proximal ones, while the surgeon elevates the vagina with the vaginal hand. In tying the sutures, one should leave a suture bridge between the vaginal wall and Cooper's ligament, so as not to place too much tension on the vaginal wall. After the sutures are tied, one can easily insert two fingers between the pubic bone and the urethra, thus avoiding compression of the urethra against the pubic bone. Vaginal fixation and urethral support depend more on fibrosis and scarring of periurethral and vaginal tissues over the obturator internus and levator fascia than on the suture material itself.

Paravaginal Defect Repair

The object of the paravaginal defect repair is to reattach, bilaterally, the anterolateral vaginal sulcus with its overlying endopelvic fascia to the pubococcygeus and obturator internus muscles and fascia at the level of the arcus tendineus fasciae pelvis. The retropubic space is entered and the bladder and vagina are depressed and pulled medially to allow visualization of the lateral retropubic space, including the obturator internus muscle, and the fossa containing the obturator neurovascular bundle. Blunt dissection can be carried dorsally from this point until the ischial spine is palpated. The arcus tendineus fasciae pelvis is frequently visualized as a white band of tissue running over

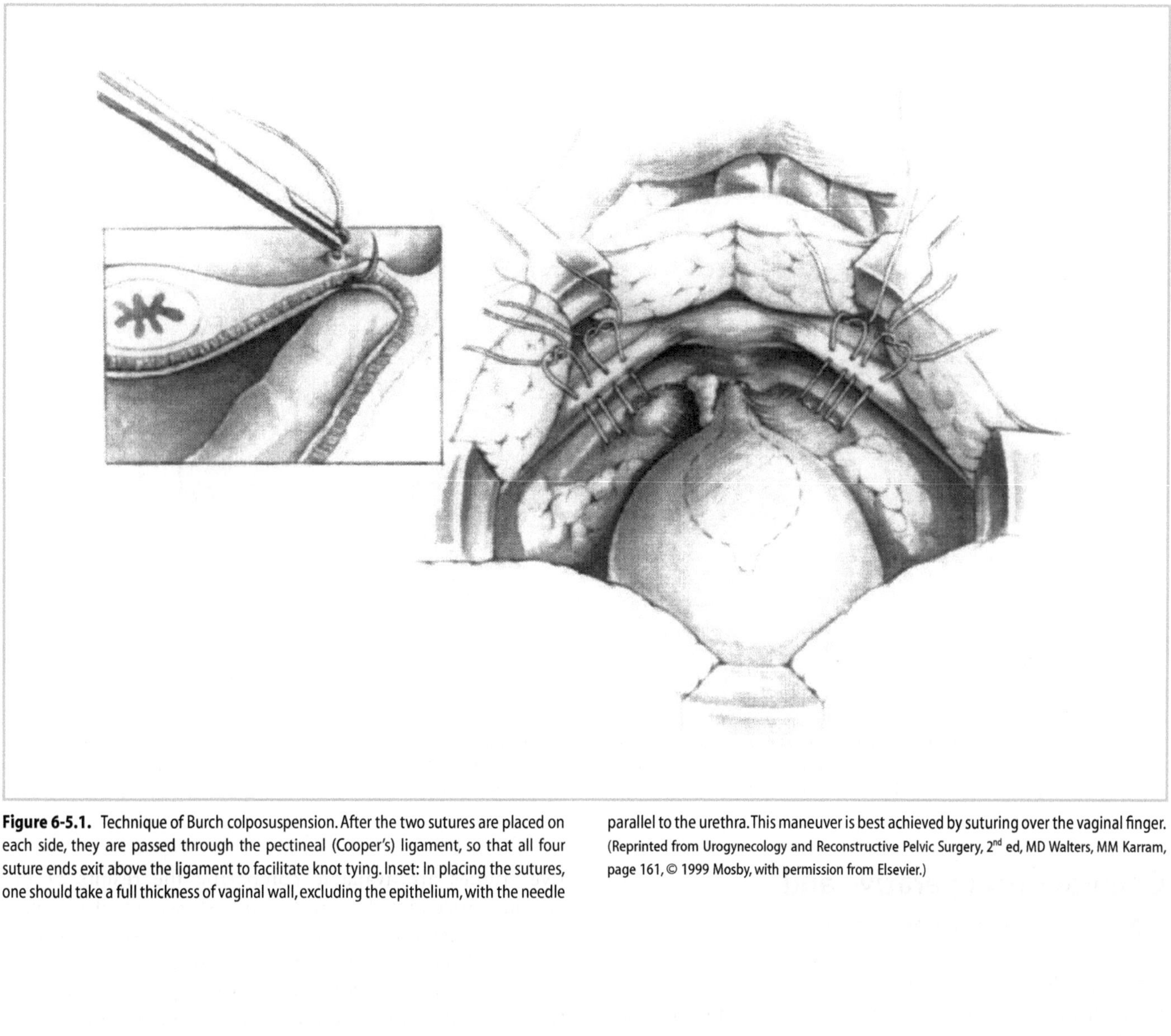

Figure 6-5.1. Technique of Burch colposuspension. After the two sutures are placed on each side, they are passed through the pectineal (Cooper's) ligament, so that all four suture ends exit above the ligament to facilitate knot tying. Inset: In placing the sutures, one should take a full thickness of vaginal wall, excluding the epithelium, with the needle parallel to the urethra. This maneuver is best achieved by suturing over the vaginal finger. (Reprinted from Urogynecology and Reconstructive Pelvic Surgery, 2nd ed, MD Walters, MM Karram, page 161, © 1999 Mosby, with permission from Elsevier.)

the pubococcygeus and obturator internus muscles from the back of the lower edge of the symphysis pubis toward the ischial spine. A lateral paravaginal defect representing avulsion of the vagina off the arcus tendineus fasciae pelvis or of the arcus tendineus fasciae pelvis off the obturator internus muscle may be visualized (Figure 6-5.2).

The surgeon's nondominant hand is inserted into the vagina. While gently retracting the vagina and bladder medially, the surgeon elevates the anterolateral vaginal sulcus. Starting near the vaginal apex, a suture is placed, first through the full thickness of the vagina (excluding the vaginal epithelium), and then into the obturator internus and levator fascia or arcus tendineus fasciae pelvis, 1 to 2 cm anterior to its origin at the ischial spine. After this first stitch is tied, additional (four or five) sutures are placed through the vaginal wall and overlying fascia and then into the levator muscle and fascia at about 1-cm intervals toward the pubic ramus (Figure 6-5.2). The most distal suture should be placed as close as possible to the pubic ramus, into the pubourethral ligament. We use No. 2-0 nonabsorbable suture on a medium-sized, tapered needle for the paravaginal repair. If a Burch procedure is combined with a paravaginal defect repair, it is easier to place and tie the most cephalad paravaginal sutures first, then place the Burch colposuspension sutures at the level of the urethra and bladder neck.

The procedure leaves free space between the symphysis pubis and the proximal urethra, but secures support so that rotational descent of the proximal urethra and bladder base is prevented with sudden increases in intraabdominal pressure. The paravaginal defect repair avoids overcorrection and fixation of the paraurethral endopelvic fascia, which might compromise the functional movements of the urethra and bladder base and lead to obstruction and voiding difficulty. This principle may explain why the paravaginal defect repair usually results in spontaneous voiding on the first or second postoperative day.

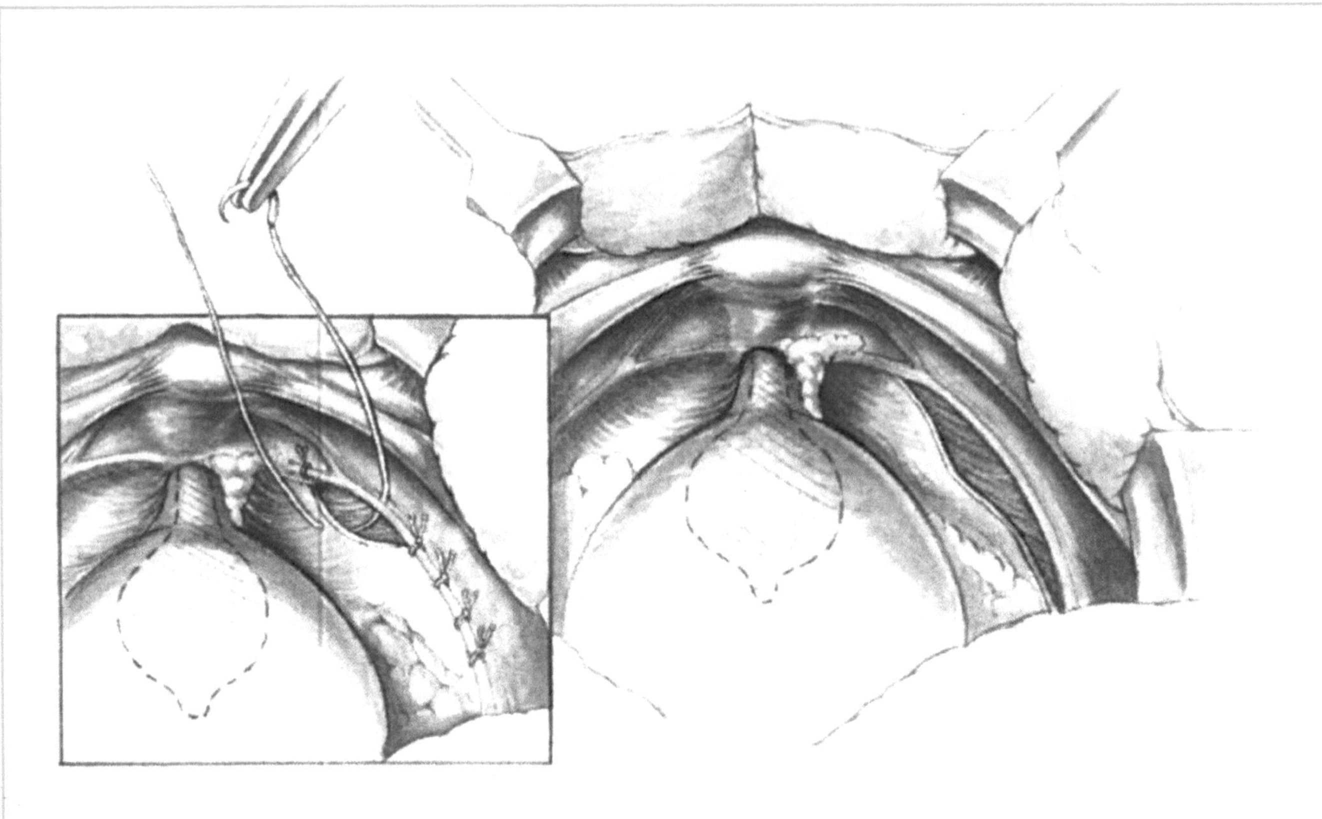

Figure 6-5.2. Lateral paravaginal defect and technique of paravaginal defect repair. Four to six sutures are placed, first through the full thickness of the vagina (excluding the vaginal epithelium), and then into the obturator internus and levator fascia or arcus tendineus fasciae pelvis, 3 to 4 cm below the obturator fossa. (Reprinted from Urogynecology and Reconstructive Pelvic Surgery, 2nd ed, MD Walters, MM Karram, page 162, © 1999 Mosby, with permission from Elsevier.)

General Intraoperative and Postoperative Procedures

We routinely perform cystoscopy, either transurethrally or through the dome of the bladder, or cystotomy to document ureteral patency and absence of intravesical sutures after retropubic procedures.

Closed suction drains in the retropubic space are used only as necessary when hemostasis is incomplete and there is concern about postoperative hematoma. The bladder is routinely drained with a suprapubic or transurethral catheter for 1 to 2 days. After that time, the patient is allowed to begin voiding trials, and postvoid residual urine volumes are checked, either with the suprapubic catheter or by intermittent self-catheterization.

Clinical Results

Many studies have reported clinical experiences with retropubic urethral suspension procedures for stress urinary incontinence. Although most of these studies are methodologically flawed, increasing numbers of quality studies, including prospective randomized trials, have been or are being conducted. Currently, however, few prospective studies are available comparing the results of the various procedures for urodynamic stress incontinence.

Only a few studies have been done assessing the paravaginal defect repair for stress incontinence. Early studies using subjective outcome measures reported that more than 90% of women were continent after this procedure. However, in a prospective randomized trial, Colombo et al.[6] found that only 61% of women were continent 3 years after a paravaginal defect repair compared with 100% of women continent after a Burch colposuspension. We currently believe that the paravaginal defect repair should be used for anatomic correction of anterior vaginal wall prolapse (cystocele), but not as primary treatment of urodynamic stress incontinence.

The Burch colposuspension is the best studied of the retropubic procedures. From 1980 to 1990, 18 studies were reported using the Burch colposuspension in women with urodynamically proved stress incontinence and with objective measures of cure.[7] Follow-up times in most studies ranged from 1 to 7 years. At 3 to 24 months after surgery, 59% to 100% of patients became continent, for an overall average cure rate of 84%. At 3 to 7 years, continence rates ranged from 63% to 89%, for an average rate of 77%. Although objectively incontinent, a small percentage of additional patients were judged to be improved and satisfied with their surgical results. The overall reported absolute failure rate was 14% at 3 to 24 months and 14% at 5 to 7 years.

Several studies have now been done that have assessed women more than 10 years after undergoing a Burch procedure. Alcalay et al.[8] followed a cohort of 109 women (out of a group of 366 eligible women) who underwent Burch colposuspension between 1974 and 1983. The mean follow-up interval was 13.8 years. Both subjective and objective outcome measures were collected during the follow-up period. The cure of incontinence was found to be time-dependent, with a decline for 10 to 12 years and then a plateau at 69%. Cure rates were significantly lower in woman who had had previous bladder neck surgery. Approximately 10% of patients required at least one additional surgery to cure her stress incontinence.[8]

Black and Downs[9] published an excellent systematic review in 1996 describing the effectiveness of surgery for stress incontinence in women. The methodologic quality of studies was assessed especially including all of the randomized controlled trials to that time. Only two randomized controlled trials of colposuspension were available. The study noted that different methods of performing colposuspension (e.g., Burch colposuspension vs. MMK procedure) have not been shown to be associated with significant differences in outcome. There is preliminary evidence that laparoscopic colposuspension as well as open paravaginal defect repair may have somewhat lower cure rates than open Burch procedures. Colposuspension seems to be more effective than both anterior colporrhaphy and needle urethropexy procedures in curing and improving stress incontinence. Approximately 85% of women can expect to be continent 1 year after colposuspension compared with 50% to 70% after anterior colporrhaphy and needle suspension. Primary procedures are generally more effective than repeat procedures. The benefit of Burch colposuspension is maintained for at least 5 years, whereas the benefits from anterior colporrhaphy and needle suspension diminish quite rapidly. Of the four prospective studies comparing Burch colposuspension and sling procedures, none has reported a difference in cure, however defined, regardless of whether the operations were being performed as primary or secondary operations.

In a recent prospective multicenter randomized trial of Burch colposuspension and tension-free vaginal tape (TVT) procedure for urodynamic stress incontinence, no significant difference was found between the surgeries for objective cure rates: 66% in the TVT group and 57% in the colposuspension group.[10] Bladder injury was more common during the TVT procedure; delayed voiding, operation time, and return to normal activity were all longer after colposuspension.

Risk of Failure and Complications

Clinical conditions that increase the risk of surgical failure for retropubic urethropexy include obesity, menopause, prior hysterectomy, and prior anti-incontinence procedures. Advanced age does not seem to be associated with lower rates of cure after colposuspension. Urodynamic findings that increase the risk of surgical failure include signs of intrinsic urethral sphincter deficiency, abnormal perineal electromyography, and concurrent detrusor overactivity. Patients with intrinsic sphincter deficiency probably are better treated with a sling procedure if the urethral is hypermobile, or with periurethral injections of a bulking agent, if the urethra is nonmobile.

Detrusor overactivity or urge incontinence may coexist in up to 30% of patients with stress incontinence. The term mixed incontinence has been used to describe this condition. In addition, approximately 15% of patients with stress incontinence who have a preoperative stable cystometrogram, develop de novo overactive bladder after a colposuspension. The course of the detrusor overactivity after a retropubic repair in patients with mixed incontinence is unpredictable. Interestingly, as many as 50% to 60% of patients with mixed incontinence are cured of their detrusor overactivity by surgical support of the bladder neck. A much smaller percentage (approximately 5%–15%) have worsening of their overactivity with the remainder (20%–30%) having persistence of their overactivity. Women with high-pressure detrusor overactivity or poor bladder compliance are more likely to have persistent urge incontinence after incontinence surgery. In general, women with mixed incontinence should initially receive nonsurgical therapy, followed by surgery if they have persistent stress incontinence that remains bothersome.

Detrusor overactivity is a recognized postoperative complication of retropubic procedures. Postoperative detrusor overactivity is more common in patients with previous bladder neck surgery and in those with mixed detrusor overactivity and stress incontinence preoperatively. In a study of 148 patients with urodynamic stress incontinence and stable bladders preoperatively, Steel et al.[11] reported that 24 patients (16.2%) had postoperative detrusor overactivity on cystometrogram 6 months after surgery. Ten of the 24 patients with detrusor overactivity were completely asymptomatic. Of the 14 symptomatic patients, four were improved with drugs aimed at correcting the overactivity. The remaining 10 patients (6.8%) remained symptomatic with detrusor overactivity 3 to 5 years after surgery.

Wound complications and urinary infections are the most common surgical complications after retropubic colposuspension. Direct surgical injury to the urinary tract occurs relatively infrequently. Bladder lacerations occur in approximately 1% of patients; the risk increases in women who have had previous bladder neck suspension. Accidental placement of sutures into the bladder during the Burch colposuspension or paravaginal repair, resulting in vesical stone formation, painful voiding, recurrent cystitis, or fistula can occur but is rare. Ureteral obstruction occurs rarely (0%–1.2%) after Burch colposuspension and results from ureteral stretching or kinking after elevation of the vagina and bladder base. No cases of transected ureters have been reported.

The incidence of voiding difficulties after colposuspension varies widely, although patients rarely have urinary retention after 30 days. In my hands, the mean number of days to complete voiding after open Burch procedure is 7 days.[12] Colposuspension can change the original micturition pattern and introduce an element of obstruction that can disturb the balance between voiding forces and outflow resistance, resulting in immediate postoperative as well as late voiding difficulties. Urodynamic findings that may occur after colposuspension include decreased flow rate, increased micturition pressure, and increased urethral resistance. In a study conducted at our institution, risk factors for prolonged voiding after a bladder neck suspension included advanced age, previous incontinence surgery, increased first sensation to void, high postvoid residual volume, and postoperative cystitis.[12] Abdominal straining during voiding was not associated with prolonged voiding after surgery.

Burch[2] first reported that enteroceles occurred in 7.6% of cases after the Burch procedure, but only two-thirds of these patients required surgical correction. Langer et al.[13] reported that 13.6% of patients who had undergone Burch procedures, but no hysterectomy or cul-de-sac obliteration, developed an enterocele 1 to 2 years postoperatively. Alcalay et al.[8] noted that 26% of patients during a 10- to 20-year follow-up period after Burch colposuspension underwent a rectocele repair and 5% underwent an enterocele repair. Whenever possible, a cul-de-sac obliteration in the form of uterosacral suspension or plication should be considered at the time of retropubic suspension to prevent enterocele formation, although the true efficacy of this prophylactic maneuver is unknown. Rectocele repair should be performed as indicated for symptomatic or large rectoceles.

Role of Hysterectomy in the Treatment of Incontinence

Gynecologists frequently perform hysterectomies at the time of retropubic or vaginal surgery for stress incontinence although few data support this practice. Langer et al.[13] assessed the effect of concomitant hysterectomy during Burch colposuspension on the cure rate of stress incontinence. Forty-five patients were randomly assigned to receive colposuspension only or colposuspension plus abdominal hysterectomy and cul-de-sac obliteration. Using urodynamic investigations 6 months after surgery, the rate of cure for stress incontinence between the two groups did not differ statistically (95.5% and 95.7% for the no-hysterectomy and hysterectomy groups, respectively). This study clearly showed that hysterectomy adds little to the efficacy of Burch colposuspension in curing stress incontinence. In general, we perform hysterectomies only for specific uterine pathology or for the treatment of uterovaginal prolapse.

References

1. Marshall VF, Marchetti AA, Krantz KE. The correction of stress incontinence by simple vesicourethral suspension. Surg Gynecol Obstet 1949;88:509.
2. Burch JC. Urethrovaginal fixation to Cooper's ligament for correction of stress incontinence, cystocele, and prolapse. Am J Obstet Gynecol 1961;81:281–290.
3. Tanagho EA. Colpocystourethropexy: the way we do it. J Urol 1976; 116:751–753.
4. Richardson AC, Edmonds PB, Williams NL. Treatment of stress urinary incontinence due to paravaginal fascial defect. Obstet Gynecol 1981;57:357–362.
5. Shull BL, Baden WF. A six-year experience with paravaginal defect repair for stress urinary incontinence. Am J Obstet Gynecol 1989; 160:1432–1439.
6. Colombo M, Milani R, Vitobello D, Maggioni A. A randomized comparison of Burch colposuspension and abdominal paravaginal defect repair for female stress urinary incontinence. Am J Obstet Gynecol 1996;175:78–84.
7. Walters MD, Karram MM. Urogynecology and Reconstructive Pelvic Surgery. 2nd ed. St. Louis: Mosby; 1999.
8. Alcalay M, Monga A, Stanton SL. Burch colposuspension: a 10–20 year follow up. Br J Obstet Gynaecol 1995;102:740–745.
9. Black NA, Downs SH. The effectiveness of surgery for stress incontinence in women: a systematic review. Br J Urol 1996;78: 497–510.
10. Ward K, Hilton P. Prospective multicentre randomized trial of tension-free vaginal tape and colposuspension as primary treatment for stress incontinence. BMJ 2002;325:1–7.
11. Steel SA, Cox C, Stanton SL. Long-term follow-up of detrusor instability following the colposuspension operation. Br J Urol 1986;58: 138–142.
12. Kobak WH, Walters MD, Piedmonte MR. Determinants of voiding after three types of incontinence surgery. Obstet Gynecol 2001; 97:86–91.
13. Langer R, Ron-El R, Neuman M, Herman A, Bukovsky I, Caspi E. The value of simultaneous hysterectomy during Burch colposuspension for urinary stress incontinence. Obstet Gynecol 1988;72:866–869.

6-6

Laparoscopic Surgery for Urodynamic Stress Incontinence

Marie Fidela R. Paraiso

Laparoscopic Burch colposuspension is a minimally invasive alternative for the treatment of urodynamic stress incontinence. This procedure has been modified by some investigators by using mesh and staples or reducing the number of colposuspension sutures because of technical difficulty of laparoscopic suturing. At our institution, we believe that the laparoscopic Burch colposuspension should be identical to the conventional open Burch procedure. The laparoscopic access for the Burch procedure had been gaining popularity since its inception in 1991 by Vancaillie and Schuessler.[1] However, with rapid adoption of the mid urethral sling procedures and their similar cure rates when compared with open colposuspension,[2] more pelvic surgeons are choosing to forego the steep learning curve associated with laparoscopic suturing. Furthermore, many reconstructive pelvic surgeons prefer the vaginal route alone rather than a combined vaginal–laparoscopic or vaginal–abdominal approach for concurrent treatment of pelvic organ prolapse and stress incontinence. Surgeons who are experienced and adept at laparoscopic suturing may convert all of their pelvic reconstructive procedures, such as sacral colpopexy and paravaginal defect repairs, to the laparoscopic route, thus converting all open Burch colposuspensions to this route.

The advantages of laparoscopic surgery are: 1) the improved visualization of anatomy of the space of Retzius and peritoneal cavity because of laparoscopic magnification, insufflation effects, and improved hemostasis; 2) shortened hospitalization; 3) decreased postoperative pain, rapid recovery, and return to work; and 4) cosmetic appearance of smaller incisions. Disadvantages of laparoscopic surgery include: 1) technical difficulty of retroperitoneal dissection and acquiring suturing skills; 2) increased operating time early in the surgeon's experience especially if concomitant procedures are laparoscopically performed; and 3) increased hospital cost secondary to increased operating room time and the use of disposable surgical instruments.[3]

Indications

After the diagnosis of urodynamic stress incontinence has been made and the patient has opted for surgical management, the choice of laparoscopic versus open retropubic colposuspension depends on several factors: previous pelvic or anti-incontinence surgery; history of severe abdominopelvic infection or known extensive abdominopelvic adhesions; patient ability to undergo general anesthesia; need for concomitant abdominal, pelvic, or vaginal surgery; patient preference, and operator experience and preference.[3] Most laparoscopic colposuspensions have been done for primary stress incontinence because of difficulty in dissecting retropubic adhesions; however, dissection of the space of Retzius seems easier even when compared with open access once a surgeon has gained sufficient experience. I prefer the laparoscopic Burch colposuspension rather than mid urethral slings for the treatment of stress-provoked detrusor overactivity.

Anatomy

I will briefly summarize anatomy of the anterior abdominal wall, which is essential for safe and effective trocar insertion. The anatomy of the space of Retzius has been summarized in the chapter of anatomic correlates. The umbilicus is at the L3–4 level in normal-sized women. In obese women, the umbilicus is caudal to the L4–5 level, which marks the aortic bifurcation. Therefore, the intraumbilical trocar should be introduced at a more acute angle relative to the abdominal wall and toward the pelvis in thin women and closer to 90 degrees in obese women. The left common iliac vein lies 4 to 7 cm inferior to the umbilicus. The superficial epigastric artery branches from the femoral artery, coursing cephalad, and can be transil-

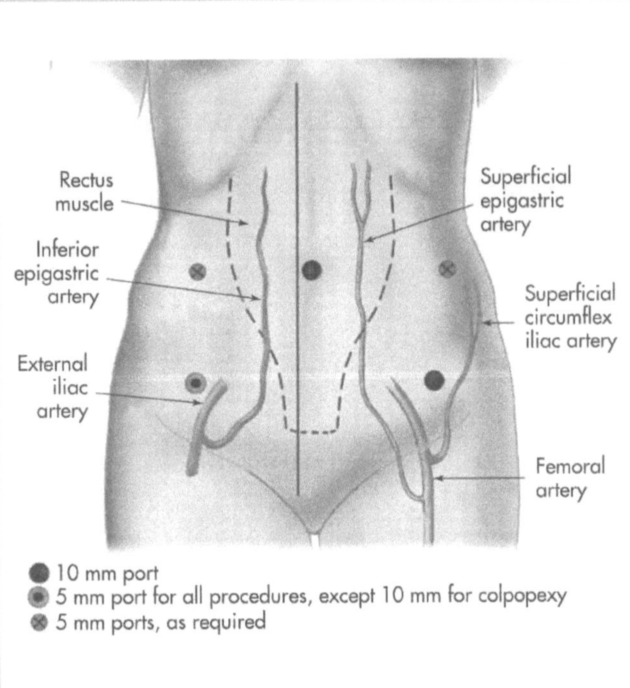

Figure 6-6.1. Anatomy of the anterior abdominal wall and suggested laparoscopic trocar positions. (Reprinted from Urogynecology and Reconstructive Pelvic Surgery 2nd ed, MD Walters, MM Karram, 190–198, © 1999 Mosby, with permission from Elsevier.)

luminated. The inferior epigastric artery branches from the external iliac artery at the medial border of the inguinal ligament and runs lateral to and below the rectus sheath at the level of the arcuate line. It is accompanied by two inferior epigastric veins (Figure 6-6.1).[3]

Operative Technique

A consistent operating room setup and a nursing staff, dedicated and trained in minimally invasive surgery, are essential in laparoscopic reconstructive pelvic surgery. The monitor screens should be placed lateral to the legs in direct view of the surgeon standing on the opposite side of the table. Ideal stirrups for combined laparovaginal cases are the Allen stirrups (Allen Medical, Garfield, OH).

For standard suturing technique, needle-holder preference is determined by comfort of the surgeon. The Storz Scarfi needle-holder and notched assistant needle-holder (Karl Storz Endoscopy, Culver City, CA) are most like conventional needle-holders; however, the tips tend to become magnetized, which makes needle placement difficult at times. The handles will pop as a result of overuse; therefore, regular maintenance is recommended. Conventional and 90-degree self-righting German needle-holders (Ethicon Endo-Surgery, Inc., Cincinnati, OH) have ratchet spring handles and are very sturdy. Disposable suturing devices have been introduced, which include the Endostitch (U.S. Surgical Corp., Norwalk, CT) and the Capio (Microvasive Boston Scientific, Inc., Natick, MA). Extracorporeal knot-tying is preferred because of technical facility and the ability to hold more tension on the suture. The choice of an open-ended or close-ended knot pusher for extracorporeal knot-tying depends on surgeon preference. Our suture of choice is the double-armed No. 0 Ethibond 30-in suture on a CT-2 needle (Ethicon, Inc., Somerville, NJ). Our alternate choice for suture is No. 0 Gore-Tex (W.L. Gore and Associates, Inc., Phoenix, AZ). Forty-eight-inch suture is preferred when suturing from ports at the level of the umbilicus. Sterile steel thimbles may be used by the surgeon or assistant when elevating the vagina while the surgeon is placing the stitches in the vaginal wall.

Skin Incisions for Port Sites

Intraumbilical or infraumbilical incisions are made depending on the anatomy of the umbilicus. I use two additional trocars: a 10/12-mm disposable port with reducer in the right and left lower quadrants lateral to the inferior epigastric vessels (Figure 6-6.1).[3] Trocars are placed lateral to the rectus muscle, approximately 3 cm medial to and above the anterior superior iliac spine. A recent anatomic investigation by Whiteside et al.[4] at our institution revealed that the accessory ports should be placed at least 2 cm above the anterior superior iliac spine to avoid iliohypogastric or ilioinguinal nerve entrapment or injury. An additional 5-mm port may be placed on the principal surgeon's side so that he or she can operate with two hands. Both reusable and disposable ports may be secured with circumferential screws to prevent port slippage.

Laparoscopic Burch Colposuspension

The choice of an extraperitoneal or intraperitoneal approach depends on whether concomitant pelvic procedures are being performed, whether the patient has had previous abdominal surgery, or by surgeon's preference. Previous retropubic surgery is a contraindication for extraperitoneal approach, and low transverse or midline incisions make the dissection more difficult and prone to failure. We prefer the intraperitoneal approach because it allows a larger operating space for safe, secure, and comfortable handling of the suture. Furthermore, other intraperitoneal surgery can be performed concomitantly. If a surgeon has performed a vaginal hysterectomy before laparoscopic concomitant colposuspension, hemostasis of the operative site can be evaluated. I will only describe the intraperitoneal route in this chapter for brevity.

The intraperitoneal approach begins with insertion of the 0-degree laparoscope through a 5- or 10-mm intra- or infraumbilical cannula followed by intraabdominal insufflation. We perform the direct puncture trocar technique either in a blind manner or with the use of an optical trocar. If infraumbilical adhesions are suspected, open laparoscopy or left upper quadrant puncture are applied.

Two to three additional trocars are placed under direct vision, as previously noted.

The bladder is filled retrograde with a three-way Foley catheter with 200 to 300 mL of sterile water or normal saline (indigo carmine or methylene blue is optional). Using sharp dissection with prudent use of electrocautery or harmonic scalpel, a transverse incision 2 cm above the bladder reflection between the medial umbilical folds is made. Identification of the loose areolar tissue at the point of incision confirms a proper plane of dissection. Blunt and sharp dissection aiming toward the posterior superior aspect of the pubic symphysis decreases risk of bladder injury. Blunt dissection is then performed inferolaterally on both sides to identify the pubic symphysis, Cooper's ligaments, and bladder neck.

After the space of Retzius is exposed, the surgeon places two fingers in the vagina and identifies the urethrovesical junction by placing gentile traction on the Foley catheter. With elevation of the vaginal fingers, the vaginal wall lateral to the bladder neck is exposed by using a laparoscopic blunt-tipped dissector or a suction irrigator tip. No dissection is performed within 2 cm of the bladder neck to avoid bleeding and damage to the periurethral musculature and nerve supply.

I prefer to use two disposable 10/12-mm trocars through which 5- to 10-mm instruments are introduced. These sites are used specifically for introduction and removal of the needles and suture. We place stitches in the vaginal wall excluding the vaginal epithelium at the level of, or just proximal to, the mid urethra and bladder neck (Figure 6-6.2). I prefer to introduce the double-armed No. 0 nonabsorbable suture in the left lower quadrant thus taking one bite through the endopelvic fascia and subsequently through the ipsilateral Cooper's ligament. I then remove this needle through the right lower quadrant port. The second arm of the suture is then introduced in the left port and thrown at an angle to the first throw so that a double throw results. This suture is thrown through Cooper's ligament and is subsequently removed through the right lower quadrant port. The stitch is then tied extracorporeally above Cooper's ligament. This technique is one I developed for the sake of time efficiency and avoidance of locking sutures and suture dragging. Some surgeons prefer to backload suture through 5-mm ports and introduce and remove needles through the skin incisions, which is easily accomplished in thinner patients. Trauma to the subcutaneous and inferior epigastric vessels may result with this technique. Other surgeons prefer to use Gor-Tex suture because of ease in sliding through the tissue. Because this is not double-armed, the resultant knot is located in the suture bridge.

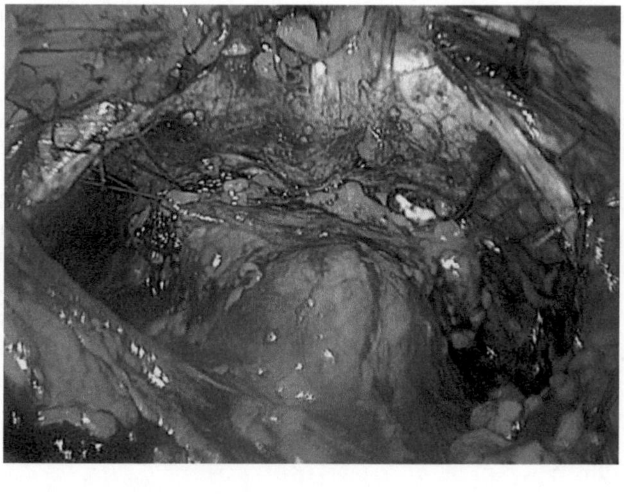

Figure 6-6.2. Laparoscopic panoramic view of a completed Burch colposuspension.

We place Gelfoam (Pharmacia and Upjohn, Inc., Kalamazoo, MI) between the vaginal wall and the obturator fascia before knot-tying to promote fibrosis. With simultaneous vaginal elevation, the suture is tied with six extracorporeal square knots. Two granny half hitches (equivalent to a surgical knot) and a flat square knot will secure the stitch. Our technique for laparoscopic Burch is illustrated in Figure 6-6.2. Sutures are tied as they are placed in order to avoid tangling. Mid urethral stitches are placed first, although this is a matter of preference. With port placement close to the anterior superior iliac spine it is easier to place stitches from the contralateral port.

If the lower quadrant ports are placed higher (at or slightly below the level of the umbilicus), placement of ipsilateral stitches is facilitated because the angle to the ipsilateral vaginal wall and Cooper's ligament is less acute. The appropriate level of bladder neck elevation is estimated with the assistant's vaginal hand. The assistant elevates the vaginal wall in order to place the urethra and bladder neck in a high retropubic position, which does not result in kinking or compression of the urethra. The goal is to elevate the vaginal wall to the level of the arcus tendineus fasciae pelvis bilaterally so that the bladder neck is supported and stabilized by the vaginal wall, which acts as a hammock between both arcus tendineus fasciae. In tying the sutures, the surgeon should not reapproximate the vaginal wall to Cooper's ligament or place too much tension on the vaginal wall. A suture bridge of 1.5 to 2.0 cm is common.

After all sutures are placed and tied, transurethral cystoscopy or suprapubic teloscopy is done to document ureteral patency and absence of sutures in the bladder. A suprapubic catheter is placed, if desired. The surgeon must reinspect the space of Retzius for bleeding while reducing the carbon dioxide insufflation. Routine closure of the peritoneum is not performed. All ports are removed under direct visualization and the peritoneum and fascia of all 10/12-mm incisions are reapproximated with the Endoclose device (U.S. Surgical Corp.) or the Grice needle (New Ideas in Medicine, Inc., Clearwater, FL).

Technical Skill Development for Laparoscopic Colposuspension

Before doing laparoscopic retropubic procedures, the surgeon should have adequate experience in performing these procedures by laparotomy and experience in performing operative laparoscopic procedures. A surgeon can begin to develop suturing skills on inanimate models in pelvic trainers using laparoscopes, cameras, and video monitors in order to simulate operative conditions and improve depth perception. The next step involves performing laparoscopic retropubic procedures in animal laboratories. Cadaver laboratories are ideal but less accessible. A surgeon should perform initial cases of laparoscopic bladder neck suspension in patients with an experienced advanced laparoscopist. A surgeon gains the most proficiency during the first 20 cases.

Clinical Outcomes

Reported continence rates for laparoscopic Burch colposuspension vary from 69% to 100%.[3] Only half of the studies presented objective outcomes. Most authors reported less blood loss, shorter hospitalization, and less-frequent postoperative voiding dysfunction and de novo detrusor instability when compared with the abdominal route.

A recent systematic review[5] of laparoscopic colposuspension summarized the findings of five randomized[6–10] and three "quasi-randomized" trials.[11–13] Analysis of subjective perception of cure showed no difference between open and laparoscopic colposuspension. When one poor-quality trial[9] was excluded from the analysis for objective cure, the relative risk of a positive stress test at follow-up was less in the open colposuspension group but not significantly so when compared with the laparoscopic group. Based on a single trial, two stitches are better than one.[11] The authors concluded that the evidence is limited by short-term follow-up and small numbers; therefore, valid conclusions are difficult to make. A current randomized clinical trial at our institution comparing laparoscopic Burch colposuspension and the tension-free vaginal tape procedure demonstrates superior subjective and urodynamic cure in the tension-free vaginal tape group at a median of 24.5 months. However, the sample size of our study is relatively small with 36 patients in each group and 2-year follow-up of all subjects is not yet complete. Cost was similar for both procedures.

Conclusion

Laparoscopy is a means of surgical access, not a unique procedure, and its use is expanding rapidly in all surgical specialties because of a trend toward minimally invasive procedures. More prospective clinical trials with identical technique of laparoscopic and open colposuspension and follow-up of 3 or more years are warranted.

References

1. Vancaillie TG, Schuessler W. Laparoscopic bladderneck suspension. J Laparoendosc Surg 1991;1:169–173.
2. Ward K, Hilton P. Prospective multicentre randomized trial of tension-free vaginal tape and colposuspension as primary treatment for stress incontinence. BMJ 2002;325:67–70.
3. Paraiso MF, Falcone T. Laparoscopic treatment of incontinence and prolapse. In: Walters MD, Karram MM, eds. Urogynecology and Reconstructive Pelvic Surgery. 2nd ed. St. Louis: Mosby; 1999: 197–209.
4. Whiteside J, Barber MD, Walters MD, Falcone T. Anatomy of ilioinguinal and iliohypogastric nerves in relation to trocar placement and low transverse incisions. Am J Obstet Gynecol 2003;189(6): 1574–1578.
5. Moehrer B, Carey M, Wilson D. Laparoscopic colposuspension: a systematic review. Br J Obstet Gynaecol 2003;110:230–235.
6. Burton G. A three year prospective randomized urodynamics study comparing open and laparoscopic colposuspension [abstract]. Neurourol Urodyn 1997;16:353–354.
7. Carey M, Rosamilia A, Maher C, et al. Laparoscopic versus open colposuspension: a prospective multicentre randomized single-blind trial. Neurourol Urodyn 2000;19:389–391. Abstract 8.
8. Fatthy H, El Hao M, Samaha I, Abdallah K. Modified Burch colposuspension: laparoscopy vs laparotomy. J Am Assoc Gynecol Laparosc 2001;8:99–106.
9. Su TH, Wang KG, Hsu CY, Wei HJ, Hong BK. Prospective comparison of laparoscopic and traditional colposuspensions in the treatment of genuine stress incontinence. Acta Obstet Gynecol Scand 1997;76:576–582.
10. Summitt Rl, Lucente VL, Karram MM, Shull BL, Bent AE. Randomised comparison of laparoscopic and transabdominal Burch urethropexy for the treatment of genuine stress incontinence [abstract]. Obstet Gynecol 2000;95(4 suppl 1):S2.
11. Persson J, Wollner-Hanssen P. Laparoscopic Burch colposuspension for stress urinary incontinence: a randomized comparison of one or two sutures on each side of the urethra. Obstet Gynecol 2000;95: 151–155.
12. Ross J. Two techniques of laparoscopic Burch repair for stress incontinence: a prospective randomized study. J Am Assoc Gynecol Laparosc 1996;3:351–357.
13. Wallweiner D, Grischke EM, Rimbach S, Maleika A, Bastert G. Endoscopic retropubic colposuspension: "Retziusscopy" versus laparoscopy—a reasonable enlargement of the operative spectrum in the management of recurrent stress incontinence? Endosc Surg Allied Technol 1995;3:115–118.

6-7

Other Therapies for Stress Urinary Incontinence

Tara L. Frenkl and Sandip P. Vasavada

Whereas simple and definitive sling, retropubic, or injectable therapy will suffice for the overwhelming majority of incontinent patients, it is often extremely difficult to manage stress urinary incontinence that is of a refractory nature. Accordingly, less often used modalities may be required to cure the condition, including the use of obstructing or circumferential slings, artificial urinary sphincters (AUS), or even transvaginal closure of the bladder neck.[1] These methods are certainly retained for a "last resort" of therapy, yet still maintain a prominent position in care, because all patients will not be successfully managed with standard therapies.

Obstructing Sling Procedures

In the severely incontinent patient whose condition is refractory to a standard anti-incontinence procedure, an obstructing sling is a good option. The basics of the approach may be as simple as a standard sling procedure that is ultimately fixed or tied down with more tension than in usual circumstances with the intent to obstruct the outlet. It should be noted that extreme degrees of tension may create more problems and irritation, so a balanced approach to the nature of outlet resistance should be maintained. We typically prefer the use of biological tissues when placing them under more tension. In cases in which a synthetic sling may be preferred, our choice has been the use of the circumferential or spiral sling. We initially begin by performing a complete urethrolysis and obtain access to the supraurethral space. We then use a large strip of polypropylene mesh trimmed to 2.0 × 30 cm and create a slit inside it so as to be able to pass the sling through itself (Figure 6-7.1). The needle passers are passed from one side suprapubically to the *opposite* side vaginally so the sling spirally encompasses the entire urethra (Figure 6-7.2). A pressure test or cough test is performed to assure good coaptation and lack of leakage with Valsalva/cough/Credé maneuvers. It remains imperative that the patient and physician realize that the patient will likely require intermittent postoperative self-catheterization, permanently in the majority of these types of cases. The goal of the procedure is truly not to achieve spontaneous balanced voiding at low pressures, but rather, a continent and well-coapting urethral mechanism.

Artificial Urinary Sphincters

The AUS is currently not approved for use in females in the United States.[2] Although it is not approved, it is used for circumstances similar to those for which the spiral or obstructing slings are used; in other words, a severely incompetent outlet. Often, these patients have had multiple, failed anti-incontinence procedures and have a "pipestem" urethra that does not coapt well even with injectable therapies. One must realize that the successes with this procedure have been detailed in two approaches, and there exists no consensus as to which approach is best or most efficacious.

Transvaginal placement of the AUS entails a transvaginal incision and placement of the cuff via this route. All tubing and connections can be transvaginally done before pump and reservoir placement into the labia and perivesical spaces, respectively. The typical size of cuff used for female AUS is approximately 5.0 cm. The main risks of complications after abdominal/retropubic placement are erosion and infection. The vaginal incision does allow the potential for nonhealing of the space under the cuff, thereby allowing it to be exposed and resulting in either an erosion or infection. By placement of the device via a retropubic approach, erosion may be minimized, if not eliminated. One may adjunctively use a Martius flap or omental flap to place underneath the cuff, between it and the vaginal wall to minimize the potential for erosion. As in other surgical procedures for complex, refractory incontinence, many of these patients have had multiple failed procedures, and the tissue planes for dissection may, in fact, be fused, thus creating an inadvertent incision and increased risk for erosion potential. Similarly, as in the

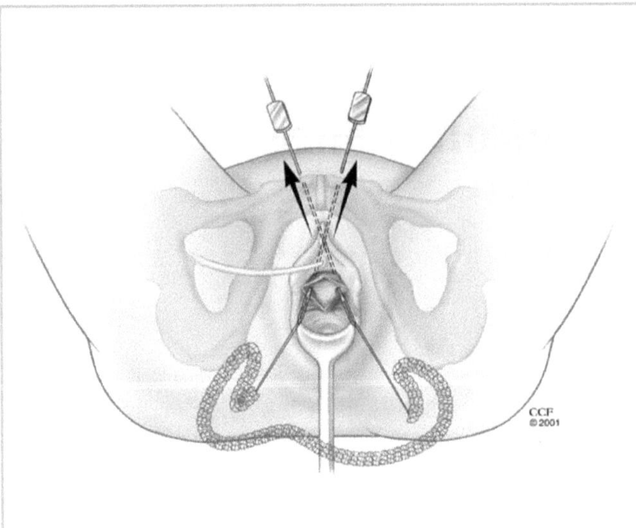

Figure 6-7.1. Synthetic sling material being passed from one side to opposite side. (Reprinted with the permission of The Cleveland Clinic Foundation.)

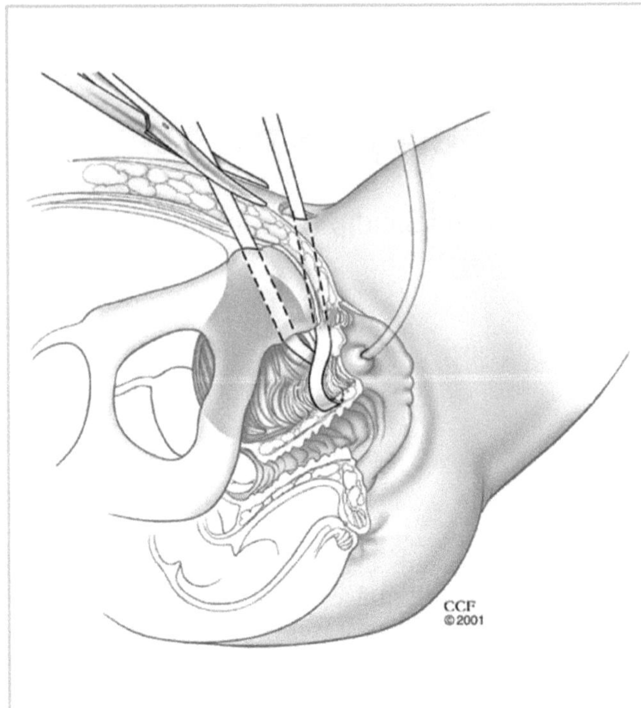

Figure 6-7.2. Synthetic sling in cross-sectional view after sling has been transferred suprapubically. (Reprinted with the permission of The Cleveland Clinic Foundation.)

male patient with an AUS, infection, tissue atrophy, and mechanical failure may occur.

Transvaginal Closure of the Bladder

Transvaginal closure of the bladder neck represents an uncommon surgical procedure, because its main indication is in patients who have urethral destruction secondary to prolonged catheter drainage in a neurogenic bladder.[3] The resultant effect of the catheter and Foley balloon on the urethra is erosion and pressure necrosis as well as spontaneous urethral extrusion from severe bladder spasms. The usual course of events is progressive worsening of leakage around the catheter, necessitating larger catheters and more fluid placement in the balloon. These sequential increases in catheter size along with the concomitant urethral destruction result in a wide, patulous, and severely damaged urethra. Accordingly, proper management of these patients requires transvaginal closure of the bladder neck and simultaneous urinary diversion with a continent catheterizable augmentation, incontinent ileovesicostomy, or suprapubic catheter. Typically, an obstructing sling or AUS will not be possible because of insufficient length of urethra for coaptation caused by erosion.

Most patients that would be candidates for transvaginal closure of the bladder neck are those who have a neurogenic bladder and urethral destruction secondary to long-term catheter drainage. Other indications include refractory urethrovaginal fistula, severely shortened urethra, and severe refractory intrinsic sphincteric deficiency.

Preoperative Work-up

Visual inspection of the urethra demonstrates a large, patulous urethra that is shortened and may even permit visualization of the bladder neck or bladder itself. Cystoscopy is performed to confirm the absence of other pathology in the bladder such as stones, neoplasms, or diverticuli that may accompany long-term catheter drainage. Simultaneous study of the upper urinary tracts may be indicated to evaluate for hydronephrosis or ureteric obstruction. The patient should also be evaluated for manual dexterity or have an adequate caregiver to facilitate catheterization of the augmented segment. If this is not feasible, one must plan for an incontinent augmentation (ileovesicostomy) and subsequent stomal drainage or suprapubic catheter drainage. Urine cultures should be routinely assessed preoperatively, because these patients are prone to infections. Broad-spectrum antibiotics are essential before surgery.

Surgical Technique

Initial preparatory steps include placement of the patient in the dorsal lithotomy position, and prepping and draping the lower abdomen and vagina in the usual sterile manner. If a suprapubic tube placement is necessary, it is placed at this time, typically with use of a Lowsley retractor elevating the bladder to the anterior abdominal wall. Normal saline or vasoactive substances are injected into the vaginal wall to develop a subepithelial plane around the urethra. A

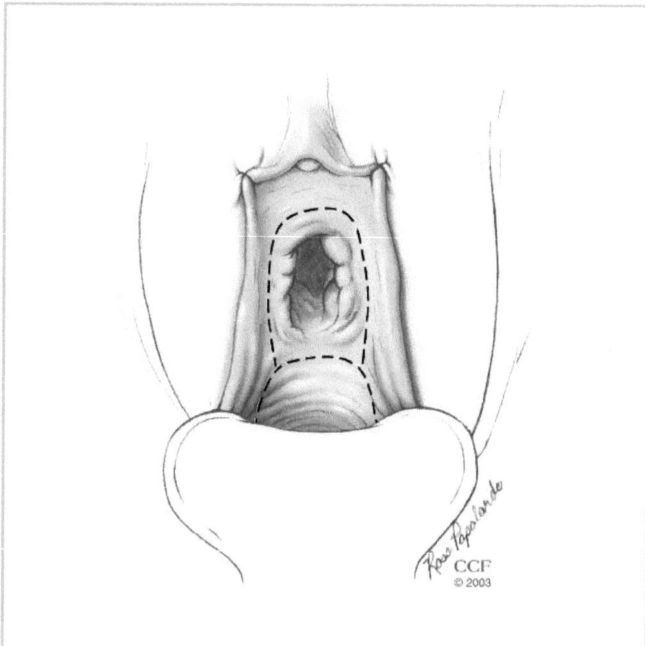

Figure 6-7.3. Open bladder neck and eroded urethra from long-term catheter use. (Reprinted with the permission of The Cleveland Clinic Foundation.)

circumscribing incision is made around the urethra and an inverted U incision is made inferior to this in the anterior vaginal wall (Figure 6-7.3). A vaginal flap is created by sharp dissection in the anterior vaginal wall separating the epithelium from the underlying perivesical fascia. This dissection is carried laterally around the open bladder neck (Figure 6-7.4). The dissection is then continued toward the bladder neck, freeing the bladder from its attachments to the symphysis pubis and lateral pelvic sidewall (Figure 6-7.5). This necessitates entering the retropubic space by detaching the urethropelvic ligaments on either side of the bladder neck and transecting the pubourethral ligaments above the urethra. Typically, the attenuated urethra requires some degree of excision; however, we often use it to create an additional layer of closure (Figure 6-7.5). Indigo carmine is intravenously administered to facilitate identification of the ureteral orifices to prevent injury during bladder neck closure. The bladder neck is then closed using a vertical closure of 2-0 polyglycolic acid suture in an interlocking manner (Figure 6-7.6). This is followed by a second layer of interrupted 2-0 polyglycolic acid suture incorporating the bladder neck and anterior bladder wall to prevent overlapping suture lines and transfer the closed bladder neck into the retropubic space, high behind the pubic symphysis (Figure 6-7.7). After this second layer, placement of the attenuated posterior urethral plate allows a third layer of support of the closure. If necessary, a Martius flap may be placed at this time to advance yet another layer over the repair. When no simultaneous continent urinary augmentation is planned, a separate Penrose drain is advanced through a suprapubic incision. Finally, the anterior vaginal wall flap is advanced as the final layer of closure, and it covers the original urethral opening (Figure 6-7.8).

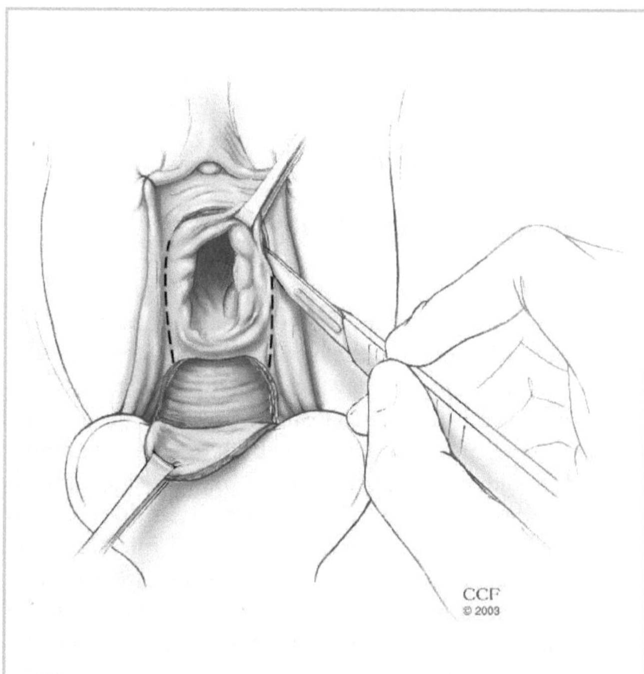

Figure 6-7.4. Incision around urethra and open bladder neck to initially free it from vaginal wall and vaginal wall flap inferiorly. (Reprinted with the permission of The Cleveland Clinic Foundation.)

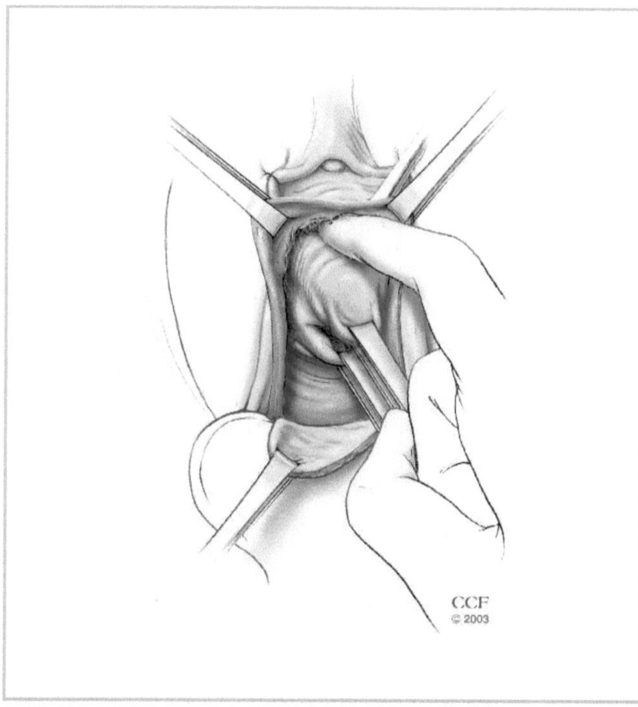

Figure 6-7.5. Bladder neck dissection performed circumferentially to free it from surrounding attachments. Any remaining urethra should be excised except at the base to prevent a bladder contraction from being propagated down the urethra and risk opening the incisional closure. (Reprinted with the permission of The Cleveland Clinic Foundation.)

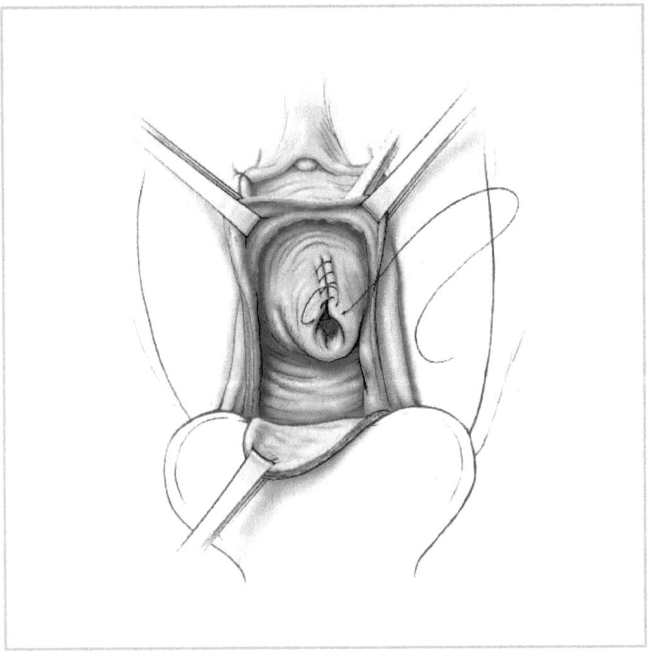

Figure 6-7.6. The bladder neck is then closed using a vertical closure of 2-0 polyglycolic acid suture in an interlocking manner to assure a watertight first closure layer. (Reprinted with the permission of The Cleveland Clinic Foundation.)

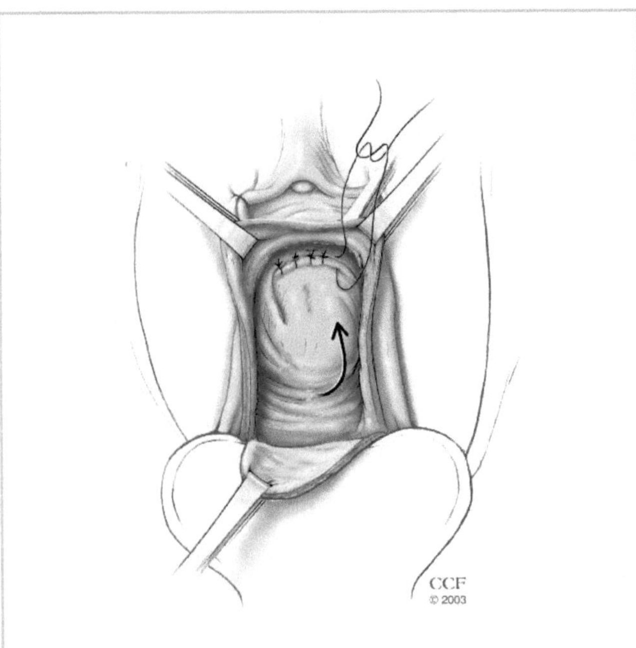

Figure 6-7.7. A second layer of interrupted 2-0 polyglycolic acid suture incorporating the bladder neck and anterior bladder wall is placed to prevent overlapping suture lines and transfer the closed bladder neck into the retropubic space, high behind the pubic symphysis. If the closure is tenuous or one thinks it necessary, one would place a Martius flap at this time after the second layer closure. (Reprinted with the permission of The Cleveland Clinic Foundation.)

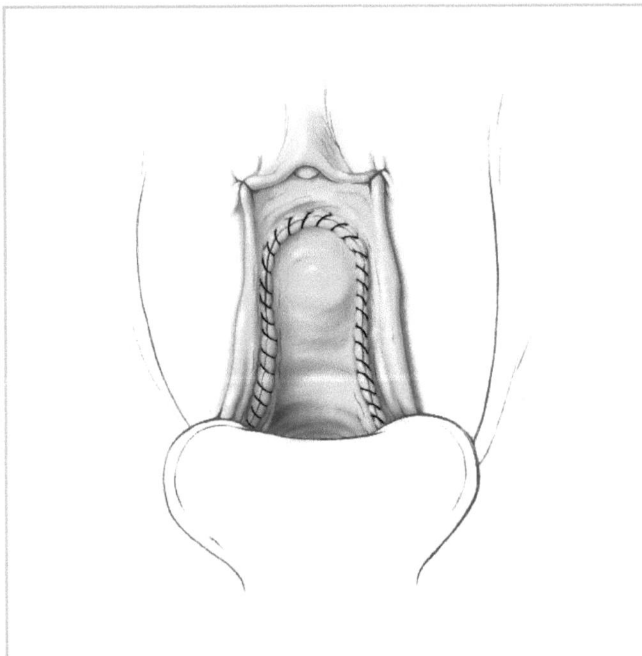

Figure 6-7.8. The vaginal wall flap is placed over the entire closure and the vagina is packed overnight. (Reprinted with the permission of The Cleveland Clinic Foundation.)

Postoperative Care

A vaginal pack is placed in the vagina to prevent bleeding. The suprapubic tube is irrigated and confirmed to function properly to allow for adequate drainage and healing of the incisions. Antibiotics are used postoperatively for 2 weeks during the healing period. Depending on whether or not a continent catheterizable augment is performed, a cystogram may be performed to ensure proper healing, and the suprapubic tube may be allowed to remain or removed at that time.

Postoperative Complications

The most significant complication of transvaginal bladder neck closure is postoperative vesicovaginal fistula. This may occur early or late in the healing period. Steps one can perform to prevent the development of fistulas include: 1) use of several layers of closure; 2) watertight closure; 3) positioning the bladder neck high in the retropubic space; 4) avoidance of postoperative bladder spasms with anticholinergics; 5) use of a Penrose drain; 6) use of a Martius flap when necessary. Bleeding and infection may require surgical drainage, but should be readily controlled perioperatively. If fistula formation occurs, then secondary repair may be performed after a prudent period of observation. Often a rotational flap may be required to assist in the closure. Alternatively, an abdominal approach may be used with interposition of an omental flap to cover the repair.

References

1. Thomas K, Venn SR, Mundy AR. Outcome of the artificial urinary sphincter in female patients. J Urol 2002;167(4):1720–1722.
2. Wilson TS, Lemack GE, Zimmern PE. Management of intrinsic sphincteric deficiency in women. J Urol 2003;169(5):1662–1669.
3. Zimmern PE, Hadley R, Leach G, Raz S. Transvaginal closure of the bladder neck and placement of a suprapubic catheter for destroyed urethra after long-term indwelling catheterization. J Urol 1985; 134(3):554–557.

6-8

Anal Sphincter Repair

Susan M. Cera and Steven D. Wexner

Fecal continence relies on normal sensory, motor, and reflex activity of the colon, rectum, and anus. Stool consistency, colonic transit, rectal sensation, neural integrity, and sphincter function all have a role in individual control of stool and gas. The principal component in facilitating continence is the anal sphincter, whereas the hemorrhoidal cushions, sensory epithelium of the anal mucosa, intrinsic anorectal reflexes, and pelvic floor muscles contribute to its function. The anal sphincter complex is composed of the internal anal sphincter (IAS), the external anal sphincter (EAS), and the puborectalis muscle. The pudendal nerve innervates the EAS and puborectalis whereas the innervation of the rectum, pelvic floor muscles, and IAS is a complex system of sympathetic and parasympathetic neurons supplied by the pelvic and sacral nerves. Damage to any muscular or neural component of the sphincter mechanism may result in fecal incontinence and possible need for surgical intervention. The two most common causes of fecal incontinence are sphincter dysfunction and neuropathy.

Physical disruption of the sphincter may occur secondary to surgical, obstetric, or other trauma. Surgical trauma most often occurs after anorectal surgical procedures used in the treatment of anal fistulas, hemorrhoids, tumors, or anal fissures. Each of these surgical injuries is similar, in that the IAS, the EAS, or both may be injured. Anal fistulotomy inherently causes some form of damage to the sphincter mechanism, and the risk of incontinence may be proportional to the amount of muscle divided. Alternatives to fistulotomy involve controlled transection of the muscle with cutting setons, instillation of fibrin glue, or endorectal advancement flaps.

Chronic fissures are frequently treated by lateral internal sphincterotomy in which a portion or all of the internal sphincter is transected. The internal sphincter, which consists of smooth muscle under autonomic involuntary control, is responsible for the resting pressure in the anal canal. When divided, the resting pressure is lowered, allowing the anal fissure to heal. The vast majority of patients have no untoward effects from this technique, but a small minority may develop transient incontinence to gas. A few individuals will develop chronic incontinence, which may be more severe and associated with involuntary loss of liquid or solid stool. Alternatives to this procedure have been developed and include medications, such as nitrate paste and calcium channel blocking cream, to relax the internal sphincter. These therapies are now often the first line treatment in the management of anal fissures.

Sphincter injury resulting from hemorrhoid surgery is fortunately rare and usually avoided by careful attention to the anatomy.

Trauma can injure the anal sphincter mechanism; impalement injuries may result in sphincter disruptions, rectal perforations, and soft tissue damage to the perineum. Their treatment is often staged with initial debridement and fecal diversion followed by delayed definitive repair when the sepsis is resolved and the injury has healed.

Obstetric injury is the most common cause of sphincter damage and occurs as the result of a tear in the perineum or an episiotomy. Parous women or women who undergo assisted evacuations are at the highest risk for sphincter laceration. The sphincter damage may be occult for many years; it is common for this dormant injury to clinically manifest as fecal incontinence years to decades postpartum.

The surgical treatment of fecal incontinence is categorized into two groups: reparative procedures that focus on repair or augmentation of the sphincter muscle, versus replacement procedures in which the goal is reconstruction, replacement, or external modulation of the sphincter apparatus. The choice of procedure is dependent on the underlying etiology of the incontinence as well as on complete evaluation of the anorectal and pelvic floor anatomy. The type and severity of injury found on evaluation and physiologic testing directs therapy to the individual's problem. If possible, repair of the native sphincter mechanism is undertaken before resorting to major reconstructive surgery, artificial sphincters, neurostimulation, or stoma. In particular, patients who are candidates to undergo sphincter repair often have injuries from physical disruption of the EAS complex.

Types of anal sphincter repairs include direct apposition sphincteroplasty, overlapping sphincteroplasty, anterior plication (reefing), Parks' postanal repair, and total pelvic floor repair. Sphincteroplasty involving direct apposition is traditionally reserved for the acute setting of traumatic sphincter laceration. However, recent data have challenged this dogma. The overlapping sphincteroplasty is the most widely accepted procedure for secondary repairs in patients with isolated anterior defects. Anterior plication, postanal repair, and total pelvic floor repair do not involve direct repair of the muscle but theoretically attempt to augment the function of the sphincter mechanism by restoring the lax muscular architecture to its original anatomic configuration. These procedures are seldom performed today but remain important surgical options in patients who have failed conservative measures but do not desire more complex forms of surgical therapy.

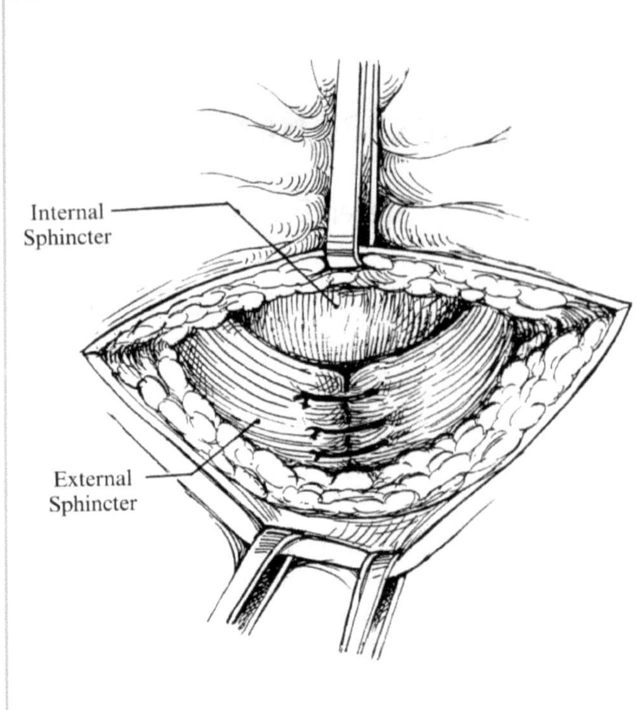

Figure 6-8.1. Direct apposition sphincter repair. (Reprinted from Wexner SD, Beck DE. Fundamentals of Anorectal Surgery. 2nd ed. p 133–136, Copyright 2001 WB Saunders, with permission from Elsevier.)

Direct Apposition Sphincter Repair

The majority of fecal incontinence is attributed to anatomic defects in the EAS. This type of defect most often occurs as a result of midline perineal tears sustained during vaginal delivery. Overt sphincter rupture occurs during 5% of vaginal deliveries and increases to an incidence of 50% with the use of midline episiotomy and/or operative vaginal delivery.[1] At the time of delivery, any documented trauma to the anal sphincter (third or fourth degree tear) should be given due regard. Even if waiting 12 to 18 hours is required, performing the repair in the operating room with adequate lights and instruments optimizes the repair, as judged by ultrasound after the repair has healed. The most common technique used in the acute setting is direct apposition of the sphincter ends in which the cut ends of the IAS and EAS are sutured together in layers. Direct apposition of the sphincters without tension in the acute setting has been reported to give satisfactory results in 47% to 100% of cases.[2] Failures are attributed to tearing of the sutures through the tissue with resultant sphincter repair disruption. Hematoma formation, wound infection, technical considerations, and an unrecognized second sphincter injury also predispose to poor outcome requiring second repair in at least 5% of patients.[1,3] In addition, in women with severe traumatic lacerations and significant perineal contamination, debridement and formation of a colostomy may be necessary until definitive sphincter repair can be performed. Secondary sphincter repairs should only be undertaken after all pelvic sepsis has resolved, contaminated perineal wounds have healed, the inflammation has completely resolved, and the tissues are soft and pliable. For most failed primary sphincter repairs, a period of waiting of at least 3 months before performing an overlapping sphincteroplasty is prudent. Overlapping sphincteroplasty is generally undertaken as the technique for secondary repairs because, historically, direct apposition has demonstrated disappointing long-term results, although it yields good results in the acute setting (Figure 6-8.1).

Overlapping Sphincter Repair

For most women with fecal incontinence secondary to obstetric trauma, sphincter damage is often occult, persistent, or associated with pudendal nerve damage leading to fecal incontinence many years after delivery. For women with fecal incontinence secondary to a functional yet anatomically disrupted anterior EAS, overlapping sphincteroplasty is the operation of choice. This technique is the result of an evolution of a variety of modifications of the initial descriptions of sphincter repair. Initial reports involved identification and mobilization of the EAS, excision of all scar tissue, and direct apposition of muscle ends. Further modifications associated with significant improvement in functional outcome included overlapping of the ends of the muscle and avoidance in excising any associated scar tissue. Additionally, it was found that sphincteroplasty could be safely performed without fecal diversion if an appropriate bowel preparation was used. Other modifications including levatorplasty and repair of the IAS have further been incorporated.[4]

Candidacy for this procedure is based on clinical history, physical examination, and physiologic evaluation. A bowel habit history with regard to the stool consistency, the number of bowel movements in a 24-hour period, symptoms of urgency, the ability to defer defecation, staining of the undergarments, and any loss of control of bowel movements is obtained. In women, an obstetric history should be obtained with specific questions regarding episiotomies, tears, and assisted delivery with suction or forceps. A surgical history should be inquired concentrating on surgical procedures for fissures, hemorrhoids, fistulas, and tumors. Any history of pelvic irradiation or success with previous therapies is important.

The physical examination begins with inspection of the anorectal region. Important clinical findings include appearance of the anus at rest (open versus closed), visible contraction with squeeze, and the presence of any scars. Digital rectal examination provides additional information with particular regard to resting (IAS) and squeeze pressures (EAS). The thickness of the rectovaginal septum and any associated sphincter defects should be noted. Sensation is evaluated by touching the perianal skin with a sharp object. Anoscopic and endoscopic evaluation often complete the examination.

The incontinence score guides both the intensity of the investigation and any subsequent therapeutic decisions because its takes into account the severity, type, and degree of incontinence with consideration of its effect on lifestyle. Conservative measures, including medical intervention and biofeedback, are reserved for patients with low incontinence scores or who are poor surgical candidates. Only moderately or severely incontinent patients whose lifestyles are significantly affected should be offered surgical intervention. Physiologic evaluation is important in not only determining candidacy for these procedures, but also in assigning prognosis. The typical physiologic evaluation involves endoanal ultrasound, anal manometry, electromyography, and pudendal nerve terminal motor latency (PNTML) testing. Discovery of isolated anterior sphincter defects warrants surgical correction in appropriate patients before implementation of other techniques.

Initial physiologic evaluation of patients suspected of having an isolated sphincter defect involves endoanal ultrasound (EAUS). At our institution, EAUS is performed in the office and gives structural information about the integrity of the IAS, EAS, and puborectalis as well as the presence and location of any defects or scars (see Chapter 3-5). This study will ultimately help determine the type of surgery from which the patient may most benefit. If the EAUS reveals a single defect in the IAS and EAS or the EAS alone, the patient may be a candidate for an overlapping sphincteroplasty. If the ultrasound reveals multifocal defects or an intact muscle, the patient may benefit from either biofeedback or alternative surgical intervention such as sacral nerve stimulation or sphincter reconstruction or replacement. Anal magnetic resonance imaging provides similar information when performed with an anal coil, but this procedure is more expensive and less comfortable for the patient.

Discovery of an isolated IAS defect poses a challenging problem because it is not the sole cause of incontinence in symptomatic patients. Repair of isolated defects in the IAS is controversial because studies have not clearly shown benefit in incontinent patients without EAS damage. However, some surgeons may repair the defect in conjunction with overlapping repair of the EAS, although this method has not yielded better results than overlapping the EAS alone. An isolated defect in the IAS should be followed by evaluation for other causes of incontinence and for rectoanal intussusception (cinedefecography) with concomitant pudendal neuropathy (PNTML). An IAS defect alone without intussusception or pudendal neuropathy is usually treated with biofeedback therapy. However, newer therapies such as submucosal injection of carbon-coated beads and the administration of endoanal radiofrequency energy may be useful in these cases.

Further physiologic testing includes anal manometry and anal electromyography. Anal manometry is the least useful in determining the etiology of the incontinence; however, it provides important information about pressures, compliance, sensation, and reflexes of the anorectal region (see Chapter 3-4). Anal manometry confirms any sphincter damage by decreased resting (function of the IAS) and/or squeeze (function of the EAS) pressures. Manometry will also identify other abnormalities that may not be corrected by sphincter repair such as decreased rectal compliance (inflammatory bowel disease, radiation proctitis), intrinsic nerve impairment/loss in the form of absent rectoanal inhibitory reflex (adult Hirschsprung's, Chagas' disease), or decreased anal or rectal sensation (aging). These abnormalities may be improved with nonoperative therapy in the form of biofeedback with sensory retraining or other forms of surgery (proctectomy, stoma) in addition to or instead of sphincter repair.

Electromyography (EMG) records action potentials derived from motor units within a contracting muscle. It allows functional correlation of the EAUS in a quadrant-by-quadrant manner. Before EAUS, EMG was used for "mapping" sphincter defects using needle electrodes. An anterior defect of the anal sphincter on EAUS should have an appropriate decrease in the level of muscle function on EMG in the anterior quadrant. Electromyography will also provide information about the functional quality of the intact sphincter in the remaining three muscle quadrants (posterior, left, and right). Increased fiber density, or polyphasia, and/or the absence of action potentials is suggestive of either neural damage or significant muscular degeneration. This inadequate neuromuscular function may argue against a successful outcome after overlapping sphincteroplasty. Overlapping sphincteroplasty may still be recommended as the initial surgical therapy in this situation after thorough counseling. However, these patients may require additional therapy in the form of biofeedback, sacral nerve stimulation, or neosphincter procedures if an inadequate result is obtained.

To complete the physiologic evaluation of the patient with incontinence, measurement of the PNTML is routinely performed. Damage to the EAS may be accompanied by injury to the pudendal nerve that manifests as prolonged pudendal nerve latency (see Chapter 3-6). A PNTML of more than 2.2 milliseconds is considered abnormal at our institution. Pudendal neuropathy corresponds on EMG to a decreased or absent action potentials or polyphasia. Pudendal nerve damage may contribute to incontinence after obstetric trauma and is found to be a poor prognostic factor after overlapping sphincteroplasty. However, prolonged PNTML may not preclude sphincteroplasty after extensive counseling.[5] A waiting period of 6 to 12 months may allow time to recover after a birthing injury; however, if the injury is remote, recovery is unlikely.

At our institution, the technique of overlapping sphincteroplasty for the treatment of isolated anterior EAS defects is generally performed without the use of a stoma. Preoperative management includes full mechanical bowel preparation (45 mL of Fleets Phospho-Soda per os at 4:00 PM and at 9:00 PM, each followed by three 8-oz glasses of water) and oral antibiotics (1 g of neomycin and 500 mg of metronidazole at 1:00, 2:00, and 11:00 PM) the day before surgery. In addition, 2 g of cefotaxime and 1 g of metronidazole are administered intravenously as well as 5000 units of heparin subcutaneously at the start of the operation. The patient is positioned in the prone jackknife position on a Kraske roll with PAS stockings. The procedure is performed under general anesthesia although regional (caudal or spinal) anesthesia can be substituted. After sterile preparation of the perianal area, vagina, and perineum, a circumanal and bilateral nerve block is achieved with a mixture of 0.25% Marcaine (bupivacaine) and 0.5% Xylocaine with 1:400,000 units of epinephrine. A 120-degree curvilinear incision is made anterior to the anus, approximately 0.5 cm distal and parallel to the anal verge. The anterior portion of the sphincter muscle is identified and dissected with the index finger of the nondominant hand inserted into the vagina. This maneuver aids in the prevention of inadvertent vaginal wall injury. In addition, care must be taken not to extend the dissection through the mucosa into the anal canal or rectum, particularly in areas where the sphincter muscle is disrupted or absent or the rectovaginal septum is extremely thin. Such an injury predisposes to the development of an anovaginal or rectovaginal fistula. Dissection of the EAS is initiated laterally where the muscle anatomy is usually intact and can aid the surgeon in identifying the proper plane of dissection. Additionally, care must be taken not to injure the pudendal nerve bundles that enter the EAS bilaterally in the posterolateral positions. The intersphincteric space is dissected from lateral to medial on each side and the internal and external sphincters are separated (Figure 6-8.2). If scar tissue is connecting the two ends of the disrupted external sphincter, the scar is divided in the midline, but not excised. It is important to preserve all scar tissue in order to better anchor the sutures. If the ends are separated, the scar is identified and preserved and the muscle is mobilized.

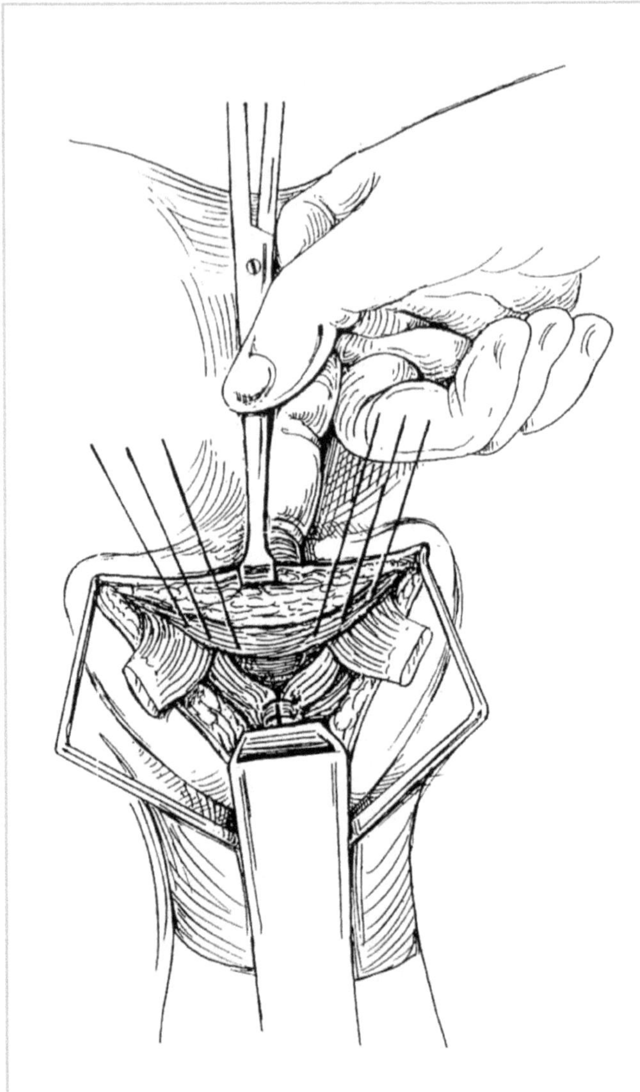

Figure 6-8.2. Dissection of the external anal sphincter and division of the anterior scar. (Reprinted from Wexner SD, Beck DE. Fundamentals of Anorectal Surgery. 2nd ed. p 133–136, Copyright 2001 WB Saunders, with permission from Elsevier.)

The repair starts with the imbrication of the levator muscles, found just beneath the two ends of divided scar tissue, using interrupted 2-0 polypropylene sutures. The IAS is plicated with interrupted 2-0 polydioxane acid sutures (Figure 6-8.3). These sutures are placed far enough laterally for a snug repair that is then verified by inserting an index finger into the anal canal. At this point in the procedure, the ends of the external sphincter muscle should overlap without significant tension and are secured with interrupted 2-0 polydioxane mattress sutures, using the scar tissues to provide a significant portion of the suture fixation (Figure 6-8.4). All retractors and buttocks tapes are removed before tying of the sutures of the overlapped ends to avoid a lax repair. The sutures are tied snugly but not so tight as to induce muscular ischemia. The wound is closed from lateral to medial on each side using 3-0 polyglactin sutures through the skin, leaving the central-most portion of the wound open for drainage.

Postoperatively, the patient receives 2 g of cefotaxime every 12 hours and 500 mg of metronidazole every 6 hours for 2 days, followed by oral ciprofloxacin and metronidazole for 7 days. Ambulation is encouraged and long periods of sitting are discouraged. The wound is not packed, although patients may need a pad to protect clothing. Bowel confinement has not been used since completion of a prospective, randomized trial at our institution in which no advantage was found over starting a regular diet.[6]

Short-term outcome of overlapping sphincteroplasty yields extremely good short-term functional results with an overall reported early success rate of 71% to 86% during a mean follow-up of 10 to 29 months.[2] At our institution, 55 women who underwent overlapping sphincteroplasty were analyzed over a period of 29 months.[5] More than 95% had a history of previous vaginal delivery and 55% had had a previous sphincter repair. The outcome was excellent in 71% and fair in the remaining. Several important conclusions were gained from this study. A significant increase in mean squeeze pressure and high-pressure zone length correlated with better functional outcome. In other words, restoring the integrity of the sphincter muscle resulted in improved sphincter function and associated decrease in incontinence. Age was not predictive of success because this procedure was as effective in women older than age 60 as those younger than 60. Previous sphincter repairs did not predict failure with future overlapping repairs because repeat repairs were shown to demonstrate significant improvements in continence in those with residual anterior sphincter damage after a previous sphincter surgery. The major factor found to be associated with poor outcome was prolonged PNTML, a finding that has been separately confirmed in several other studies. A review of factors predictive of good outcome in overlapping sphincteroplasty revealed that the optimal conditions for this type of repair are no previous repair, preservation of scar during dissection, bilaterally intact pudendal nerves, normal rectal sensation, and an asthenic young patient with an isolated anterior EAS defect.[2] However, long-term results are less

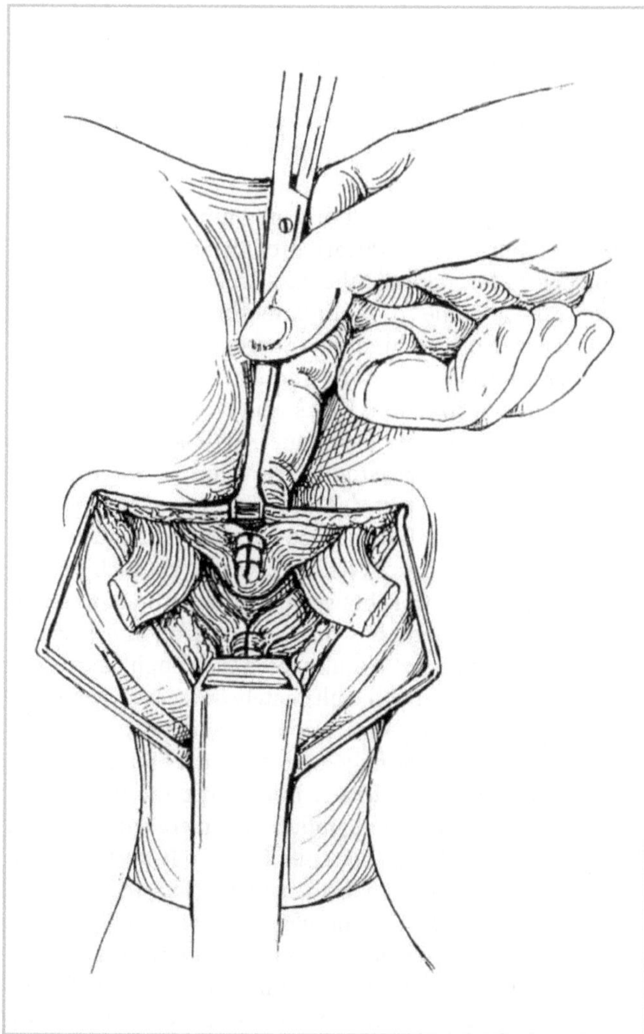

Figure 6-8.3. Apposition of the levator and anterior plication of the internal anal sphincter. (Reprinted from Wexner SD, Beck DE. Fundamentals of Anorectal Surgery. 2nd ed. p 133–136, Copyright 2001 WB Saunders, with permission from Elsevier.)

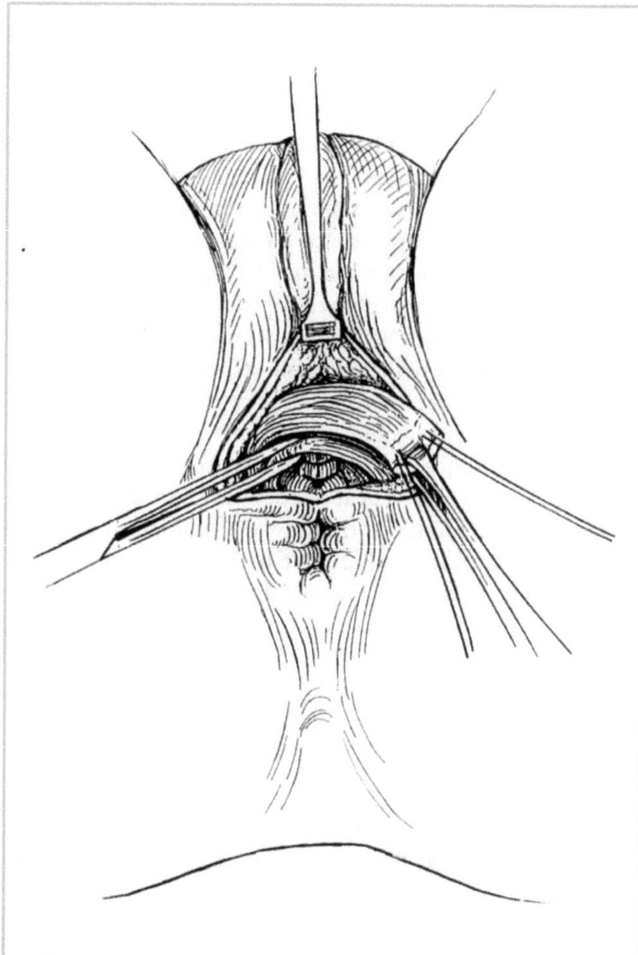

Figure 6-8.4. Overlapping repair. (Reprinted from Wexner SD, Beck DE. Fundamentals of Anorectal Surgery. 2nd ed. p 133–136, Copyright 2001 WB Saunders, with permission from Elsevier.)

gratifying, which is one of the reasons why other procedures may have a role.

As many as 60% of patients present with recurrent symptoms of fecal incontinence.[7] Early or late failure of overlapping sphincteroplasty warrants repeat physiologic evaluation to determine the integrity of the repair. Identification of a persistent anterior sphincter defect should lead to contemplation of a repeat repair. In a study performed at our institution on 36 women who had a history of overlapping sphincteroplasty, patients had the same significant improvement as women with no history of repair.[7] In addition, the outcome seemed similar regardless of the number of previous repairs, although no definitive conclusion could be drawn with regard to this matter because only five patients had more than one previous repair. However, repeat overlapping sphincteroplasty should be considered the treatment of choice in patients with fecal incontinence that have had one or more previous sphincter repairs in whom residual anterior sphincter damage is found. However, a functional failure despite anatomic success may be better treated with biofeedback therapy, postanal or total pelvic floor repair, stimulated graciloplasty, artificial bowel sphincter, sacral nerve stimulation, or may necessitate a permanent stoma.

Anterior Plication, Postanal Repair, and Total Pelvic Floor Repair

Anterior, postanal, and total pelvic floor repairs are not frequently performed and are no longer promoted as first line procedures in patients with incontinence. When the sphincter mechanism itself is anatomically intact but laxity in the pelvic floor or damage to pelvic innervation is implicated as the cause of fecal incontinence, the resulting dysfunction has been referred to as perineal descent syndrome, idiopathic incontinence, or neurogenic incontinence. If the symptoms are minimal, conservative treatment in the form of dietary modification or medical intervention is warranted. However, if symptoms are severe or the patient has pudendal neuropathy after an unsuccessful trial of sphincteroplasty, the postanal or total pelvic floor repair, sacral nerve stimulation, a neosphincter procedure, or a stoma all become options. Although postanal sphincter repairs have limited success, they demonstrate a low morbidity and remain a valuable option in patients with neurogenic incontinence for whom conservative measures have failed, are not candidates for sacral nerve stimulation (because of exclusion criteria), or neosphincter procedures (stimulated graciloplasty or artificial bowel sphincter), and do not want a stoma.

Anterior plication (also called anterior reefing procedure) involves plication of the anterior perineal musculature. The procedure has been described with multiple variations but usually encompasses reapproximation of the anterior puborectalis with or without simultaneous approximation of the levators and S-shaped plication of the EAS. The theoretical outcome of this technique is a tightening of the anterior perineal musculature and, when employed, is used in combination with other procedures. In the early twentieth century, this technique was used as an alternative to reconstruction of thin or damaged sphincters. Although some studies have shown good initial results with this procedure, the long-term benefits significantly decrease over time. Consequently, this procedure is rarely used today in the current treatment of fecal incontinence.

Postanal repair (also called posterior sphincteroplasty or posterior plication) was devised by Parks et al.[8] for the treatment of idiopathic and neurogenic incontinence. The goal of the procedure was to increase the length of the anal canal with the intent of restoring the anorectal angle. At our institution, after induction of general anesthesia, the patient is placed in the prone jackknife position on a Kraske roll and a curvilinear incision is made 5 cm posterior to the anus.[9] An intersphincteric approach between the internal and EAS is used to expose Waldeyer's fascia and the levator ani muscle complex. Interrupted 2-0 polypropylene sutures are then placed from one side of the pelvis to the other through the two limbs of the iliococcygeus muscle. Because of the intervening distance, the two halves of the iliococcygeus muscle cannot be approximated, rather a lattice of suture is formed that supports the posterior rectal wall. Interrupted sutures are then placed in the pubococcygeus and puborectalis muscles in a similar manner (Figure 6-8.5). Plication of the EAS may also be performed before closing the skin in a V-Y manner. The theoretical outcome of approximating the levator ani muscles posteriorly behind the anorectal junction is to displace these muscles anteriorly and increase the angulation of the anorectal junction. Parks achieved the best results (>80% success) for this technique in his series. A review of several subsequent studies revealed a success rate of between 32% to 87%.[2] Although this operation was designed to improve the anorectal angle and lengthen the anal canal, anatomic analysis has revealed no such results. In addition, correlation between successful outcome and reduction of this angle has not been proven and remains theoretical. Long-term results of this repair are poor because successful outcome ultimately decreases over time. The only predictor of adverse outcome is pudendal neuropathy, which, interestingly, is present in a majority of the patients considered candidates to undergo the repair.

Postanal repair was more popular in an era when routine EAUS was not available. Currently, it is rarely performed in the United States, but may be considered in select patients with no sphincter defect who have failed biofeedback and conservative measures or in whom an overlapping sphincter repair has achieved anatomic but not functional success. A study of this technique at our institution involved the analysis of results from 20 postanal repairs with a follow-up of 23 months.[10] The success rate was 35% as indicated by a significant improvement in the inconti-

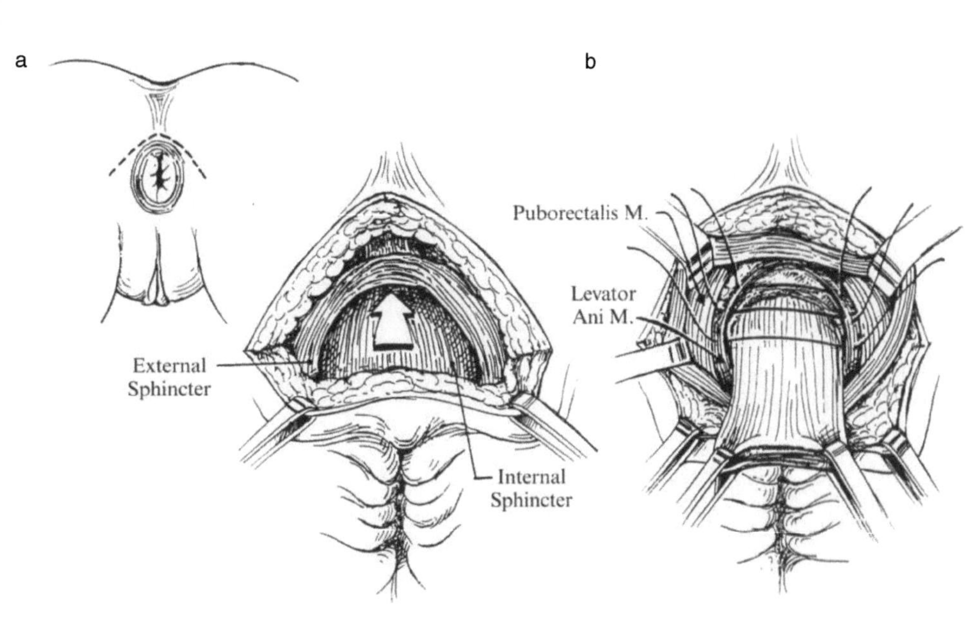

Figure 6-8.5. Postanal repair with dissection in the intersphincteric space (a) and posterior plication of the muscles of the pelvic floor (b). (Reprinted from Wexner SD, Beck DE. Fundamentals of Anorectal Surgery. 2nd ed. p 133–136, Copyright 2001 WB Saunders, with permission from Elsevier.)

nence score in 22 months of follow-up. The morbidity rate was minimal (5%) with no mortality. No clinical or physiologic variables were found to be predictors of success. Although the success rate of this operation is poor, it is relatively free of morbidity and remains an important option for those patients who may have few alternatives other than high-risk complicated operations (neosphincter procedures) or stomas. Nonetheless, it is rarely used by most surgeons because of its poor long-term success.

Total pelvic floor reconstruction refers to a combination of postanal repair, anterior levatorplasty, and can include sphincteroplasty. Indications include complicated injuries to the EAS and puborectalis in one or more sites with associated pudendal neuropathy. This approach was designed to provide a comprehensive repair, which also reinforces any perineal deficiency such as rectocele. It can be performed in stages or in one step with or without colostomy. This procedure also purportedly increases the angle of the anal canal and corrects the degree of pelvic floor descent at rest and during straining, but does not influence resting or squeeze anal canal pressures. As with postanal repair, studies have shown that clinical improvement did not correlate with change in the anorectal angle. Short-term results are encouraging, although long-term success is rare. This repair has not gained widespread popularity.

Conclusion

Fecal incontinence secondary to sphincter damage is best managed by overlapping sphincteroplasty. The most successful outcomes are found in patients with isolated sphincter defects without evidence of pudendal neuropathy. Multifocal defects and unilateral or bilateral pudendal neuropathy argue for poor outcome. These patients may be better served with postanal or total pelvic floor repair, sacral nerve stimulation, or sphincter augmentation or replacement before embarking on a permanent stoma. All types of surgical intervention for fecal incontinence should simultaneously address other disorders of the pelvic floor and urinary incontinence.

References

1. Venkatesh KS, Ramanujam PS, Larson DM, Haywood MA. Anorectal complications of vaginal delivery. Dis Colon Rectum 1989;32: 1039–1041.
2. Baig MK, Wexner SD. Factors predictive of outcome after surgery for fecal incontinence. Br J Surg 2000;87(10):1316–1330.
3. Jacobs PP, Scheuer M, Kuijpers JHC, Vingerhoets MH. Obstetric fecal incontinence. Role of pelvic floor denervation and results of delayed sphincter repair. Dis Colon Rectum 1990;33:494–497.
4. Wexner SD, Machetti F, Jagelman D. The role of sphincteroplasty for fecal incontinence reevaluated: a prospective physiologic and functional review. Dis Colon Rectum 1991;34:22–30.
5. Olivera L, Pfeifer J, Wexner SD. Physiological and clinical outcome of anterior sphincteroplasty. Br J Surg 1996;83:502–505.
6. Nessim A, Wexner SD, Agachan F, et al. Is bowel confinement necessary after anorectal reconstructive surgery? A prospective randomized surgeon-blinded trial. Dis Colon Rectum 1999;42(1):16–23.
7. Giordano P, Renzi A, Effron J, et al. Previous sphincter repair does not affect outcome of repeat repair. Dis Colon Rectum 2002;45(5): 635–640.
8. Parks AG, Porter NH, Hardcastle J. The syndrome of descending perineum. Proc R Soc Med 1996;59:477–482.
9. Schmitt SL, Wexner SD. Anterior and posterior sphincter repair. Semin Colon Rectal Surg 1992;3(2):92–97.
10. Matsuoka H, Mavrantonis C, Wexner SD, Oliviera L, Gilliland R, Pikarsky A. Postanal repair for fecal incontinence – is it worthwhile? Dis Colon Rectum 2001;43(11):1561–1567.

6-9

Fecal Diversion

Benjamin Person, James Doty, and Steven D. Wexner

Causes of Fecal Incontinence

Fecal continence is maintained by the integrated action of the external and internal anal sphincters, the levator ani musculature, and by intact neural pathways. It depends on the consistency of the stool, the capacity of the rectum, the anorectal sampling reflex, and normal resting anal tone. If any of these mechanisms malfunction, incontinence may result.[1] There are other numerous causes of fecal incontinence as outlined below.

Normal pelvic floor
- Diarrheal states (infectious diarrhea, inflammatory bowel disease, short gut syndrome, laxative abuse, radiation enteritis)
  - Overflow (impaction, rectal tumor)
  - Neurologic conditions (congenital anomalies, multiple sclerosis, dementia, strokes, neuropathy, injuries to brain, spinal cord)

Abnormal pelvic floor
- Congenital anorectal malformation
- Trauma (accidental injury, anorectal surgery, obstetric injury)
- Pelvic floor denervation (vaginal delivery, chronic straining at stool, rectal prolapse, descending perineum syndrome)

Definitions and Epidemiology

Fecal incontinence is defined as the impaired ability to control gas or stool to allow evacuation at a socially acceptable time and place. It has been shown in a number of studies that incontinence affects self-confidence and personal image, and can lead to social isolation causing a significant impact on quality of life.[1–5] The true prevalence of fecal incontinence is unknown because it is thought to be greatly underreported. In a community survey of 7000 patients in the United States, 2.2% of patients reported incontinence to liquid or solid stool or gas.[2] In other population studies, the prevalence of some form of incontinence has been reported to approach 60%, particularly higher in women than men (4 to 1) because of the trauma of vaginal delivery.[1,6] Fecal incontinence is the second leading cause of nursing home placement, and up to 45% of patients in nursing homes are estimated to have some fecal incontinence.[6]

There are many different methods to measure the severity of fecal incontinence; the Cleveland Clinic Florida Fecal Incontinence (CCFFI) score is the most popular and widely cited.[7] This scale is a measure of frequency of incontinence to gas, liquid, and solid stool, alteration in lifestyle, and use of a protective pad. In one of several validation studies of this scoring system, it was shown that a CCFFI score greater than 9 is consistent with a significant alteration in quality of life and can be used as an indication for surgical therapy.[4] This score is equivalent to losing stool more than once a week, needing a diaper, and feeling significantly restricted in daily functioning.[4]

Besides being a social burden, fecal incontinence places a significant financial burden on United States health care. In a study of 63 patients, treatment charges were $559,341 over an average of 10 years.[8]

Treatment

Treatment of fecal incontinence is divided into medical therapy and surgical therapy. In general, medical therapy should be pursued first because there are less risks involved. However, the risks and benefits of treatments must be weighed for each patient based on their degree and etiology of incontinence. For example, a patient with a documented sphincter defect with only occasional incontinence may not want surgery and may benefit from biofeedback and dietary modification. Alternatively, a patient with moderate incontinence and a sphincter defect who is willing to accept the risks of surgery may benefit from an overlapping sphincter repair rather than being chronically dependent on antidiarrheal medications.

The mainstay of medical therapy is altering the stool consistency through dietary changes and antidiarrheal medications. Concurrently or subsequently, one can pursue a course of biofeedback, because it is relatively noninvasive and has essentially no side effects. There has been some variation in the results obtained with biofeedback for fecal incontinence depending on severity and etiology. In a study of 25 patients with more than five episodes of fecal incontinence per day, biofeedback improved sphincter endurance and sphincter contraction. Forty-four percent became completely continent and 48% had significant improvement in their incontinence.[9]

Surgical Therapy

The key to deciding what kind of surgical therapy should be used depends on whether or not the anal sphincters are intact. This evaluation and treatment are discussed elsewhere in this book in greater detail. Simply stated, those patients in whom surgical therapy is planned who have sphincter defects should have an overlapping sphincter repair. For those patients with intact sphincters, other interventions such as submucosal injection of bulking agents – Durasphere™ (formerly known as ACYST™; Boston Scientific Corp., Boston MA), radiofrequency energy delivery to the anal canal – the Secca® procedure (Curon Medical, Sunnyvale, CA) (Chapter 6-10, Figure 6-10.7, a and b), sacral nerve stimulation (Chapter 6-10, Figure 6-10.9), stimulated or nonstimulated unilateral or bilateral graciloplasty (Chapter 6-10, Figures 6-10.1 and 6-10.2), or artificial bowel sphincter implantation (Chapter 6-10, Figure 6-10.4, a and b) may be considered.

Finally, one must consider the role of fecal diversion in the management of fecal incontinence. Fecal diversion can be used as a temporary or permanent procedure, or as primary or secondary therapy. In a select group of patients, permanent fecal diversion may be the best therapeutic option. For those patients with severe incontinence and altered mobility or mental status, a permanent stoma may significantly improve quality of life.[10] In patients with spinal cord injury, ostomies have been shown to improve the quality of life and decrease the amount of weekly time spent on bowel care from 10.3 to 1.9 hours on average .[11] Other patients in whom permanent fecal diversion may be considered are those with unreparable defects, patients who have failed other attempts at surgery, or those in whom other forms of surgery are undesirable or too risky. Fecal diversion may be the best choice for elderly, debilitated, and institutionalized patients or those who do not wish to have other procedures.[10] In appropriately selected cases, ostomies are easier to care for and may in some patients be preferable to a life of restricted social activities, especially because enterostomal therapy has advanced so significantly.[12]

The other major role for fecal diversion in the management of incontinence is that of a temporary stoma. Patients who have had multiple or complex repairs, severe trauma, and/or perineal contamination (perineal sepsis), associated pelvic injuries, or Crohn's disease may benefit from diversion before repair.[13] The decision to divert is dependent on the type of surgery used to restore continence, the patient's comorbidities, and the quality of the tissue used to affect the repair. For example, studies have shown that routine diversion in the setting of overlapping sphincteroplasty is not necessary.[1,14] The goal of diversion is to allow the tissues to heal with less risk of sepsis, and to close the stoma at a later date. Closure of the stoma is individualized based on the healing process and is generally performed within 3 to 6 months after the restorative procedure. Experience at Cleveland Clinic Florida has shown that laparoscopic creation of colostomies and ileostomies can be performed safely and with a more rapid recovery than if performed by laparotomy.[15] Creation of a stoma, although usually considered a relatively simple procedure, is not entirely free of complications. The overall complication rate varies between 25% and 75%[16–20] and is usually divided into early complications (less than a month after the procedure) and late complications. Early complications include skin irritation, ischemia, and partial necrosis of the stoma, retraction, dehydration caused by high output (mainly in ileostomies), and bleeding. Late complications occur more than a month after the procedure and may include parastomal hernias, stenosis, and prolapse as well. Obesity and inflammatory bowel diseases predispose to complications whereas enterostomal nursing care may help prevent them.[17] There is no consensus as to which is the best method to divert the fecal stream. Some authors prefer colostomies[19] whereas others demonstrated the superiority of ileostomies.[18] The method of diversion that is used most often in the Cleveland Clinic Florida is a loop ileostomy, preferably performed laparoscopically.[21]

Conclusion

The role of fecal diversion in the management of fecal incontinence is diverse and therefore must be individualized based on the severity of the problem, the nature of the surgery, and the patient's condition and desires.

References

1. Mavrantonis C, Wexner SD. A clinical approach to fecal incontinence. J Clin Gastroenterol 1998;27(2):108–121.
2. Nelson R, Norton N, Cautley E, Furner S. Community based prevalence of anal incontinence. JAMA 1995;274(7):559–561.
3. Rockwood TH, Church JM, Fleshman JW, et al. Fecal Incontinence Quality of Life Scale: quality of life instrument for patients with fecal incontinence. Dis Colon Rectum 2000;43(1):9–16.
4. Rothbarth J, Bemelman WA, Meijerink WJ, et al. What is the impact of fecal incontinence on quality of life? Dis Colon Rectum 2001; 44(1):67–71.
5. Sangwan YP, Coller JA. Fecal incontinence. Surg Clin North Am 1994; 74(6):1377–1398.

6. Whitehead WE, Wald A, Norton NJ. Treatment options for fecal incontinence. Dis Colon Rectum 2001;44(1):131–142.
7. Jorge JM, Wexner SD. Etiology and management of fecal incontinence. Dis Colon Rectum 1993;36(1):77–97.
8. Mellgren A, Jensen LL, Zetterstrom JP. Long term cost of fecal incontinence secondary to obstetric injuries. Dis Colon Rectum 1999; 42(7):857–865.
9. Ko CY, Tong J, Lehman RE, Shelton AA, Schrock TR, Welton ML. Biofeedback is effective therapy for fecal incontinence and constipation. Arch Surg 1997;132(8):829–833.
10. Rotholtz N, Wexner SD. Surgical treatment of constipation and fecal incontinence. Gastroenterol Clin North Am 2001;30(1):131–166.
11. Branagan G, Tromans A, Finnis D. Effect of stoma formation on bowel care and quality of life in patients with spinal cord injury. Spinal Cord 2003;41(12):680–683.
12. Wexner SD, Schmitt SL. Fecal incontinence: surgical therapy. In: Brubaker L, Sacclarides T, eds. The Female Pelvic Floor: A Multidisciplinary Approach. Philadelphia: FA Davis; 1996:220.
13. Baig MK, Wexner SD. Factors predictive of outcome after surgery for faecal incontinence. Br J Surg 2000;87:1316–1330.
14. Hasegawa H, Yoshioka K, Keighley MR. Randomized trial of fecal diversion for sphincter repair. Dis Colon Rectum 2000;43(7):961–964.
15. Oliveira L, Reissman P, Nogueras J, Wexner SD. Laparoscopic creation of stomas. Surg Endosc 1997;11(1):19–23.
16. Bagi P, Jendresen M, Kirkegaard P. Early local stoma complications in relation to the applied suture material: comparison between monofilament and multifilament sutures. Dis Colon Rectum 1992; 35(8):739–742.
17. Duchesne JC, Wang YZ, Weintraub SL, Boyle M, Hunt JP. Stoma complications: a multivariate analysis. Am Surg 2002;68(11):961–966.
18. Edwards DP, Leppington-Clarke A, Sexton R, Heald RJ, Moran BJ. Stoma-related complications are more frequent after transverse colostomy than loop ileostomy: a prospective randomized clinical trial. Br J Surg 2001;88(3):360–363.
19. Park JJ, Del Pino A, Orsay CP, et al. Stoma complications: the Cook County Hospital experience. Dis Colon Rectum 1999;42(12):1575–1580.
20. Shellito PC. Complications of abdominal stoma surgery. Dis Colon Rectum 1998;41(12):1562–1572.
21. Wexner SD, Taranow DA, Johansen OB, et al. Loop ileostomy is a safe option for fecal diversion. Dis Colon Rectum 1993;36(4):349–354.

6-10

ACYST, Secca, Sacral Nerve Stimulation, Artificial Bowel Sphincter, and Stimulated Graciloplasty

Susan M. Cera and Eric G. Weiss

Patients with severe fecal incontinence unresponsive to conservative measures can be divided into two broad categories of surgical approach. The first group includes those patients with an identifiable anatomic sphincter defect who can expect a 70% to 80% surgical success rate with overlapping sphincteroplasty.[1] The absence of pudendal neuropathy yields the best outcome in these patients.[2] The second group includes those patients with extensive sphincter damage, muscle loss, or pudendal neuropathy not amenable to direct sphincter repair. Fortunately, because of the development of newer surgical techniques, the patient with severe fecal incontinence is not obligated to a permanent stoma.

In the first half of the century, patients who were not candidates for sphincter repair underwent sphincter reconstruction with muscle transpositions involving either the gluteus maximus or gracilis. These techniques only met with moderate success because the static, striated muscle flaps were prone to fatigue with chronic contraction. The transposed muscle did not have any involuntary tone at rest, and patients had to perform awkward movements to achieve imperfect continence.

In the 1980s, external stimulators were applied to the muscle transpositions to create dynamic neosphincters with resting muscle tone. The low-frequency electrical stimulation provided by these stimulators transforms the skeletal muscle from fast-twitch fatigue-prone (type II) muscle fibers to slow-twitch fatigue-resistant (type I) muscle fibers. However, the procedure involves many components and requires technical expertise predisposing to an array of complications that proved to undermine the advantages. Since its introduction in the 1990s, the Artificial Bowel Sphincter™ (ABS) offers an alternative, safe, and convenient option in sphincter reconstruction. In addition, the development of other techniques, including injection of submucosal beads (ACYST™), radiofrequency (the Secca® procedure), and sacral nerve stimulation (SNS), serve to augment the native continence mechanism in those who do not require neosphincter construction.

Stimulated Graciloplasty

The technique of stimulated graciloplasty involves the transposition of the gracilis muscle from the thigh to form a skeletal muscular ring around the anus with the distal portion anchored to the contralateral ischial tuberosity (Figure 6-10.1). Two phases are used in this procedure with the number of required operations dependent on the use of an optional stoma. Phase I consists of transposition of the muscle and implantation of the stimulator and the electrodes (Figures 6-10.2 and 6-10.3). Phase II involves 8 weeks of muscle conditioning with increasing levels of neuromuscular stimulation. The use of a diverting stoma requires additional operative intervention for creation and closure. Upon completion of phase II, the patient is able to control continence with the use of an external magnet. The patient can switch the neurostimulator on, causing the muscle to contract, and off, causing the muscle to relax.

The Dynamic Graciloplasty Therapy Study Group (DGTSG) has been instrumental in providing the largest prospective, multicenter data with regard to outcome of this procedure. The initial report on the efficacy revealed improvement in continence and quality of life in the majority (60%) of patients.[3] Long-term efficacy reported in a separate study revealed a 62% success rate and improvements in functional and quality-of-life variables that persisted during a 2-year duration.[4]

Although many studies proved the efficacy of the stimulated graciloplasty, concern arose about the high rates of complications and need for reoperation. In the original study by the DGTSG, the complication and reoperative rates were 74% and 40%, respectively.[3] Other studies also revealed high rates of infection, hardware failure, and postoperative evacuatory dysfunction. In a single institution

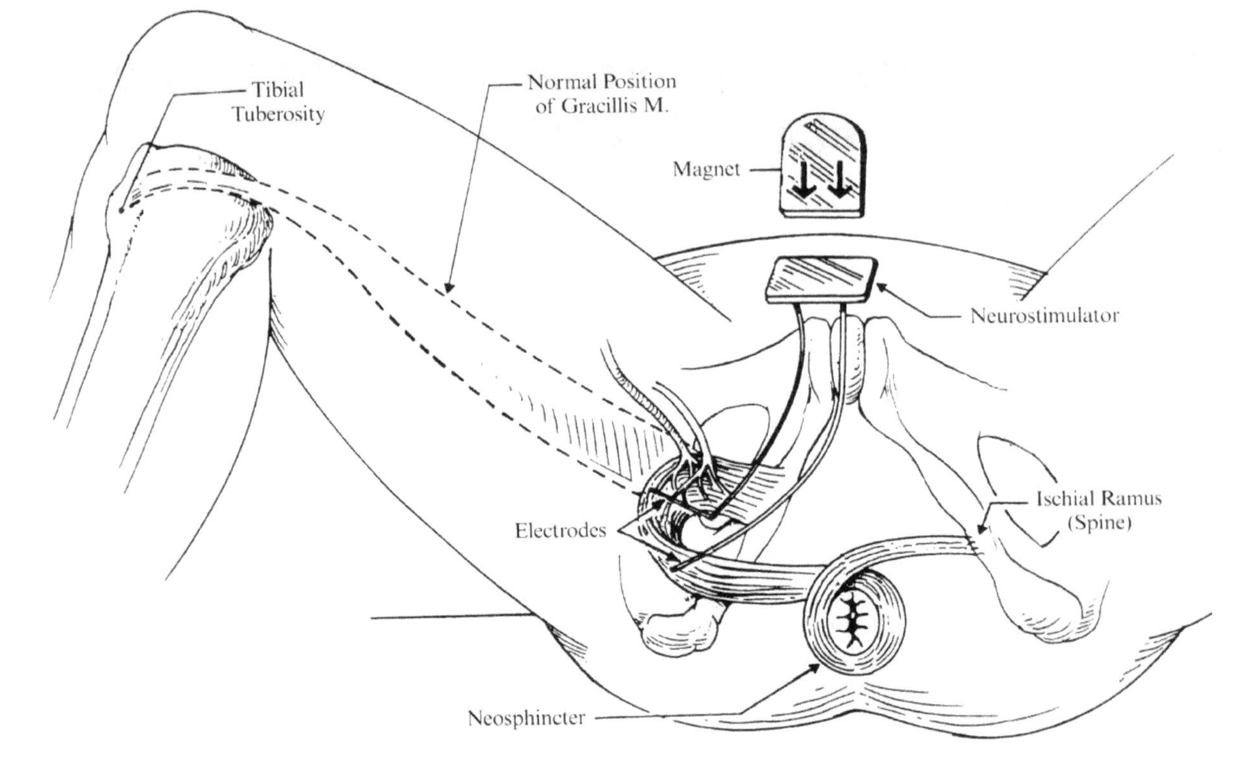

Figure 6-10.1. Schematic of the stimulated graciloplasty with lead implants. (Reprinted from Fundamentals of Anorectal Surgery, 2nd ed, DE Beck, SD Wexner, page142, © 1996, with permission from Elsevier)

study performed at Cleveland Clinic Florida, complications included lead fibrosis, seroma of the thigh incision, excoriation of the skin above the stimulator, fecal impaction, anal fissure, parastomal hernia, rotation of the stimulator, premature battery discharge, fracture of the lead, perineal skin irritation, perineal sepsis, rupture of the tendon, tendon erosion, muscle fatigue during programming sessions, electrode displacement from the nerve, and fibrosis around

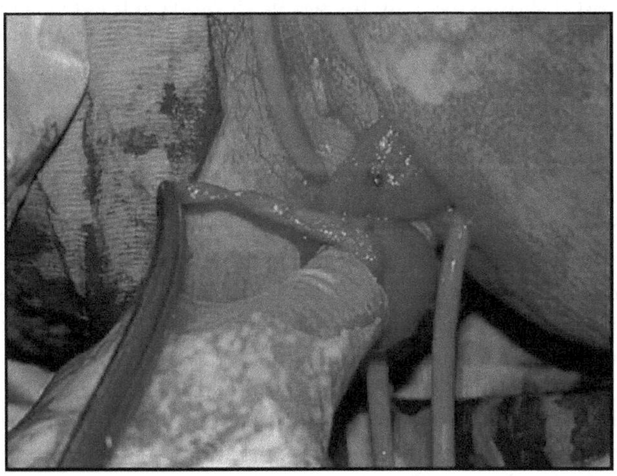

Figure 6-10.2. Intraoperative view of the gracilis transposition for sphincter reconstruction.

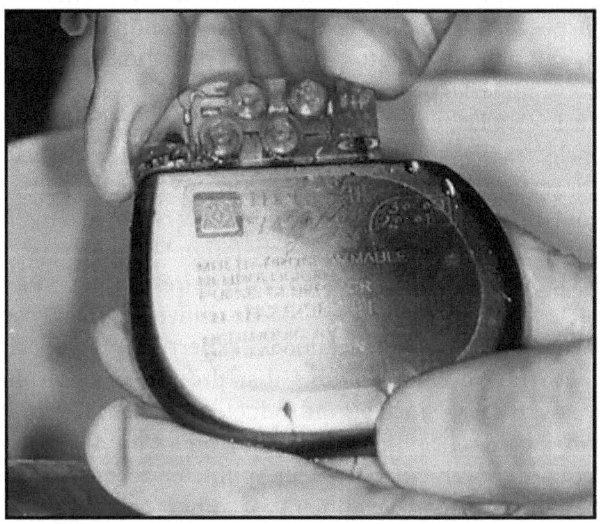

Figure 6-10.3. Stimulator for the stimulated gracilis sphincter reconstruction.

the nerve.[5] Some of these complications led to stoma creation or death. Consequently, the DGTSG investigated the etiology and impact of these complications.[6] In this report, 211 complications occur in 93 cases of dynamic graciloplasty. Forty-two percent had severe complications, although recovery was achieved in 92%. Of all the complications, only major infections adversely affected outcome, leading the authors to conclude that although the complication rate was high, most of the complications were treatable and did not adversely affect outcome. Because of the complexity of the procedure and the high rate of complications, the stimulator device has been removed form the United States market, although it remains a viable option in other countries. Currently in the United States, gracilis transpositions are being used to augment sphincter mass before placement of the ABS in those patients with significant muscle loss from trauma or from congenital atresia.[7]

Artificial Bowel Sphincter™

Augmentation of the sphincter with a prosthetic device was first reported for fecal incontinence in 1992 after the idea was borrowed from urology, where artificial sphincters are used for urinary incontinence.[8] The current device used for fecal incontinence, the Acticon™ NeoSphincter (American Medical Systems, Minnetonka, MN), consists of three silastic components: an inflatable cuff, a pressure-regulating balloon, and a control pump that allows activation or deactivation of the device (Figure 6-10.4). The inflatable cuff is implanted around the anus and is connected by silastic tubing to the control pump placed in the scrotum of males or in the labium major of females. The control pump is also connected to the pressure-regulating balloon implanted in the space of Retzius. When activated, the cuff is distended and the anus is occluded. The pressure-regulating balloon maintains the cuff pressure. To defecate, the patient compresses the control pump several times, and the fluid is displaced out of the cuff and into the regulating balloon.

The artificial sphincter is placed with the patient under general anesthesia in the lithotomy position after having undergone a mechanical and antibiotic bowel preparation and rectal irrigation with Betadine solution. Attention is given to prevention of contact of the silastic components with lint and powder because these materials tend to easily adhere with potential for contribution to infection. Either through an anterior perineal incision or bilateral perianal incisions, blunt dissection is used to create the circumferential tunnel around the anal canal several centimeters deep in the ischiorectal fossa. The occlusive cuff is appropriately sized and placed with the connection tubing on the same side as the patent's dominant hand. A suprapubic incision is made and the pressure-regulating balloon placed in the space of Retzius. Blunt dissection creates a dependent pouch in the scrotum or labia into which the control pump is placed. The tubings are connected but the device is left deactivated for the first 6 weeks postoperatively.

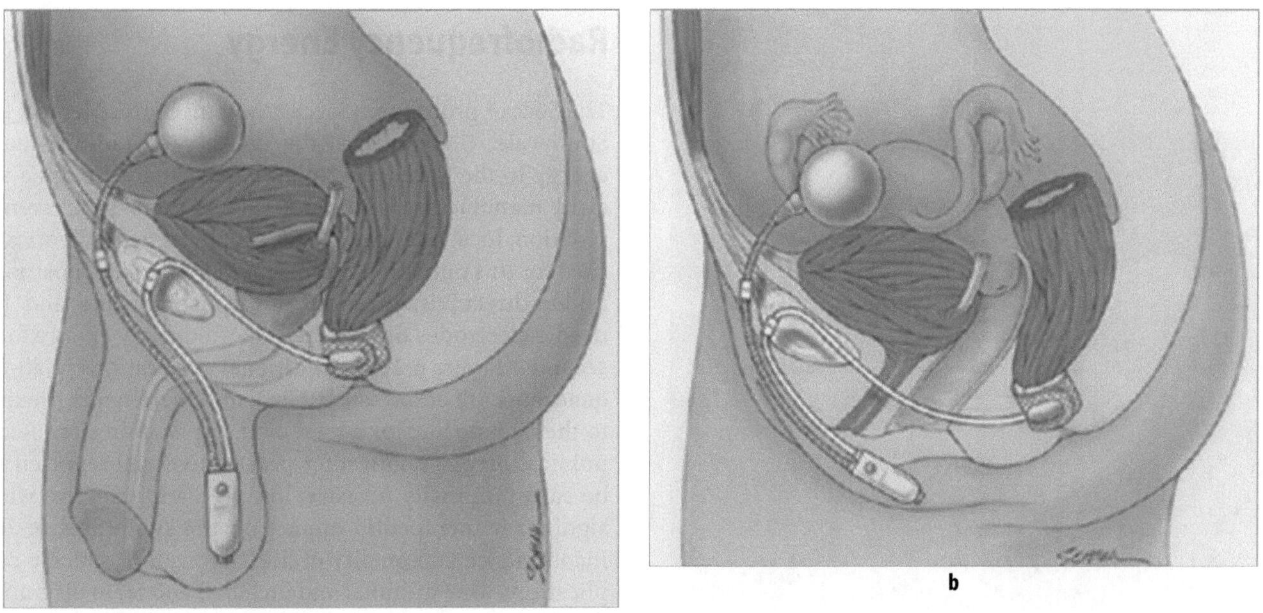

a b

Figure 6-10.4. Schematic of the Artificial Bowel Sphincter™ in male (**a**) and female (**b**).

The Acticon™ NeoSphincter received Food and Drug Administration (FDA) approval in 1999. A recent, multicenter, nonrandomized trial revealed the device to have a significant rate of clinical success (85%), enhancement in quality of life, and a high degree of safety.[9] However, similar to the stimulated graciloplasty, limitations in the use of this technique are related to the high rate of complications, most of which are related to infections of the foreign material with subsequent need for surgical revision. Other complications are related to erosion of the components into adjacent structures or device malfunction with a device explant rate of 36%.[9] The cost involved and the morbidity from this device and the stimulated graciloplasty are approximately the same. In a recent prospective comparison of eight cases of dynamic graciloplasty and eight implantations of the ABS followed over 3 years, there was no difference in complications, wound healing problems, or explantation rates, although the ABS was found to be more effective in lowering the fecal incontinence score.[10] Nonetheless, this remains an important alternative for patients with end-stage fecal incontinence when no other surgical or medical options exist except stoma. At this point, long-term studies are still needed to determine the longevity of the device.

Carbon-coated Bead Injection

Injection of carbon-coated beads (ACYST™ or Durasphere FI®; Carbon Medical Technologies, St. Paul, MN) into the submucosa of the upper anal canal and lower rectum serves to provide additional bulk in the anal canal (Figures 6-10.5 and 6-10.6). This bulk provides resistance to the passage of stool and allows for improved sensation and discrimination. It is an outpatient office procedure that is well tolerated. A recent prospective, open-label trial study performed at our institution on 20 patients revealed symptomatic improvement of the fecal incontinence in 75% of cases at 2-year follow-up.[11] This procedure has proven to be a safe, simple, inexpensive, and effective technique in the treatment of moderate to severe fecal incontinence. Currently, a prospective, randomized trial further evaluating this method is beginning.

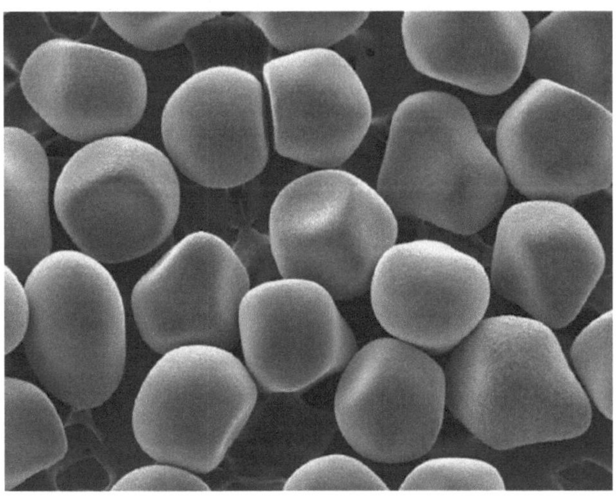

Figure 6-10.5. Microscopic picture of the carbon-coated beads in a gel suspension. The beads are 212 to 500 μg in size.

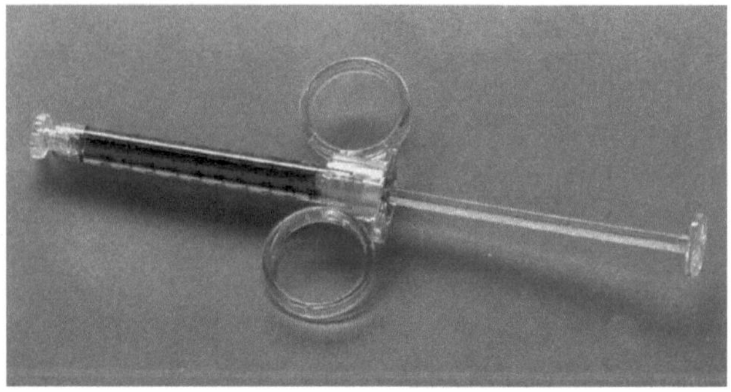

Figure 6-10.6. The beads are injected via a 1-mL syringe.

Radiofrequency Energy

The Secca® procedure (Secca® System; Curon Medical Inc., Sunnyvale, CA) involves the delivery of radiofrequency energy to the lower rectum and anal canal through a specially manufactured anoscope (Figure 6-10.7). Intravenous sedation, local anesthesia, and prophylactic antibiotics are used for this outpatient procedure. The device is positioned under direct visualization at the dentate line and four needle electrodes deliver the radiofrequency energy for 90 seconds (Figure 6-10.8). Additional applications in all four quadrants are administered in 5-mm increments proximal to the dentate line for a total of 16 application sites. Recent publication of a multicenter, prospective trial reported it to be safe, minimally invasive, low risk, and effective with a significant therapeutic impact on the symptoms of fecal incontinence and quality of life.[12] Procedure-specific complications were minimal and included anoderm ulcerations and bleeding. A prospective, randomized, blinded, sham-controlled trial is underway to further study the procedure. The device is not currently FDA approved except for use in clinical trials.

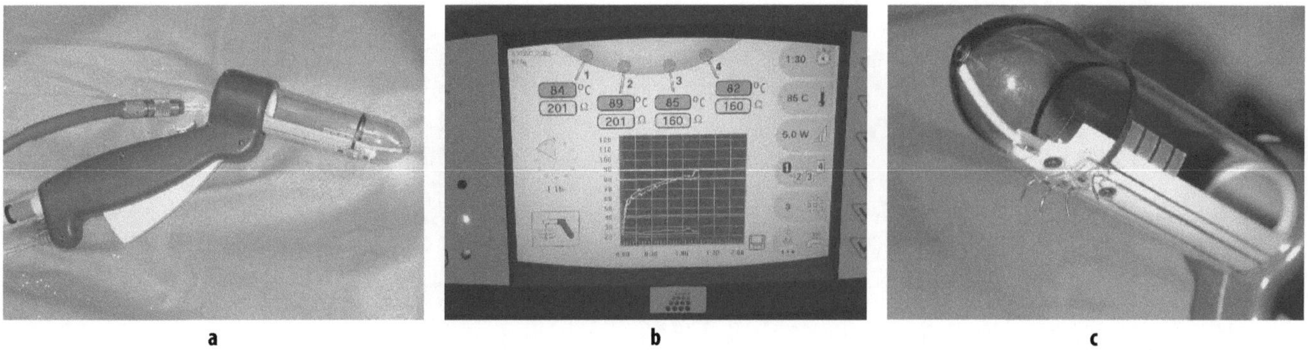

Figure 6-10.7. Anoscope (a) and radiofrequency (b) device of the Secca® system. The anoscope has four electrodes that deliver the radiofrequency energy (c).

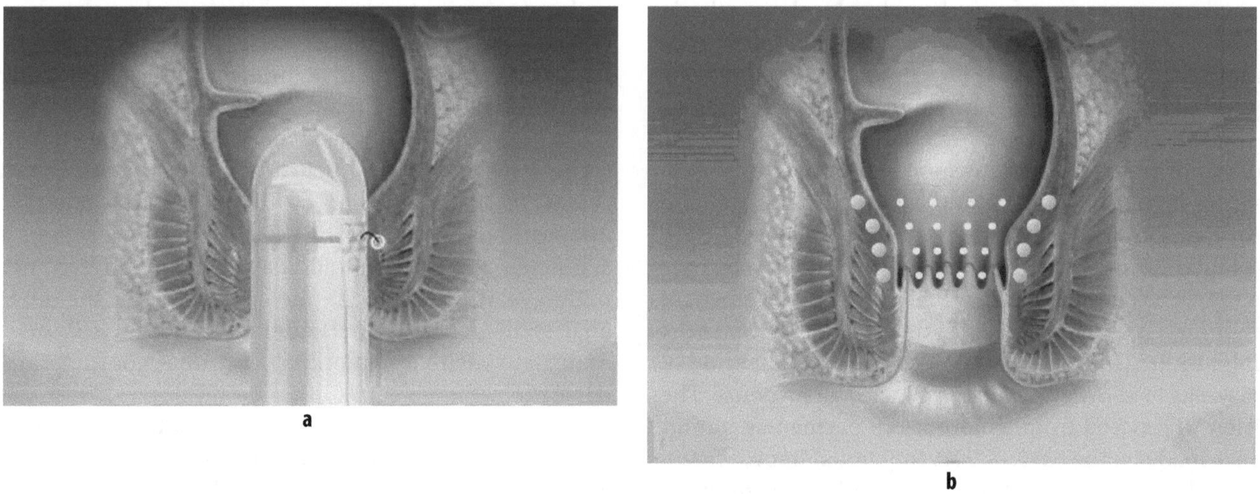

Figure 6-10.8. Probe is inserted into anal canal (**a**). Needles are deployed and radiofrequency delivered and four to five levels of radiofrequency delivered in the four quadrants at each level (**b**).

Sacral Nerve Stimulation

Sacral nerve stimulation is the most widely published "new" technique for the restoration of fecal incontinence (Figure 6-10.9). Once again, the technology has been adapted from urology techniques where SNS is widely accepted in the current treatment of urinary incontinence (see Chapter 7-2). Patients with bladder dysfunction and concomitant fecal incontinence were noted to have improvement in both symptoms with external stimulation to the pelvic neuroplexus. It was first reported for treatment of fecal incontinence in 1995 by Matzel et al.[13] Since then, it has been performed in several hundred patients in Europe. In the United States, it is currently under investigation including at our institution.

To be a candidate, a patient must have an intact sphincter without substantial defects or loss of muscle, reduced or absent sphincter function (by anal manometry), and intact residual reflex function (confirmed by pudendal stimulation) demonstrating an intact nerve–muscle connection. Performance of this technique consists of two stages: peripheral nerve evaluation (PNE) (the diagnostic stage) and the permanent implant (the therapeutic stage). Peripheral nerve evaluation of the sacral roots (S2, S3, S4) is divided into two phases: an acute phase to test the functional integrity of each spinal nerve to striated anal sphincter function and a chronic phase to assess the therapeutic potential of sacral spinal nerve stimulation in individual patients. For PNE, the patient is placed in the prone position using either local or general anesthesia and the sacral

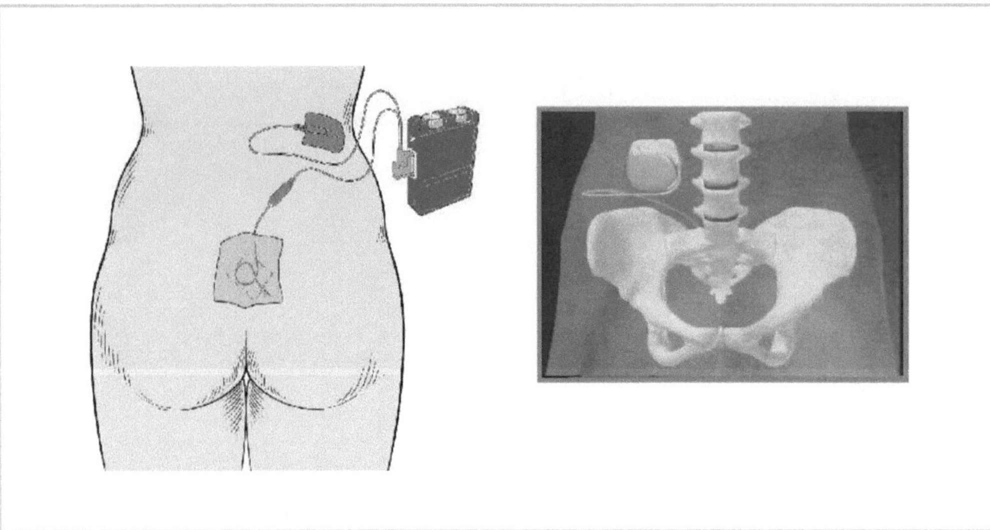

Figure 6-10.9. Sacral nerve stimulation.

foramina are located using bony landmarks. The acute phase test is performed under local anesthesia using a 20-gauge spinal insulated needle (Medtronic Inc., Minneapolis, MN) attached to an external neurostimulator (Medtronic Inc.). The needle is placed in the sacral foramina, and an electrical current is gradually applied to the needle until a visual muscle response is obtained. Muscle responses include movement of the external sphincter and lateral rotation of the leg (S2), contraction of pelvic floor and plantar flexion of the big toe (S3), or contraction of the anus (S4). The chronic phase of PNE involves placement of a temporary stimulator lead into the same position as the testing needle. This lead is left in place for a trial period of 1 to 2 weeks to allow evaluation of functional response. The decision to proceed from temporary to permanent stimulation is made on the basis of 50% functional improvement in either the number of episodes or incontinence-free days. For placement of the permanent stimulator, the patient is placed in the prone position. The previous scar in the upper buttocks area is opened and the temporary stimulator removed. A permanent stimulator is placed in a subcutaneous pocket and the wound is reclosed. Perioperative antibiotics are continued for 24 hours for each of the procedures.

In the patients that have undergone this procedure since 1995, all but one have had functional benefit and these results have remained consistent over a course of follow-up of up to 8 years.[14–19] In addition, SNS clearly improved quality of life. The complication rate was between 0% to 33% and consisted of pain at the site of the pulse generator, electrode migration, and infection.

Patients with fecal incontinence from a wide variety of causes have been treated successfully including: deficits resulting from anal or rectal procedures, those with obstetric or neurologic trauma, scleroderma, systemic sclerosis, primary internal anal sphincter degeneration, and even sphincter disruption. The exact mechanism of action is unknown; however, the sacral nerve roots are the most proximal point of the combined dual nerve supply, somatic and autonomic, to the pelvic floor and anal sphincter mechanism. Stimulation of these sacral nerves augments the neural input to a native mechanical apparatus consisting of the muscular architecture, its associated system of neural connections involved in function of striated muscle, sacral reflexes, and the intrinsic nervous system. The effect seems to be the result of direct efferent stimulation and contraction of the muscles of the pelvic floor and sphincter and from its modulation of the afferent neural pathways is involved in the activation of the internal anal sphincter, rectal relaxation, and sacral reflexes that regulate sensitivity, motility, and the coordination of defecation.

Sacral nerve stimulation has many potential advantages over sphincter repairs, reconstruction, and replacement. The main advantage is that it is minimally invasive because it involves placement of electrodes at a proximal source of the nerve supply with no manipulation of the rectum, anus, or pelvic floor. Consequently, it has a very low complication rate, and the need for discontinuation of treatment is rare. Revisions, repeat surgeries, and removal of the apparatus do not necessarily obligate the patient to a stoma because the stimulation device can be reimplanted again if temporary removal is necessary. Other advantages include the ability to perform the temporary stimulation phase as a screening method for appropriate patients before permanent electrode placement, the absence of required bowel preparation, the performance of procedures in an outpatient setting, lack of decline in efficacy over an 8-year period, and use in a variety of causes of fecal incontinence. Despite the fact that the exact mechanism of action remains to be elucidated, satisfying clinical results have been achieved with this technique. It is an exciting treatment option in a population in whom conservative measures have failed and traditional surgical approaches are conceptually questionable, have limited success, or are considered too high risk.

Choosing a Technique

The recent development of multiple surgical techniques for fecal incontinence offers physicians and patients a variety of options for treatment. Choosing the appropriate therapy is based on the patient risk factors, etiology of the incontinence, procedure-specific contraindications, and associated pelvic floor deficits.

Patient risk factors for surgery have an important role in choosing a procedure. The least invasive is the ACYST™ procedure, which can be performed during an office visit and is ideal for those who are poor surgical candidates or refuse more invasive techniques. The Secca® and SNS procedures are also minimally invasive but require local anesthesia, complicated equipment, and operating room monitoring for performance. The ACYST™, Secca®, and ABS demonstrate very low risk of complications and do not preclude the subsequent use of other techniques that can be used in the case of their failure. The ABS is an invasive procedure reserved for those in whom other techniques have failed and requires a motivated and otherwise healthy patient that is physically fit for possible multiple surgical revisions. It has a high rate of complications, is usually reserved as a last resort, and failure of this technique usually obligates that patient to a stoma.

Another patient factor that needs consideration before choice of therapy is mental capacity with respect to the more complicated techniques. The SNS, stimulated graciloplasty, and ABS procedures require patients with the cognitive ability to understand the technique, incorporate it into daily life, maintain the components, and understand potential complications and failure. In addition, those who undergo the stimulated graciloplasty and ABS should be prepared for possible repeat surgical interventions. Those with psychiatric conditions or emotional instability may not be suitable for these high-maintenance procedures and are probably better candidates for either the ACYST™ or Secca® procedure, which do not require routine maintenance or follow-up.

Etiology is important to the consideration of the type of therapy chosen. The ACYST™ and Secca® procedures do not attempt to correct the underlying etiology of the incontinence. Instead, these procedures involve augmenting the resistance of fecal passage by bulking or tightening the mucosa of the anal canal and rectum. Consequently, these techniques can be used in all forms of fecal incontinence, but are more effective in milder forms. Sacral nerve stimulation has also been used in a variety of causes of fecal incontinence but requires adequate neuromuscular architecture demonstrated on electromyography and pudendal nerve testing. Patients with severe neuropathy (absent bilateral pudendal motor latencies) or significant deficits/loss of sphincter muscle may not be amenable to SNS and are better candidates for sphincter replacement with the ABS.

Procedure-specific contraindications relate to the safety in performing the procedures in certain situations or in the placing of the artificial components. The presence of associated anorectal pathology precludes choice of the ACYST™, Secca®, and ABS procedures because they all involve direct manipulation of the anus and rectum. Perianal disease (i.e., fistulas, fissures, abscesses, inflammatory bowel disease, perianal infections, anorectal carcinoma) and local radiation are contraindications to these procedures because of the inherently high risk of infectious complications. In contrast, contraindications to SNS include sacral diseases, such as spina bifida or sacral agenesis, cauda equina syndrome, or skin pathology at the site of electrode placement. Patients with a cardiac pacemaker or an implantable defibrillator cannot undergo SNS or the stimulated graciloplasty because of the obvious interference of the electrical stimulators. Lastly, some patients with fecal incontinence have associated pelvic floor dysfunction in the form of concomitant urinary incontinence. These patients may benefit most from SNS because it has shown to improve both of these impairments as opposed to the placement of separate urinary and bowel sphincters.

Multiple options are now available for patients with fecal incontinence without a sphincter defect and for whom conservative measures have failed. Each offers advantages and disadvantages that can be used to tailor specific treatments to individual needs.

References

1. Oliveira L, Wexner SD. Anal incontinence. In: Beck DE, Wexner SD, eds. Fundamentals of Anorectal Surgery. 2nd ed. Philadelphia: WB Saunders; 1998:136.
2. Wexner SD, Marchetti F, Jagelman D. The role of sphincteroplasty for fecal incontinence reevaluated: a prospective physiologic and functional review. Dis Colon Rectum 1991;34:22–30.
3. Baeten CG, Bailey HR, Bakka A, et al. Safety and efficacy of dynamic graciloplasty for fecal incontinence: report of a prospective, multicenter trial. Dynamic Graciloplasty Therapy Study group. Dis Colon Rectum 2000;43:743–751.
4. Wexner SD, Baeten CM, Bailey R, et al. Long-term efficacy of dynamic graciloplasty for fecal incontinence. Dis Colon Rectum 2002;45:809–818.
5. Wexner SD, Gonzalez-Padron A, Rius J, et al. Stimulated gracilis neosphincter operation. Initial experience, pitfalls, and complications. Dis Colon Rectum 1996; 39:957–964.
6. Matzel KE, Madoff RD, LaFontaine LJ, et al. Complications of dynamic graciloplasty: incidence, management, and impact on outcome. Dis Colon Rectum 2001;44:1427–1435.
7. da Silva GM, Jorge JM, Belin B, et al. New surgical options for fecal incontinence in patients with imperforate anus. Dis Colon Rectum 2004;47:204–209.
8. Christiansen J, Sparso B. Treatment of anal incontinence by implantable prosthetic anal sphincter. Ann Surg 1992;215:383–386.
9. Wong WD, Congliosi SM, Spencer MP, et al. The safety and efficacy of the artificial bowel sphincter for fecal incontinence: results from a multicenter cohort study. Dis Colon Rectum 2002;45:1139–1153.
10. Ortiz H, Armendariz P, DeMiguel M, Solana A, Alos R, Roig JV. Prospective study of artificial anal sphincter and dynamic graciloplasty for severe anal incontinence. Int J Colorect Dis 2003;18: 349–354.
11. Weiss EG, Efron JE, Nogueras JJ, Wexner SD. Submucosal injection of carbon coated beads is a successful and safe office-based treatment for fecal incontinence [abstract]. Dis Colon Rectum 2002;45: A46–A47.

12. Efron JE, Corman ML, Fleshman J, et al. Safety and effectiveness of temperature-controlled radiofrequency energy delivery to the anal canal (Secca procedure) for the treatment of fecal incontinence. Dis Colon Rectum 2003;46:1606–1618.
13. Matzel KE, Stadelmaier U, Hohenfellner M, et al. Electrical stimulation of the sacral spinal nerves for treatment of fecal incontinence. Lancet 1995;346:1124–1127.
14. Ganio E, Luc AR, Clerico G, Trompetto M. Sacral nerve stimulation for treatment of fecal incontinence: a novel approach for intractable fecal incontinence. Dis Colon Rectum 2001;44:619–631.
15. Leroi AM, Michot F, Grise P, Denis P. Effect of sacral nerve stimulation in patients with fecal and urinary incontinence. Dis Colon Rectum 2001;44:779–789.
16. Malouf AJ, Vaizey CJ, Nicholls RJ, Kamm MA. Permanent sacral nerve stimulation for fecal incontinence. Ann Surg 2000;232:143–148.
17. Matzel KE, Stadelmaier U, Hohenfellner, Gall FP. Chronic sacral spinal nerve stimulation for fecal incontinence: Long term results with foramen and cuff electrode. Dis Colon Rectum 2001;4:59–66.
18. Rosen H, Urbarz C, Holzer B, Novi G, Schiessel R. Sacral nerve stimulation as a treatment for neurogenic and idiopathic fecal incontinence. Gastroenterology 2001;121:536–541.
19. Vaizey CJ, Kamm MA, Turner IC, Nicholls RJ, Woloszko J. Effects of short-term sacral nerve stimulation on anal and rectal function in patients with anal incontinence. Gut 1999;44:407–412.

6-11

Case Presentation: Stress Urinary Incontinence/Fecal Incontinence

Daniel H. Biller and G. Willy Davila

Chief Complaint

"I leak urine and stool."

History and Physical Examination

The patient is a 40-year-old G2P1 woman with a 4-year history of stress urinary incontinence (SUI) and fecal incontinence (FI) that began with the vaginal delivery of her now 4-year-old daughter complicated by a fourth degree laceration. Since that time, she has had progressively worsening FI to both liquid and solid stools, as well as an inability to control flatus. Her urinary incontinence is stress related with no significant urgency, frequency, enuresis, or nocturia. She voids with a normal flow. In addition, she complains of dyspareunia and vaginal dryness. Her symptoms have caused significant quality-of-life impairment.

Medical History

1. Vaginal delivery × 1
   a. 5 pounds 14 ounces
   b. Fourth degree laceration
2. Low back pain

Social History

1. Tobacco – 1/2 pack per day
2. Social alcohol consumption

Medication

None

Physical Examination

Pelvic Examination

Normal external genitalia. Urethra, bladder, and vagina were normal. First degree cervical prolapse with good vaginal vault support.

Second degree midline cystocele associated with second degree enterocele and third degree rectocele. Vaginal caliber was normal to slightly enlarged.

Bimanual examination was normal with no uterine enlargement and no adnexal masses. There was mild attenuation of the perineal body.

Reflexes and sensation were normal suggesting neuromuscular integrity. Total vaginal length was 9.2 cm.

Rectal Examination

Thin rectovaginal septum with a probable anteriorly based anal sphincter defect. Decreased resting pressures and no squeeze pressures were identified. Sensation to sharp stimuli in the perianal area was normal.

Work-up

- Urinalysis: negative
- Urine culture: negative

Endoanal Ultrasound

- Defect of both internal and external anal sphincter muscles anteriorly at the level of the mid anal canal with a thin perineal body (Figure 6-11.1).

Urodynamics

1. Postvoid residual was 75 mL and a negative urinalysis (Figure 6-11.2).

2. Cystometrogram: Capacity of 479 mL. First sensation at 83 mL and first desire at 98 mL. She did not experience any urgency throughout the cystometrogram. No uninhibited detrusor contractions. She did demonstrate leakage with cough, prompt urinary loss at capacity.

3. Urethral pressure profiles: Functional urethral length was 3.1 cm. Maximal urethral closing pressure was 79 cm H_2O at capacity.

4. Leak point pressure testing: There was no leakage at 150 mL with a pressure of 83 cm H_2O. There was also no leakage at capacity to 86 cm H_2O pressure.

5. Uroflowmetry: The patient had an interrupted pattern voiding 700 mL over 106 seconds giving a mean flow of 10.8 mL/s with a peak flow of 19.4 mL/s. The configuration was interrupted with a Valsalva pattern.

6. Cough pressure profile: Positive with equalization (Figure 6-11.2).

7. Q-tip angle was positive with a resting angle of 35 to 80 degrees with Valsalva.

8. Electromyography was appropriate.

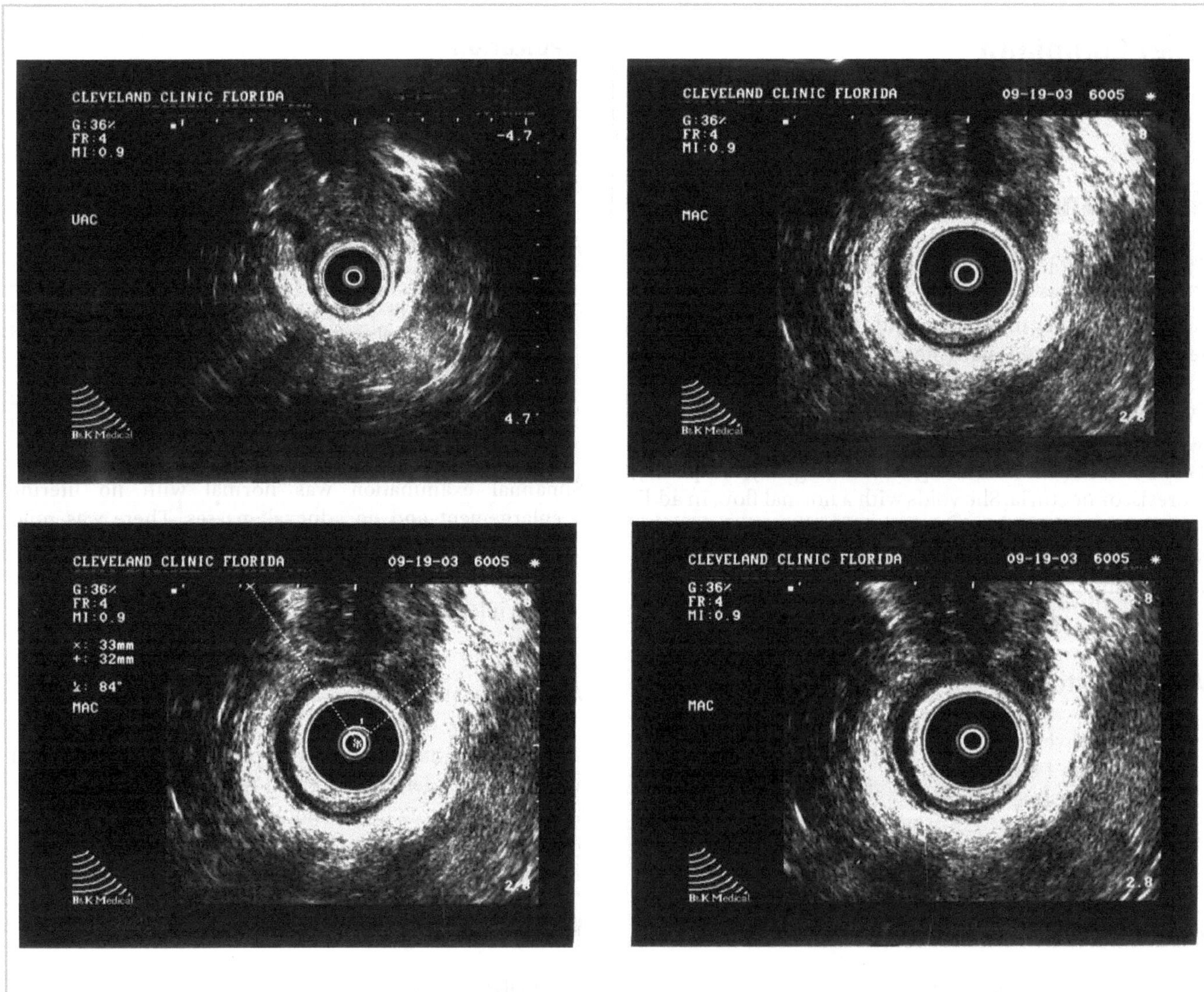

Figure 6-11.1. Endoanal ultrasounds.

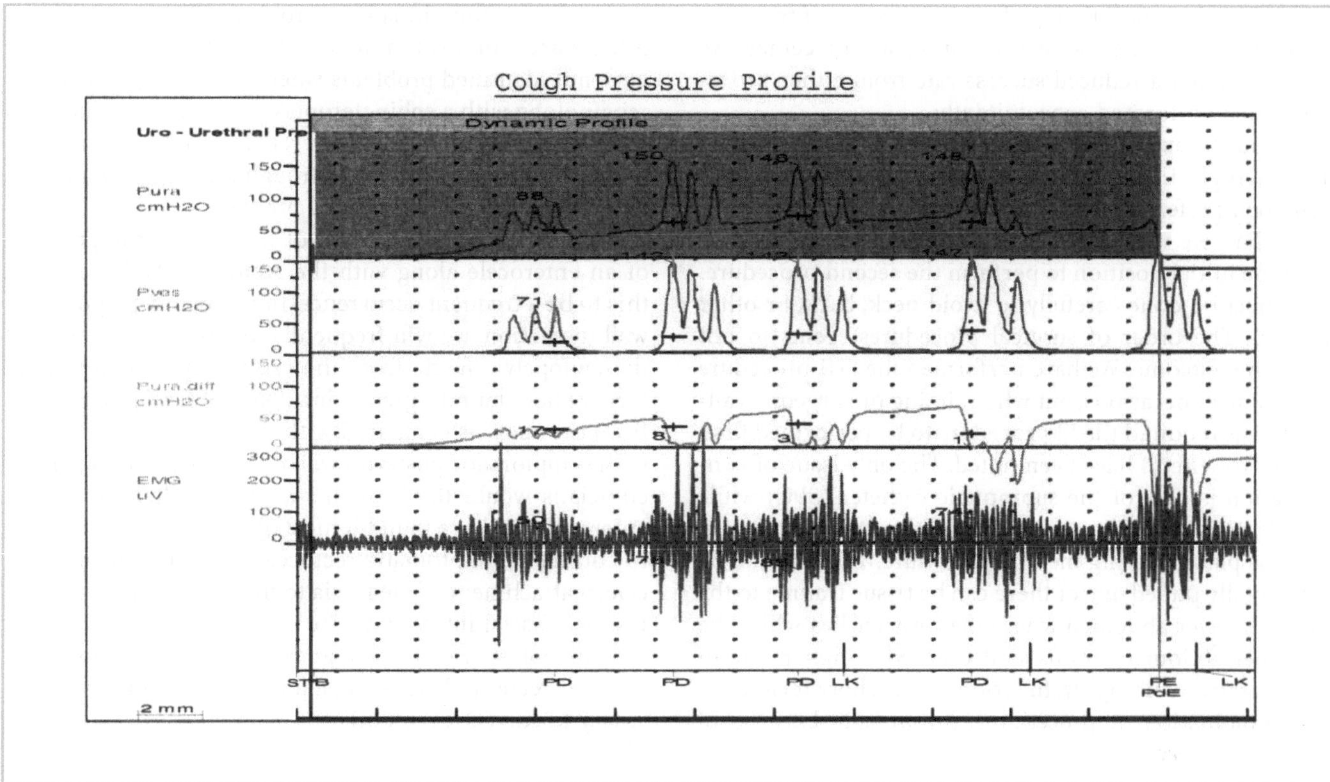

Figure 6-11.2. Urodynamics.

Assessment

Complex pelvic floor dysfunction including:

1. Stress urinary incontinence with urethral hypermobility
2. Vaginal prolapse with primarily posterior wall relaxation
3. Fecal incontinence with evident anal sphincteric defect

Concomitant Procedures Performed

1. Transobturator suburethral sling procedure
2. Vaginal enterocele and rectocele repairs
3. Anal sphincteroplasty

Six-month Follow-up

The patient said she feels "greatly improved." She reports one episode of urine leakage with cough and several episodes of fecal soilage during the first 2 months of recovery. Since, she reports rare incontinence of flatus, and no incontinence of urine or liquid or solid stool.

Commentary

This case represents a common clinical scenario of post-obstetric pelvic floor damage. Coexistent urinary and fecal incontinence caused by neuromuscular trauma to the levator and anal sphincter muscles as well as to the pudendal nerve is not uncommonly seen on a short-term basis. If the symptoms persist past 3 to 6 months postpartum, they should be evaluated and treated. Pelvic floor exercises can be instituted as initial conservative therapy, but if a sphincter muscle defect is noted on examination, FI is likely to persist. Each symptom should be comprehensively evaluated if surgery is planned. Some clinicians would suggest that multichannel urodynamics are not necessary if no evidence of intrinsic sphincter deficiency is noted, or other indications for urodynamics are not present. In this case, the patient had been performing Kegel exercises, so conservative therapy had failed. The colorectal evaluation should always include endoanal ultrasound to identify a sphincteric defect. Some clinicians would argue that a careful digital examination is sensitive enough to identify a torn sphincter muscle. However, we believe sonography will document the defect and is very helpful in localizing the extent of sphincter end separation in order to facilitate identification of the muscle ends for reapproximation.

Combined surgical therapy for SUI/FI is one of the most frequently performed joint procedures at our center. We have not noted a reduced success rate from either procedure when performed concomitantly.

Although our colorectal surgeons have performed sphincteroplasties with the patient in the lithotomy position, their preference is clear for the prone, jackknife position. Thus, we will typically need to flip the patient over from the initial position to perform the second procedure. This must be done carefully to avoid neck, back, or other damage. The order of surgical procedures seems to not affect the outcome. We have performed the SUI procedure first on most occasions, but when timing or surgeon availability has required the FI procedure to be performed first, no adverse effects have been noted. The only issue of consideration is care of the suprapubic catheter (SPC) while repositioning the patient and during the sphincter repair, if one is placed during the SUI procedure. The SPC can be accidentally pulled out, or there can be tissue trauma to the patient's lower abdomen where the catheter lies while she is on her abdomen. We have also noted a higher rate of postoperative voiding dysfunction in the short term after this combination of procedures, presumably because of perineal pain from the sphincteroplasty. Thus, we will most often place a thin suprapubic catheter in this scenario. We have not identified problems when performing a rectocele repair along with a sphincteroplasty. Care must be taken to not disrupt the anal sphincter repair sutures. Because we perform all rectocele repairs through the vaginal approach, this is typically not a technical problem.

Another noteworthy aspect of this case is the presence of an enterocele along with the rectocele. We have found this to be a frequent occurrence. In performing a posterior wall dissection, we will frequently identify a separation of the endopelvic fascia from the vaginal apex or cervix. In fact, we have found a greater incidence of fascial separation from the apex, as compared with the perineum, in women with symptomatic posterior wall relaxation. Although most clinicians would think of this anatomic alteration as an enterocele, we have found it most often during a dissection to correct a symptomatic rectocele. In correcting a rectocele, reattachment of the fascia to the vaginal cuff or cervix restores fascial integrity to the posterior vaginal wall, and a perineoplasty restores integrity to the perineal body and anterior rectal wall. As such, a levator plication is not necessary to correct a rectocele.

Section VII

Urgency/Frequency Syndromes Therapy

Section VII

Urgency/Frequency Syndromes Therapy

Gamal M. Ghoniem

Irritative diseases of the lower urinary tract share common symptoms of urinary frequency and urgency. Collectively, urgency/frequency syndromes encompass multitudes of pelvic diseases including overactive bladder (OAB). Other symptoms may help to direct investigation and help to establish a diagnosis. For example, hematuria could be a symptom of cystitis, urethral diverticulum, calculus, or malignancy. Most importantly, lower urinary tract malignancy, especially carcinoma in situ, should be considered by the treating physician. Pelvic pain can arise from any organ in the pelvis or its musculofascial support. Historically, fibroids of the uterus have been erroneously blamed for the symptoms of urgency and frequency, and it was used as an indication for hysterectomy. Similarly, constipated patients have significant lower urinary tract symptoms, and successful treatment of constipation is usually accompanied by resolution of urinary symptoms. More recently, treatments that address the pelvic floor as a unit have success for multiorgan symptoms. At Cleveland Clinic Florida, some patients with bowel and urinary symptoms experienced relief from both after neuromodulation therapy.

The new definition of OAB as urinary frequency, urgency with or without urge incontinence, has helped to define the problem, and subsequent epidemiologic data showed striking similarity of a 17% incidence of OAB in both Europe and North America. Suddenly, a new condition/disease emerged that has a high prevalence compared with chronic diseases such as asthma and arthritis. Direct QOL impact on this large population with OAB has led to feverish research, driven mostly by industry, with the introduction of compounds that are as effective as old ones with fewer side effects, such as dry mouth and constipation and subsequent negative effects on patient compliance. This was accompanied by both public and medical education increasing the awareness of the condition among health care professionals and the community and driving more patients to seek help. Dr. Davila's chapter explores the current pharmacologic therapies for OAB.

Unfortunately, the present pharmacologic therapies have their limitations. Blocking M receptors whether nonspecific or specific (M2), has reached its maximum tolerable efficacy. The search for newer compounds that block adenosine 5′-triphosphate (ATP)-regulated receptors such as P2X, K and N channel openers, and others may lead to improved efficacy. Until then, alternative therapies for refractory OAB should be tried. In such refractory cases, a minimally invasive technique is presented in my chapter on neuromodulation, and Dr. Rackley's chapter, which reviews other alternatives ranging from minimally invasive office procedures such as Botox injections to the more extensive surgical procedures of bladder augmentation with bowel. Similar irritative conditions causing bowel dysfunction and their treatment are described by Drs. Efron and Weiss.

7-1

Overactive Bladder: Pharmacologic Therapy

Daniel H. Biller and G. Willy Davila

The term *overactive bladder* (*OAB*) encompasses a wide range of irritative bladder filling and storage symptoms including the symptoms of *urinary frequency*, *urgency*, and *urge incontinence*, alone or in combination. Overactive bladder represents a health condition of increasing public and medical recognition, and because of the wide variation in reported symptoms of OAB, accurate estimates of prevalence and incidence are difficult to calculate.

The underlying pathophysiology of OAB has not been completely elucidated but is believed to be multifactorial. The voiding process is a complex mechanism that integrates the nervous system, including the brain and spinal cord, with the smooth muscle in the bladder and urethra. These circuits control the two integral functions of the bladder: storage and elimination. Theories concerning the pathogenesis of OAB include abnormal excitability or loss of inhibition at various levels of the peripheral and central nervous system and the bladder smooth muscle. Overactive bladder can develop as a result of various neurologic disorders, aging, and intrinsic detrusor abnormalities resulting in detrusor overactivity, urgency, and urge incontinence.

Dysfunction at the receptor level has traditionally received the most attention with the major pharmacologic interventions focusing on the muscarinic receptor behavior. The urinary bladder is a heterogeneous organ possessing the high density of muscarinic receptors. At least five receptor subtypes are based on molecular cloning and four receptor subtypes are based on pharmacology. M1, M2, and M3 receptor subtypes have been identified in the human bladder, noting that M2 receptors predominate, but M3 receptors mediate the cholinergic smooth muscle contraction. M2 receptors may have a role in pathologic states, because it has been postulated that the coactivation of M2 receptors enhances the response to M3 stimulation. The muscarinic receptors are widely distributed throughout the central nervous system, and clearly have a role in OAB whether in regulation of detrusor function or in modulating anticholinergic side effects. The exact mechanism of detrusor function and dysfunction remains undefined, with the most likely explanation incorporating disorders at multiple sites within the nervous system and bladder.[1]

Management of Overactive Bladder

Overactive bladder management should be individualized, because the impact on quality of life is relative to that patient. Most published treatment algorithms include a combination of behavioral modification and pharmacologic therapies.

Behavioral Therapies

Behavioral therapy includes bladder drills, bladder training, habit training, timed voiding, pelvic floor muscle training, and biofeedback. It is a means for the patient to regain cortical control over the detrusor muscle and bladder outlet. Bladder training incorporates timed voiding habits and bladder function, whereas pelvic floor training emphasizes control of the bladder outlet and increased sphincter resistance. Current approaches include combinations of voiding schedules, bladder diaries, fluid and diet management, and pelvic floor exercises.

In a prospective, randomized study comparing behavioral pelvic floor therapy to oxybutynin therapy and placebo, behavioral therapy provided an 80.7% reduction (30% cure) in incontinence episodes; it was significantly more effective than drug treatment (mean 68.5% reduction; $P = .04$), and both treatments together were more effective than placebo (mean 39.4% reduction; $P < .001$ and $P \pm .009$, respectively).[2] In a later randomized, controlled trial with a modified crossover design, it was demonstrated that combining drug and behavioral therapy in a stepped program can produce added benefit for patients with urge incontinence. In this study, single therapy resulted in a mean reduction of incontinence episodes of 57.5%, whereas addition of oxybutynin chloride to behavioral therapy reduced incontinence episodes by 88.5% ($P = .034$), and similarly, addition of pelvic floor therapy to isolated drug therapy reduced incontinence episodes from 72.7% to 84.3% ($P = .001$).[3]

Pharmacologic Therapy

The anticholinergic/antimuscarinic agents are the most commonly used drugs to treat OAB and urge incontinence (Table 7-1.1). This class of drugs remains the standard of care, and is recommended as the first line pharmacologic therapy for patients with detrusor overactivity by the Agency for Health Care Policy and Research (AHCPR) (www.ahrq.gov/clinic/uiovervw.htm). These agents suppress muscarinic receptors on the bladder smooth muscle, but the relative lack of selectivity for the bladder over other organ systems accounts for significant adverse side effects such as xerostomia, constipation, and blurred vision. These adverse effects result in a high rate of discontinuation that has led to a search for receptor subtype-specific antimuscarinic agents with improved tolerability profiles. When reduction of urge incontinence episodes is used as the primary outcome, anticholinergics have been shown to be significantly efficacious in 50% to 80% of cases.

As with all pharmacologic therapy, medication should be initiated with the lowest dose of a single agent and titrated to the desired clinical effect whereas minimizing adverse effects. The two most frequently prescribed pharmacologic agents for the treatment of OAB are oxybutynin (available in immediate- and extended-release oral formulations and a transdermal delivery system), and tolterodine (available in extended-release oral formulation). Both agents can be effective in the treatment of OAB.

Oxybutynin

Oxybutynin is an antimuscarinic agent with pronounced antispasmodic effects and local anesthetic activity, and it has served as the gold standard for treatment of OAB for 3 decades. Several double-blind, controlled studies have demonstrated its efficacy with reports of 50% symptomatic improvement in 61% to 86% of subjects at 15 mg daily.[4]

The original formulation (OXY-IR) is an immediate-release oral administration dosed at 5 mg three to four

Table 7-1.1. Available anticholinergic agents for treatment of OAB

Name of Drug	Class	Dosage	Route	Side effects	Contraindications
Oxybutynin	Antimuscarinic Local anesthetic Smooth muscle relaxant	IR = 2.5 mg BID–5 mg QID ER = 5 mg–30 mg QD TD = 3.9 mg 2×/week	PO TD	Xerostomia Blurred vision Constipation tachycardia	Closed-angle glaucoma Gastric retention Arrhythmia Myasthenia gravis
Tolterodine	Antimuscarinic Smooth muscle relaxant	IR = 1–2 mg BID ER = 2–4 mg QD	PO	antincholinergic	Closed-angle glaucoma Gastric retention Arrhythmia Myasthenia gravis
Hyoscyamine	Anticholinergic Antispasmodic	0.125–0.25 mg Q4h or 0.375–0.75 mg BID	PO SL	Anticholinergic	Closed-angle glaucoma Gastric retention Arrhythmia Myasthenia gravis
Imipramine	Tricyclic antidepressant Anticholinergic Alpha-adrenergic Antihistaminic	25–100 mg QHS	PO	Orthostatic hypotension Urinary retention anticholinergic	Recent MI MAO inhibitor usage
Propantheline	Anticholinergic	15–30 mg TID/QID	PO	anticholinergic	Closed-angle glaucoma Gastric retention Arrhythmia Myasthenia gravis
Solifenacin	Antimuscarinic	5–10 mg QD	PO	Anticholinergic	Closed angle glaucoma Urinary retention Gastric retention
Darifenacin	Antimuscarinic	7.5–15 mg QD	PO	Anticholinergic	Closed angle glaucoma Urinary retention Gastric retention Hepatic impairment
Trospium	Antimuscarinic Antispasmodic	20 mg BID	PO	Anticholinergic	Closed angle glaucoma Urinary retention Gastric retention

IR=immediate release, ER=extended release, TD=transdermal, SL=sublingual times a day with the maximum dosage of 20 mg per day. Thuroff et al.[5] demonstrated a mean decrease in incontinence episodes of 52%, and in urinary frequency of 33%. Significant levels of discontinuation were observed despite the high subjective improvement rates (74%) secondary to the incidence of reported side effects (70%), most notably dry mouth. Other placebo-controlled studies demonstrated reduced incontinence frequency by 19% to 58%.

Approximately 80% of patients taking OXY-IR 5 mg four times a day report intolerable adverse side effects. Oral administration of oxybutynin results in the metabolism of oxybutynin to *N*-desethyloxybutynin (DEO), its primary metabolite. A strong correlation has been identified between DEO levels and the severity of dry mouth. Thus, reducing small bowel and hepatic production of DEO results in lessened anticholinergic side effects. This led to the release of an extended-release oral oxybutynin and a transdermal delivery system. In 1999, oxybutynin extended release – OXY-ER (Ditropan XL) was approved. It is a once daily, controlled-release formulation using a push–pull osmotic system (OROS) allowing for consistent release of oxybutynin, affording a lower mean degree of fluctuation in plasma levels, and avoiding enterohepatic circulation by increased delivery in the large intestine, thus reducing the side effect profile whereas maintaining clinical efficacy. As a result, the incidence of moderate to severe xerostomia is reduced to 23%.[6]

Various clinical studies have demonstrated the effectiveness of oxybutynin in controlling OAB. In a randomized study that compared the immediate- and extended-release formulations of oral oxybutynin, the mean number of weekly urinary incontinence episodes was reduced by 88% and 84%, respectively. Although effectiveness was high, so was the incidence of dry mouth – 68% for extended-release oxybutynin and 87% for immediate-release oxybutynin.[7]

Transdermal Oxybutynin

Approved by the U.S. Food and Drug Administration (FDA) in 2003, transdermal oxybutynin is a matrix-type patch system (Oxytrol™; Watson Pharma, Inc.) that provides 3.9 mg of oxybutynin per day via twice-weekly applications.

In a phase III double-blind, placebo-controlled study comparing 3.9 to 1.3 and 2.6 mg/d transdermal oxybutynin as well as placebo patch, the 3.9-mg daily dose of transdermal oxybutynin significantly decreased weekly incontinence episodes and urinary frequency, increased voided volumes, and significantly improved quality of life (Incontinence Impact Questionnaire total score, $P = .0327$) compared with placebo.[1]

The most common adverse event with the transdermal system is application-site pruritus (transdermal oxybutynin, 16.8%; placebo, 6.1%).[1] The incidence of dry mouth in the transdermal group was similar to placebo (9.6% vs. 8.3%).[1] The dosage of oxybutynin is dependent on the surface area of the applied patch. We have noted that many patients who benefit from 10 mg of Ditropan XL may need 1 to 2 patches to achieve a similar reduction in OAB symptoms.

When we initiate oxybutynin therapy, we will offer patients either 10 mg of Ditropan XL or a 3.9-mg patch. Depending on the response, tolerability, and other factors, some patients may increase the dosage to 15 mg of Ditropan XL or 1.5 to 2 patches weekly. Alternatively, patients may supplement the oxybutynin dosage by adding 2.5 to 5 mg of immediate-release oxybutynin as necessary for a social event or travel. As with any anticholinergic agent, it is contraindicated in patients with narrow-angle glaucoma, severe constipation, or an allergy to the medication.

Tolterodine

Tolterodine is a competitive muscarinic receptor antagonist that was developed for the management of OAB. There are currently two oral formulations: an immediate-release (Detrol) available in 1- and 2-mg tablets given twice per day, and an extended-release pill available as 4 mg given once per day (Detrol LA; Pfizer, Inc., New York, NY).

Tolterodine immediate-release has consistently demonstrated a 20% reduction in micturition episodes, and a 40% to 60% reduction in weekly urge incontinence episodes versus placebo in randomized, double-blind, placebo-controlled studies.[8–10]

The mechanism of action is similar to oxybutynin, but the adverse side-effect profile is improved. The better tolerability profile of tolterodine compared with immediate-release oxybutynin has been confirmed in two randomized studies of detrusor overactivity. In a meta-analysis including 1120 patients, Appell[11] reported severe xerostomia in 6% of placebo, 4% of the 2-mg tolterodine arm, 17% of the 4-mg tolterodine arm, and 60% of the 15-mg immediate-release oxybutynin arm.

Extended-release tolterodine (Detrol LA) is a once-daily dosage that utilizes a microsphere system to deliver a more uniform serum concentration. Van Kerrebroeck et al.[12] demonstrated in a phase III study that Detrol LA was statistically superior to Detrol in terms of tolerability, namely, xerostomia ($P < .02$), whereas maintaining similar efficacy in significantly reducing mean urge incontinence episodes per week and mean micturition frequency.

Oxybutynin Versus Tolterodine

There have been multiple studies comparing oxybutynin-IR to tolterodine-IR. Because the extended-release formulations are more often used, we will focus on trials involving either of these two agents. Two studies have compared oxybutynin-ER to tolterodine-ER. The Antimuscarinic Clinical Effectiveness Trial (ACET) compared tolterodine-ER (2 and 4 mg) with oxybutynin-ER (5 and

Table 7-1.2. Date in perspective: Anticholinergic AEs (%)

	OXY-IR[6] (n = 52)	OXY-ER[14]	TOL-IR[8]	TOL-ER[8] (n = 507)	OXY-TDS[1]
Dry mouth	87	28	30	23	9.6
Constipation	31	7.0	7.0	6.0	0.8
Dizziness	38	4.9	2.0	2.0	0.8
Somnolence	40	4.3	3.0	3.0	1.6
Abnormal vision	—	2.2	1.0	1.0	0
Headache	—	8.1	4.0	6.0	0

OYX, oxybutynin; IR, immediate release; ER, extended release; TOL, tolterodine, AE, adverse events; TDS, transdermal system.

10 mg) in a prospective, nonrandomized, nonblinded, open-labeled study. Primary outcomes were measured by physician and patient questionnaires regarding drug efficacy pretreatment and at 8 weeks. The patient perception of efficacy was greatest for tolterodine-ER [4 mg every day (q.d.)], reporting 70% efficacy, compared with 60% efficacy for tolterodine-ER (2 mg q.d.), and 60% efficacy for oxybutynin-ER (10 mg q.d.).[13]

The Overactive Bladder: Performance of Extended Release Agents (OPERA) trial compared tolterodine-ER to oxybutynin-ER in a prospective, randomized, double-blind study.[14] The tolerability profile was similar for both arms, but total xerostomia was greater in the oxybutynin-ER arm. Both agents demonstrated similar reduction in weekly urge urinary incontinence episodes, but oxybutynin-ER was significantly more effective than tolterodine-ER in reducing micturition frequency ($P = .003$), and it reported a higher percentage of patients reaching total dryness with oxybutynin-ER (23.0%) versus tolterodine-ER (16.8%) ($P = .029$). Because it seems to have a lower rate of dry mouth, we prefer to use tolterodine-ER (Detrol LA) in patients who are more susceptible to side effects (i.e., polypharmacy) or who already have anticholinergic side-effect symptoms. Side-effect profiles must be kept in mind when selecting an anticholinergic agent (Table 7-1.2).

Propantheline Bromide

Propantheline bromide (Pro-Banthine) is also an anticholinergic agent that inhibits involuntary detrusor contractions and increases bladder capacity. The side effects are xerostomia, constipation, tachycardia, and blurry vision. We use this agent in patients refractory to other medicines at a dosage of 30 mg orally three times a day. Careful monitoring of side effects is necessary, and the dosage should be titrated to the maximum benefit with limited side effects.

Imipramine Hydrochloride

Imipramine hydrochloride (Tofranil) is a tricyclic antidepressant that facilitates urine storage by decreasing bladder contractility and increasing urethral sphincteric function. These effects result from the anticholinergic, antihistaminic, local anesthetic, and alpha-adrenergic properties of the drug. Patients with mixed incontinence often benefit from this drug. It can be very useful for OAB patients whose prior treatments have failed. We start at a dosage of 25 mg orally at bedtime, and may gradually increase the dose to 100 mg at night. We give the dosage at night, because nocturia is frequently a significant complaint, and somnolence is common after administration. Care must be taken in treating the elderly patient, because imipramine has a narrow side-effect profile, including potential cardiovascular toxic effects and orthostatic hypotension. Its effectiveness in enhancing sphincteric function is increased by concomitant administration of local estrogen cream. Thus, we frequently prescribe both medications together.

Hyoscyamine

Hyoscyamine (Levsin/Levbid) is an anticholinergic/antispasmodic used to relieve irritable bowel syndrome, which frequently coexists with OAB. It also suppresses detrusor overactivity by inhibiting muscarinic cholinergic receptor sites. Similar to other nonspecific anticholinergics, potential side effects limit its use. The dosage is 0.125 to 0.25 mg orally or sublingually every 4 hours for Levsin, or twice a day for the extended-release formulation, Levbid.

Intravesical Therapy

Success of intravesical administration of anticholinergics for overactive bladder and detrusor hyperreflexia has been documented in many studies.[15] However, secondary to the cumbersome nature of administration and the relatively short duration of action requiring repeated catheterization, intravesical anticholinergic drug therapy is not widely accepted.

Intravesical administration of local anesthetics such as lidocaine has been shown to be effective for suppressing overactive bladder, but secondary to its short-term effectiveness, the use of intravesical lidocaine is limited to its diagnostic use for differentiating detrusor hyperactivity caused by lesions of the spinal cord versus lesions of the brain.

Intravesical peppers or vanilloids such as capsaicin and resiniferatoxin (RTX) activate nociceptor sensory nerve fibers known as vanilloid receptor subtype 1. These receptors function as transducers of painful thermal stimuli, and are located predominantly on C fiber bladder afferent nerves. Activation of these receptors initially excites these nerves, but ultimately results in desensitization. C fiber hyperactivity has been identified in patients with idiopathic detrusor hyperactivity, multiple sclerosis, and spinal cord injuries.[15]

Intravesical capsaicin and RTX are still considered experimental, but in a recent meta-analysis of published studies, the beneficial clinical effects of increased bladder capacity on cystometrics and mean symptomatic improvement were noted in approximately 72%, and may last up to 1 year.[15] The side effects are acute burning and pain, as initial instillation causes stimulation of the unmyelinated C fibers. Resiniferatoxin is markedly more potent than capsaicin as a vanilloid receptor agonist, but causes significantly less local pain and inflammatory neurotransmitter release.

Botulinum toxin (Botox) inhibits acetylcholine release at the presynaptic membrane, thereby decreasing muscle contractility and causing muscle atrophy at the site of injection. This action is temporary because the nerves regenerate in 3 to 6 months. Traditionally, botulinum toxin has been used to treat detrusor-sphincter dyssynergia because the toxin produces a chemical denervation at the neuromuscular junction of the external sphincter. Schurch et al.[16] demonstrated increases in mean maximum bladder capacity of 296 to 480 mL ($P < .016$) and a decrease in mean maximum detrusor voiding pressure of 65 to 35 cm H_2O ($P < .016$) measured 6 weeks after injection in spinal cord–injured patients with detrusor hyperreflexia who received intravesical injection at up to 30 sites.

In our experience, patients respond well to Botox injections, but symptoms are likely to recur within 6 months. Limited clinical data are available on Botox for OAB, and it is not FDA approved for this indication.

Recently Available Therapy

Darifenacin

Darifenacin (Enablex) is an M3 receptor antagonist that has an 11-fold-higher affinity for M3 than for M2 receptors. It has similar potency for blocking acetylcholine-induced contractions of the bladder, but has a lower affinity than atropine for the M3 receptor in the salivary glands. Phase III studies are currently in progress.

Solifenacin

Solifenacin (Vesicare) is an antimuscarinic agent also selective for the M3 receptor, and has demonstrated efficacy in OAB symptoms relative to placebo, with low dry mouth rates.

Trospium Chloride

Trospium chloride (Sanctura) is an antimuscarinic agent with atropine-like effects that has no selectivity for the M1 to M5 receptors subtypes. It has been used in many forms in Europe for several years, and was recently FDA approved in the United States. It is different from traditional antimuscarinics in its unique large molecular size that prevents it from crossing the blood–brain barrier, thus minimizing central anticholinergic side effects. Additionally, 80% of an oral dose of the drug is excreted as the active compound in the urine, thus directly contacting the urothelium and possibly accounting for the favorable effects on the bladder and the limited effects on the salivary glands.

Controlled trials of trospium indicated a statistically significant improvement in volume to first contraction and bladder capacity compared with placebo with a subjective symptomatic improvement threefold greater.[17]

Hofner et al.[18] compared trospium to oxybutynin in a multicenter trial, and found that maximum bladder capacity, volume at first unstable contraction, and volume at first desire to void, were improved without any statistical difference between the arms. Adverse events were less frequent in the trospium group (47.9% vs. 58.9% for trospium and oxybutynin, respectively; $P = .002$) and xerostomia was significantly decreased (33% vs. 50%, trospium and oxybutynin, respectively; $P < .01$).[18,19]

Summary

Multiple effective agents are available orally, transdermally, and intravesically for the treatment of OAB. Most patients will respond to either of the available agents. In those patients with a greater degree of complexity, or a negative quality-of-life impact, we prefer to use an oxybutynin-based agent, either orally or transdermally, because the circulating level can be individualized via a combination of doses and routes, realizing that expected dry mouth severity may limit patient compliance, although tolterodine may be preferable in the elderly poly-pharmacy patient. The future of OAB therapy looks promising as new agents increase the available treatment options for patients with OAB.

References

1. Dmochowski RR, Davila GW, Zinner NR, et al. Efficacy and safety of transdermal oxybutynin in patients with urge and mixed urinary incontinence. J Urol 2002;168(2):580–586.
2. Burgio KL, Locher JL, Goode PS, et al. Behavioral vs drug treatment for urge urinary incontinence in older women: a randomized controlled trial. JAMA 1998;280:1995–2000.
3. Burgio KL, Locher JL, Goode PS, et al. Combined behavioral and drug therapy for urge incontinence in older women. J Am Geriatr Soc 2000;48:370–374.
4. Yarker YE, Goa KL, Fitton A. Oxybutynin: a review of its pharmacodynamic and pharmacokinetic properties, and its therapeutic use in detrusor instability. Drugs Aging 1995;6(3):243–262.
5. Thuroff JW, Chartier-Kastler E, Corcus J, et al. Medical treatment and medical side effects in urinary incontinence in the elderly. World J Urol 1998;(16 suppl 1):S48–S61.

6. Yoshimura N, Chancellor MB. Current and future pharmacological treatment for overactive bladder. *Journal of Urology* 2002;168:1897–1913.
7. Anderson RU, Mobley D, Blank B, Saltzstein D, Susset J, Brown JS. Once daily controlled versus immediate release oxybutynin chloride for urge urinary incontinence. OROS Oxybutynin Study Group. J Urol 1999;161(6):1809–1812.
8. Larsson G, Hallen B, Nilvebrant L, Tolterodine in the treatment of overactive bladder: an analysis of the pooled phase II efficacy and safety data. *Urology* 1999;53(5):990–998.
9. Van Kerrebroeck P, Amarenco G, Thuroff JW, et al. Dose-ranging study of tolterodine in patients with detrusor hyperreflexia. *Neurourol Urodyn* 1998;17(5):499–512.
10. Rentzhog L, Stanton SL, Cardozo L, Nelson E, Fall M, Abrams P. Efficacy and safety of tolterodine in patients with detrusor instability: a dose-ranging study. *Br J Urol* 1998;81(1):42–48.
11. Appell RA. Clinical efficacy and safety of tolterodine in the treatment of overactive bladder: a pooled analysis. Urology 1997;50(6A suppl):90–96; discussion 97–99.
12. Van Kerrebroeck P, Kreder K, Jonas U, Zinner N, Wein A, on behalf of the Tolterodine Study Group. Tolterodine once-daily: superior efficacy and tolerability in the treatment of the overactive bladder. Urology 2001;57(3):414–421.
13. Sussman D, Garely A. Treatment of overactive bladder with once-daily extended release tolterodine or oxybutynin: the Antimuscarinic Clinical Effectiveness Trial (ACET). Curr Hosp Res Opin 2002;18(4):177–184.
14. Diokno AC, Appell RA, Sand PK, et al. Prospective, randomized, double-blind study of the efficacy and tolerability of the extended release formulations of oxybutynin and tolterodine for overactive bladder: results of the OPERA trial. Mayo Clin Proc 2003;78:687–695.
15. Chancellor MB, de Groat WC. Intravesical capsaicin and resiniferatoxin therapy: spicing up the ways that we treat the overactive bladder. J Urol 1999;162:3.
16. Schurch B, Stöhrer M, Kramer G, et al. Botulinum: a toxin for treating detrusor hyperreflexia in spinal cord injured patients: a new alternative to anticholinergic drugs? Preliminary results. J Urol 2000;164:692.
17. Cardozo L, Chapple CR, Toozs-Hobson P, et al. Efficacy of Trospium chloride in patient with detrusor instabiligy: a placebo-controlled, randomized, bouble blind, multicentre clinical trial. *BJU Int* 2000;85:659–664.
18. Hofner K, Oelke M, Machtens S, Grunewald V. Trospium chloride: an effective drug in the treatment of overactive bladder and detrusor hyperreflexia. World J Urol 2001;19:336–343.
19. Appell RA, Sand P, Dmochowski R, et al. Prospective randomized controlled trial of extended-release oxybutynin chloride and tolterodine tartrate in the treatment of overactive bladder: results of the OBJECT Study. Mayo Clin Proc 2001;76(4):358–363.

7-2

Sacral Nerve Stimulation

Joanna M. Togami and Gamal M. Ghoniem

Sacral nerve stimulation (SNS) is a useful tool to have in the armamentarium for the treatment of voiding dysfunction. Since Food and Drug Administration approval, SNS has evolved to attain a major role in treating refractory overactive bladder (OAB) and patients with difficult pelvic floor dysfunction. It has become a viable option in those patients who failed conservative therapy, but who desire alternative therapy before more invasive procedures are performed. The technique and the technology have been improved, increasing the safety and efficacy whereas decreasing patient morbidity.

Functional Electrical Stimulation

Electrical stimulation has been used for a variety of lower urinary tract symptoms including those of OAB and painful bladder syndrome. There are mainly two types of electrical stimulation: chronic, in which low current is used for many hours daily, and acute submaximal functional electrical stimulation, which is applied up to the patient's tolerance for 15 minutes, once or twice a day, three times a week, or daily. The stimulation can be applied transvaginally, transrectally, or transcutaneously. There are a variety of home units that can be used by the patient.

The proposed mechanism of action is alteration of lower urinary tract function by stimulation of the sacral autonomic or somatic nerves. There may also be a direct stimulatory effect on pelvic floor muscles and sphincters causing muscular hypertrophy and a change from fast-twitch to slow-twitch fiber types, which can maintain muscle tone more effectively.

The stimulation probe is placed in the vagina at the level of the pelvic musculature in the mid-vagina. During stimulation, the probe is held by the patient to prevent any migration and possible discomfort. Functional electrical stimulation current is pulsed, i.e., short periods of stimulation are alternated with longer rest periods. The strength of the stimulus is adjusted to avoid pain. Low frequency (10–20 Hz) is used for OAB, mid-frequency (50–100 Hz) for stress incontinence, and high frequency (200 Hz) for urinary retention. A review of trials of functional electrical stimulation for stress urinary incontinence showed cure in 18% and improvement in 34% of patients. In the treatment of OAB, maximal electrostimulation cured 20% and improved 37% of women with urodynamic detrusor overactivity incontinence.[1] Functional electrical stimulation is of limited value in the treatment of stress incontinence.

Mechanism of Action

The precise mechanism of action by which electrical stimulation works remains to be proven. It is thought that the efficacy is dependent on stimulation of the afferent nerves that modulate sensory processing of the voiding reflex and its pathways to the central nervous system. Several voiding reflexes are affected by neuromodulation, and it is useful to review these. The *guarding reflex* serves to maintain continence with increased bladder volume. The detrusor muscle relaxes, the levator ani muscles contract, and urinary leakage is prevented. The *micturition reflex* facilitates voiding. Normally, a voluntary signal to empty the bladder originates from the cerebral cortex. The pontine micturition center acts as a switch between the guarding reflex and the micturition reflex. After sensory information is processed in the pontine micturition center, micturition is initiated under voluntary control. This is mediated by afferent delta nerve fibers and is composed of two reflexes. The *bladder-bladder reflex* is a positive reflex that activates the full bladder allowing a sustained contraction until the bladder empties. The *bladder-urethra reflex* allows smooth muscle of the proximal urethra to remain open while the bladder contracts. One theory is that these reflexes become pathological with voiding dysfunction. For bladder overactivity, SNS may work by inhibition of several neurons: spinal tract neurons involved in the micturition reflex, interneurons involved in spinal segmental reflexes, and postganglionic neurons, thus inhibiting the primary afferent pathway.[2] For nonobstructed urinary retention, SNS may suppress the guarding reflex by turning off bladder afferent input to the internal sphincter sympathetic or external urethral sphincter interneurons, and activation of postganglionic neurons and including bladder activity.[2] Stimulation is not curative for the underlying condition, and symptoms will recur when stimulation has been turned off. However, by modulating these reflexes that have become aberrant, symptoms improve.

Indications

Sacral neuromodulation is indicated in urgency frequency, urge incontinence, mixed incontinence with a significant urge component, and unobstructed urinary retention. More recently, it is indicated for interstitial cystitis. The work-up includes a complete history and physical examination, urinalysis, urine culture, cytology, and voiding diary. Cystoscopy, urodynamics, and upper tract imaging are performed as indicated. These diagnostics serve three purposes: 1) to identify an underlying disorder that should be treated by other means (calculus or carcinoma in situ), 2) to clarify the etiology of the lower urinary tract dysfunction, and 3) to serve as a baseline for later comparison.

Contraindications

Mental incapacity is a contraindication, because the patient must be able to adjust the patient programmer used to turn the stimulator on and off. In addition, the patient must be able to track and understand the response to therapy. Electromagnetic interference can result in injury to the patient from heating of the implanted components of the system, resulting in damage to the tissues. System damage may require surgical replacement. Operational changes to the neurostimulator from a magnetic source can potentially turn the system off, resulting in resumption of symptoms. This may require reprogramming. Unexpected changes in stimulation can result in a shock or jolt-type sensation. Sources of electromagnetic interference include defibrillators, cardioverters, electrocautery, high-output ultrasonics or lithotripsy, magnetic resonance imaging and radiofrequency, or microwave ablation. Use of magnetic resonance imaging and radiofrequency or microwave ablation can cause tissue damage.[3] Diathermy, or energy from the diathermy (shortwave, microwave, or ultrasound) can be transferred through the implanted system and cause tissue damage at the site of the implanted electrodes. Diathermy is contraindicated in patients with InterStim.

Patient Selection

Sacral neuromodulation is ineffective for the following patient populations: those with end-stage, small, contracted bladders; those with peripheral nerve damage, e.g., sacral spinal cord injury, radical pelvic surgery, spinal cord tumors, and those with areflexic bladders from myogenic damage secondary to chronic overdistention.

Nerve Evaluation

Before placement of the permanent pulse generator, it is useful to evaluate the sacral nerve to assess its suitability for stimulation. The purpose is to confirm the integrity of the peripheral nerves. Also, it can identify the optimal site for a trial of temporary stimulation. This may be performed by one of two methods – percutaneous nerve evaluation (PNE) or staged implantation of the permanent lead. We use the PNE in patients who have urgency and frequency. Patients who are being treated for nonobstructed urinary retention generally undergo a staged procedure. This allows for a longer period of evaluation, because retention generally improves over a week to 10 days, whereas with urgency and frequency, the results are generally apparent within several days. The advantages of the PNE are that it is performed as an office procedure, it is completely percutaneous, and the wires can be removed simply by pulling them out. The drawbacks are that lead migration occurs in 12% of patients and there is a potential difference of approximately 15%, in clinical response when the permanent lead is placed with the internal pulse generator. The advantage of the staged procedure is that the permanent lead is placed during the first stage; therefore, the clinical response when the internal pulse generator is inserted is the same, because lead migration is minimal. The disadvantages are that a small open incision is needed to implant the lead, and a separate procedure is necessary to explant the lead should the test period be unsuccessful.

Technique

The method for implantation of the lead and the wires is similar. The patient is placed in the prone position with a 30-degree flexion at the hips. This position places S3, S4, and S5 into the vertical plane. The patient is given local anesthesia and, under fluoroscopic guidance, the S3 sacral foramen is located using a crossed-hair technique developed by Chai and Mamo.[4]

The horizontal axis is defined at the inferior borders of the sacroiliac joints, and the vertical axis is the midline (Figure 7-2.1). Once these are marked on the skin, points that are 2 cm lateral and cephalad on both sides are marked. This is the skin location of S3. We identify a point 2 to 3 cm above this mark and infiltrate with 2% lidocaine. The needle is introduced at an angle of 60 degrees (Figure 7-2.2).

Once the foramen is located, on both anterior/posterior and lateral views, the needle is stimulated. Both the motor and sensory responses are monitored. The lead is

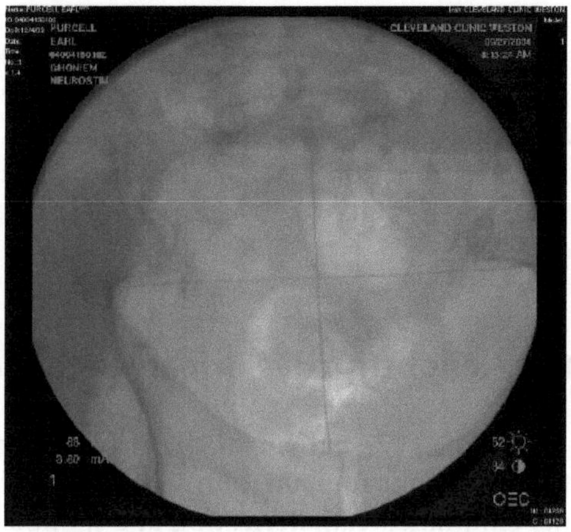

Figure 7-2.1. Crossed-hair technique.

Table 7-2.1. Motor and sensory responses observed with stimulation of the S2, S3, and S4

	S2	S3	S4
Motor	• Lateral rotation of the leg • Contraction of the toe and foot • Contraction of the calf • Contraction of the superficial pelvic fbor	• Bellows (contraction of the levator ani muscles causing a deepening and fbttening of the buttocks groove) • Plantar flexion of the hallux caused by sciatic nerve stimulation	• Bellows activation of the posterior levator ani muscles • No motor response of the leg or foot
Sensory	• Pulling sensation of the vagina or base of the penis	• Paresthesia of the perineal skin, external genitalia • Pulling sensation of the vagina, rectum, or bladder base	• Pulling sensation in the rectum

quadripolar, and each one is stimulated separately to determine the lead that has the most optimal placement. Table 7-2.1 is a guide for motor and sensory responses observed with stimulation of the S2, S3, and S4 nerves.

Once the placement is confirmed, either the wire or tined lead is placed. The wire is fed through the foramen needle. The needle is removed, and the wires are then taped carefully to the back. In the case of the staged implant, the lead is placed through a sheath, and its placement is verified by fluoroscopy. A small incision is created at the future neurostimulator pocket site in the upper buttock. A trocar is used to tunnel the lead to the neurostimulator pocket, the percutaneous lead extension is connected, and the connection is buried at the site. The wires are brought out at a separate site, usually the contralateral side, and the connection site is closed in two layers.

During the second stage, the patient is brought back to the operating room and given local anesthesia with sedation. The pocket is opened, and the lead is disconnected from the extension wire. The wires are pulled out from the externalized side to prevent the exposed wires from being brought through the subcutaneous tissues. The

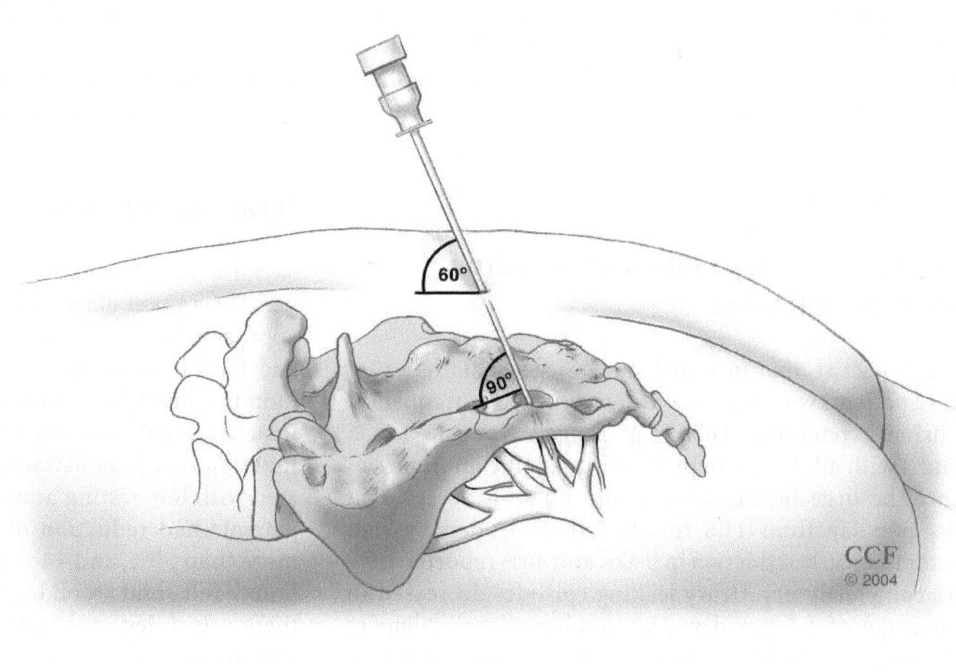

Figure 7-2.2. Appropriate needle orientation within S3 foramen. The needle entry is 60 degrees to the horizontal. (Reprinted with the permission of The Cleveland Clinic Foundation.)

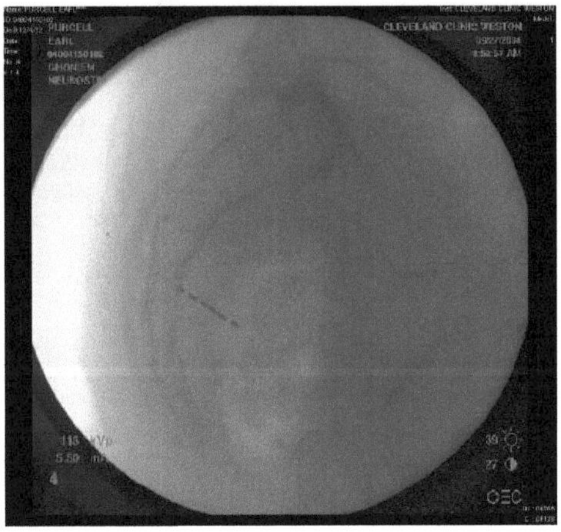

Figure 7-2.3. Lateral view of radiograph showing the implanted lead.

pocket is developed for the neurostimulator. The lead is connected to the neurostimulator. Connections are sealed with silicone glue and the boot is placed over the connection and tied with prolene suture. Care is taken not to create a trigger point by placing the wires around the neurostimulator. The wound is closed in two layers (Figure 7-2.3).

Because a foreign body is being introduced into the patient, prevention of wound infections is imperative. We administer preoperative antibiotics to cover coliforms and skin flora. For the first stage and complete implantation, the patient is placed in an operating room equipped with laminar airflow. The pocket developed for the neurostimulator is irrigated copiously with neomycin irrigation. It is important not to immerse the neurostimulator in a liquid medium because this may cause it to malfunction. A two-layer closure is desirable should the skin open, so that there is a closed layer beneath.

Clinical Uses

Urgency/Frequency Syndromes and Urge Urinary Incontinence

In a prospective, multinational study, Siegel et al.[5] reported on 112 patients with urge incontinence, urgency/frequency, or urinary retention. Follow-up ranged from 18 to 36 months with all three groups showing significant improvement. The urge incontinence group noted a decrease in leaks per day from 11.6 to 5.0. Fifty-nine percent had greater than 50% reduction in leaks, and 46% reported they were completely dry. Heavy leaking episodes decreased to an average of 1.3 per day. The number of pads/diapers replaced was reduced from 6.7 pads to 3.4 pads per day. In the urgency/frequency group, average reported voids per day decreased from 17 to 10.6. Improvement in the degree of urgency was reduced as demonstrated by 69% of patients. Volume of urine per void increased from 132 to 225 mL. Adverse effects were lead migration (18.2%), pain at the implantation site (15.3%), new pain (9.0%), suspected lead migration (8.4%), infection (6.1%), adverse change in bowel function (3.0%), suspected device problems (1.6%), nerve injury (0.5%), device rejection (0.5%), and other (9.5%).

Idiopathic Nonobstructive Urinary Retention

Jonas et al.[6] studied 177 patients with urinary retention, of which 68 qualified for implantation. They found that 69% of patients eliminated catheterization at 6 months and an additional 14% had a 50% or greater reduction in catheterized volume per catheterization. Compared with the control group with 9% improvement, 83% of patients achieved successful results. Temporary inactivation of sacral stimulation led to increased residual volumes. Success was measured as a 50% or greater improvement in baseline voiding symptoms (number of catheterizations per day, total catheter volume per day, maximum catheter volume, number of voids, and total volume of voids).

Interstitial Cystitis

Maher et al.[7] studied 15 women prospectively and found increases in mean voided volume (90–143 mL), frequency decreased from 20 to 11, nocturia decreased from 6 to 2 times, and mean bladder pain decreased from 8.9 to 2.4 points on a scale of 1 to 10. Quality-of-life measures, namely, the UDI-6 (the short form of urogenital distress inventory) and SF-36 Health Survey also showed improved parameters during the stimulation period. Sacral neuromodulation shows promise for both the urgency/frequency and pain components of interstitial cystitis.

Fecal Incontinence

Sacral nerve stimulation has been used for fecal incontinence. Ganio et al.[8] studied 23 patients with a two-staged approach, five of whom went on to be implanted. Patients had fecal incontinence at least twice every 2 weeks for 2 months, failed biofeedback therapy, and had structurally intact external and internal anal sphincters. Sphincter dysfunction was demonstrated with fecal incontinence associated with low resting and/or squeeze pressures. Seventeen patients had reduction of liquid or solid stool losses by more than 50%, and 14 were completely continent to both liquid and solid stool. In patients with urge incontinence, there was a decrease in losses per week from five to one. Manovolumetric findings showed improvements in maximum resting pressure, rectal sensitivity, first sensation, and the distention pressure for urge threshold. Two patients who also had voiding complaints had complete resolution after SNS. Sacral nerve stimulation for fecal incontinence is currently being studied at our institution.

Complications

Reported adverse events include lead migration, new pain (at the site or from other nerve stimulation), persistent skin irritation, change in bowel function, change in bladder function, local infection, technical problem, device problem, and transient electric shock.[9] If the lead is placed as a first stage, it is recommended that the lead be dissected out and removed from the original insertion point rather than from the tunneled skin site because lead breakage has been reported. Very few patients have had surgical intervention with adverse effects of SNS. In the original series, only one patient required explantation.

References

1. Berghmans LC, Hendriks HJ, DeBie RA, van Waalwijk van Doom ES, Bo K, van Kerrebroeck PE. Conservative treatment of urge urinary incontinence in women: a systematic review of randomised clinical trials. BJU Int 2000;85:254–263.
2. Chancellor MB, Leng W. The mechanism of action of sacral nerve stimulation in the treatment of detrusor overactivity and urinary retention. In: Jonas U, Grünewald V, eds. New Perspectives in Sacral Nerve Stimulation for Control of Lower Urinary Tract Dysfunction. London: Martin Dunitz; 2002.
3. Medtronic Inc. medical technology company official company Web site. Available at: http://www.medtronic.com. Accessed May 6, 2004.
4. Chai TC, Mamo GJ. Modified techniques of S3 foramen localization and lead implantation in S3 neuromodulation. Urology 2001;58:786–790.
5. Siegel SW, Catanzaro F, Dijkema HE, et al. Long-term results of a multicenter study on sacral nerve stimulation for treatment of urinary urge incontinence, urgency-frequency and retention. Urology 2000; 56(S1):87–91.
6. Jonas U, Fowler CJ, Chancellor MB, et al. Efficacy of sacral nerve stimulation for urinary retention: Results 18 months after implantation. J Urol 2001;165:15–19.
7. Maher CF, Carey MP, Dwyer PL, Schluter PL. Percutaneous sacral nerve root neuromodulation for intractable interstitial cystitis. J Urol 2001;165:884–886.
8. Ganio E, Luc AR, Clerico G, Trompetto M. Sacral nerve stimulation for treatment of fecal incontinence. Dis Colon Rectum 2001;44: 619–631.
9. Jonas U, van den Hombergh U. Complications of sacral nerve stimulation. In: Jonas U, Grünewald V. New Perspectives in Sacral Nerve Stimulation for Control of Lower Urinary Tract Dysfunction. London: Martin Dunitz; 2002.

7-3

Surgical Management of the Overactive Bladder: Evacuation Disorders

Raymond R. Rackley and Joseph Abdelmalak

Overactive bladder (OAB) is a term describing the complex symptoms and conditions of urinary frequency and urgency, with or without urge incontinence and may affect more than 17 million Americans. Although OAB may be successfully managed in a variety of ways, the majority of cases (50%–80%) presenting to the physicians respond to the triad of measures integrating the individual merits of behavior modification, pelvic muscle physiotherapy, and pharmacotherapy. Awareness by patients and primary care physicians of new advances in this trivium of OAB management has resulted in an increasing number of referred cases to the pelvic health specialist for this condition in which the prevalence is not only much higher than previously expected, but also increasing with the aging of our population. The growing awareness and increasing prevalence of this manageable condition not only increases the absolute number of referred cases, but also the number of cases that will be refractory to the trivium of standard management. However, continued advances in patient evaluation as well as behavioral modification, physiotherapy, pharmacology, and surgical interventions continue to evolve our multimodal strategies for the management of refractory cases of OAB. The merits of evolving surgical management strategies for refractory cases of OAB are reviewed below.

Surgical Management Options

Neuromodulation

Neurotoxins: Botulinum Toxin Injections

The diversity and multidimensional application of pharmacologic agents have resulted in this form of therapy being the most widely adopted form of managing the OAB. Specialists have mastered both individual and synergistic benefits of topical estrogen replacement of the vagina, anticholinergic therapies for the hyperactivity of the detrusor, and judicious use of medications such as diuretics, calcium channel blockers, and DDAVP (desmopressin acetate) for managing the timing of urine production for refractory OAB conditions. Improved agents with better clinical effectiveness based on drug delivery, as well as different mechanisms of action, continue to be developed.

Botulinum toxin inhibits the calcium-mediated release of acetylcholine at the presynaptic neuromuscular junction. By creating a reversible field denervation at the site of injection and subsequent muscle atrophy, botulinum toxin injections have proven to be a safe and effective therapy for a variety of somatic and autonomic disorders and are being used in a variety of clinical applications such as the refractory OAB. Botulinum toxin injection in the office setting under a topical bladder and urethral anesthetic addresses not only the hyperactivity of the bladder muscle from motor efferent innervation, but may also address the hypersensitivity of the bladder's sensory afferent nerves that may contribute to refractory OAB symptoms. In the United States (US), the procedure is done on an outpatient basis under local anesthesia provided by instilling 90 mL of a 1% lidocaine solution in the bladder for 10 to 15 minutes. Under cystoscopic visualization, 300 IU of botulinum A toxin (Botox) is injected into the detrusor muscle with a standard 5-French flexible injection needle into 30 detrusor muscle sites (10 U per 1 mL per site) (Figure 7-3.1). Trigonal injections have not been performed in European protocols; however, in the US, physicians have been safely placing 20% (six injections) into the trigonal area without complications of ureteral reflux. Another note of difference between European and US studies is the injection volume used for this therapy. In the US, most investigators use a total injection volume of 3 mL compared with 30 mL used elsewhere. Not only is the smaller injection volume easier to tolerate by the patient when done under a local instillation anesthetic, but this smaller volume leads to less extravasation of the toxin

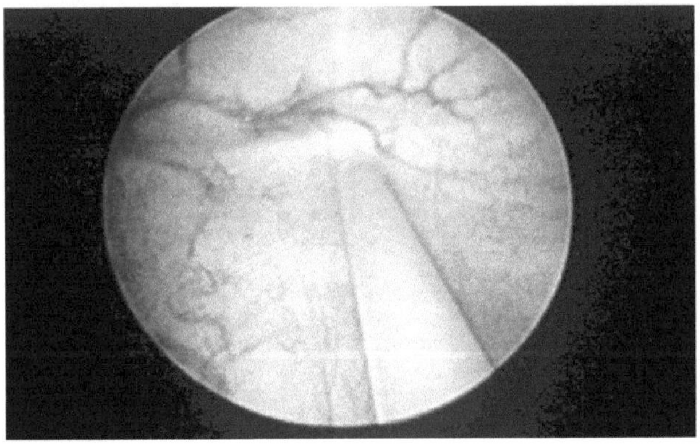

Figure 7-3.1. Cystoscopic view of a transurothelial botulinum toxin injection into the detrusor muscle.

material that would explain the rare reports of systemic hypoesthesia associated with the larger volume. Because the onset of toxin activity gradually takes place over 7 to 14 days, patients are asked to progressively decrease their anticholinergic treatment within 1 week and should be asked to complete a micturition diary, perception of bladder condition, and quality-of-life questionnaires to objectively monitor their results in comparison to their baseline voiding evaluations.

We have reported on 29 patients with refractory OAB injected with 300 IU of Botox. Three-month follow-up was completed in 18 of the 29 patients. Nine of the 18 patients completed a 6-month follow-up. The average frequency per day at the baseline was 10.7; at 3 months, it was 8.8 ($P = .01$); and at 6 months, it was 10.3 ($P = .06$). The average number of leakage episodes at the baseline was 4.2; at 3 months, it was 2.5($P = .01$); and at 6 months, it was 2.5 ($P = .46$). The average number of urgency events at baseline was 6; at 3 months, it was 4 ($P = .08$); and at 6 months, it was 4.2 ($P = .23$). The average score for the bladder perception at baseline was 3.6; at 3 months, it was 2.5 ($P = .001$); and at 6 months, it was 2.4 ($P = .02$). The average score for the IIQ7 at baseline was 14.8; at 3 months, it was 10.4 ($P = .01$); and at 6 months, it was 11 ($P = .1$). The average score for the UDI6 at baseline was 10; at 3 months, it was 8.1 ($P = .23$); and at 6 months, it was 8.7 ($P = .2$). No short- or long-term side effects or systemic complications have been noted.

Reimbursement for covering the cost of the botulinum toxin medication in the US is currently hampered by the lack of a specific Food and Drug Administration indication for use of this drug in treating OAB. This has obligated the application of this office-based therapy to self-pay/off-labeled use or clinical studies outlining the long-term safety and cost effectiveness; however, clinical success with reported 6 to 12 months of follow-up in both Europe and the US suggests that this effective and reversible therapy will complement our options for a large subset of people with refractory OAB conditions.

Intravesical Therapy: Resiniferatoxin

Evidence suggests that unmyelinated C fibers become predominant in mediating OAB symptoms in cases of neurogenic and idiopathic detrusor hyperactivity. Through desensitization of these sensory fibers by capsaicin or its potent analog, resiniferatoxin, the micturition reflex mediated by C fibers can be suppressed and detrusor overactivity can be successfully managed in cases of anticholinergic treatment failures. Resiniferatoxin has a clinical effect similar to capsaicin, but with less of a neuronal excitatory effect that is suitable for office instillation therapy with local bladder anesthesia in patients who have normal bladder sensation and OAB. Although not yet available in the US for routine clinical use, non-US–based studies suggest that this form of intravesical therapy will be effective in treating a significant subset of refractory cases of OAB including non-neurogenic conditions.

Nerve Stimulation: Applied Electrical Stimulation

Sacral nerve neuromodulation represents an outpatient therapeutic option that may be strategically positioned between botulinum toxin injections and laparoscopic augmentation cystoplasty. This form of therapy provides electrical stimulation of sacral nerves that modulate the bladder, sphincter, and pelvic floor muscle functions that influence the OAB. As percutaneous or implantable, programmable neurostimulation systems represent minimally invasive treatments that have clinically significant therapeutic success without major complication rates. For the urologist seeking a therapeutic way to address individual or combinations of efferent nerve hyperactivity, afferent nerve hypersensitivity, or central nervous system facilitation of peripheral nerve pathways, neuromodulation becomes an instrumental strategic option. This is discussed in detail in chapter 7-2.

Hydrodistention: Transient Peripheral Nerve Ischemia

Bladder hydrodistention has been used in the past to reduce the irritative symptoms of OAB, hypersensitivity of bladder afferent nerves, and the pain associated with bladder filling in interstitial cystitis patients. The exact mechanism of action for the perception of clinical benefit is unknown, as is the exact etiology(ies) for this condition; however, mechanical damage and local ischemia of the suburothelial nerve plexus is thought to provide temporary benefit. Potential but rare complications of significant hematuria and bladder rupture associated with only a transient (1–6 months) clinical benefit for a procedure requiring a regional or general anesthesia have dampened the use of this treatment for refractory OAB.

Denervation Procedures

Because the etiologic mechanisms for the development of the OAB may lie in the neural control of the detrusor muscle, denervation procedures of the overactive bladder muscle has been tried with varying success using both central and peripheral approaches. Surgical or chemical denervation can be used at central or peripheral levels to interrupt motor and/or sensory reflex pathways.

In general, the central denervation procedures involve the levels of sacral roots S3 and S4, which are responsible for the parasympathetic and somatic innervations to the lower urinary tract and pelvic organs. Unfortunately, achieving stabilization of bladder filling and increases in reflex trigger volumes can be obtained with a posterior root rhizotomy, yet the attendant complications of induced sensory loss and sexual dysfunction at this sacral level restricts these procedures to only people with complete spinal cord injuries resulting in neurogenic detrusor dysfunction. Our experience with central deafferentation through posterior root rhizotomy procedures are currently reserved for those cases in which the implantation of the Brindley device for driving bladder voiding in spinal cord patients is accomplished through anterior sacral root stimulation.

Peripheral nerve transection, cystolysis, bladder transection, myomectomy, autoaugmentation, and chemical nerve ablation procedures with phenol or alcohol have all been performed over the last 50 years with controversial short-term reporting and no significant improvement in long-term outcomes that warrant the complications associated with these procedures. Although Ingelman-Sundberg described a transvaginal approach for partial bladder denervation in 1959 that was later modified by Cespedes and McGuire[1] and reported in 1996, the consistent, reproducibility of this minimally invasive procedure remains to be tested against other evolving procedures providing similar 1-year outcomes in a refractory group of OAB conditions. What remains a challenge to the pelvic health specialist is the inability to reproducibly induce urinary retention and continence with peripheral and central denervation procedures when colleagues from neurosurgery, gynecology, and colorectal surgery invariably refer their patients to us with these "complications" after routine procedures on the spine or pelvic organs.

Laparoscopic Augmentation Cystoplasty

Augmentation cystoplasty remains the most widely accepted reconstructive technique for creating a compliant, large-capacity bladder that protects the upper urinary tract and provides urinary continence for people with bladder dysfunction secondary to noncompliance or reduced functional capacity. This form of bladder reconstruction may even be combined with developing a continent catheterizable stoma for use as an accessible port for bladder emptying. Since 1888, the standard enterocystoplasty has classically evolved as a procedure performed through an open laparotomy incision using various segments of well-vascularized segments of the gastrointestinal system that are reconfigured before anastomosis with the urinary bladder. One hundred years have passed since the original open approach for this procedure has become an established reconstructive technique that can be performed laparoscopically. No matter which approach is chosen, the use of any bowel segment for augmentation is associated with advantages and disadvantages, but the versatility of choosing a particular bowel segment, both open and laparoscopic, provides a variety of clinical options based on an individualized set of objectives for the person requiring this form of bladder reconstruction.

Enterocystoplasty is effective in providing a durable increase in bladder capacity and compliance; however, the morbidity and postoperative discomfort associated with the open laparotomy incision are major deterrents. For patients with preexisting debilitating neurologic and other comorbid conditions, the open procedure may significantly prolong the hospital stay, increase the metabolic needs for wound healing, and delay postoperative recovery. Laparoscopy has distinct advantages when compared with open surgical procedures in regard to postoperative pain and morbidity, improved cosmesis, and a shorter hospital stay and decreased convalescence period. Recent studies have indicated that postoperative intraabdominal adhesions are significantly reduced after laparoscopic surgery when compared with open surgery.

Despite the established role of laparoscopy in diagnostic and ablative urologic surgery, the use of laparoscopic techniques in reconstruction has been limited because of the technical complexity of the procedures involved. The technical steps in performing a laparoscopic bladder augmentation are designed to emulate the open surgical counterpart in every aspect, thereby producing similar functional results with an improved recovery.

For laparoscopic or open approaches, the surgical technique of enterocystoplasty is based on the following fundamentals: a) selection of an optimal segment of bowel based on a broad, well-vascularized mesenteric pedicle, b) isolation of the bowel segment, c) reestablishment of bowel continuity and closure of the mesenteric defect, d) detubularization and reconfiguration of the bowel segment without peritoneal soiling of bowel contents, e) bladder mobilization with formation of an adequate-sized cystotomy, f) creation of a tension-free, water-tight, full-thickness, circumferential anastomosis of the bowel to the bladder, and g) confirmation of adequate postoperative urinary drainage (Figure 7-3.2). As is true in open entero-

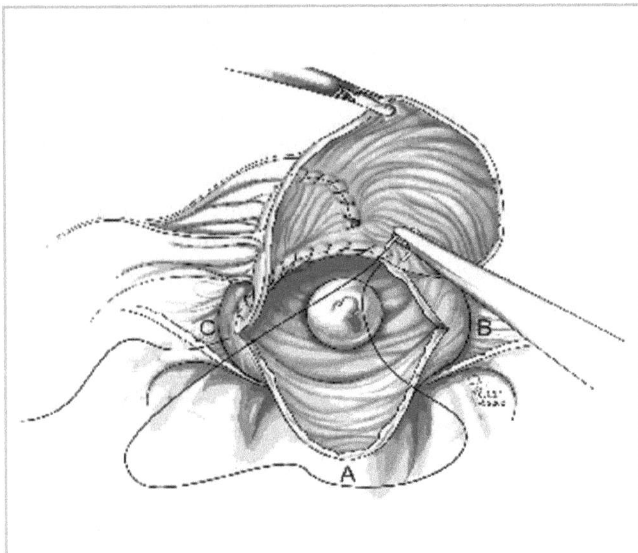

Figure 7-3.2. Laparoscopic augmentation enterocystoplasty. (Reprinted with the permission of The Cleveland Clinic Foundation.)

cystoplasty, various segments of the gastrointestinal system may be used for the procedure depending on the clinical requirements of the patient. A length of 20 cm of bowel is usually desirable to attain an adequate augmented bladder capacity. An appropriate segment of bowel is identified based on the following criteria: 1) the bowel segment will reach the area of the bladder neck without tension, and 2) a well-defined arterial arcade should be present in the bowel mesentery.

We have reported on the largest series of laparoscopic enterocystoplasty in 17 patients with functionally reduced bladder capacities attributed to neurogenic causes. Procedures included ileocystoplasty (5), sigmoidocystoplasty (3), colocystoplasty (1), and cecocolocystoplasty with a continent catheterizable ileal stoma (8). Total surgical time (including cystoscopy, stent placement, extracorporeal bowel anastomosis and refashioning of the isolated loop, and laparoscopic dissection and suturing) ranged from 5.3 to 8 hours (average, 7.0 hours). The time for laparoscopic suturing ranged from 1.7 to 3.1 hours (average, 2.4 hours). Blood loss was minimal, and did not exceed 250 mL during any of the cases (average, 175 mL). Oral feeding was resumed by 24 hours in 11 of 12 patients. Although most of the study group had moderate to severe forms of neurologic dysfunction because of multiple sclerosis, the average hospital stay until the patient was discharged home was 5.7 days and ranged from 3 to 7 days. Most notable was the absence of long-term or extended-care needs for the patients with multiple sclerosis because of the demands of wound healing or functional neurologic loss that typically occur with an open procedure in this subset of patients.

All patients consented to participate in a long-term outcome study using preoperative and postoperative validated questionnaires concerning bladder control (BLCS) and bowel control (BWCS). In regard to the quality-of-life (QOL) measure using the BLCS, there was significant clinical improvement. In regard to the potential risk of causing bowel dysfunction by harvesting various bowel segments for augmentation cystoplasty, there was no clinically significant difference in the bowel control score before or after the procedure.

Laparoscopic augmentation enterocystoplasty is technically feasible and successfully emulates the established principles of open enterocystoplasty whereas minimizing operative morbidity and maximizing clinical effectiveness. As is true in open surgery, various bowel segments can be fashioned and anastomosed to the bladder laparoscopically. The increased costs associated with laparoscopy and with minimally invasive surgery in general have been a significant disadvantage; however, a previous report on the costs of laparoscopic procedures concluded that increased surgical experience reduces the surgical time and length of the hospital stay, thereby decreasing costs. Furthermore, the increased use of reusable instruments results in considerable economic benefits. Implementation of appropriate cost-saving strategies will ultimately result in decreased expenses associated with laparoscopy. Although laparoscopic enterocystoplasty is currently a lengthy procedure lasting twice as long as open surgery, further technical modifications and increasing experience will continue to reduce the surgical time involved.

For patients with complex comorbid illnesses desiring an improved quality of life associated with traditional augmentation cystoplasty, the reduced morbidity observed in our series of patients undergoing a laparoscopic procedure makes this approach an attractive option to consider. A clinically significant positive impact on their postoperative QOL related to their bladder control compared with their preoperative status will be achieved using a laparoscopic approach. Furthermore, this benefit in their QOL from improvement of their bladder control can be achieved without a negative impact on their bowel control. Our experience suggests that laparoscopic enterocystoplasty has become a viable alternative to open enterocystoplasty and is a surgical option to consider in people with refractory OAB conditions who have failed other management options.

Urinary Diversion

Several minimally invasive surgical techniques have been developed to effectively provide urinary diversion for management of the refractory OAB when bladder and urethral function is compromised to the point that reconstruction or management options described above fail to provide a meaningful improvement in the patient's quality of life and impact on their condition. Although options that preserve the bladder and upper urinary tract with minimally invasive techniques that allow for urinary diversion in cases of evacuation disorders are covered in a later chapter, some consideration must be given to the merits of urinary diver-

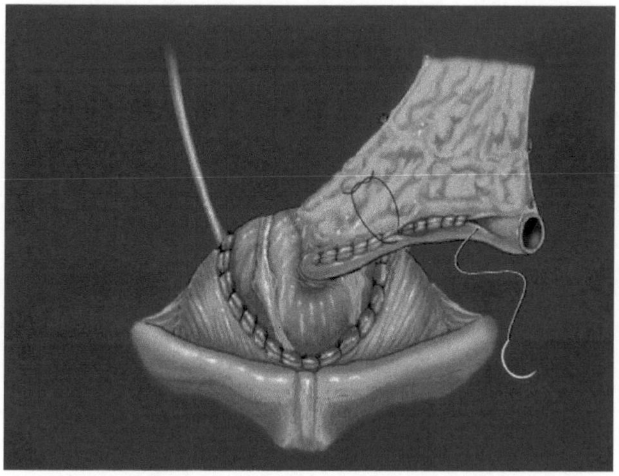

Figure 7-3.3. Extracorporeal construction of a right colon reservoir using a reinforced ileocecal valve continence mechanism. (Reprinted with the permission of The Cleveland Clinic Foundation.)

sion with or without cystectomy as the ultimate solution for management of severe cases of refractory OAB conditions. These cases include both bladder hypersensitivity as well as hyperactivity disorders. Furthermore, the choice for cystectomy at the time of urinary diversion must be made on an individual case basis because the long-term risk of developing pyocystitis or bladder malignancy must be weighed against the short-term morbidities associated with this part of the extirpative procedure.

Although the appliance-dependent urinary ileostomy with or without cystectomy is still considered an acceptable practice for urinary diversion, the optimal urinary diversion should replicate bladder function as closely as possible. For this reason, the continent urinary diversions such as the Studer (orthotopic) or Indiana (nonorthotopic) reservoirs have evolved to a point of reliability and simplicity that allows one to consider these options as first-line diversion therapies for the majority of people with severe conditions of refractory OAB. Because they offer a significantly better quality of life, self-image, and sexual function, and have complication rates comparable to those of the standard ileal-conduit, we have been successfully developing laparoscopic approaches for achieving even better results with these forms of continent urinary diversion.

Currently, the Indiana pouch (a right colon reservoir using a reinforced ileocecal valve continence mechanism) offers the best documented outcomes for continence for a continent catheterizable urinary reservoir in a female requiring cystectomy with possible removal or damage to the bladder neck/urethral continence mechanism (Figure 7-3.3). A diversion reservoir such as a Studer pouch (small bowel reservoir without construction of a continence mechanism) can be anastomosed directly to a functional bladder neck/urethral continence mechanism. However, if the continence mechanism is damaged or the reservoir is anastomosed to the remnant urethral or native urethral meatal site, then an artificial sphincter or sling must be placed around the catheterizable bowel segment for continence formation; such techniques are extremely complex and are prone to complications.

Summary

For evolving strategies to effectively address the growing number of refractory cases of OAB, they must have widespread, practical application that translates therapeutic effectiveness to the patient as defined by clinical efficacy, tolerability, and persistence. Although many options seem limited in scope or complex in application, most are inspirational to specialists seeking optimal outcomes for people with refractory and complex conditions. The available strategies outlined above are a reflection of our current understanding of the pathophysiology of OAB and attendant refractory conditions. The promise for future successful strategies lies not only in the successful translation of our basic science knowledge of the OAB to date, but advancement of our basic and clinical science research of refractory conditions for the future.

References

1. Cespedes RD, Cross CA, McGuire EJ. Modified Ingelman-Sunberg bladder denervation procedure for intractable urge incontinence. J Urol 1996;156(5):1744–1747.
2. Abdelmalak JB, Rackley RR, Vasavada SP, et al. Botulinum a toxin (Botox) therapy for the treatment of refractory overactive bladder. J Urol 2004;171(4):141(abstract 531).
3. Kuo HC. Effectiveness of intravesical resiniferatoxin for anticholinergic treatment refractory detrusor overactivity due to nonspinal cord lesions. J Urol 2003;170:835–839.
4. Rowland RG, Mitchell ME, Bihrle R, Kahnoski RJ, Piser JE. Indiana continent urinary reservoir. J Urol 1987;137:1136–1139.

7-4

Irritable Bowel Syndrome

Ronnie R. Pimentel

Irritable bowel syndrome (IBS) is the most common functional gastrointestinal disorder representing up to 25% of outpatient visits in gastrointestinal practices.[1] It is twice as common in females as in males and more prevalent in young patients of high socioeconomic status. Interestingly, only one-fourth of all affected patients will seek medical attention and they seem to be those with a greater degree of somatization and psychiatric disorders. Association with other functional disorders such as migraine or fibromyalgia is not uncommon.

Irritable bowel syndrome is a chronic condition that appears early in life and is characterized by periods of exacerbation and remission. Exacerbations are frequently associated with stressful situations or "life events." The most common symptoms are those of abdominal pain, altered bowel habits, bloating, and sensation of incomplete evacuation. The characteristics and severity of the abdominal pain vary from patient to patient but keep a particular pattern over time. The pain is often described as crampy, rarely severe, and temporarily relieved by bowel movements. Nocturnal pain, weight loss, fever, and other constitutional symptoms are typically absent.

Based on the pattern of bowel alteration, IBS can be described as diarrhea- or constipation-predominant; nevertheless, an alternating pattern in which patients go from one extreme to the other is also seen. Diarrhea-predominant IBS is more common in women. Nocturnal occurrence is quite rare and suggestive of organic etiologies. Fecal incontinence is not uncommon. In constipation-predominant cases, bloating, incomplete evacuation, and changes in stool shape, along with the presence of mucus is a typical scenario.

In an effort to standardize the clinical definition of IBS, the Rome II criteria were created. Although quite useful and widely utilized, these criteria failed to include a subgroup of IBS patients with postprandial exacerbation and those with painless, functional diarrhea.

Rome II Criteria

Recurrent symptoms, two of three, for at least 12 weeks in the preceding year:

- Abdominal pain relieved by defecation and/or
- Onset associated with change in frequency of stool and/or
- Onset associated with change in stool appearance

Symptoms supportive of diagnosis of IBS:

- Abnormal stool frequency
- Abnormal stool form
- Abnormal stool passage
- Passage of mucus
- Bloating

Evaluation

Because we do not have a single, objective confirmatory test, the diagnosis of IBS is based on symptomatology, the absence of organic disease, and a comprehensive medical evaluation.

Besides a detailed history and physical examination, diagnostic studies should include a complete blood count, comprehensive metabolic panel, and thyroid function tests. The use of imaging studies should be tailored to each patient, taking into consideration family history, age, and type and severity of symptoms. Colonoscopy is often indicated but, in selected cases, flexible sigmoidoscopy, double contrast barium enema, or virtual colonoscopy are appropriate alternatives.

The presence of "red flag" signs such as weight loss, fever, anorexia, rectal bleeding, and laboratory abnormalities should raise doubts about the diagnosis of IBS and prompt a detailed evaluation for alternative conditions.

The differential diagnosis of IBS is quite broad and includes gastrointestinal neoplasia, inflammatory or infectious colitis, malabsorptive processes, systemic diseases that may affect the gastrointestinal system such as thyroid disorders, and others as listed below.

- Inflammatory bowel disease
- Collagenous or lymphocytic colitis
- Infectious enteropathies
- Eosinophilic gastroenteritis
- Lactose intolerance
- Malabsorption syndromes
- Thyroid dysfunction
- Colonic inertia
- Pelvic floor dysfunction
- Urogenital diseases in women

Treatment

In the management of IBS, reassurance and patient education are of great value. Reasonable and clear expectations regarding treatment response should be established.

Our preference is to approach patients and direct treatment based on the dominant complaint. Symptom severity and overlapped features of psychiatric alterations should also help us guide therapy.

The treatment of constipation-predominant IBS should be initiated by behavioral modification, including a rich fiber diet and toilet training. Stool softeners should be initiated and stimulant laxatives avoided. If needed, fiber supplements should be added; however, with higher doses, a significant number of patients will experience worsening bloating. We tend to use a light osmotic laxative such as magnesium hydroxide as a rescue measure no more than once or twice a week. Tegaserod (Zelnorm), a 5-HT4 receptor agonist, was recently approved by the Food and Drug Administration for women with constipation-predominant IBS. This agent has been shown to relieve the global symptoms of constipation, bloating, and abdominal pain in up to two-thirds of the cases. Most common adverse effects are diarrhea and headaches. Few cases of ischemic colitis have been reported. The recommended dose is 6 mg twice a day.

In cases in which the major complaint is diarrhea or alternating patterns of bowel dysfunction, the use of fiber supplements can improve stool consistency and reduce incontinence without leading to constipation. At times, cholesterol-binding resin such as cholestyramine or colestipol can be quite effective. In refractory cases, more potent antimotility agents such as diphenoxylate or loperamide may be needed. Alosetron, a 5-HT3 receptor agonist, is proven to improve the global symptoms of abdominal pain, bloating, and diarrhea in approximately 60% of the cases. Although withdrawn from the market in 2000 because of its association with ischemic colitis, it is now available under close patient surveillance and only for women with diarrhea-predominant IBS. Alosetron should not be used in patients with constipation or an alternating pattern of bowel dysfunction. The recommended dose is 1 mg once or twice daily.

Antispasmodic agents such as dicyclomine and hyoscyamine are the most common drugs used for the management of abdominal pain in IBS. Clinical trials have shown marginal benefit when compared with placebo.[2,3] In clinical practice, their benefit is modest and we prefer to use them in mild IBS cases and for short periods of time. Undesirable anticholinergic effects are common.

In cases of severe IBS, a comprehensive approach is pivotal. These patients tend to have greater levels of somatization, depression, and anxiety disorders. Sleep disturbances are common. The patient often claims multiple drug allergies and sensitivities. In women in particular, a history of sexual abuse should be identified and addressed, likely needing the help of a therapist. Other functional chronic pain syndromes such as fibromyalgia and migraines are not rare. Several and extensive past medical evaluations are rather the norm. These patients benefit from education, reassurance, and behavioral modification. Fiber and antispasmodic agents are of minimal benefit. Psychotropics are the mainstay form of treatment in this subgroup of patients. Tricyclic agents at low doses are of great benefit, especially in those patients with diarrhea-predominant IBS. These agents can target the symptoms of insomnia, diarrhea, emotional disturbance, and abdominal pain. We tend to use those with less anticholinergic effect, in particular amitriptyline, desipramine, and doxepin. Adverse effects include anticholinergic effects, weight gain, and cardiac dysrhythmias. Selective serotonin reuptake inhibitors can be used in cases with predominant symptoms of constipation.

Summary

Our approach to medical therapy for mild cases of IBS starts with a high fiber diet, fiber supplements, and antispasmodics. Stool softeners, osmotic laxatives, and antidiarrheal agents should be used as indicated by the predominant pattern of bowel alteration. In moderate cases, the use of the new serotonin receptor agents should be considered. Patients presenting with severe IBS should be initially managed with psychotropic agents and intensive behavioral modification. Concomitant obstructed defecation and symptoms of constipation should be addressed.

References

1. Drossman DA, Camilleri M, Mayer EA, Whitehead WE, Camilleri MI. AGA technical review on irritable bowel syndrome. Gastroenterology 2002;123:2108–2131.
2. Jailwala J, Imperiale TF, Kroenke K. Pharmacologic treatment of the irritable bowel syndrome: a systematic review of randomized, controlled trials. Ann Intern Med 2000;133:136–147.
3. Talley NJ. Pharmacologic therapy for the irritable bowel syndrome. Am J Gastroenterol 2003;98:750–758.

7-5

Irritable Bowel-Anismus

Wael Solh and Eric G. Weiss

The causes of constipation and altered defecation are multifactorial, and the manifestations are varied. Etiologies of constipation or altered defecation can be divided into two categories – slow transit constipation, and pelvic outlet obstruction. Pelvic outlet obstruction includes etiologies such as paradoxical or nonrelaxation of the puborectalis muscle or anismus (nonrelaxation of the "anal canal"), rectal prolapse or intussusception, and nonemptying rectoceles. Associated findings may include perineal descent and solitary rectal ulcer syndrome.

In normal defecation, the pelvic floor muscles and external anal sphincter are voluntarily inhibited resulting in an increase in the anorectal angle with increasing intraabdominal pressure. Patients with nonrelaxing puborectalis syndrome (NRPS) or anismus are unable to voluntarily inhibit contraction of the pelvic floor. When inappropriate function of the puborectalis muscle (inappropriate contraction or failure to relax) results in the inability to evacuate the rectum, the condition is termed anismus. Unlike levator syndrome, anismus is painless.

Diagnosis

The diagnosis of NRPS is confirmed with a combination of anorectal manometry, electromyography (EMG), and cinedefecography. Cinedefecographic findings that suggest NRPS include obstructed defecation, a long and persistently closed anal canal, and the presence of a rectocele. Electromyography studies can confirm persistent contraction during defecation and straining. The balloon expulsion test is another modality that can be used. This simple and inexpensive test has shown high reliability in diagnosing pelvic outlet obstruction resulting from nonrelaxation of the puborectalis muscle.[1] However, these diagnostics should be interpreted with caution. In a prospective study, we found the sensitivity and specificity of EMG and cinedefecography to be suboptimal if applied individually to diagnose anismus.[2] Others argue that paradoxical contraction of the puborectalis muscle on EMG analysis is not a specific finding, being found in patients with solitary ulcer syndrome and idiopathic perineal pain.[3] Therefore, the diagnosis of NRPS must be made based on the patient's clinical findings, supported by more than one physiologic investigation.

Treatment

Because NRPS is a behavioral rather than an anatomic abnormality, biofeedback is the standard therapy. In biofeedback training, patients are allowed to view their own EMG or manometric tracings on a video monitor while attempting to relax the pelvic floor and sphincter muscles. Numerous reports have demonstrated success rates ranging from 37% to 100%.[4] In the largest series to date, Gilliland and Wexner[5] found only one variable to be predictive of a successful outcome: patients who self-discharged from therapy had a success rate of only 29%, whereas patients who remained in therapy until discharged by the therapist had a success rate of 63%.

Where biofeedback therapy has failed to relieve a patient's symptoms, botulinum toxin type A injection under EMG guidance can be offered in selected cases. We found high initial success rates with moderate long-term results.[6] Another therapeutic modality available is electrogalvanic stimulation (EGS). Hull et al.[7] reported poor long-term outcome (19% had their symptoms relieved, 24% had partial relief at 28 months follow-up) for the treatment of levator syndrome. More recently, Chiarioni et al.[8] reported a 50% success rate when they treated 30 patients with pelvic floor "dyssynergia" with EGS. Overall, we believe EGS may represent a useful adjunct in the treatment options for these patients because of its simplicity and low morbidity.

Surgical Approaches

Surgery has been universally unsuccessful in the treatment of NRPS. A review of the literature provides numerous examples of failures to surgically alleviate these patients' symptoms. These futile attempts have included division of the puborectalis muscle either posteriorly or laterally, anorectal myectomy, rectopexy, and progressive anal dilation. Not only were these attempts unsuccessful, but some resulted in unacceptable levels of temporary or permanent incontinence. Another dilemma the clinician may be confronted with is how to approach the constipated patient with both slow colonic transit and pelvic outlet obstruction. In short, the key to success in this population is appropriate patient selection. In fact, a significant proportion of constipated patients have NRPS.[9] It is imperative to accurately identify these patients using a combination of transit studies and pelvic floor function tests, before proceeding with colectomy and ileorectal anastomosis. The use of such comprehensive preoperative investigations resulted in a 97% success rate in those patients who underwent surgical therapy.[10] It is our practice to recommend biofeedback therapy before any surgical intervention in those patients with concomitant slow transit and NRPS. In severe cases of refractory pelvic outlet obstruction, we perform a diverting stoma (loop ileostomy or sigmoid colostomy depending on colonic function). It is our preference to use the laparoscopic approach to divert patients with benign pathology.

References

1. Fleshman JW, Dreznik Z, Cohen E, Fry RD, Kodner IJ. Balloon expulsion test facilitates diagnosis of pelvic floor outlet obstruction due to non-relaxing puborectalis muscle. Dis Colon Rectum 1992;35:1019–1025.
2. Jorge JM, Wexner SD, Ger GC, Salanga VD, Nogueras JJ, Jagelman DG. Cinedefecography and electromyography in the diagnosis of nonrelaxing puborectalis syndrome. Dis Colon Rectum 1993;36:668–676.
3. Jones PN, Lubowski DZ, Swash M, Henry MM. Is paradoxical contraction of puborectalis muscle of functional importance? Dis Colon Rectum 1987;30:667–670.
4. Corman ML. Colon and Rectal Surgery. 4th ed. Philadelphia: Lippincott Williams & Wilkins; 1998:383–385.
5. Gilliland R, Wexner SD. Outcome and predictors of success of biofeedback for constipation. Br J Surg 1997;84:1123–1126.
6. Joo JS, Wolff B, Wexner SD. Initial North American experience with botulinum toxin type A for treatment of anismus. Dis Colon Rectum 1996;39:1107–1111.
7. Hull TL, Milsom JW, Church J, Oakley J, Lavery I, Fazio V. Electrogalvanic stimulation for levator syndrome: how effective in the long run? Dis Colon Rectum 1993;36(8):731–733.
8. Chiarioni G, Chistolini F, Menegotti M, et al. One year follow-up study on the effects of electrogalvanic stimulation in chronic idiopathic constipation with pelvic floor dyssynergia. Dis Colon Rectum 2004;47(3):346–353.
9. Wexner SD, Nogueras JJ, Jagelman DG. Physiologic assessment of colorectal functional disorders: use or abuse of technology? Dis Colon Rectum 1992;35:10–11.
10. Nyam DC, Pemberton JH, Ilstrup DM, Rath DM. Long term results of surgery for chronic constipation. Dis Colon Rectum 1997;40:273–279.

Section VIII

Prolapse Syndromes Therapy

Section VIII

Prolapse Syndromes Therapy

G. Willy Davila

Herniation of a pelvic floor organ through the levator hiatus or into an adjacent compartment is a very commonly found anatomic disturbance. Our understanding of the etiologic basis for development of these herniations has improved as our knowledge of pelvic support anatomy has increased. We encourage all pelvic floor surgeons to utilize cadaveric specimens in attaining a better understanding of pelvic floor anatomy. The use of fresh frozen cadavers has allowed clinicians to better identify causes of genital prolapse, as well as increase their understanding of the mechanism of action for reparative surgery.

All pelvic floor anatomic and functional defects should be addressed at one surgical setting, whenever possible. It is therefore important for the pelvic surgeon to have a clear understanding of the identification of pelvic support defects, their underlying anatomic defects, and preferred surgical repair techniques. Although there are nonsurgical means for addressing pelvic organ prolapse, surgical therapy remains the primary means of addressing these problems, even in the elderly patient.

8-1

Vaginal Pessaries

Minda Neimark

Pessaries have long been used for the treatment of pelvic floor dysfunction in women. Initial descriptions of a vaginal device entailed intravaginal placement of objects to support genital prolapse and/or administer therapeutic chemicals. Pessaries were originally designed to treat genital prolapse. Application of pessaries for women with stress urinary incontinence (SUI) are rather recent and will be discussed elsewhere. With the high coexistence of genital prolapse and urinary incontinence, it is likely that many women who were fit with a pessary for genital prolapse noted an improvement in their urinary incontinence. In addition, women with exteriorized prolapse may have occult stress incontinence, which is uncovered upon being fit with a vaginal pessary for reduction of the prolapse.

As our population ages, with an increased incidence of pelvic floor dysfunction, alternatives to surgery become more desirable. There are many urogynecologic indications for pessary usage, as mentioned below:

- Patient unable to safely undergo pelvic reconstructive surgery
- Desire to avoid surgery for genital prolapse
- Evaluation of continence mechanism in moderate to severe vaginal prolapse
- Pregnant patient with incompetent cervix/premature delivery/multiple gestation
- Genital prolapse in the neonate
- Low back/pelvic pain secondary to genital prolapse
- Postoperative usage after pelvic reconstructive surgery to prevent recurrence

Historical Pessary Applications

In the 1800s, Hugh Lenox Hodge designed the lever pessary to treat uterine retroversion thought to be a cause of pelvic pain.[1] As modifications of the lever pessary were made, other indications were proposed for its use. In 1961, Vitsky[2] suggested that cervical incompetence was attributable to a lack of central uterine support. Uterine retroversion has also been associated with infertility and pelvic pain. Placing a lever pessary would displace the cervix posteriorly, thus lifting the weight of the uterus off of the incompetent cervix. Women diagnosed with an incompetent cervix were treated during pregnancy with a Hodge pessary from 14 to 38 weeks' gestation, with an 83% success rate.[3] Currently, cervical cerclage is the treatment of choice for women with cervical incompetence. There is great controversy regarding the possible causative role of uterine retroversion in many gynecologic conditions including pelvic pain, infertility, and sexual dysfunction.

Pessaries for Genital Prolapse

Use of vaginal devices for prolapse reduction and administration of chemicals is documented as early as the Egyptian civilization. Written documentation of efforts to reduce genital prolapse with vaginal objects dates back as far back as the fifth century. Modernization of the pessary came with the discovery of vulcanization of rubber and a better understanding of female anatomy. Since then, multiple modifications in pessary design and material selection for manufacture have been made.[4]

Advances in gynecologic surgery and anesthesia over the last several decades have reduced the need for pessary usage in the treatment of prolapse and incontinence. However, the recent increase in the elderly population requiring conservative treatment of prolapse and incontinence has led to a resurgence of pessary use.[5] In addition, there remains a very acceptable role for therapeutic use of a vaginal pessary in the premenopausal patient. In a recent survey administered to members of the American Urogynecologic Society, 77% of the respondents used pessaries as a first-line therapy for prolapse, and only 12% reserved pessaries for women who were not surgical candidates. In addition, 92% of the physicians surveyed believed that pessaries relieved symptoms associated with pelvic organ pro-

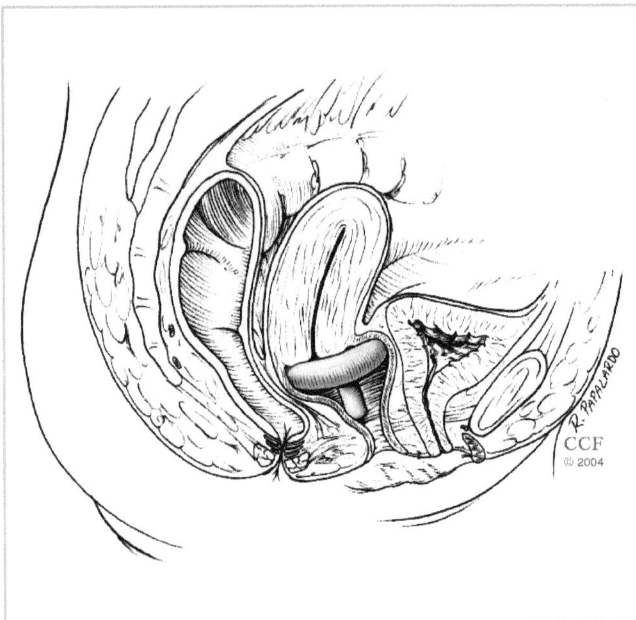

Figure 8-1.1. Gellhorn pessary. (Reprinted with the permission of The Cleveland Clinic Foundation.)

lapse, and 48% believed they had a therapeutic benefit in addition to relieving the symptoms.[6]

In a simple prospective protocol for pessary management, patients with symptomatic pelvic prolapse were given the option of pessary use versus surgery or expectant management. If the vaginal pessary was chosen as the method of treatment, the patient was fitted with a ring pessary or a pessary that could be retained without difficulty. The patient then followed up at scheduled intervals to evaluate pessary effectiveness. Sixty-six percent of those who used a pessary for more than 1 month remained users after 12 months. Fifty-three percent of the patients continued to wear the pessary after 36 months.[7] A retrospective series of 107 patients who were fit with a Gellhorn, cube, or ring pessary for symptomatic vaginal prolapse for various indications including medically unfit for surgery, awaiting surgery, or desired conservative management, confirmed that at least 50% of the women continued use of their pessary without complications at follow-up.[8] Of those who continued pessary use, 20% were patients who initially desired surgery but later declined because of satisfaction with the vaginal pessary (Figure 8-1.1).

Pessary Types

Many different types of pessaries are currently available, most made of medical grade silicone or rubber. Although the majority were designed for specific types of prolapse, many have the indication for prolapse as well as SUI. Clinical use of the pessary is typically based on a "best fit" choice (Table 8-1.1). The best pessary is the one that elevates the prolapse and remains in place during ambulation. In addition, it should sit in the vaginal canal without causing pain or becoming obstructive in terms of voiding and bowel evacuation. The health care provider must be prepared to try several different types and sizes until the correct pessary is found for each individual patient. Because of variations in pelvic anatomy, optimal fitting of a pessary may be challenging. In addition, the large number of pessary types, each with various sizes, further complicates the process. Because pessary fitting is a trial and error process, a clinician must have a large inventory of pessaries of varying sizes in order to accomplish appropriate fitting of patients (Figures 8-1.2–8-1.8).

Table 8-1.1. Pessary types and indications

		Indications					
		Cystocele					
Pessary	Sizes	Small	Large	Enterocele	Rectocele	Vault Prolapse	Stress Incontinence
Smith-Hodge	0–9	X		X			X
Risser	0–9	X		X			X
Marland	2–8	X					X
Incontinence dish (mm)	55–85	X					X
Incontinence ring	0–10	X					X
Ring (with or without support)	0–9	X	X	X			X
Gerhrung (with or without knob)	0–9	X			X		X
Gellhorn (in)	$1\frac{1}{2}$–$3\frac{1}{2}$	X	X	X	X	X	
Schaatz (in)	$1\frac{1}{2}$–$3\frac{1}{2}$	X		X	X	X	
Doughnut (in)	2–$3\frac{3}{4}$	X	X	X	X	X	
Inflatable (in)	2–$3\frac{3}{4}$	X	X	X	X	X	
Cube	0–7	X	X	X	X	X	

Source: Adapted from Oster and Javert[3] (Table 1).

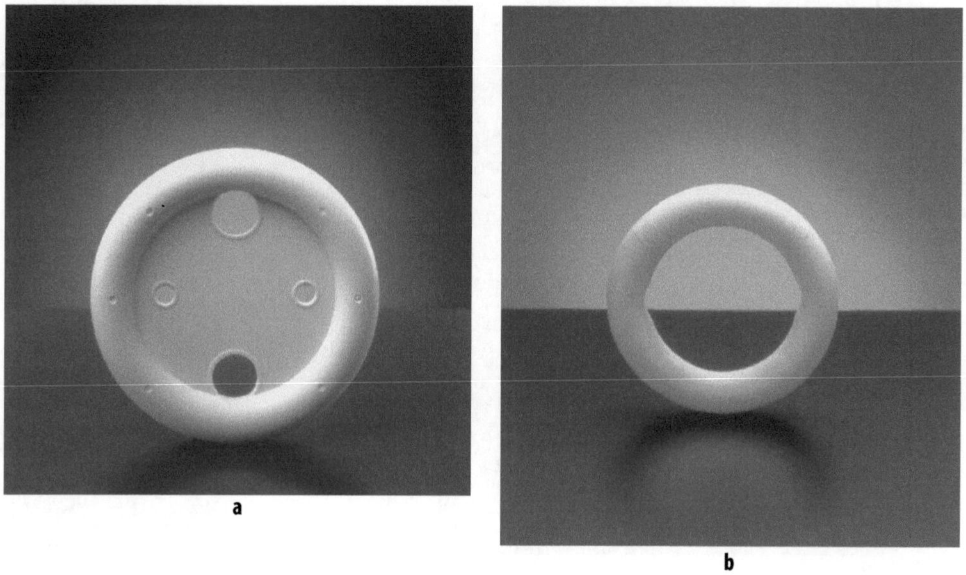

Figure 8-1.2. Ring pessary (**a**) with and (**b**) without support. (Courtesy of Milex Products, a division of Cooper Surgical.)

Figure 8-1.3. Shaatz pessary. (Courtesy of Milex Products, a division of Cooper Surgical.)

Figure 8-1.4. Donut pessary. (Courtesy of Milex Products, a division of Cooper Surgical.)

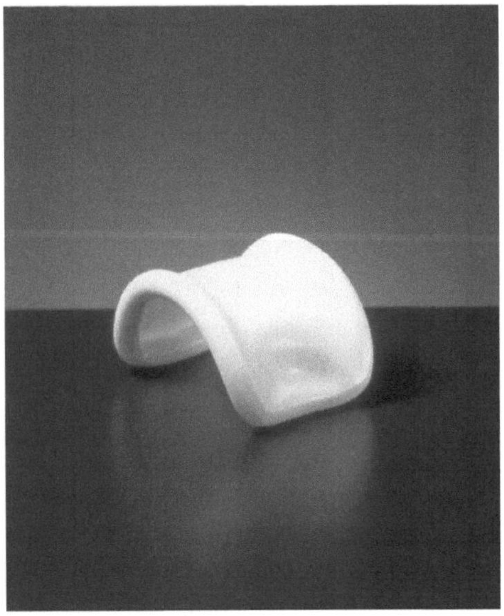

Figure 8-1.5. Gehrung pessary. (Courtesy of Milex Products, a division of Cooper Surgical.)

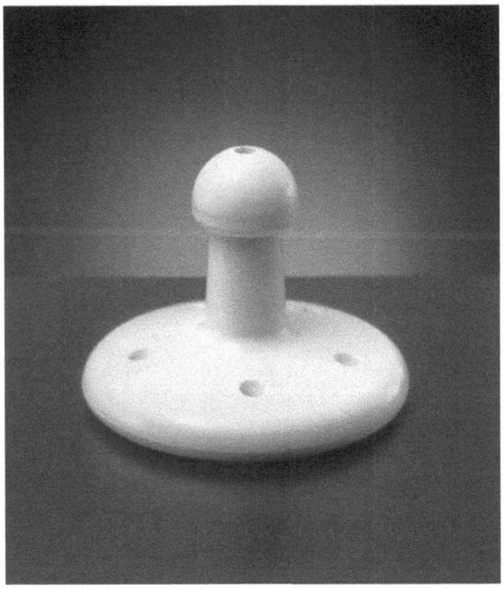

Figure 8-1.6. Gellhorn pessary. (Courtesy of Milex Products, a division of Cooper Surgical.)

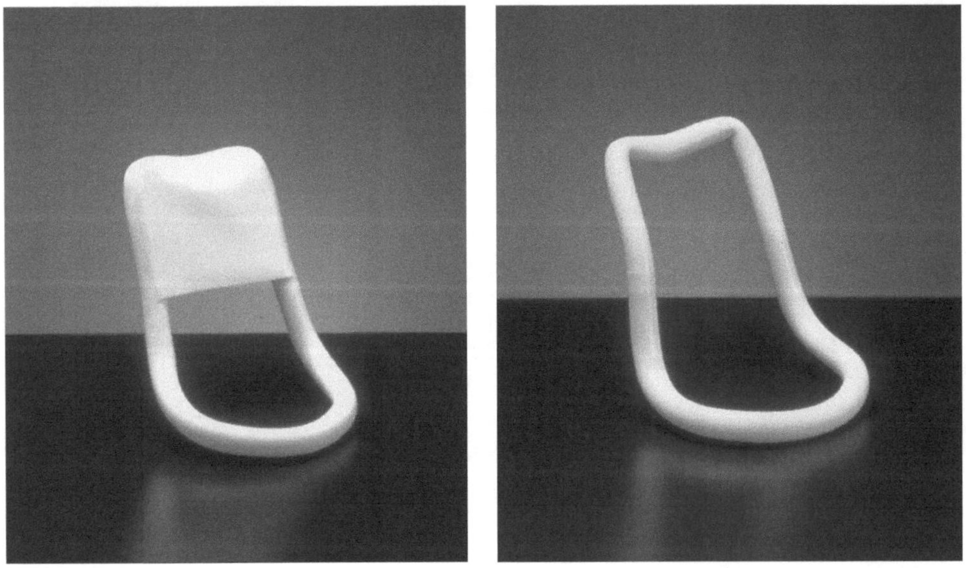

a b

Figure 8-1.7. Smith-Hodge pessary (**a**) with and (**b**) without support. (Courtesy of Milex Products, a division of Cooper Surgical.)

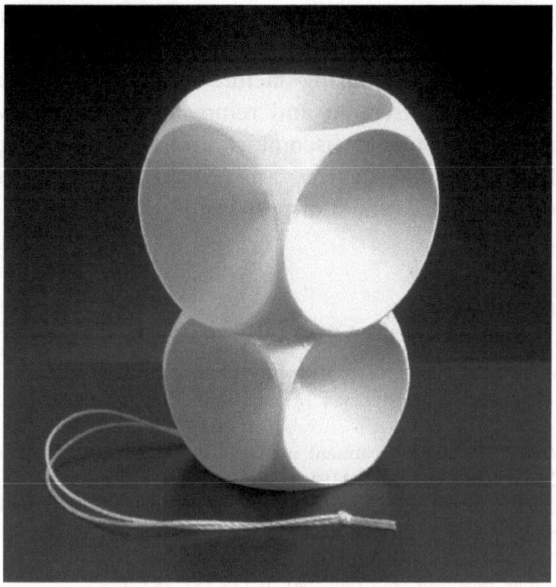

Figure 8-1.8. Cube pessary. (Courtesy of Milex Products, a division of Cooper Surgical.)

Pessary Care

Although the pessary can initially be effective in reduction of the prolapse, with further weakening of the pelvic floor caused by continued progression of pelvic floor neuropathy as well as an increase in the vaginal hiatus, or decreased vaginal hiatus diameter due to levator muscle hypertrophy the size and type of pessary may need to be changed. Pessary size may also need to be changed over time because of patient weight loss. For those patients who are able to easily remove and place the pessary themselves, follow-up with the clinician should occur every 6 months to assure proper fit and placement. Following is a list of recommended care procedures; the patient who cannot self-care for the pessary must be seen by the clinician on a regular basis. Pessary care recommendations should be followed closely for safe long-term use.

- Remove at least two nights per week
- Leave out overnight
- Insert 1 to 2 g of estrogen cream during the night while pessary is out
- Wash pessary with soap and water
- Reinsert using water-soluble lubricant
- Report any unusual discharge, bleeding, or discomfort
- Report any changes in bladder or bowel function
- Have pelvic examination every 6 to 12 months

For those unable to provide self-care, we recommend scheduled office visits for pessary care every 6 to 12 weeks. Patients must be encouraged to use intravaginal estrogen cream regularly to help prevent vaginal ulceration from foreign-body use within the vagina. This is especially true in the postmenopausal patient. Some pessaries that are used, such as the Gellhorn, may be difficult for the patient to remove herself. This can be facilitated by attaching a string (dental floss or suture) to the pessary. For those patients who are sexually active, the pessary may either be removed or left in place provided there is no discomfort to either the patient or her partner.

Many young patients with prolapse believe that pessaries are designed for the elderly women with prolapse, dissuading them from considering their use. Young women of reproductive age are often ideal candidates for pessary treatment of prolapse, as well as SUI. However, most women of reproductive age desire surgical correction of their prolapse, and they may use a vaginal pessary only to relieve the symptoms of genital prolapse until the desired time to undergo surgical treatment.

Other ideal candidates include those:

- Of reproductive age
- Comfortable with genital contact (i.e., tampon user)
- With adequate manual dexterity
- Compliant with safe usage recommendations
- With a well-estrogenized vagina
- With an unscarred vagina

The Patient Who Cannot Be Fit with a Pessary

Not all female candidates can be successfully fit with a pessary. Typical reasons for inability to be fit include vaginal scarring with loss of vaginal caliber or length from previous surgery, severe urogenital atrophy, vaginal pain, and markedly restricted or enlarged vaginal introitus. In those with significantly increased vaginal caliber, performing a perineoplasty and subsequent refitting with a pessary should be considered. For elderly women with severe genital prolapse who are not, and will not become, sexually active, and cannot be fit with a pessary, consideration should be given to a colpocleisis performed under regional or local anesthesia.

Contraindications to Pessary Use

Although the pessary can be a valuable tool in the treatment of genital prolapse, there are certain patients for whom pessary use may be considered contraindicated.

- Severe atrophic tissues
- Erosive ulcerative changes in the vaginal mucosa
- Undiagnosed vaginal bleeding
- Undiagnosed vaginal discharge

- Vaginal or cervical cancer
- Noncompliance in follow-up
- Impaired mental capacity

At our center, pessaries are frequently used as first-line therapy for prolapse and incontinence. For advanced degrees of prolapse, our most commonly used pessary is the Gellhorn, which is available with either a short or long stem depending on the patient's vaginal length. Cube pessaries should only be used with extreme caution in women who will be compliant with follow-up, as the suctioned adherence to the vaginal sidewalls may lead to significant mucosal erosions and ulcerations. Shaatz pessaries are also used in patients with advanced prolapse, but may be more difficult to remove, because of the absence of the stem.

For those elderly women with severe genital prolapse as well as SUI who cannot or do not want surgery, a pessary may be placed and a bulking agent such as collagen used to treat the incontinence. Many women who wear pessaries for SUI or prolapse eventually become frustrated with the efforts required for safe pessary use, and opt for surgical management. Some women will postoperatively use a pessary upon return to increased physical activity to help provide additional support to the pelvic floor and reduce the risk of prolapse recurrence.

Cost issues must also be considered. In many cases, reimbursement by insurance companies does not cover the entire cost of a pessary. Additionally, no optimal distribution systems through pharmacies or durable medical supply houses exist. The physician is thus typically at financial risk when fitting pessaries.

Motivated patients who wish to avoid surgical therapy are the optimal candidates for vaginal pessary use. Although several limiting factors may arise, including difficulty with insertion and removal, interference with sexual activity, pelvic discomfort, and associated vaginal discharge, vaginal devices are a valuable conservative option for the treatment of prolapse.

References

1. Miller DS. Contemporary use of the pessary. In: Sciarra JJ, ed. Gynecology and Obstetrics. Vol 1. Philadelphia: Lippincott-Raven; 1992: 1–12.
2. Vitsky M. Simple treatment of the incompetent cervical os. Am J Obstet Gynecol 1961;81:1194–1197.
3. Oster S, Javert CT. Treatment of the incompetent cervix with the Hodge pessary. Obstet Gynecol 1966;28:206–208.
4. Davila GW. Vaginal prolapse: management with nonsurgical techniques. Postgrad Med 1996;99:171–176, 181, 184–185.
5. Poma PA. Nonsurgical management of genital prolapse: a review and recommendations for clinical practice. J Reprod Med 2000;45:789–797.
6. Cundiff GW, Weidner AC, Visco AG, Bump RC, Addison WA. A survey of pessary use by members of the American Urogynecologic Society. Obstet Gynecol 2000;95:931–935.
7. Wu V, Farrell SA, Baskett TF, Flowerdew G. A simplified protocol for pessary management. Obstet Gynecol 1997;90:990–994.
8. Sulak PJ, Kuehl TJ, Shull BL. Vaginal pessaries and their use in pelvic relaxation. J Reprod Med 1993;38:919–923.

8-2

Vaginal Vault Prolapse Surgery

G. Willy Davila

Identification of weakness of vaginal vault support is crucial for the surgical correction of vaginal prolapse. In the presence of a large cystocele or enterocele, accurate identification of a prolapsed vaginal apex can be challenging, even to an experienced examiner. However, it is not uncommon to see a patient referred with recurrent vaginal prolapse after an anterior and posterior repair who likely had preexisting vaginal vault prolapse before the initial reparative procedure. Recent estimates have calculated that approximately 14% of women in their peri- and postmenopausal years may have uterine prolapse. Of those who have been hysterectomized, approximately 38% will have some form of prolapse. It is estimated that 10% of this age group of women will have vaginal vault prolapse.

Anatomy

Full details of the anatomy of vaginal vault support are discussed in Chapter 4-2. It is important to realize that there are multiple components to vaginal vault support, including uterosacral and cardinal ligaments, endopelvic fascial envelopes, and lateral paravaginal attachments. Many surgical techniques for vault prolapse are designed to utilize one, or a combination of those structures. It is possible to restore attachment of the vaginal apex to the uterosacral ligaments. Specific restoration of cardinal ligament support of the vaginal apex is less likely and not currently within our armamentarium. As such, most surgical approaches to vaginal vault prolapse rely on compensatory techniques rather than restoration of previously existing support mechanisms. The creation of a central attachment site at the vaginal apex is crucial to achieving vault support and integrity of the anterior and posterior vaginal fascial envelopes (Figure 8-2.1). In the nonhysterectomized woman, the cervix acts as the central attachment site. In a woman undergoing a hysterectomy, the vaginal cuff anatomy should be restored by attaching the uterosacral ligaments firmly to the vaginal apex during the course of the procedure. Prevention of vaginal vault prolapse or development of a posterior enterocele should be a goal during every hysterectomy by performing a McCall culdoplasty or other form of attachment of the uterosacral ligaments to the vaginal apex.

Surgical Approaches

The goals of vaginal vault prolapse surgery include restoration of vaginal apical support and maintenance of normal vaginal length, axis, and caliber. To achieve a long-term successful outcome, correction of all pelvic floor defects should be achieved at the time of a vaginal vault suspension. Care must thus be taken to achieve integrity of the fascial envelope of the anterior and posterior vaginal walls and reattaching any fascial tears off of the vaginal apex during the surgical procedure.

Vaginal vault support procedures are divided into abdominal approaches and vaginal approaches, as well as reconstructive and obliterative procedures as shown below:

- Vaginal
  - McCall culdoplasty
  - Ileococcygeus suspension
  - Sacrospinous fixation
  - Uterosacral ligament suspension
  - Posterior intravaginal slingplasty (IVS) vault suspension
- Abdominal
  - Abdominal sacrocolpopexy
  - Reattachment of uterosacral ligaments
- Laparoscopic
  - Abdominal sacrocolpopexy
  - Reattachment of uterosacral ligaments
- Obliterative vaginal
  - Le Fort colpocleisis
  - Colpectomy

Choice of the surgical approach to a patient's vault suspension procedure is dependent on multiple individual

Figure 8-2.1. The vaginal cuff or the cervix act as a central attachment site for suspensory ligaments (uterosacral and cardinal) and anterior and posterior vaginal wall endofascial layers. 1, Vaginal apex. 2, perineal body. (Reprinted with the permission of The Cleveland Clinic Foundation.)

variables. We will typically perform an abdominal sacrocolpopexy in a patient with advanced prolapse in which the vaginal apex reaches significantly above the ischial spines on vaginal examination, for whom maintaining normal sexual activity is very important, someone who will require another abdominal procedure for their vaginal prolapse such as an abdominal paravaginal repair, and someone with significant vaginal wall scarring, large apical fascial defects, or significant foreshortening of the vaginal canal. Vaginal approaches are typically performed in postmenopausal women for whom sexual activity may not be as important and in whom other reconstructive procedures can be appropriately performed vaginally. Obliterative procedures are performed in elderly women who are not, and will not, be sexually active, and who request the least invasive procedure for advanced vaginal prolapse. Laparoscopic approaches are reserved for surgeons request laparoscopic skills adequate enough to safely achieve a good anatomic result. At our center, we restrict our laparoscopic surgeries to those patients who are good candidates for reattachment of the uterosacral ligaments to the vaginal apex or shortening of the uterosacral ligaments in enhancing vaginal vault support.

The importance of prevention of posthysterectomy vaginal vault prolapse cannot be overemphasized. Performance of a McCall culdoplasty at the time of a vaginal hysterectomy has proven value in maintaining vaginal vault support. We encourage surgeons to perform this procedure at the time of any vaginal hysterectomy and to be very methodical in restoration of vaginal vault support by reattaching uterosacral and cardinal ligaments to the vaginal apical fascia and mucosa. This is of special importance in women who are undergoing vaginal hysterectomy for preexisting vaginal vault prolapse. We perform a McCall culdoplasty by using a permanent suture (i.e., 0-silk) to incorporate both uterosacral ligaments to full thickness of the apical vaginal wall. It is important to tag the uterosacral ligaments during the hysterectomy procedure, so as to be able to easily identify the ligaments for suture placement. In women with vault prolapse, we measure the number of centimeters the vault has prolapsed [pelvic organ prolapse quantitation (POP-Q) TVL minus point D], and place the sutures that same distance up each ligament from the vaginal cuff. In addition, during cuff closure, we approximate the cardinal and uterosacral ligaments to each other in the midline with individual figure-of-8 sutures. A McCall culdoplasty will prevent the development of an enterocoele unless a posterior fascial separation from the cervix was preexistent. Such a defect may be identified during a posterior colporrhaphy dissection, and the fascia should be reattached to the newly formed cuff.

Vaginal Procedures for Vaginal Vault Prolapse

Ileococcygeus Suspension

This is likely the least complex means for restoring vaginal vault support. Permanent, monofilament sutures are placed bilaterally between the apical lateral vaginal wall and the ileococcygeus muscle anterior to the ischial spine. Sutures can be placed either submucosally after dissection, or transmucosally without dissection. We typically reserve this for patients who are not significantly physically active and are candidates for a rapid vaginal procedure for isolated vaginal vault prolapse. A requirement for this procedure is sufficient vaginal length and caliber such that the vaginal sidewall reaches the ileococcygeus muscle where the suture will be placed. We will typically place 1 0-Prolene suture at either apical sulcus and tie transmucosally. We are always concerned about the development, or persistence, of central vault prolapse. Other procedures are more effective in restoring appropriate vaginal depth. Nevertheless, this procedure has been demonstrated in a randomized manner to be an effective means for restoring vaginal vault support.[1] The need for associated reconstructive or anti-incontinence procedures may limit the presumed simplicity of this procedure.

Uterosacral Ligament Suspension

Reattaching the uterosacral ligaments to the vaginal vault represents a very attractive means of normalizing vaginal vault support and anatomy – similarly to a McCall culdoplasty at the time of vaginal hysterectomy. In the hysterectomized patient, it can be challenging to identify the actual uterosacral ligaments. In addition, there is a reported risk of ureteral trauma of 10% to 15% in published series.[2] As such, we do not favor this procedure to restore vaginal vault support in a hysterectomized patient. However, the popularity of this procedure is growing, as is the number of published series.

Normalization of vaginal axis is the most attractive feature of this procedure. To identify the uterosacral ligaments, most surgeons recommend palpation of the cul-de-sac peritoneum overlying the ischial spine. This should be inferior to the location of the ureters. In a patient with a large, bulbous prolapse, this may be somewhat difficult because of redundancy of peritoneum. Typically, 1-3 monofilament permanent sutures are used to perform this procedure. Long-term results are not yet available.

Sacrospinous Ligament Fixation

The sacrospinous fixation is likely the most popular vaginal approach to vaginal vault suspension in the United States (US). Initially described in Germany and popularized in the US during the 1980s, this procedure has a high degree of effectiveness and longevity in restoring vaginal vault support. The vaginal apex is attached to the sacrospinous ligaments with permanent or delayed absorbable sutures. The procedure can be performed via a posterior vaginal dissection or anterior paravaginal dissection, and unilaterally or bilaterally. The main drawback of this procedure is the well-recognized recurrence of anterior vaginal wall defects. Because of the downward deviation of the vaginal axis, the anterior vaginal wall is more subject to transmission of intraabdominal pressure and creation of a significant cystocele. In our experience, the resultant cystocele typically does not extend beyond the vaginal introitus and is typically not symptomatic. In addition, when a bladder neck suspension procedure or mid-urethral sling is performed concomitantly, there is a well-recognized risk of postoperative voiding dysfunction.

Over the past few years, the number of sacrospinous fixations performed at our center has decreased with the advent of new vaginal vault suspension techniques and technology. When we perform a sacrospinous fixation, we will use a posterior vaginal dissection. The patient is placed in high stirrups and the vaginal apices are marked with marking sutures. This will help identify both apices because our preference is for a bilateral sacrospinous fixation as a means of normalizing vaginal support. If the procedure is performed unilaterally, we believe there is a higher risk of cystocele or opposite-side enterocele development. We have found that most patients can undergo a bilateral procedure without difficulties.

Once the posterior vaginal wall has been infiltrated with a hemostatic agent, a vertical incision is made from the introitus to an area approximately 2 cm below the vaginal apex. The rectum and connective tissue are then dissected off of the vaginal mucosa bilaterally to the level of the lateral vaginal sulcus, and the pararectal space is entered. The dissection is followed bluntly to the ischial spine. This is typically a fairly straightforward dissection. The loose connective tissue overlying the ischial spine and sacrospinous ligaments are then digitally cleared and the ligament is very clearly identified. We then place two CV-2 Gore-Tex sutures through each vaginal apex. Because we use Gore-Tex suture, we avoid full-thickness penetration of the vaginal mucosa with the suture material. If a woman's vaginal mucosa is significantly atrophic, we will use a small (2 × 3 cm) piece of Prolene mesh along the underside of either apex to enhance vaginal support. We then place each Gore-Tex suture through the mid portion of the ipsilateral sacrospinous ligament with a Miya hook. Other instruments are available for this purpose, but we have found the Miya hook to be quite acceptable and have not modified our instrument of choice. Once the sutures have all been placed, the upper half of the vaginal mucosal incision is closed. The Gore-Tex sutures are then to be tied, elevating the vaginal apex. Care must be taken to maintain appropriate vaginal symmetry when tying the sutures. Additional reconstructive surgical procedures can then be performed. Because these patients typically have a posterior enterocele as well, we will reattach any endopelvic fascia that has separated from the vaginal apex to the vaginal apex with permanent sutures to restore posterior wall fascial integrity.

Care must be taken in passing the hook through the sacrospinous ligaments. It is important to pass the hook through the ligament and not behind it, to avoid any damage to pudendal, sciatic, or other vascular or nerve structures. In addition, the sutures should be placed at least a finger breadth medial to the ischial spine.

In our series of 89 patients, our success rates have been quite positive.[3] Recurrence of vaginal vault prolapse has been identified in only 2.2% of patients. When this occurs, it is typically because of the de novo occurrence of an apical midline enterocoele between the two sites of sacrospinous ligament suspension sutures. The placement of a submucosal reinforcing piece of mesh has not been demonstrated to enhance our outcomes. In addition, our reoperative rate for any recurrent prolapse was 4.5%, and cystocele development rate was 14.6%.

Posterior Intravaginal Slingplasty Vaginal Vault Suspension

The posterior IVS (Tyco/US Surgical) procedure was initially described as a means of enhancing posterior pelvic floor support as a component of the "Integral Theory" described by Petros. As initially described, this procedure entails placing a piece of multifilament Prolene mesh through the pararectal space onto the vaginal apex to help provide apical support. The tape is inserted along the medial aspect of each buttock approximately 2 to 3 cm posterior and lateral to the anus using a blunt tunneler. In our initial experience, and in cadaveric dissections, we have

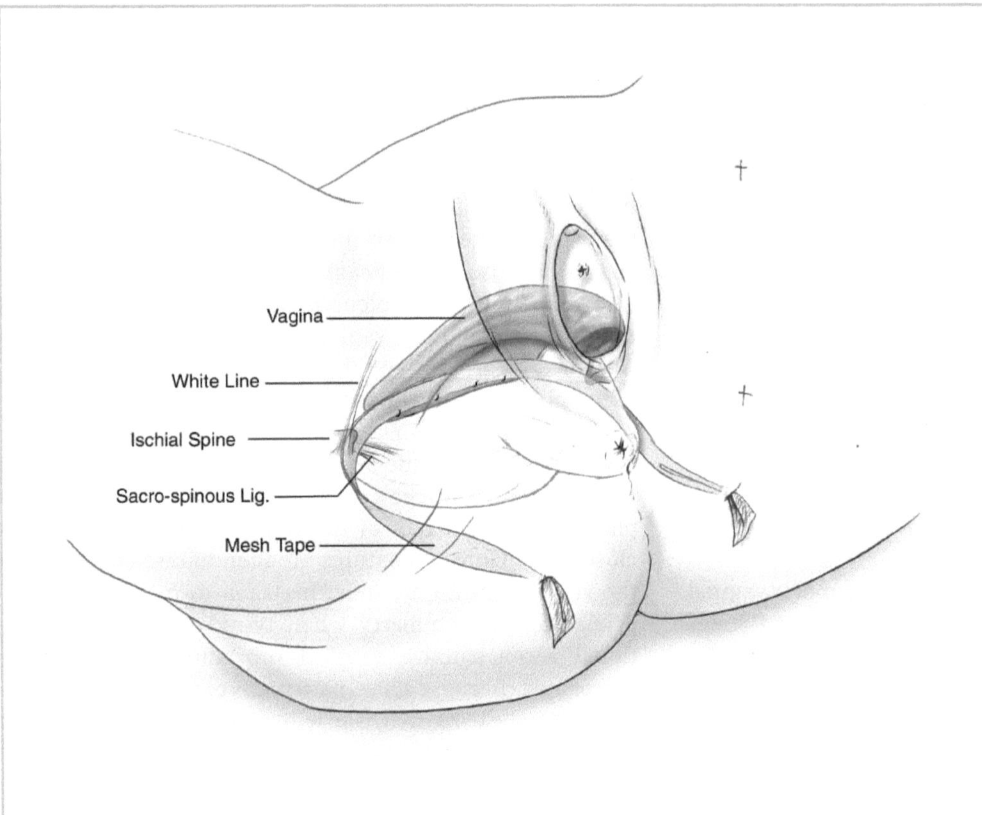

Figure 8-2.2. The tape from the posterior IVS procedure lies from pelvic sidewall to pelvic sidewall, allowing the vaginal cuff to be supported by neo-"Cardinal" ligaments. (Reprinted with the permission of The Cleveland Clinic Foundation.)

found the procedure, as originally described, to result in significant vaginal foreshortening. Therefore, we have modified the technique to place the tape further lateral along the pelvic sidewall in order to attain more cephalad support of the vaginal apex. We guide the IVS tunneler through the pararectal space and direct the tip along the lateral pelvic sidewall with the resultant placement immediately anterior to the ischial spine deep to the arcus tendineus. Once this is performed bilaterally, the tape essentially lies from ischial spine to ischial spine, analogous to Cardinal ligaments. The vaginal apex is then sutured to the tape to restore vaginal vault support. The ends of the tape are then pulled at their perianal ends, elevating the tape (and attached cuff) to the desired position. The dissection required is less than that for a sacrospinous fixation, and the resultant vaginal axis is more physiologic because it is anterior to the ischial spines, and not as horizontal as after a sacrospinous fixation (Figure 8-2.2).

The surgical procedure is performed via a near identical procedure to a sacrospinous fixation. One of the most attractive features of this procedure is the fact that the tape serves as an attachment site for the endopelvic fascia along the anterior and posterior vaginal walls. Fascia can be sutured to the tape such that when the tape is adjusted by pulling the ends of the tape paraanally, the vaginal apex and its fascial attachments are then elevated to their physiologic position. The presence of the tape also prevents the development of a central enterocele, which can occur after a bilateral sacrospinous fixation. In our current series of 77 patients treated with a posterior IVS, we have not had any vaginal vault prolapse recurrences and have only had one vaginal mucosal erosion of the tape.[4] Resultant anatomy includes a vaginal length of 7.6 cm (range, 5–11) and a mean POP-Q point C of 7.0 cm.

Radiographic evaluation of these patients has demonstrated that the vaginal axis is less horizontal than that found with a sacrospinous fixation and approximates more the axis resultant from an abdominal sacrocolpopexy.

Abdominal Approaches to Vaginal Vault Prolapse

Abdominal Sacrocolpopexy

Abdominal sacrocolpopexy is considered by many to be the gold standard procedure for repair of vaginal vault prolapse. The resultant vaginal axis is the most physiologic of all reconstructive procedures. Because this procedure is performed through an abdominal incision, its associated morbidity is greater than with the vaginal approaches. Therefore, it is typically reserved for women in whom sexual activity is of great importance. The goal of the procedure is to elevate the vaginal apex with a mesh bridge to the sacral promontory (Figure 8-2.3). The procedure is begun through a transverse or vertical laparotomy incision. We place the sacral promontory sutures first, before

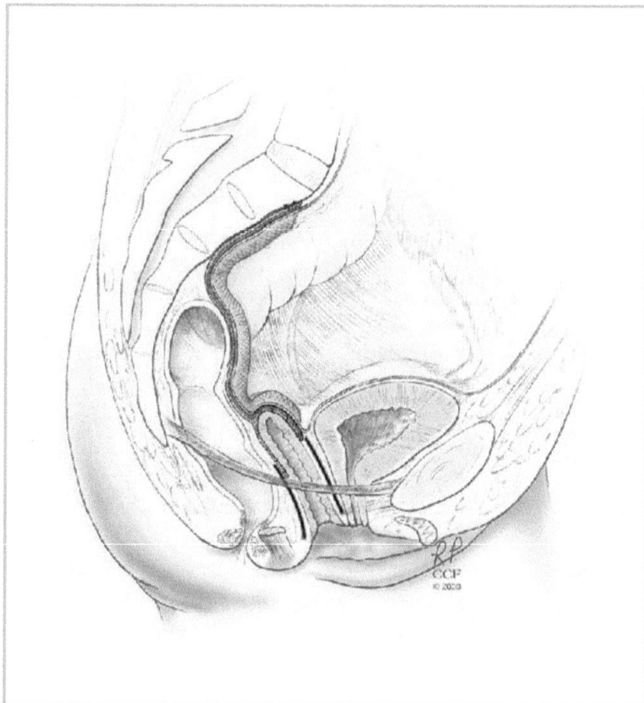

Figure 8-2.3. Abdominal sacrocolpopexy involves elevation of the vaginal apex to the sacral promontory with a mesh bridge. (Reprinted with the permission of The Cleveland Clinic Foundation.)

any vaginal manipulation. We open the peritoneum overlying the sacral promontory using the right ureter and the colon as the lateral landmarks for identifying the peritoneum to be incised. The dissection is then followed vertically to an area above the sacral promontory and down to the pelvic floor musculature inferiorly. Sharp and blunt dissection is then used to dissect fatty and connective tissue off of the sacral promontory. Once the anterior vertebral ligament is seen, the dissection is complete. Multiple vessels including the middle sacral artery and periosteal perforators are typically seen overlying the sacral promontory. To avoid significant bleeding, we use bone anchors for attachment of the sacral suspensory sutures. Life-threatening hemorrhage has occurred during suture placement at this location. Therefore, rather than elevating portions of the ligament with sutures, bone anchors minimize trauma because of the presence of a single puncture site (Figure 8-2.4). Either a drilled or a pressed in bone anchor can be used for this purpose. The currently available bone anchors have a monofilament #1 Prolene suture attached to the bone anchor. Once we have placed the two bone anchors on the sacral promontory, we will then address the vaginal apex. Rather than using a vaginal obturator, I prefer to place my fingers within the vaginal canal in order to be able to identify both vaginal apices. This is important for later approximation of the uterosacral ligaments in an abdominal McCall fashion to prevent development of an enterocele. We use Prolene mesh folded over and fashioned into a "Y" shape such that the longer arm extends along the posterior vaginal wall to the level of the rectal reflection. Anteriorly, the bladder is dissected off of the vaginal wall for a distance of a few centimeters for attachment of the short arm of the mesh. It is of great importance to identify any fascial tears along the anterior or posterior vaginal walls before placement of the mesh. These are typically seen as segments of fascial tissue that have separated from the vaginal apex. The edges should be sutured to the vaginal apex with permanent suture before placing the suspensory mesh. We will then use three rows of two 2-0 Prolene sutures along the posterior vaginal wall and one row of two 2-0 Prolene sutures along the anterior vaginal wall. Once the Prolene mesh has been secured to the vaginal walls, the abdominal McCall procedure is performed approximating the uterosacral ligaments in the midline. This part is important in prevention of the development of a posterior enterocele. The mesh is then elevated to the sacral promontory such that it is placed in a tension-free manner to suspend the vaginal apex. Once the mesh has been attached to the sacral promontory, the area is reperitonealized making sure to cover the entire mesh for the prevention of any internal hernias. If additional reconstructive procedures are performed, they are then performed at this time; typically including abdominal paravaginal and vaginal rectocele repairs.

It is important to maintain continuity of the endopelvic fascia with the Prolene mesh used for the sacrocolpopexy. This will allow for continuity of fascial envelope or mesh from the perineum to the vaginal apex both anteriorly and posteriorly (Figure 8-2.3).

It is important to note specific pre- and postoperative care required by the patient undergoing a sacrocolpopexy. We use preoperative mechanical bowel preps in all of our patients who undergo this surgical procedure to facilitate bowel packing and improve visualization of the sacral promontory. Postoperatively, we advance the patient's diet very slowly because there is an underlying risk of ileus,

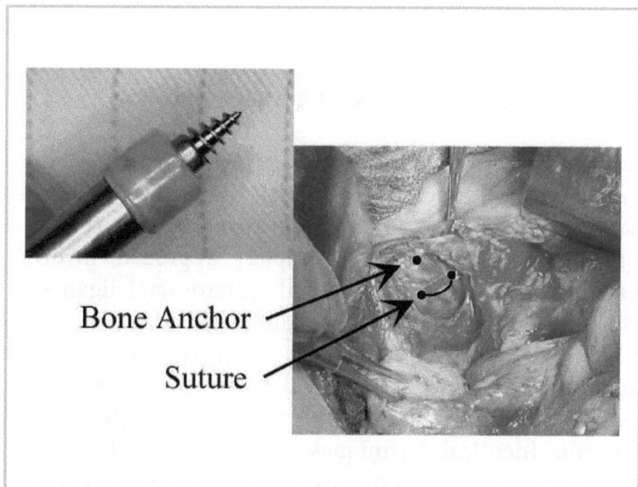

Figure 8-2.4. Use of bone anchors for abdominal sacrocolpopexy minimizes the amount of periosteal trauma and risk of hemorrhage.

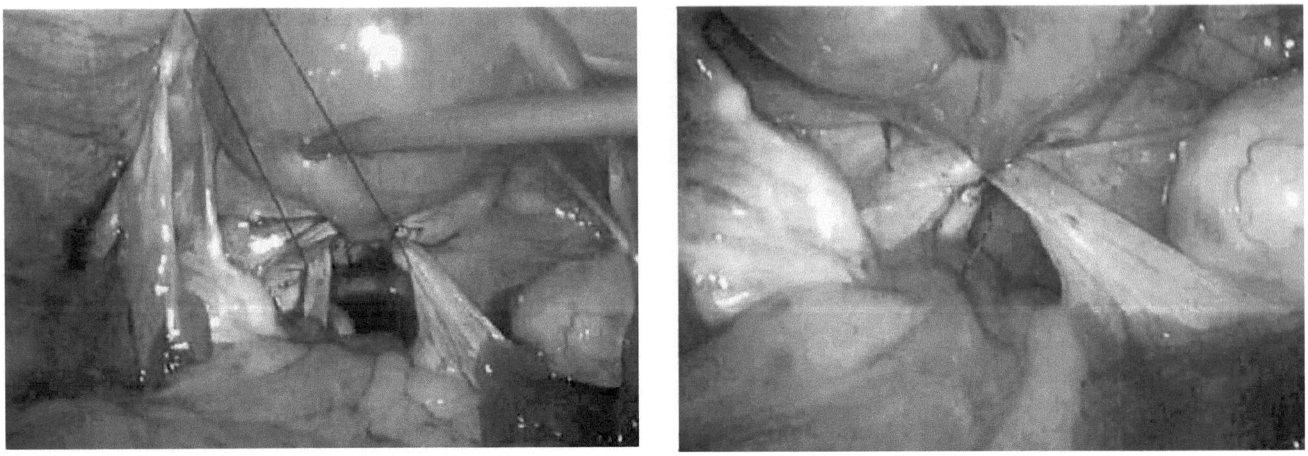

Figure 8-2.5. Laparoscopic vault suspension can be performed using the uterosacral ligaments and extracorporeal knot tying.

which can be quite problematic if the patient is fed too quickly. Our patients typically stay in the hospital for 3 to 4 postoperative days.

We attach the suspensory mesh to the sacral promontory rather than to S2 or S3 as has been described by others. This results in easier placement of the bone anchors through the anterior sacral ligament. This does not seem to result in an excessively vertical vagina, or any significant long-term difficulties.

In a recent series of 46 patients who underwent bone-anchored sacrocolpopexy using Prolene mesh at our center, we had no recurrences of vaginal vault prolapse.[5] We noted one rectocele, but no cystoceles. Mean blood loss was 213 mL (range, 100–500). We had one mesh erosion, which responded to conservative therapy. We use local estrogen cream postoperatively in all of our patients to maintain mucosal integrity.

Laparoscopic Vaginal Vault Suspension

Laparoscopic approaches to vault prolapse include laparoscopic sacrocolpopexy and uterosacral ligament suspension. A significant amount of laparoscopic skill is required for either procedure. The simplest approach involves reattachment or shortening of the uterosacral ligaments using a monofilament suture and extracorporeal knot tying (Figure 8-2.5). Modifications of the traditional sacrocolpopexy have been reported. However, success rates cannot be equated unless the procedure is performed with the identical techniques. We do not perform this surgical procedure at our institution, although it is often performed by our colleagues at the Cleveland Clinic in Cleveland.

Obliterative Procedures

LeFort Colpocleisis

As a result of the aging of our population, there has been a reported increase in the number of obliterative vaginal procedures being performed for vaginal vault prolapse. Many women of advanced age are not, and will not be, sexually active, and therefore, the simplicity, effectiveness, and low associated morbidity of the LeFort colpocleisis offers a very attractive treatment option. We perform this procedure under spinal or local anesthesia, thus reducing morbidity rates. Preoperative surgical clearance is of great important in these elderly patients.[6] Preparation with local estrogen cream is also important.

Our surgical technique entails denuding a rectangular segment of the anterior and posterior vaginal walls. The rectangles are then approximated to each other such that when the prolapse is reinserted into the vaginal canal, the rectangles are sutured to each other (Figure 8-2.6). If an anti-incontinence procedure is necessary, it is performed before completion of closure of the vaginal defect. A perineoplasty is a very important part of the surgical procedure in order to decrease the strain on the superior sutures. We perform a tight perineoplasty with resultant exposure limited to the urethral meatus.

In a multicenter series of 224 patients with a mean age of 78.3 years (range, 61–92), the success rate for vaginal prolapse was 96.4%, with four wound breakdowns and four recurrent rectoceles.[7] There was no perioperative mortality. This surgery seems to carry the highest degree of satisfaction for patients who undergo the procedure. However, preoperative counseling is of critical importance because of the postoperative inability to have sexual activity.

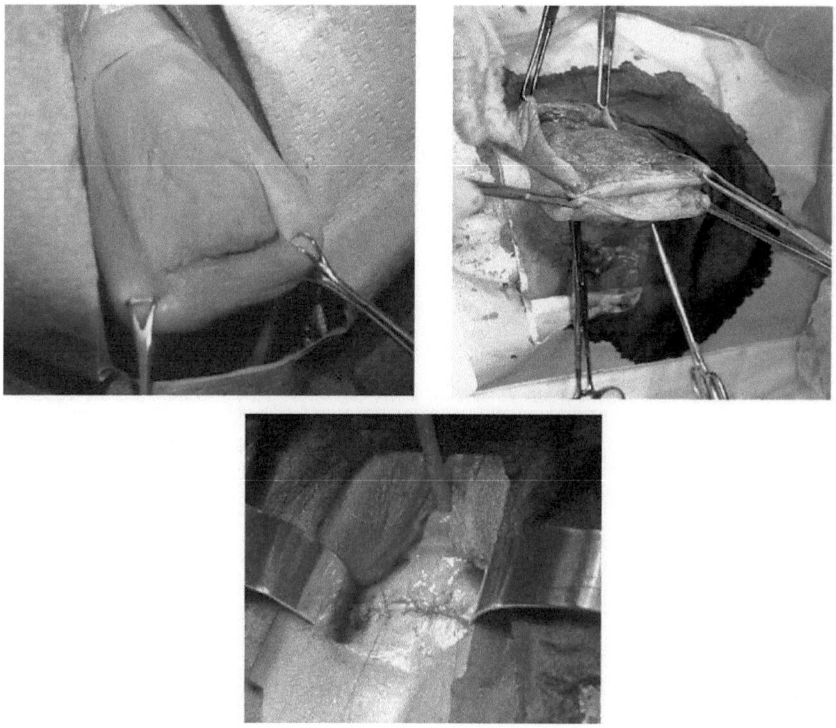

Figure 8-2.6. LeForte colpocleisis entails suturing the posterior and anterior vaginal walls to each other in order to obliterate the vaginal canal.

Summary

Multiple techniques are available for restoration of vaginal vault support. The primary challenge is identification of prolapse of the vaginal vault in women with advanced degrees of prolapse. The surgical approach is then individualized based on the patient's anatomy, visceral function, age, health status, and additional required procedures. Success rates of more than 95% are reported for all the mentioned techniques.

References

1. Maher CF, Murray CJ, Carey MP, Dwyer PL, Ugoni AM. Iliococcygeus or sacrospinous fixation for vaginal vault prolapse. Obstet Gynecol 2001;98:40–44.
2. Barber MD, Visco AG, WeidnerAC, Anubdsen CL, Bump RC. Bilateral uterosacral ligament vaginal vault suspension with site-specific endopelvic fascia defect repair for treatment of pelvic organ prolapse. Am J Obstet Gynecol 2000;183:1402–1411.
3. Guerette NL, Davila GW. Bilateral sacrospinous fixation: is vault reinforcement necessary? Int Urogynecol J 2003;14:S69.
4. Davila GW, Miller D. Vaginal paravaginal vault suspension with the posterior IVS technique. Presented at the Joint meeting of the American Urogynecologic Society and Society of Gynecologic Surgeons. July 2004.
5. Guerette NL, Davila GW. Abdominal sacrocolpopexy utilizing bone screws. Int Urogynecol J 2003;14:S11.
6. Pollak JT, Davila GW, Kopka SL, Ciocon J. Urogynecological and reconstructive pelvic surgery in women aged 80 and older. J Am Geriatr Soc 2004;52(5):851–852.
7. Guerette NL, Mattox F, Neimark M, Contreras-Ortiz O, Davila GW. Le Fort colpocleisis: a safe and effective procedure for elderly women. Int Urogynecol J 2003;14:S52.

8-3

Anterior Vaginal Prolapse

Sandip P. Vasavada and Matthew D. Barber

The successful management of anterior vaginal wall prolapse remains one of the greatest challenges in female pelvic reconstructive surgery. Prolapse of the anterior vaginal wall is the most common presentation of pelvic organ prolapse. Recent studies suggest that, of the various segments of the vagina that may be involved in prolapse (anterior, posterior, or apical), the anterior vaginal wall is the segment most likely to demonstrate recurrent prolapse after reconstructive surgery. Additionally, normal anterior vaginal support has an important role in supporting the urethra, and loss of this support can contribute to the development of stress urinary incontinence. It is, therefore, important that the pelvic reconstructive surgeon understand the normal support mechanisms of the anterior vaginal wall and the full spectrum of techniques for correction of anterior vaginal prolapse and cystoceles.

The vagina is a hollow, flattened fibromuscular tube lined with nonkeratinized stratified squamous epithelium. Underlying the vaginal epithelium is the vaginal muscularis, a well developed layer made up primarily of smooth muscle along with collagen and elastin. Some have labeled this layer of the anterior vaginal wall *pubocervical fascia* or *vesicopelvic fascia*. Although these terms are widely used, the use of the term "fascia" is a misnomer, because it does not accurately reflect the histology of the vagina. Advanced anterior vaginal prolapse can result from defects in several areas of pelvic support including:

1. The attenuation or defects of the vaginal muscularis in the midline (central defect)
2. Loss of lateral attachments from the anterior vagina to the pelvic sidewall (lateral or paravaginal defect)
3. Loss of bladder neck support
4. Separation of the cardinal-uterosacral ligament complex from the vaginal apex (superior defect)

These four defects accompany most large cystoceles and must be corrected to achieve good support, elevation of the bladder base, and a solid, durable repair. A detailed description of the anatomy of anterior vaginal support can be found in Chapters 4-1 and 4-2.

Evaluation

History

As with all patients with symptomatic pelvic floor dysfunction, a detailed history of bladder, bowel, and sexual function should be elicited. Patients with anterior vaginal prolapse often complain of symptoms directly related to the prolapse as well as symptoms of bladder dysfunction. Patients with prolapse at or beyond the level of the hymen often complain of pelvic pressure and bulging. Stress urinary incontinence often occurs in association with anterior vaginal prolapse and approximately one-third of women with stage II or greater prolapse will complain of symptoms of urinary urgency, frequency, and/or urge incontinence. Symptoms of voiding dysfunction such as a feeling of incomplete emptying, intermittent or reduced urine flow, or the need to splint to complete urination are common in women with prolapse beyond the hymen. Sexual difficulty is also a common complaint of women with advanced prolapse. In addition to this functional assessment, a detailed history of previous prolapse or anti-incontinence surgery should also be elicited.

Physical Examination

A pelvic examination should be performed with the patient in the lithotomy position. A detailed assessment of the support of all segments of the vagina should be made using the International Continence Society's pelvic organ prolapse quantitation (POP-Q) system. In addition, an assessment for central, lateral, and superior anterior vaginal support defects should be performed. If physical findings do not correspond to symptoms or if the maximum extent of the prolapse cannot be confirmed, the woman should be examined in the standing position. Use of a birthing chair tilted so the patient is 45 degree above the horizontal can also be useful for obtaining the maximal degree of prolapse while maintaining vaginal access to make the POP-Q measurements.

A method for clinically identifying paravaginal defects in women with prolapse was originally described by Richardson and subsequently standardized by Baden and Walker. A curved ring forceps is placed in the lateral vaginal sulci and directed toward the ischial spines along the course of the arcus tendineus fasciae pelvis (ATFP) to reproduce the lateral support of the vagina. A Sims speculum is placed posteriorly to expose the anterior vaginal wall and reduce any posterior prolapse. The patient is asked to strain maximally and, if complete reduction of the anterior prolapse occurs with the ring forceps in this position, she is thought to have paravaginal defects. If, when she strains, the anterior vagina continues to bulge between the arms of the ring forceps without any evidence of reduction, then this suggests a midline loss of support. If supporting the lateral vagina results in partial reduction of the prolapse, then she is thought to have lost both midline and lateral support. Unilateral elevation of each vaginal sulcus allows differentiation of bilateral and unilateral paravaginal defects.

Current evidence suggests that this technique of physically assessing anterior vaginal support may not accurately reflect findings at the time of surgery. A study of 117 women found a discrepancy between the prevalence of paravaginal defects noted at a standardized preoperative clinical examination (63%) and the prevalence of discrete paravaginal detachment noted at surgery (42%).[1] The clinical finding of paravaginal defects in this study was sensitive (sensitivity 92%) but not specific (specificity 52%) and had an adequate negative predictive value (91%), but low positive predictive value (61%). Subjects in whom normal paravaginal support was found at physical examination usually had intact paravaginal support confirmed at surgery, but less than two-thirds of women who were thought to have paravaginal defects based on physical examination actually had them at the time of surgery.

Diagnostic Tests

Urodynamics

Symptoms of urinary incontinence and voiding dysfunction are common in women with advanced vaginal prolapse. Additionally, because significant anterior vaginal prolapse often results in urethral kinking that may mask underlying stress incontinence it is our practice to obtain preoperative urodynamics with vaginal packing or pessary placement to evaluate the lower urinary tract in these patients.

Magnetic Resonance Imaging

Single-shot turbo fast spin echo (SSFSE) dynamic magnetic resonance imaging (MRI) sequences can be useful to further define the nature of the prolapse and evaluate for potential enteroceles, rectoceles, or uterine prolapse. Many patients with severe anterior vaginal wall prolapse have defects in pelvic floor relaxation such that they are predisposed to formation of these other forms of prolapse. The MRI also allows simultaneous viewing of the distal ureters in order to demonstrate ureteral dilation that may accompany high-grade cystoceles with kinking of the trigone. This information is invaluable as a preoperative tool because it allows one to properly define the exact nature of the prolapse (see chapter Imaging of the Genitourinary Tract in Females).

Cystoscopy

Preoperative office cystoscopy is useful in the evaluation of the prolapse patient with lower urinary tract symptoms such as urinary urgency, hematuria, and obstructed voiding in order to detect concurrent bladder pathology. Additionally, we have found it useful to use a "light test" to more accurately differentiate a cystocele from other types of vaginal prolapse. This is performed during cystoscopy with the tip of the cystoscope aimed downward. Although this test is quite simple, it can yield important information as well.

Surgical Techniques

Surgical correction of anterior vaginal prolapse can be accomplished through a vaginal, abdominal, or laparoscopic approach. The appropriate route of surgery is determined by several factors including the preoperative evaluation of anterior vaginal support defects, the presence of other vaginal support defects, history of previous reconstructive or anti-incontinence surgery, the patient's surgical risk, and surgeon skill and preference. In the following section, several techniques for correcting anterior vaginal prolapse are described. The procedures discussed all address correction of anterior vaginal support defects. In our experience, it is rare that a patient with advanced anterior vaginal prolapse has an isolated anterior support defect; almost universally, an apical support defect exists when the prolapse is beyond the hymen. For this reason it is essential to include a vaginal vault suspension procedure as part of the surgical correction of anterior vaginal prolapse. Techniques for vaginal vault suspension are detailed elsewhere in this book. Additionally, because of the contribution of anterior vaginal support toward the urinary continence mechanism, preferential support of the bladder neck using a pubovaginal sling, retropubic urethropexy, bladder neck plication, or some similar technique should be performed routinely as part of any anterior vaginal reconstruction.

Anterior Colporrhaphy

Anterior vaginal prolapse resulting from a central defect is best corrected through a transvaginal approach. The anterior colporrhaphy was popularized by Howard Kelly in 1912,[2] and although no longer an acceptable treatment for stress urinary incontinence, it remains a commonly used technique for transvaginal correction of anterior vaginal prolapse today. Although many variations of this technique have been described in the last century, the basic approach is still similar to that originally described by Kelly.

In recent years there has been a growing interest in the use of synthetic mesh or biologic grafts in the surgical repair of cystoceles to reduce the risk of prolapse recurrence. The role of mesh or tissue grafts in the surgical correction of prolapse is currently unknown because few randomized trials investigating these techniques exist. However, the limited evidence that is available remains promising. Below we describe both our technique for the traditional anterior colporrhaphy and the mesh or graft patch repair.

Traditional Approach

The patient is placed in dorsal lithotomy position in candy-cane or Allen stirrups. A Foley catheter is placed to dependent drainage. A weighted speculum is placed in the vagina and a midline vaginal incision is made with a scalpel. Sterile saline or local anesthetic can be injected into the anterior vaginal wall before incision if desired. If performed along with a vaginal hysterectomy, it is useful to complete the hysterectomy before beginning the anterior vaginal dissection. Sharp dissection is carried laterally to remove the vaginal epithelium from the vaginal muscularis up to the lateral sulcus. The retropubic space is entered sharply with the curved Mayo scissors to allow palpation of the pubic bone if simultaneous pubovaginal sling is required. Alternatively, it can be preserved if a tension-free or other mid urethral sling is to be performed. Also, one may curtail the incision immediately below the bladder neck to allow a tension-free mid urethral sling to be placed through a separate incision.

After completing the dissection, the vaginal muscularis is plicated in the midline using several interrupted stitches of delayed absorbable or permanent suture thereby repairing the central defect and elevating the bladder base and anterior vagina. If a sling is not performed, the bladder neck can be preferentially supported by placating the periurethral tissue underneath the bladder neck (Kelly plication). After completing the anterior colporrhaphy, repair of lateral and/or apical support defects is then performed. The excess vaginal epithelium is then trimmed and the incision is closed with use of a 2-0 polyglycolic acid suture. A vaginal pack is then placed for postoperative hemostasis. Cystoscopy should then be performed to ensure bladder and ureteral integrity.

Mesh or Graft Patch Repair

A central defect repair using mesh or tissue graft patch begins similarly to the traditional repair. After completion of the vaginal dissection, the cardinal ligaments are isolated and plicated with two 0-polyglycolic acid sutures to correct their separation and laxity. The sutures are placed into the levator fascia on each side, thereby preventing the sliding defect herniation of the bladder base. Additionally, the cardinal ligaments form the base of the cystocele repair and anchor the posterior portion of the patch (Figure 8-3.1). The vaginal muscularis is plicated over the central defect cystocele which may be reduced with use of an absorbable mesh. The fascia is reapproximated with horizontal mattress sutures of delayed absorbable or permanent suture. The tissue is brought to the midline, over the mesh to facilitate reapproximation without tension and this maneuver also helps to reduce the incidence of ureteric injury. Before this, intravenous indigo carmine dye is administered to ensure patency of the ureters. A separate

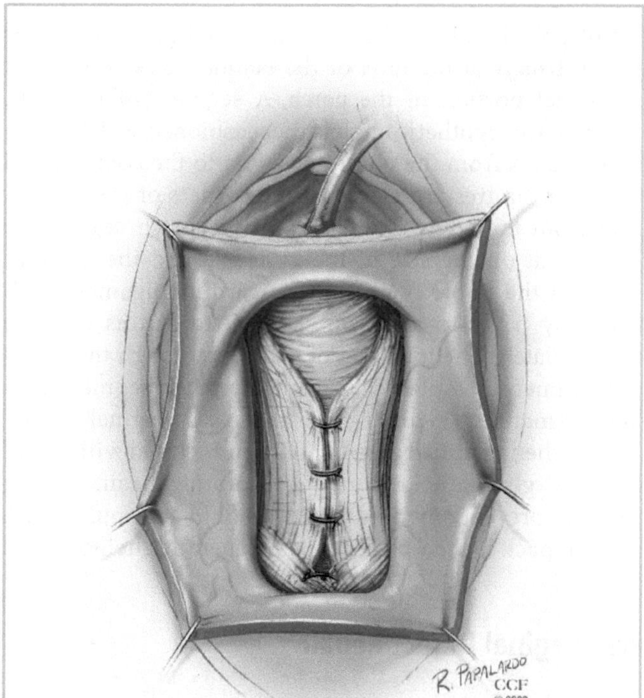

Figure 8-3.1. Cystocele repair. The vaginal muscularis is reapproximated with horizontal mattress sutures of delayed absorbable or permanent suture. (Reprinted with the permission of The Cleveland Clinic Foundation.)

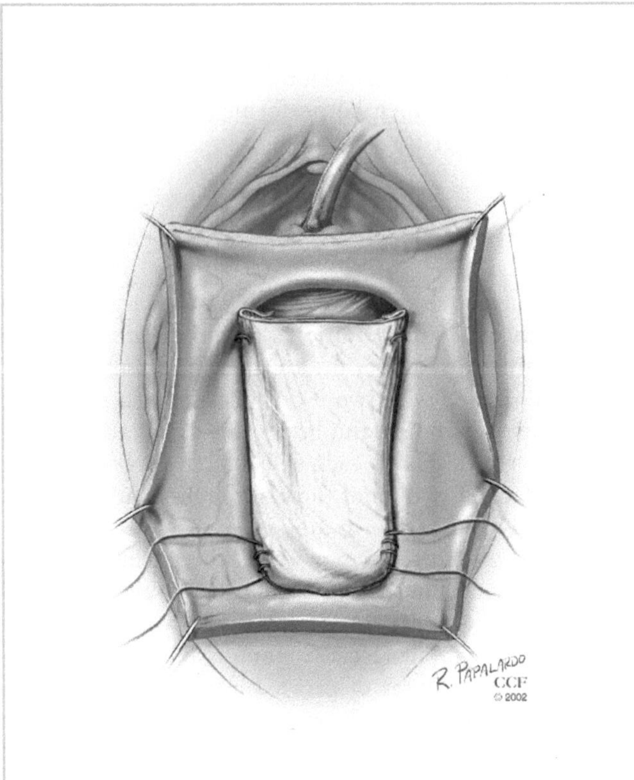

Figure 8-3.2. A segment of allograft, xenograft, or synthetic mesh is then fashioned such that the width spans from the obturator fascia to the contralateral side. The sutures are placed in the corners of the patch 5 mm from the edge. (Reprinted with the permission of The Cleveland Clinic Foundation.)

set of polyglycolic acid sutures is placed into the levator fascia distally at the level of the bladder neck to support the distal portion of the patch. A segment of allograft, xenograft, or synthetic mesh is then fashioned such that the width spans from the obturator fascia to the contralateral side. The sutures are placed in the corners of the patch 5 mm from the edge. The length of the patch segment is dependent on the size of the cystocele and the distance between the pubic bone and the cardinal ligaments. We routinely trim a 5- to 7-cm segment to fit this distance appropriately. The lower set of sutures (through the cardinal ligaments) is placed through the patch segment in a similar manner (Figure 8-3.2). The excess vaginal epithelium is then trimmed and the incision is closed with use of a 2-0 polyglycolic acid suture incorporating the underlying mesh to prevent any dead space for fluid accumulation. A vaginal pack is then placed for postoperative hemostasis.

Paravaginal Defect Repair

The goal of the paravaginal defect repair is to correct anterior vaginal wall prolapse that results from loss of lateral support by reattaching the lateral vaginal sulcus to its normal lateral attachment site. The lateral vagina attaches to the levator ani muscle on each side along a line from the anterior pubic rami to the ischial spine known as the "white line" or arcus tendineus fasciae pelvis (ATFP). The ATFP is formed from a condensation of the obturator internus and levator ani fascia and is composed primarily of organized fibrous collagen, making the lateral connective tissue attachment of the vagina more dense than the superior/apical connective tissue support of the cardinal and uterosacral complex. The paravaginal defect repair can be performed retropubically or vaginally.

Retropubic Approach

The retropubic approach for the repair of paravaginal defects can be performed via laparotomy or laparoscopy. In our opinion, the surgical technique for these two approaches, other than the means of access to the retropubic space, should be identical.

The patient is placed in modified lithotomy position using low leg holders such as Allen stirrups and is draped to allow both abdominal and vaginal access. The bladder is drained with a Foley catheter. The abdomen may be entered through either a transverse or vertical abdominal incision or via laparoscopy. The retropubic space is entered, and the bladder is retracted medially to expose the lateral aspect of the retropubic space. The pubic bone, obturator muscle, obturator fossa, and neurovascular bundle are identified. Blunt dissection is used to identify the lateral vagina, urethra, and ischial spine. The normal site of lateral vaginal attachment on the pelvic sidewall from the interior aspect of the superior pubic ramus to the ischial spine is then identified. If the lateral vagina is avulsed from this attachment site, then a paravaginal defect is present. Because physical examination is less than perfect in identifying paravaginal detachment, a bilateral assessment of paravaginal support should be made in all patients in whom paravaginal defects are suspected.

The surgeon's nondominant hand is then placed into the vagina and used to elevate the lateral superior vaginal sulcus to its site of normal attachment along the course of the ATFP. Typically, four to six simple interrupted stitches of nonabsorbable suture (No. 0 or 2-0) are used to reattach the lateral vagina to the ATFP. The first suture is placed through full thickness (excluding the vaginal epithelium) of the lateral vaginal apex and then through the ATFP and the aponeurosis of the levator ani muscle just distal to the ischial spine. This suture is tied and cut. Additional sutures are placed at 1-cm intervals through the lateral vaginal wall and into the levator ani aponeurosis along the entire course of the ATFP. In patients with a cystocele and genuine stress incontinence, many surgeons combine an abdominal paravaginal repair with a Burch colposuspension (Figure 8-3.3). At the end of the procedure, cystoscopy should be performed to document ureteral patency and the absence of intravesical sutures. Closed-suction drainage of the retropubic space is rarely indicated. Postoperatively, the bladder is drained with either a transurethral or suprapubic catheter until normal voiding occurs.

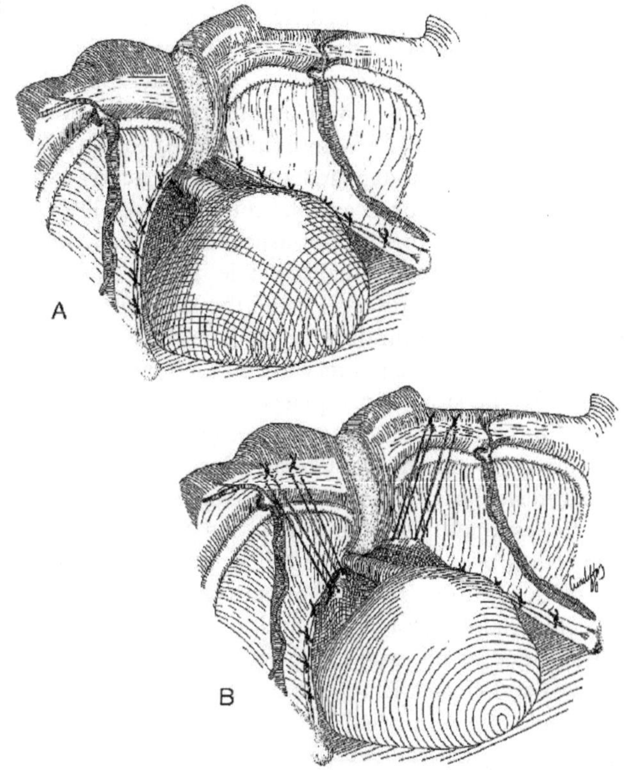

Figure 8-3.3. **A**, Retropubic approach to the paravaginal defect repair. The avulsed vaginal sulcus has been sutured to the arcus tendineus fascia pelvis bilaterally. **B**, The paravaginal repair has been combined with a Burch colposuspension to provide preferential support to the urethrovesical junction. (Reprinted from Cundiff GW, Addison WA. Management of Pelvic Organ Prolapse in Obstetrics and Gynecology Clinics of North America 1998;25:914, Copyright © 1998, with permission from Elsevier.)

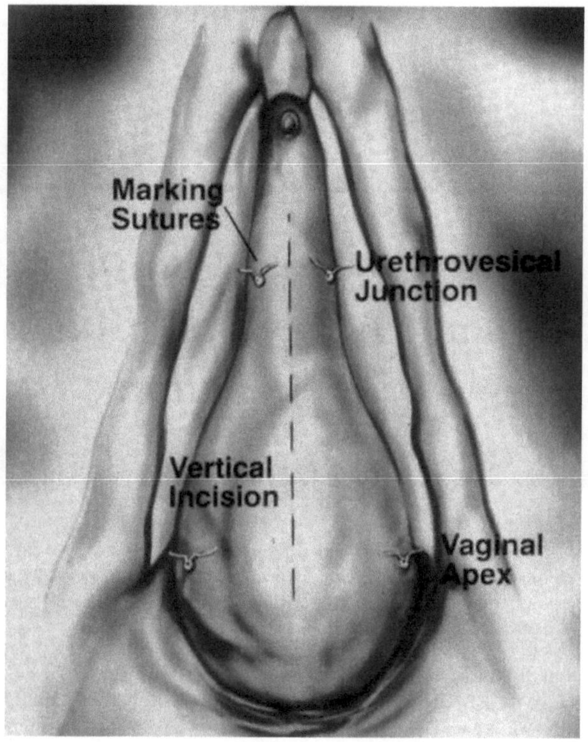

Figure 8-3.4. Vaginal paravaginal repair. Marking sutures placed at urethrovesical junction and vaginal apices. (Reprinted from Shull BL, Benn SJ, Kuehl TJ. Surgical management of prolapse of the anterior vaginal segment: an analysis of support defects, operative morbidity and anatomic outcomes. Am J Obstet Gynecol, 1994;171:1429–39, Copyright © 1994, with permission from Elsevier.)

Vaginal Approach

Paravaginal defect repair using the transvaginal approach can be more challenging than the retropubic approach, but offers the advantage of avoiding an abdominal incision and facilitating a concurrent central defect repair for those women with loss of midline as well as lateral anterior vaginal support. The technique begins similar to that of the central defect repair described above (Figures 8-3.4–8-3.7). The patient is placed in dorsal lithotomy position using candy-cane stirrups. A Foley catheter is used to drain the bladder. A weighted speculum is placed into the vagina. A midline vertical incision is made through the vaginal epithelium from the mid urethra to the vaginal apex. The vaginal epithelium is then sharply dissected off the underlying vaginal muscularis and the dissection is continued laterally to the pelvic sidewall from immediately behind the pubic rami to the level of the ischial spine. Visualization of the adipose tissue of the retropubic space from this transvaginal approach confirms the presence of a paravaginal defect, because normal lateral attachment of the anterior vaginal wall would preclude this. If visualization is limited,

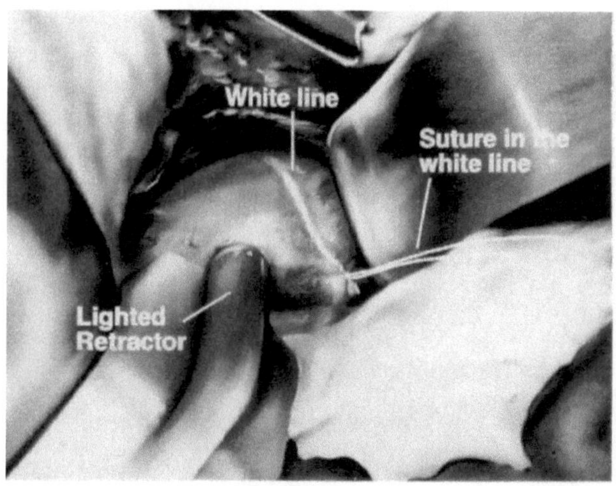

Figure 8-3.5. Vaginal paravaginal repair. A right-angle retractor is in the retropubic space, retracting the bladder medially. The suture is through the arcus tendineus fasciae pelvis approximately 2 cm ventral to the ischial spine. (Reprinted from Shull BL, Benn SJ, Kuehl TJ. Surgical management of prolapse of the anterior vaginal segment: an analysis of support defects, operative morbidity and anatomic outcomes. Am J Obstet Gynecol, 1994;171:1429–39, Copyright © 1994, with permission from Elsevier.)

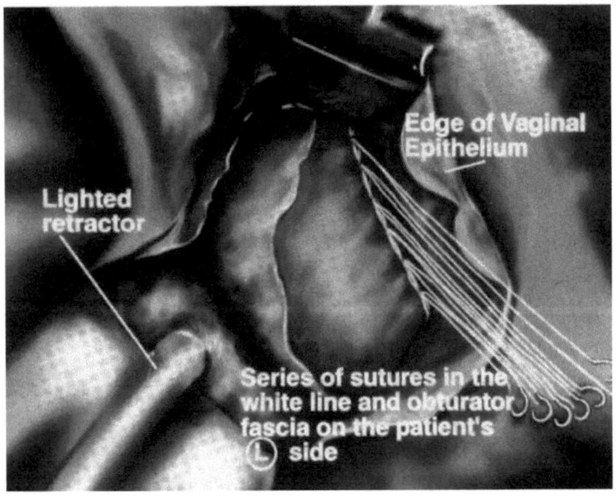

Figure 8-3.6. Vaginal paravaginal repair. A series of sutures has been placed in the arcus tendineus fasciae pelvis from a point ventral to the ischial spine to the back of the pubic bone. (Reprinted from Shull BL, Benn SJ, Kuehl TJ. Surgical management of prolapse of the anterior vaginal segment: an analysis of support defects, operative morbidity and anatomic outcomes. Am J Obstet Gynecol, 1994;171:1429–39, Copyright © 1994, with permission from Elsevier.)

gentle palpation of the lateral attachment site can be used. The ability of the examining finger to enter the retropubic space indicates a paravaginal defect. Every precaution should be taken to avoid iatrogenic creation of paravaginal defects with dissection, palpation, or retractors.

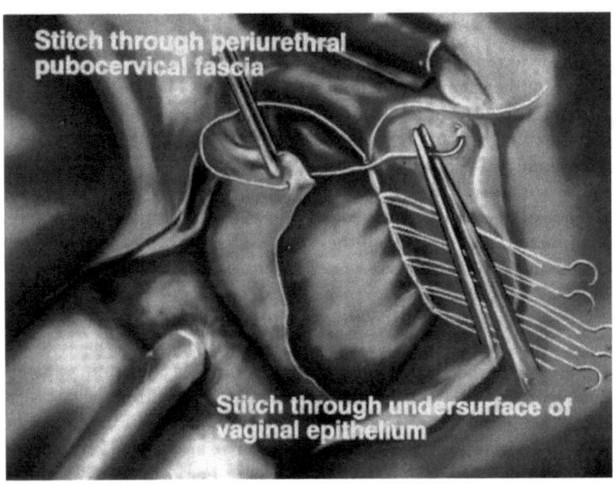

Figure 8-3.7. Vaginal paravaginal repair. The suture in the arcus tendineus fasciae pelvis near the pubic bone is also sewn into the lateral margin of the pubocervical fascia periurethrally at the site of the marking suture at the urethrovesical junction. (Reprinted from Shull BL, Benn SJ, Kuehl TJ. Surgical management of prolapse of the anterior vaginal segment: an analysis of support defects, operative morbidity and anatomic outcomes. Am J Obstet Gynecol, 1994;171:1429–39, Copyright © 1994, with permission from Elsevier.)

Once a paravaginal defect is identified, the normal site of lateral attachment of the vagina should be clearly visualized. This can be facilitated by placing a gauze sponge through the paravaginal defect into the retropubic space and using a narrow Deaver retractor to retract the sponge, underlying adipose tissue and lateral bladder anteriorly. A Briesky-Navratil retractor can then be used to retract the remainder of the bladder medially, thereby clearly exposing the levator ani muscle and the course of the ATFP from the ischial spine to the inferior aspect of the pubic ramus. Four to six interrupted nonabsorbable sutures (No. 0 or 2-0) are placed through the ATFP and the aponeurosis of the levator ani muscle from the level of the ischial spine to the pubic symphysis at 1-cm intervals and their needles left on. Once all of the stitches are placed through the ATFP, the sponge in the retropubic space is removed. Each stitch is then placed through the lateral edge of the detached vaginal muscularis (*pubocervical* or *vesicopelvic fascia*) at their corresponding level and then tied. Once paravaginal support has been assessed and, if necessary, restored on both sides, a midline plication of pubocervical fascia (anterior colporrhaphy), a bladder neck plication, or sling, can be performed, as necessary. The vaginal epithelium is then trimmed and closed. If a vaginal hysterectomy is needed, this should be performed before the paravaginal defect repair. If a vaginal vault suspension or culdoplasty is necessary, these sutures should be placed before the paravaginal defect repair, but not tied until the paravaginal defect repair has been completed. As with the retropubic approach, cystoscopy should be performed at the end of the procedure to confirm ureteral patency and the absence of intravesical sutures. Transurethral or suprapubic bladder drainage should continue until normal voiding occurs.

Outcomes

Anterior Colporrhaphy

Outcomes of the traditional anterior colporrhaphy are largely limited to retrospective reviews and case series. Reporting of outcomes has been extremely subjective and, before the advent of the POP-Q system, pre- and postoperative staging have been quite variable between surgeons. Nonetheless, the reported recurrence rates after anterior vaginal wall prolapse repair have been quite high (range, 0%–59%) (Tables 8-3.1–8-3.3). A recent randomized trial performed by the Cleveland Clinic compared traditional anterior colporrhaphy, an ultra lateral anterior colporrhaphy technique, and a traditional anterior colporrhaphy with polyglactin 910 mesh for treatment of symptomatic anterior vaginal prolapse. After a median follow-up of 23.3 months, the recurrence rate (defined as prolapse that extends 1 cm proximal to the hymen or greater) was 30% to 46% and there was no difference in efficacy among the three groups.[3]

Table 8-3.1. Outcome of anterior colporrhaphy alone for treatment of anterior vaginal prolapse

Authors (Year)	No. of Patients	Study Design	Mean Follow-up	Recurrence (%)
Macer[4] (1978)	76	Retrospective cohort	5–20 y	20
Walter et al.[5] (1982)	86	Prospective cohort	1–2.5 y	0
Stanton et al.[6] (1982)	73	Prospective cohort	2 y	15
Porges and Smilen[7] (1994)	486	Retrospective cohort	2.6 y	3
Smilen et al.[8] (1998)	245	Retrospective cohort	6 mo	9.4
Weber et al.[3] (2001)	74*	Randomized controlled trial	23.3 mo	54–70

*A total of 114 subjects were enrolled in this trial. Seventy-four subjects underwent either a traditional anterior colporrhaphy ($n = 39$) or an "ultralateral" anterior colporrhaphy ($n = 33$) without graft augmentation.

Table 8-3.2. Anterior colporrhaphy and needle suspension for treatment of anterior vaginal prolapse

Authors (Year)	No. of Patients	Study Design	Mean Follow-up (y)	Recurrence (%)
Raz et al.[9] (1989)	120	Retrospective cohort	0.5–5	2.5
Miyazaki and Miyazaki[10] (1994)	27	Retrospective cohort	3.5–4	59
Bump et al.[11] (1996)	32	Randomized controlled trial	0.5	50
Dmochowski et al.[12] (1997)	47	Retrospective cohort	1.25–6.75	17–40
Migliari and Usai[13] (1999)	15	Retrospective cohort	1.5–3.25	7

Insofar as patch augmentation of standard anterior colporrhaphy is concerned, there exist very few peer-reviewed manuscripts on the subject. This is largely attributed to the fact that follow-up is short and often postoperative evaluations have not rigorously adhered to pre- and postoperative POP-Q measurements to ascertain differences in the repairs. Nonetheless, the few reports in the literature are encouraging and suggest that routine use of patch augmentation with a variety of materials may decrease the alarmingly high rate of recurrent anterior vaginal wall prolapse and do so with minimal morbidity (Table 8-3.4).

Paravaginal Repair

The paravaginal defect repair has been widely used for correction of anterior vaginal prolapse thought to result from lateral vaginal detachment. Similar to anterior colporrhaphy, current evaluation of the success rate of this procedure for treatment of anterior vaginal prolapse is limited to case series and retrospective reviews. Reports of retropubic paravaginal repair demonstrate anatomic success rates ranging from 92% to 97% for the abdominal approach (Table 8-3.5). There are few long-term data on the efficacy of laparoscopic paravaginal repair for the treatment of anterior vaginal prolapse. Reports of vaginal paravaginal repair demonstrate success rates of 76% to 100% after variable lengths of follow-up (Table 8-3.6). Failure or recurrence of anterior vaginal prolapse after paravaginal defect may occur laterally, centrally, or both. Studies that have differentiated lateral from central recurrence have revealed that central recurrence (22%–25%) is more common than a lateral recurrence (2%–8%). To date, there are no studies comparing paravaginal defect repair with or without midline anterior repair to traditional anterior colporrhaphy alone.

Table 8-3.3. Anterior colporrhaphy and simultaneous sling/anti-incontinence procedure for treatment of anterior vaginal prolapse

Authors (Year)	No. of Patients	Study Design	Mean Follow-up (y)	Recurrence (%)
Cross et al.[14] (1997)	42	Retrospective cohort	1–3.25	8–17
Safir et al.[15] (1999)	130	Retrospective cohort	0.5–3.5	4–5

Table 8-3.4. Anterior colporrhaphy with patch augmentation of repair for treatment of anterior vaginal prolapse

Authors (Year)	Patch Material	No. of Patients	Study Design	Mean Follow-up	Recurrence (%)
Groutz et al.[16] (2001)	Cadaver fascia	21	Retrospective cohort	20 mo	0
Kobashi et al.[17] (2000)	Cadaver fascia	50	Prospective cohort	6 mo maximum	0
Migliari et al.[18] (2000)	Polypropylene mesh	12	Retrospective cohort	20.5 mo	25% grade I cystocele
Weber et al.[3] (2001)	Polyglactin 910 mesh	35	Randomized controlled trial	23.3 mo	58

Table 8-3.5. Outcomes of abdominal paravaginal repair for treatment of anterior vaginal prolapse

Authors (Year)	No. of Patients	Study Design	Mean Follow-up	Cure (%)
Richardson et al.19 (1976)	60	Retrospective cohort	20 mo	97
Richardson et al.[20] (1981)	233	Retrospective cohort	Not reported	95
Shull and Baden[21] (1989)	149	Retrospective cohort	48 mo	95
Bruce et al.[22] (1999)	52	Retrospective cohort	17 mo	92
Scotti et al.[23] (1998)	40	Prospective cohort	39 mo	97

Table 8-3.6. Outcomes of vaginal paravaginal repair for treatment of anterior vaginal prolapse

Authors (Year)	No. of Patients	Study Design	Mean Follow-up	Cure (%)
White[24] (1909)	19	Retrospective cohort	Not reported	100
Shull et al.[25] (1994)	62	Retrospective cohort	1.6 y	76
Farrell and Ling[26] (1997)	27	Retrospective cohort	8 mo	80
Nguyen and Bhatia[27] (1999)	10	Retrospective cohort	1 y	100
Elkins et al.[28] (2000)	25	Retrospective cohort	Not reported	76
Mallipeddi et al.[29] (2001)	35	Retrospective cohort	20 mo	97
Young et al.[30] (2001)	100	Retrospective cohort	11 mo	78

Complications

Intraoperative complications after repair of anterior vaginal prolapse are, for the most part, infrequent. Febrile morbidity occurs in 6% to 20% of patients and is most often self-limited. Lower urinary tract injury occurs in 0% to 4%, similar to other pelvic reconstructive procedures. Excessive blood loss may occur, requiring blood transfusion. This is infrequent with anterior colporrhaphy and retropubic paravaginal repair. Although paravaginal repair through the vaginal approach offers some potential advantages over the retropubic approach including shorter recovery time, the ability to simultaneously correct midline anterior defects, and the avoidance of an abdominal incision, it may have a higher rate of intraoperative hemorrhage and blood transfusion, as high as 9% to 12% in some series. This is in contrast to a transfusion rate of 0% to 4% in series of abdominal paravaginal defect repair. The limited exposure and technical challenge of the vaginal approach likely explains this difference.

Postoperative complications from the repair of advanced anterior vaginal prolapse are not unlike those of other vaginal reconstructive procedures. These may include prolonged urinary retention, de novo urinary urge or stress incontinence, recurrent prolapse, or vaginal shortening. Urinary retention is usually a transient phenomenon and often resolves on its own. Rarely, one may require a prolonged course of suprapubic catheter drainage or intermittent catheterization until satisfactory spontaneous voiding occurs. De novo urinary incontinence (urge or stress) may occur in a small portion of patients, quite possibly those with higher-grade prolapse and long-term obstruction. Urge incontinence may subside with time but often requires behavioral therapeutic modifications as well as anticholinergic therapy to assist in controlling symptoms. New-onset stress incontinence may result from inadequate bladder neck and urethral support or because of intrinsic sphincteric dysfunction. Accordingly, some advocate routine placement of a sling or some form of bladder neck support to minimize this potential.

In our experience, patients with stage III to IV cystoceles often require concomitant surgical procedures to correct other often severe defects in pelvic floor support in the form of enterocele, rectocele, or uterine prolapse. Simply repairing the cystocele without addressing these other potential defects in pelvic support may lead to recurrent vaginal bulges and require secondary procedures. As the anterior vaginal wall is transferred superiorly, this may allow a weakened cul-de-sac and posterior vaginal wall to prolapse in the form of an enterocele or rectocele. Vaginal shortening can be avoided by minimizing the amount of anterior vaginal wall tissue that is excised; thus, this should be a rare complication.

Ureteric obstruction may occur postoperatively, despite patency being demonstrated on cystoscopic examination, as kinking of the ureters results from the support sutures. If this occurs, one must address the obstruction by placement of either a stent or percutaneous nephrostomy tube and passage of a glidewire down the narrowed channel. After a period of observation, if no patency ensues, one may proceed with ureteric reimplantation. Our preference is to not disturb the repair site, therefore, avoid transvaginal exploration.

References

1. Barber MD, Cundiff GW, Weidner AC, Coates KW, Bump RC, Addison WA. Accuracy of clinical assessment of paravaginal defects in women with anterior vaginal wall prolapse. Am J Obstet Gynecol 1999;181: 87–90.
2. Kelly HA. Incontinence of urine in women. Urol Cutaneous Rev 1913; 17:291–293.
3. Weber AM, Walters MD, Piedmonte MR, Ballard LA. Anterior colporrhaphy: a randomized trial of three surgical techniques. Am J Obstet Gynecol 2001;185:1299–1306.
4. Macer GA. Transabdominal repair of cystocele, a 20-year experience, compared with traditional vaginal approach. Am J Obstet Gynecol 1978;13:203–207.
5. Walter S, Olesen KP, Hald T, et al. Urodynamic evaluation after vaginal repair and colposuspension. Br J Urol 1982;54:377–380.
6. Stanton SL, Hilton P, Norton C, Cardozo L. Clinical and urodynamics effects of anterior colporrhaphy and vaginal hysterectomy for prolapse with or without incontinence. Br J Obstet Gynaecol 1982; 89:459–463.
7. Porges RF, Smilen SW. Long-term analysis of the surgical management of pelvic support defects. Am J Obstet Gynecol 1994;171: 1518–1528.
8. Smilen SW, Saini J, Wallach SJ, Porges RF. The risk of cystocele after sacrospinous ligament fixation. Am J Obstet Gynecol 1998;179: 1465–1472.
9. Raz S, Klutke CG, Golombe J. Four-corner bladder and urethral suspension for moderate cystocele. J Urol 1989;142:712–715.
10. Miyazaki FS, Miyazaki DW. Raz four-corner suspension for severe cystocele: poor results. Int Urogynecol J 1994;5:94–97.
11. Bump RC, Hurt WG, Theofrastous JP, et al. Randomized prospective comparison of needle colposuspension versus endopelvic fascia plication for potential stress incontinence prophylaxis in woman undergoing vaginal reconstruction for stage III or IV pelvic organ prolapse. Am J Obstet Gynecol 1996;175:326–333.
12. Dmochowski RR, Zimmern PE, Ganabathi K, et al. Role of the four-corner bladder neck suspension to correct stress incontinence with mild to moderate cystocele. Urology 1997;49:35–40.
13. Migliari R, Usai E. Treatment results using a mixed fiber mesh in patients with grade IV cystocele. J Urol 1999;161:1255–1258.
14. Cross CA, Cespedes RD, McGuire EJ. Treatment results using pubovaginal slings in patients with large cystoceles and stress incontinence. J Urol 1997;158:431–434.
15. Safir MH, Gousse AE, Rovner ES, et al. 4-Defect repair of grade 4 cystocele. J Urol 1999;161:587–594.
16. Groutz A, Chaikin DC, Theusen E, Blaivas JG. Use of cadaveric solvent-dehydrated fascia lata for cystocele repair–preliminary results. Urology 2001;58:179–183.
17. Kobashi KC, Mee SL, Leach GE. A new technique for cystocele repair and transvaginal sling: the cadaveric prolapse repair and sling (CAPS). Urology 2000;56:9–14.
18. Migliari R, De Angelis M, Madeddu G, Verdacchi T. Tension-free vaginal mesh repair for anterior vaginal wall prolapse. Eur Urol 2000;38:151–155.
19. Richardson AC, Lyon JB, Williams NL. A new look at pelvic relaxation. Am J Obstet Gynecol 1976;126:568–571.
20. Richardson AC, Edmonds PB, Williams NL. Treatment of stress urinary incontinence due to paravaginal fascial defects. Obstet Gyencol 1981;57:357–362.
21. Shull BL, Baden WF. A six-year experience with paravaginal defect repair for stress urinary incontinence. Am J Obstet Gynecol 1989; 160:1432–1439.
22. Bruce GR, El-Galley RES, Galloway NTM. Paravaginal defect repair in the treatment of female stress urinary incontinence and cystocele. Urology 1999;54:647–651.
23. Scotti RJ, Garely AD, Greston WM, Flora RF, Olson TR. Paravaginal repair of lateral vaginal wall defects by fixation to the ischial periosteum and obturator membrane. Am J Obstet Gynecol 1998;179: 1436–1445.
24. White GR. Cystocele. JAMA 1909;21:1707–1710.
25. Shull BL, Benn SJ, Kuehl TJ. Surgical management of prolapse of the anterior vaginal segment: an analysis of support defects, operative morbidity and anatomic outcomes. Am J Obstet Gynecol 1994;171: 1429–1439.
26. Farrell SA, Ling C. Currycombs for the vaginal paravaginal defect repair. Obstet Gynecol 1997;90:845–847.
27. Nguyen JK, Bhatia NN. Transvaginal repair of paravaginal defects using the capio suturing device: a preliminary experience. J Gynecol Tech 1999;5:51–54.
28. Elkins TE, Chesson RR, Videla F, Menefee S, Yordan R, Barksdale PA. Transvaginal paravaginal repair: a useful adjunctive procedure in pelvic relaxation surgery. J Pelvic Surg 2000;1:11–15.
29. Mallipeddi PK, Steele AC, Kohli N, Karram MM. Anatomic and functional outcome of vaginal paravaginal repair in the correction of anterior vaginal prolapse. Int Urogynecol J 2001;12:83–88.
30. Young SB, Daman JJ, Bony LG. Vaginal paravaginal repair: one-year outcomes. Am J Obstet Gynecol 2001;185:1360–1367.

8-4

Vaginal Enterocele Repair

Andrew I. Sokol and Mark D. Walters

Definitions and Pathophysiology

Enterocele has been defined as a peritoneum-lined sac herniating through the pelvic floor, usually between the vagina and rectum.[1] Nichols and Randall[2] described four types, including traction, congenital, pulsion, and iatrogenic. Traction enterocele is probably the most common and occurs secondary to uterine and vaginal apical prolapse. Cystocele and rectocele usually coexist with traction enterocele. Congenital enterocele is rare and may result from connective tissue and neurologic disorders such as spina bifida. Congenital enterocele may occur independently of other types of prolapse. Pulsion enterocele results from prolonged increases in intraabdominal pressure and may be accompanied by massive prolapse. Finally, iatrogenic enterocele results after postsurgical elevation of the vaginal axis out of its normally horizontal axis toward the vertical plane, as may occur after colposuspension.

The pathophysiology of enterocele has been debated. One theory proposes a defect in the fibromuscular vaginal tube with a discreet defect of the pubocervical and rectovaginal muscularis at the vaginal apex.[3] This theory proposes that prolapse results from discrete breaks in the endopelvic fascia rather than from stretching or attenuation. Histologic samples of peritoneum directly abutting vaginal epithelium (without intervening muscularis) have been reported but are few.[4] In another histologic study of enteroceles, 13 women with posthysterectomy prolapse and enterocele were compared with 5 women undergoing hysterectomy without prolapse and 13 women undergoing radical hysterectomy. No women with enterocele had peritoneum directly in contact with vaginal epithelium. Average vaginal wall muscularis thickness was similar among the three groups, with a slightly thicker muscularis in those with enterocele.[1] This study suggested that enterocele was mainly caused by a loss of the vaginal support to the endopelvic fascia.

Evaluation

Symptoms

Women with enterocele often have concomitant vaginal support defects. Symptoms are often complex and cannot be attributed solely to the enterocele. Nonetheless, women with large enteroceles often complain of pelvic pressure, fullness, vaginal protrusion, and low backache. Additionally, women may experience irritation, spotting, and ulceration from exposure of the vaginal epithelium overlying the exposed enterocele. With severe enterocele, evisceration of bowel through the vagina has been described.[5] Rare cases of small bowel incarceration in enterocele sacs have also been reported.[2] The effects of severe enterocele on bowel function are poorly understood.

Physical Examination

Enterocele may occur in association with rectocele and may be difficult to differentiate on physical examination. Rectovaginal examination may demonstrate the rectocele as distinct from the bulging sac that arises from a point higher in the vagina, often near the apex. Visual inspection of the posterior vaginal wall may reveal a transverse furrow between the two hernias, and usually, loops of small bowel can be palpated through the vaginal wall overlying the enterocele. Occasionally, peristalsis of small bowel loops filling the enterocele sac may be seen during the examination.

Physical examination should begin with the patient in lithotomy position, as for a routine pelvic examination. Pelvic organ prolapse defects can be identified best by using a Sims speculum or the split blade of a Graves speculum. After the resting examination, the patient is instructed to strain forcefully or cough vigorously. During this maneuver, descent of the vagina is observed systematically and staged according to the pelvic organ prolapse quantification system that has been adopted by the International Continence Society, American Urogynecologic Society, and Society of Gynecologic Surgeons.[6] When the physical findings cannot be adequately evaluated in the supine position, the woman is reexamined in the standing position. The patient should be asked if the prolapse exhibited during the examination is consistent with the maximal prolapse she typically experiences.

Surgical Repair Techniques

Surgical repair of enterocele can be performed vaginally, abdominally, or laparoscopically, but few data exist comparing the various repair techniques. The approach and type of procedure performed depend on the surgeon's preference and presence or absence of concomitant vaginal or abdominal pathology. We will focus on vaginal repair techniques.

Traditional vaginal enterocele repair entails isolation of the enterocele sac, careful exploration of its contents, and closure with multiple circumferential, nonabsorbable, pursestring sutures incorporating the cardinal-uterosacral ligaments.[2] At our institution, this technique is used in combination with sacrospinous ligament fixation, ileococcygeus vault suspension, or extraperitoneal uterosacral ligament vaginal vault suspension when cul-de-sac obliteration or adhesions are present and the uterosacral ligaments are not accessible. Additional enterocele repair techniques include McCall-type culdoplasty, and more recently, uterosacral ligament vaginal vault suspension with fascial reconstruction. Whereas the latter two techniques are our preferred methods of enterocele repair, uterosacral ligament vaginal vault suspension has been described elsewhere. We will therefore present our techniques for traditional vaginal enterocele repair and McCall-type culdoplasty.

Vaginal Enterocele Repair

Enterocele usually exists with other support defects, and concurrent vaginal vault suspension, cystocele, and rectocele repair are often necessary.

The technique is as follows (Figure 8-4.1):

1. The patient is placed in dorsal lithotomy position, and the bladder is drained before incision. Prophylactic antibiotics are given routinely before the operation.

2. The vaginal epithelium at the apex is grasped with Allis clamps and a vasoconstricting solution is submucosally injected. The vaginal epithelium is incised with a scalpel vertically or in a diamond shape, the enterocele is

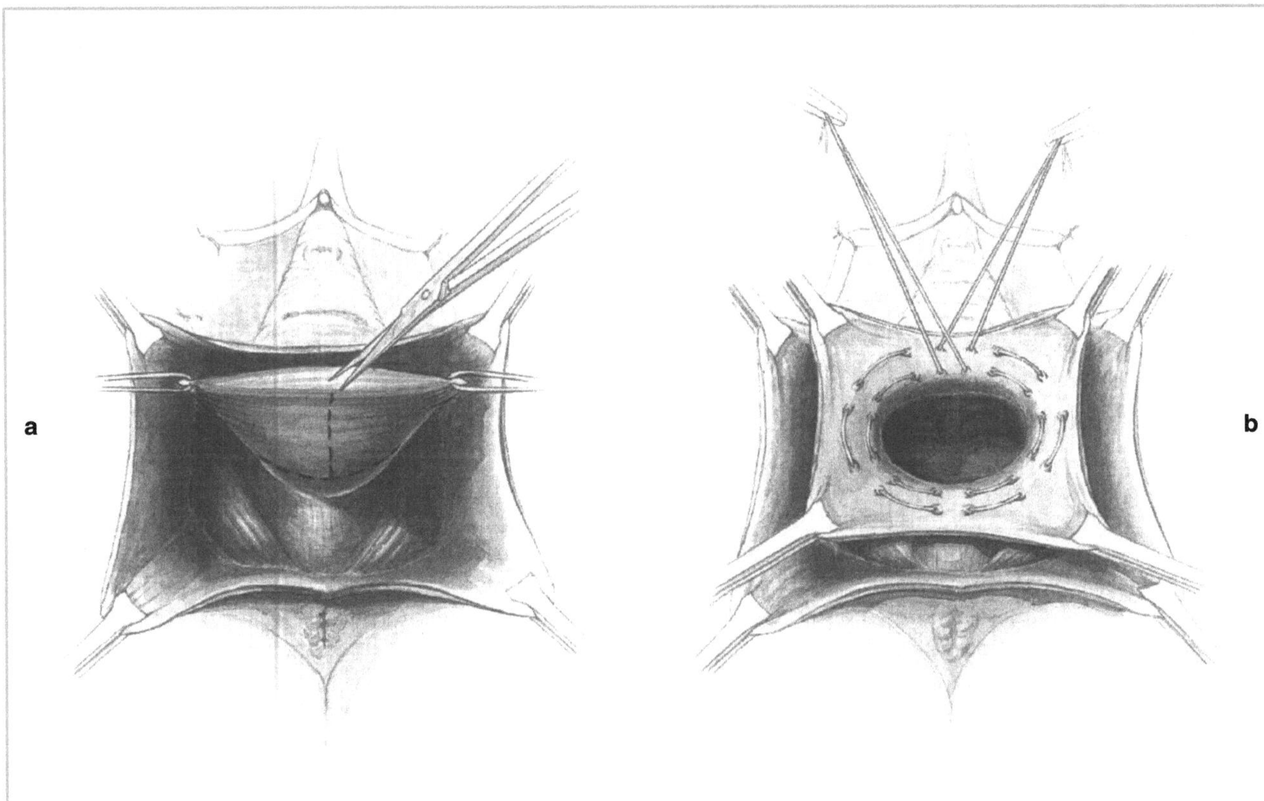

Figure 8-4.1. Vaginal enterocele repair. **a,** Isolation and entry of the enterocele sac in the rectovaginal space. **b,** Pursestring sutures are placed closed to the neck of the enterocele sac. (Reprinted from Walters MD, Karram MM. Urogynecology and Reconstructive Pelvic Surgery. 2nd ed., p 223, Copyright 1999 Mosby, with permission from Elsevier.)

identified, and the edges of the vaginal epithelium are dissected sharply away from the enterocele sac.

3. The enterocele sac should be mobilized from the vaginal walls and rectum. When the enterocele sac is difficult to distinguish from the rectum, differentiation is aided by a rectal examination with simultaneous dissection of the enterocele sac from the rectal wall. At times, distinguishing the enterocele sac from a large cystocele may prove difficult. When this occurs, placement of a curved uterine sound into the bladder or transillumination with a cystoscope may be helpful.

4. After the enterocele sac has been dissected from the vagina and rectum, traction is placed on it with two Allis clamps and the sac is entered sharply. The enterocele sac is digitally explored and adhesions of the small bowel and omentum are dissected to the level of its neck.

5. Under direct visualization, two or three circumferential, nonabsorbable, pursestring sutures are used to close the enterocele sac. The cardinal-uterosacral ligaments are incorporated as well. Once placed, the sutures are sequentially tied. Care should be taken to avoid kinking the ureters.

6. Posterior colporrhaphy and vaginal vault suspension are performed as indicated.

7. Cystoscopy is performed to ensure ureteral patency after intravenous indigo carmine is given.

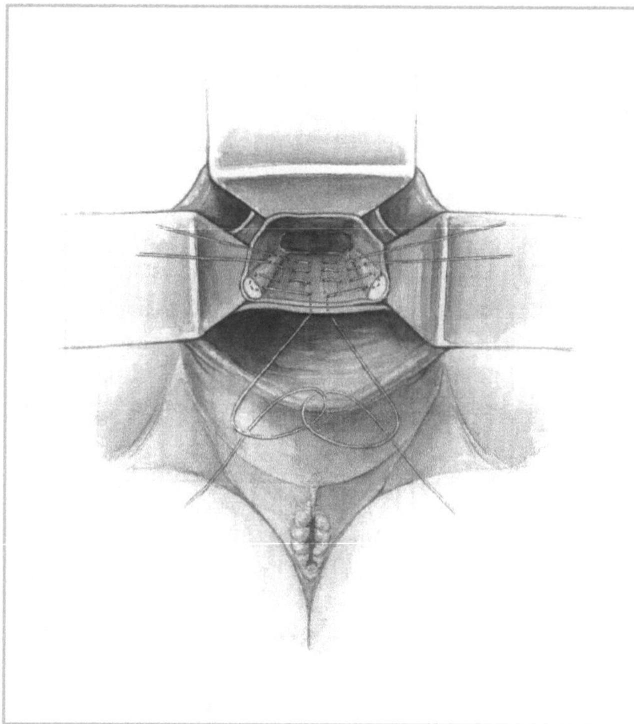

Figure 8-4.2. McCall culdoplasty technique. Note that the lowest suture incorporates the posterior vaginal wall, thus providing additional support. (Reprinted from Walters MD, Karram MM. Urogynecology and Reconstructive Pelvic Surgery. 2nd ed., p 224, Copyright 1999 Mosby, with permission from Elsevier.)

Modified McCall Culdoplasty

McCall[7] described the technique of surgical correction of enterocele and deep cul-de-sac during vaginal hysterectomy. The McCall culdoplasty closes the redundant cul-de-sac and associated enterocele, provides apical support, and lengthens the vagina. In a randomized study, Cruikshank and Kovac[8] demonstrated the superiority of McCall culdoplasty to uterosacral plication and simple peritoneal closure in the prevention of posthysterectomy enterocele. For this reason, we advocate using this procedure as a part of every vaginal hysterectomy, even in the absence of enterocele, to minimize future vaginal vault prolapse and enterocele formation. With posthysterectomy prolapse and enterocele, we may perform a modified, high McCall-type culdoplasty with a four-point attachment to the vaginal cuff.

The technique is as follows (Figure 8-4.2):

1. After the vaginal hysterectomy is completed, the surgeon places a finger into the posterior cul-de-sac to evaluate its depth and ensure accessibility of the uterosacral ligaments. Lateral traction is placed on the previously tagged uterosacral ligaments. In cases of posthysterectomy prolapse, the enterocele is identified and entered as described above.

2. With the patient in Trendelenburg position, a large, moist pack is placed in the posterior cul-de-sac and hollow of the sacrum. A wide Deaver or Breisky-Navratil retractor is used to elevate the pack and the intestines out of the operative field.

3. The ischial spines are palpated. The remnants of uterosacral ligaments are found posterior and medial to the ischial spines. They can be identified more easily by using Allis or Kocher clamps attached to the vaginal epithelium at approximately 4 o'clock and 8 o'clock (at the old hysterectomy scar) to place tension on structures of the pelvic sidewall. The clamp is elevated straight upward placing tension on the uterosacral ligament. The contralateral index finger is simultaneously used to palpate the connective tissue condensations along the side of the pelvis (uterosacral ligaments). A long Allis clamp is placed high on each uterosacral ligament.

4. A stitch of delayed-absorbable No. 0 suture is taken on each side at the level of the Allis clamp on the uterosacral ligament and held for later passage out of the apex bilaterally. The needle is passed from lateral to medial to reduce the risk of ureteral entrapment. Next, a permanent No. 0 suture is passed through the uterosacral ligament as high as possible, cephalad to the previous stitch. Successive bites are then taken at 1- to 2-cm intervals through the anterior serosa of the sigmoid colon until the opposite uterosacral ligament is reached and incorporated. This suture is left untied. Occasionally, one to three more identical sutures are placed caudally, progressing toward

the posterior vaginal cuff. The need for additional internal sutures depends on the size and depth of the enterocele or cul-de-sac. The goal is obliteration of the entire dependent portion of the cul-de-sac. After the internal permanent sutures have been placed, their ends are held laterally without tying.

5. A delayed-absorbable, No. 0 suture is placed from the vaginal lumen just below the middle of the cut edge of the posterior vaginal cuff, through the peritoneum, and through the left uterosacral ligament. Successive bites are taken across the cul-de-sac as before and into the right uterosacral ligament. This suture is passed through the peritoneum and vaginal epithelium, adjacent to the point of entry.

6. The permanent sutures are tied sequentially. A suture of delayed-absorbable, No. 0 suture is placed through the plicated uterosacral ligaments and held long, to be positioned in the anterior vaginal epithelium at the point of the new vault.

7. After the McCall sutures have been placed, the gauze pack and retractors are removed. Cystocele repair and/or bladder neck suspension or sling is performed as needed. The vaginal epithelium is trimmed as appropriate and the anterior vaginal wall closure is begun and carried to a point just short of the apex.

8. The lateral delayed-absorbable, No. 0 uterosacral sutures are then positioned bilaterally through the lateral apex. Next, the anterior delayed-absorbable, No. 0 McCall suture is positioned bilaterally through the anterior vaginal muscularis and epithelium just lateral to midline, at the level of the new apex. The anterior wall and apical closure is completed and all four delayed-absorbable, No. 0 suspension sutures are tied, bringing the anterior, posterior, and lateral vagina up to the level of the uterosacral ligaments.

9. Cystoscopy is performed after intravenous indigo carmine is given to inspect for bilateral ureteral patency after all sutures have been tied. Rectocele and perineal repair are completed as necessary.

Outcomes

Few data exist regarding long-term outcomes after vaginal enterocele repair. In a study of 48 women who had McCall culdoplasty for large enterocele, procidentia, or complete vaginal vault eversion, only two enteroceles (4%) recurred 2 to 22 years (average, 7 years) postoperatively.[9] Colombo and Milani[10] found a similar rate of recurrence, with 3 of 62 (5%) patients experiencing vault prolapse recurrence 4 to 9 years after McCall culdoplasty. However, these studies were performed with various modifications of the McCall culdoplasty, and no comparison of the varying techniques has been made.

McCall culdoplasty has been implicated as a cause of ureteral obstruction. Given[9] reported ureteral injury in 4% of McCall culdoplasty procedures. Stanhope et al.[11] also found that culdoplasty increased the risk of ureteral obstruction after vaginal surgery. Some of the reported complications of McCall culdoplasty are listed in Table 8-4.1. Because of the small but significant ureteral injury rate, we advocate cystoscopy after all vaginal reconstructive procedures to rule out bladder pathology and to document ureteral patency.

Table 8-4.1. Complications after McCall culdoplasty*

Complication	Percent of Patients ($n = 48$)
Removal of silk suture	10
Postoperative cuff infection	4
High rectocele	4
Partial prolapse of vaginal vault	4
Shortened vagina	4
Introital stenosis	2
Pulmonary emboli	2
Nerve palsy	2
Ureteral obstruction	2

* Follow-up 2 to 22 (average 7) years.
Reprinted from Given FT."Posterior culdeplasty": revisited. Am J Obstet Gynecol. 1985 15;153(2):135–9, copyright 1985, with permission from Elsevier.

Conclusion

Multiple modalities of enterocele repair are available depending on the anatomy, skill and preference of the surgeon, and need for additional abdominal or vaginal procedures. We have described our techniques for vaginal enterocele repair. Isolation and closure of the enterocele sac, along with proper repair of all of the prolapsed segments of the vagina, will result in optimal surgical results for the patient. Routine use of McCall culdoplasty after every hysterectomy would probably decrease the rate of iatrogenic enterocele and vaginal apex prolapse.

References

1. Tulikangas PK, Walters MD, Brainard JA, Weber AM. Enterocele: is there a histologic defect? Obstet Gynecol 2001;98:634–637.
2. Nichols DH, Randall CL. Vaginal Surgery. Baltimore: Williams & Wilkins; 1989.
3. Richardson AC. The anatomic defects in rectocele and enterocele. J Pelvic Surg 1995;1:214–221.
4. DeLancey JO, Starr RA, Elkins TE. Incisional hernia of the vaginal apex following vaginal hysterectomy in a premenopausal, sexually inactive woman. Obstet Gynecol 1989;73:880–881.
5. Kowalski LD, Seski JC, Timmins PF, Kanbour AI, Kunschner AJ, Kanbour-Shakir A. Vaginal evisceration: presentation and management in postmenopausal women. J Am Coll Surg 1996;183: 225–229.
6. Bump RC, Mattiasson A, Bo K, et al. The standardization of terminology of female pelvic organ prolapse and pelvic floor dysfunction. Am J Obstet Gynecol 1996;175:10–17.

7. McCall ML. Posterior culdeplasty. Obstet Gynecol 1957;10:595–602.
8. Cruikshank SH, Kovac SR. Randomized comparison of three surgical methods used at the time of vaginal hysterectomy to prevent posterior enterocele. Am J Obstet Gynecol 1999;180:859–865.
9. Given FT. "Posterior culdeplasty": revisited. Am J Obstet Gynecol 1985;153:135–139.
10. Colombo M, Milani R. Sacrospinous ligament fixation and modified McCall culdoplasty during vaginal hysterectomy for advanced uterovaginal prolapse. Am J Obstet Gynecol 1998;179:13–20.
11. Stanhope CR, Wilson TO, Utz WJ, Smith LH, O'Brien PC. Suture entrapment and secondary ureteral obstruction. Am J Obstet Gynecol 1991;164:1513–1519.

8-5

Vaginal Rectocele Repairs

Jennifer T. Pollak and G. Willy Davila

Rectocele repair represents one of the most commonly performed gynecologic pelvic reconstructive procedures. Both gynecologists and colorectal surgeons treat rectoceles. In fact, 100% of gynecologists surveyed manage rectoceles, whereas 68% of colorectal surgeons manage them.[1–3] Gynecologic surgeons perform this operation on a frequent basis by itself or in conjunction with other reconstructive procedures. Dysfunction of the posterior compartment may be managed very differently by varying clinicians, and there is a lack of consensus with regard to the indications, surgical techniques, and outcome assessment.

The restoration of normal anatomy to the posterior vaginal wall is referred to as a posterior repair or colporrhaphy. Although frequently used interchangeably with the term rectocele repair, the two operations may have vastly different treatment goals. Whereas a rectocele repair focuses on repairing a herniation of the anterior rectal wall into the vaginal canal caused by a weakness in the rectovaginal septum, a posterior colporrhaphy is designed to correct a posterior vaginal wall bulge, as well as normalize vaginal caliber by restoring structural integrity to the posterior vaginal wall and introitus.

This chapter will cover various aspects of the gynecologic approach to rectocele repair, including symptoms, anatomy, physical examination, indications for repair, surgical techniques, and treatment outcomes.

Symptoms

Symptoms of posterior wall weakness typically entail pelvic and perineal pressure, a vaginal bulge, and associated lower back pain. Many women need to digitally reduce or splint the posterior vaginal bulge or the perineum to initiate or complete a bowel movement. Accumulation of stool within the rectocele reservoir leads to increasing degrees of perineal pressure and obstructive defecation. In the absence of digital reduction, women will note incomplete emptying, which leads to a high degree of frustration. A vicious cycle of increasing pelvic pressure, need for stronger Valsalva efforts, enlargement of the rectocele bulge, and increasing perineal pressure ensues. Rectal digitation is frequently not self-reported by patients with a symptomatic rectocele unless asked by their physicians.

An enlarging rectocele will widen the levator hiatus and increase vaginal caliber. In addition, women with increasing degrees of prolapse have progressively larger genital hiatuses.[4] This may lead to sexual difficulties including symptoms of vaginal looseness and decreased sensation during intercourse. Whether this is caused by the enlargement of the vaginal introitus and levator hiatus, or coexistent damage to the pudendal nerve supply to the pelvic floor musculature is unclear. A large enterocele or rectocele may extend beyond the hymenal ring. Once exteriorized, the patient is at risk for vaginal mucosal erosion and ulceration.

Patients frequently have associated complaints of constipation. The symptom of constipation is not clearly understood by the practicing gynecologist. Its vague nature, coupled with a poor understanding of the complexity of colonic function, results in an incomplete evaluation of the symptom of constipation by the gynecologist. Unfortunately, this may result in surgical treatment of abnormal bowel function via a rectocele repair when conservative therapy for constipation may have been satisfactory. The persistence of abnormal defecation postoperatively may be responsible for the high rectocele recurrence rate.

Anatomy

Rectoceles result from defects in the integrity of the posterior vaginal wall and rectovaginal septum, and subsequent herniation of the posterior vaginal wall and anterior rectal wall into the vaginal lumen through these defects. The normal posterior vagina is lined by squamous epithelium that overlies the lamina propria, a layer of loose connective tissue. A fibromuscular layer of tissue composed of smooth muscle, collagen, and elastin underlies this lamina propria, and is referred to as the rectovaginal fascia. This is an extension of the endopelvic fascia that surrounds and supports the pelvic organs, and contains blood vessels, lymphatics, and nerves that supply and innervate the pelvic organs.

The layer of tissue between the vagina and the rectum, or rectovaginal fascia, was believed to be analogous to the rectovesical septum and became known as Denonvilliers' fascia in the female, or the rectovaginal septum. Others described the rectovaginal septum as a support mechanism of the pelvic organs, and they were successful in identifying this layer during surgical and autopsy dissections.[5–7] It is unclear whether this fascial layer extends from the vaginal cuff to the perineum or is only present along the distal vaginal wall from the levator reflection to perineum.

The normal vagina is stabilized and supported on three levels. Superiorly, the vaginal apical endopelvic fascia is attached to the cardinal-uterosacral ligament complex. Laterally, the endopelvic fascia is connected to the arcus tendineus fasciae pelvis, with the lateral posterior vagina attaching to the fascia overlying the levator ani muscles. Inferiorly, the lower posterior vagina connects to the perineal body, composed of the anterior external anal sphincter, transverse perineum, and bulbocavernosus muscles. The cervix (or vaginal cuff in the hysterectomized woman) is considered to be the superior attachment site or "superior tendon," and the perineal body the inferior attachment site or "inferior tendon." The endopelvic fascia extends between these two sites comprising the rectovaginal septum (Figure 8-5.1). A rectocele results from a stretching or actual separation or tear of the rectovaginal fascia, leading to a bulging of the posterior vaginal wall noted on examination during a Valsalva maneuver. Trauma from vaginal childbirth often leads to transverse defects above

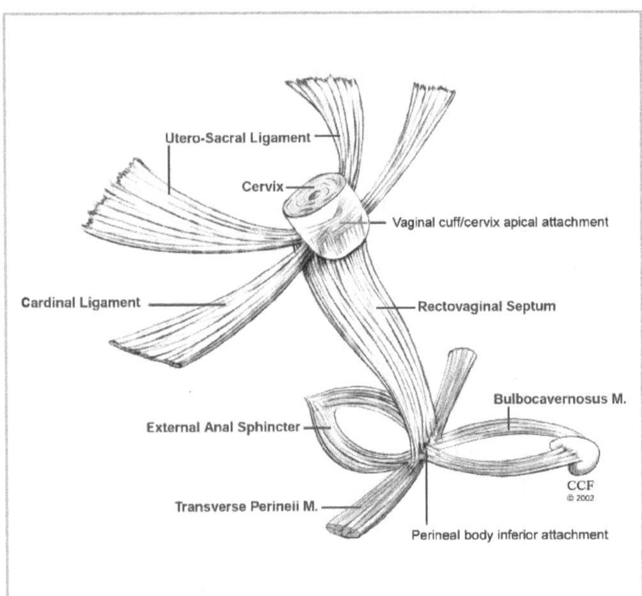

Figure 8-5.1. Diagrammatic representation of the rectovaginal septum including its attachment from vaginal apex to perineal body. (Reprinted with the permission of The Cleveland Clinic Foundation.)

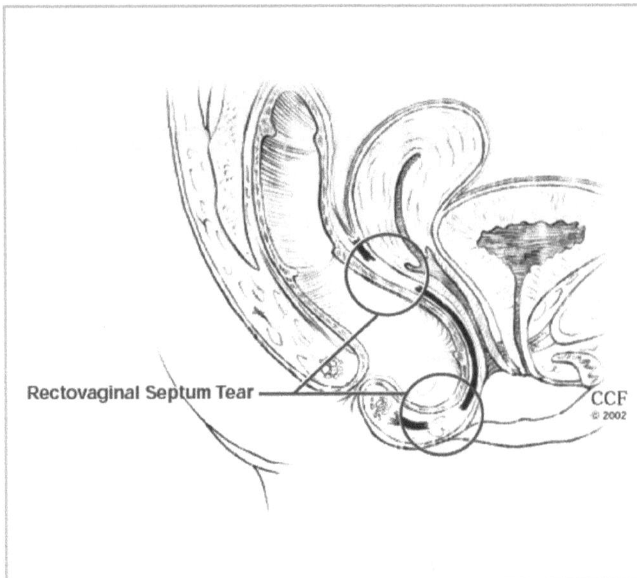

Figure 8-5.2. Fascial tears of the rectovaginal septum can occur superiorly or inferiorly at sites of attachment to a central tendon. (Reprinted with the permission of The Cleveland Clinic Foundation.)

the usual location of the connection to the perineal body (Figure 8-5.2). In addition, patients may present with lateral, midline, or high transverse fascial defects. Separation of the rectovaginal septum fascia from the vaginal cuff results in the development of an enterocele as a hernia sac without fascial lining and filled with intraperitoneal contents (Figure 8-5.2). This can present clinically as a rectocele bulge.

The levator musculature extends from the pubic bone to the coccyx and provides support for the change in vaginal axis from vertical to horizontal along the mid vagina. A rectocele typically develops at, or below, the levator plate, along the vertical vagina, weakening the fascial condensation of the attachments of the perineal musculature (Figure 8-5.3).

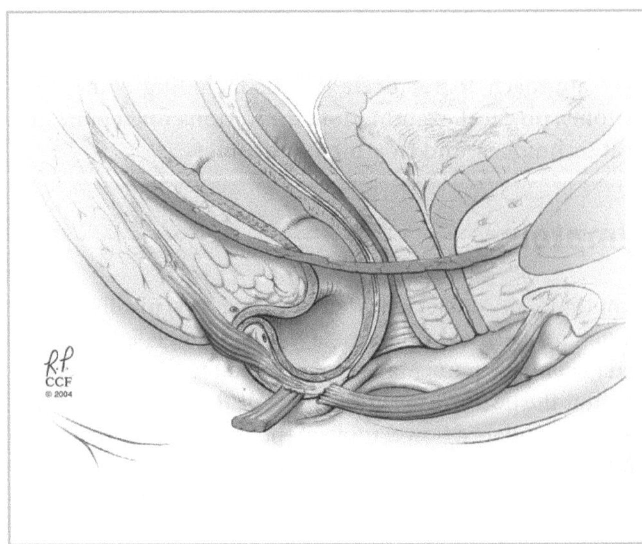

Figure 8-5.3. Rectoceles develop at or below the levator muscles, splaying the perineal musculature attachments. (Reprinted with the permission of The Cleveland Clinic Foundation.)

Physical Examination

The typical finding in a woman with a symptomatic rectocele is a lower posterior vaginal wall bulge noted on physical examination. It may extend superiorly to weaken the support of the upper, posterior vaginal wall, leading to an enterocele, or to the vaginal apex, leading to vaginal vault prolapse. In an isolated rectocele, the bulge extends from the edge of the levator plate to the perineal body. As a rectocele enlarges, the perineal body may further distend and become thinner leading to an evident perineocele. Enteroceles and rectoceles frequently coexist. The physical examination should include not only a vaginal examination but also a rectal examination, because a perineocele may not be evident on vaginal examination. At times, it can be identified only upon digital rectal examination where an absence of fibromuscular tissue in the perineal body is confirmed.

The gynecologic preoperative evaluation of a symptomatic posterior vaginal bulge typically includes only a history and physical examination. Gynecologists have not adopted the performance of defecography or other evaluation techniques to assess rectoceles. Whereas 80% of colorectal surgeons use defecography, only 6% of gynecologists use it.[1–3] In addition, differentiation between enterocele and rectocele components of posterior vaginal wall prolapse is typically performed on a clinical and intraoperative basis. It is unclear at this time whether surgical therapy outcomes are negatively impacted by the lack of a preoperative evaluation beyond a history and physical examination.

Typically, gynecologists consider repair of the rectocele indicated for obstructive defecation symptoms, lower pelvic pressure and heaviness, prolapse of posterior vaginal wall, and pelvic relaxation with enlarged vaginal hiatus. However, one should be cautioned that although repair of rectoceles may correct abnormal anatomy and alleviate these symptoms, colorectal dysfunction, including constipation, may persist.

Surgery to Correct a Rectocele

There are several goals of surgery to repair a rectocele. Endopelvic fascial integrity from the apex to the perineum and levator plate integrity should be reestablished. The anterior rectal wall should be well supported. The perineal body needs to be reinforced. The end result should be a vagina with normal caliber and length.

Posterior Colporrhaphy Technique

Posterior colporrhaphy is the most common gynecologic type of rectocele repair and is often performed in conjunction with a perineoplasty to address a relaxed perineum and widened genital hiatus. Preoperatively, the severity of the rectocele is assessed, as well as the desired final vaginal caliber. Allis clamps are placed on the inner labia minora/hymen remnants bilaterally and then approximated in the midline. The resultant vagina should loosely admit two to three fingers. A triangular incision over the perineal body is made between the Allis clamps, and sharp dissection is then performed to separate the posterior vagina from the underlying rectovaginal fascia. A midline incision is made along the length of the vagina to a site above the superior edge of the rectocele.

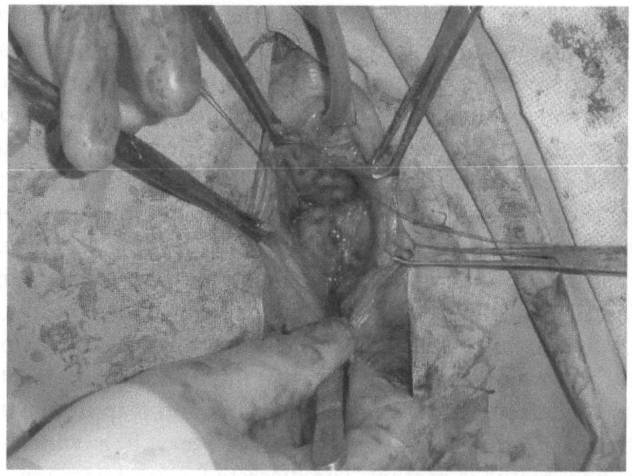

Figure 8-5.4. Multiple interrupted sutures are used to approximate the endopelvic fascia at the perineal body apex in the midline.

The dissection is carried laterally to the lateral vaginal sulcus and medial margins of the puborectalis muscles. The rectovaginal fascia with or without the underlying levator ani muscles is then plicated with interrupted sutures beginning at the level of the levator plate, while depressing the anterior rectal wall with the nondominant hand (Figure 8-5.4). Typically, thick, absorbable sutures using No. 1 Vicryl are placed along the length of the rectocele until plication to the level of the perineal body is complete. Excess vaginal mucosa is carefully trimmed and then reapproximated. A concomitant perineoplasty may be performed by plicating the bulbocavernosus and transverse perineal muscles in the midline with No. 1 Vicryl. This reinforces the perineal body and provides enhanced support to the corrected rectocele.

Discrete Fascial Defect Repair Technique

Discrete tears or breaks in the rectovaginal fascia or rectovaginal septum have been described and may contribute to the formation of rectoceles (Figure 8-5.2). Similarly to hernia repairs performed by general surgeons, the technique involves identifying the discrete fascial tears, reducing the hernia, and then closing the defect. The surgical dissection is similar to the traditional posterior colporrhaphy whereby the vaginal mucosa is dissected off the underlying rectovaginal fascia to the lateral border of the levator muscles. This dissection must be done very carefully to avoid creating iatrogenic fascial defects. Instead of plicating the fascia and levator muscles in the midline, however, the fascial tears are identified, and the edges are reapproximated with interrupted permanent sutures. Richardson[6] describes pushing anteriorly with a finger in the rectum to identify areas of rectal muscularis that are not covered by the rectovaginal septum. Thereby, the operator can locate fascial defects, identify fascial margins, and reapproximate them. A perineoplasty may be necessary if a widened vaginal hiatus is present. The discrete fascial defect repair may also be used to correct enteroceles by attaching torn endopelvic fascia to its apical attachment site at the cervix or cardinal-uterosacral ligament complex with interrupted sutures.

Modifications of the Rectocele Repair

The posterior colporrhaphy and the discrete fascial defect repair may be combined. After dissecting the rectovaginal fascia off the overlying vaginal mucosa, all fascial tears are identified and the edges are reapproximated with permanent suture, such as silk. The levator ani muscles can then be plicated in the midline, anterior to the rectovaginal fascia, using absorbable suture in the traditional manner.

Reconstructive pelvic surgeons have reported reinforcement of prolapse repairs with synthetic and biologic prostheses. Synthetic mesh is widely used for anti-incontinence surgery and abdominal sacrocolpopexy to repair vaginal vault prolapse. Although high success rates have been reported, erosion of the mesh and infection have been associated with these repairs.[8,9] Autologous grafts and allograft prostheses, including fascia lata, rectus sheath, and dermal grafts have been used for these surgeries as well as reinforcement of repairs to reduce these complications. Few complications have been associated with these grafts, and they have a comparable success rate to synthetic materials. Xenograft material, including bovine pericardium and porcine skin and small intestinal mucosa, had also been used to reinforce these repairs; however, no reports on complications and success rates exist in the literature.

When using graft material to reinforce a rectocele repair, the graft may be sutured to the lateral posterior vaginal sulcus on each side using absorbable or permanent suture. The graft should be trimmed before placement so that it lies as a flat layer between the vaginal mucosa and the newly repaired rectovaginal fascia (Figure 8-5.5). There is no evidence that soaking the graft material in antibiotic solution before placement decreases the incidence of vaginal infection or erosion.

Few studies have reported on the use of graft materials to reinforce posterior compartment defects. Sand et al.[10] reported on 132 women undergoing either standard rectocele repairs or rectocele repairs reinforced with polyglactin 910 mesh (an absorbable mesh), and found no difference in recurrence rates between the two groups. Two small observational studies on the use of Marlex mesh for rectocele repair did not report erosion or recurrence.[11,12]

Results of Surgical Repair

Although frequently performed, posterior colporrhaphy has been described as "among the most misunderstood and poorly performed" gynecologic surgeries.[13] Although many authors have reported satisfactory anatomic results, conflicting effects on bowel and sexual function postoperatively have been noted. Several authors have reported high sexual dysfunction rates of up to 50% of women reporting dyspareunia or apareunia after posterior colporrhaphy.[14] Some authors caution the performance of rectocele repair in patients with preoperative abnormal colonic transit studies secondary to continued constipation postoperatively.[15] Other authors performed preoperative defecography on all patients and found that the grade of rectocele emptying did not influence long-term outcome. In addition, pre- and postoperative defecography was reported to show an increase in maximal anal resting pressure postoperatively, suggesting that it may be caused by levator plication (Table 8-5.1).[16]

Many authors suggest that the significant rate of postoperative dyspareunia may be attributed to the plication of the levator ani muscles, and has led several authors to the popularization of the discrete fascial defect repair.[17] Several authors have reported a similar anatomic cure rate with

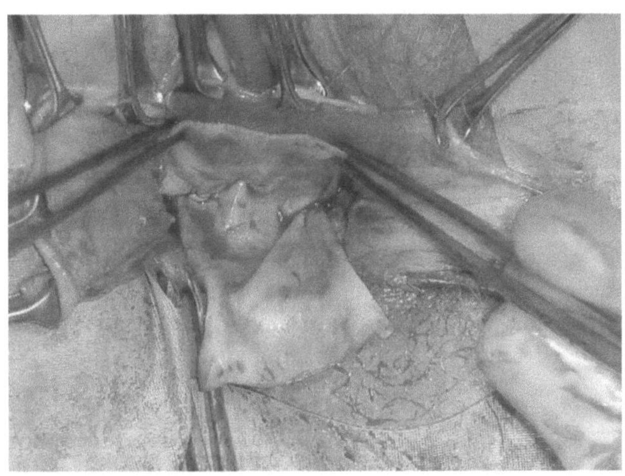

Figure 8-5.5. Posterior vaginal wall reinforcement graft in place from vaginal apex to perineum.

Table 8-5.1. Outcomes for posterior colporrhaphy

Author	No. of Patients	Mean Follow-up (mo)	Improvement in Pelvic Pressure Symptoms (%)	Anatomic Correction Rates (%)	Improvement in Evacuation Difficulty (%)	Postoperative Dyspareunia Rates (%)
Kahn and Stanton[17]	171	42.5	51	76	62	Increased from 18 to 27
Lopez et al.[16]	25	61.2	75	92	91	Increased from 6 to 33
Mellgren et al.[15]	25	12	N/A	96	88	Increased from 6 to 9
Paraiso et al.[18]	102	10	89	61	82	Increased from 2 to 12

Table 8-5.2. Outcomes for the discrete fascial defect repair

Author	No. of Patients	Mean Follow-up (mo)	Improvement in Pelvic Pressure Symptoms (%)	Anatomic Correction Rates (%)	Improvement in Evacuation Difficulty (%)	Postoperative Dyspareunia Rates (%)
Cundiff et al.[19]	69	24	87	82	63	Decreased from 29 to 19
Glavind and Madsen[20]	67	3	N/A	100	85	Decreased from 12 to 6
Kenton et al.[21]	46	12	90	77	54	Decreased from 26 to 2
Porter et al.[22]	125	18	73	82	55	Decreased from 67 to 46

this surgery, along with significant improvement in quality-of-life measures. Unlike the traditional posterior colporrhaphy, all these series report less postoperative dyspareunia (Table 8-5.2). The authors noted significant improvement in splinting, vaginal pressure, and stooling difficulties. However, rates of fecal incontinence and constipation were unchanged postoperatively. These studies show promising anatomic and functional results; however, long-term prospective studies are warranted. We have not found a significantly high rate of dyspareunia with a combination of site-specific repair and high perineoplasty. The multiple goals of normalization of anterior rectal support and vaginal hiatus size are achieved without negative consequence.

Discussion

Gynecologic indications for rectocele repair are numerous because gynecologists primarily address vaginal symptoms when repairing a rectocele. Obstructive defecation symptoms are only some of a list of accepted indications. Preoperative evaluation typically only includes clinical assessment gained from the history and physical examination, and gynecologists rarely depend on defecography to plan a reconstructive procedure for rectocele. Overall, surgical correction success rates are quite high when using a vaginal approach for rectocele correction. Vaginal dissection results in better visualization and access to the endopelvic fascia and levator musculature, which allows for a more firm anatomic correction. More comprehensive data collection is necessary to better understand the effect of various surgical techniques on vaginal, sexual, and defecatory symptoms.

References

1. Davila GW, Ghoniem GM, Kapoor DS, Contreras-Ortiz O. Pelvic floor dysfunction management practice patterns: a survey of members of the International Urogynecological Association. Int Urogynecol J 2002;13:319–325.
2. Kapoor DS, Davila GW, Wexner SD, Ghoniem GM. Posterior compartment disorders: survey of colorectal surgeons practice patterns and review of the literature. Int Urogynecol J 2001;12:S53.
3. Mizrahi N, Kapoor D, Baig MK, et al. A gynecologic perspective of posterior compartment defects. Colorectal Dis 2002;4:68.
4. Delancey JOL, Hurd WW. Size of the urogenital hiatus in the levator ani muscles in normal women and women with pelvic organ prolapse. Obstet Gynecol 1998;91:364–368.
5. Milley PS, Nichols DH. A correlative investigation of the human rectovaginal septum. Anat Rec 1969;163:443–452.
6. Richardson AC. The rectovaginal septum revisited: its relationship to rectocele and its importance in rectocele repair. Clin Obstet Gynecol 1993;36:976–983.
7. Uhlenhuth E, Wolfe WM, Smith EM, Middleton EB. The rectogenital septum. Surg Gynecol Obstet 1948;86:148–163.
8. Birch C, Fynes MM. The role of synthetic and biological prostheses in reconstructive pelvic floor surgery. Curr Opin Obstet Gynecol 2001;14:527–535.
9. Iglesia CB, Fenner DE, Brubaker L. The use of mesh in gynecologic surgery. Int Urogynecol J 1997;8:105–115.
10. Sand PK, Koduri S, Lobel RW. Prospective randomized trial of polyglactin 910 mesh to prevent recurrences of cystoceles and rectoceles. Am J Obstet Gynecol 2001;184:1357–1364.
11. Parker MC, Phillips RKS. Repair of rectocoele using Marlex mesh. Ann R Coll Surg Engl 1993;75:193–194.
12. Watson SJ, Loder PB, Halligan S. Transperineal repair of symptomatic rectocele with Marlex mesh: a clinical, physiological and radiological assessment of treatment. J Am Coll Surg 1996;183:257–261.
13. Nichols DH. Posterior colporrhaphy and perineorrhaphy: separate and distinct operations. Am J Obstet Gynecol 1991;164:714–721.
14. Francis WJ, Jeffcoate TN. Dyspareunia following vaginal operations. J Obt Soc Am 1961;68:1–10.
15. Mellgren A, Anzen B, Nilsson BY, et al. Results of rectocele repair. A prospective study. Dis Colon Rectum 1995;38:7–13.
16. Lopez A, Anzen B, Bremmer S, et al. Durability of success after rectocele repair. Int Urogynecol J Pelvic Floor Dysfunct 2001;12:97–103.

17. Kahn MA, Stanton SL. Posterior colporrhaphy: its effects on bowel and sexual function. Br J Obstet Gynaecol 1997;104:82–86.
18. Paraiso MF, Weber AM, Walters MD, Ballard LA, Piedmonte MR, Skibinshi C. Anatomic and functional outcome after posterior colporrhaphy. J Pelvic Surg 2001;7:335–339.
19. Cundiff GW, Weidner AC, Visco AG, Addison WA, Bump RC. An anatomic and functional assessment of the discrete defect rectocele repair. Am J Obstet Gynecol 1998;179:1451–1457.
20. Glavind K, Madsen H. A prospective study of the discrete fascial defect rectocele repair. Acta Obstet Gynecol Scand 2000;79:145–147.
21. Kenton K, Shott S, Brubaker L. Outcome after rectovaginal reattachment for rectocele repair. Am J Obstet Gynecol 1999;181:1360–1363.
22. Porter WE, Steele A, Walsh P, Kohli N, Karram MM. The anatomic and functional outcomes of defect-specific rectocele repairs. Am J Obstet Gynecol 1999;181:1353–1358.

8-6

Rectal Prolapse Therapy

T. Cristina Sardinha and Steven D. Wexner

The definition of rectal prolapse is the protrusion of the full thickness of the rectal wall through the anus (Figure 8-6.1). This protrusion differs from mucosal prolapse and internal intussusception. In cases of mucosal prolapse, only the inner mucosal rectal layer protrudes through the anus (Figure 8-6.2). Conversely, in cases of rectoanal internal intussusception, the prolapsed tissue remains confined within the rectal lumen. The preoperative knowledge of this type of prolapse will help direct the appropriate therapeutic option.

The etiology of rectal prolapse is unclear. However, factors involved in the development of rectal prolapse may be congenital or acquired. Moreover, there are conditions that can be associated or predispose to the development of rectal prolapse. These problems include intestinal disorders such as constipation and chronic straining, neurologic diseases, especially spinal cord abnormalities and depression, nulliparity, lack of rectal fixation to the sacrum, previous anorectal surgery, and pelvic floor defects. In addition, rectal prolapse is almost always a condition of the female gender.[1] Despite the implication of multiple pregnancies in the etiology of rectal prolapse, this pathology is more frequent in nulliparous patients.

The true incidence of rectal prolapse is unknown. However, rectal prolapse is more common in the fourth and seventh decades of life. The majority of these patients are elderly females. Moreover, a small number of children, usually younger than 3 years of age, are affected with rectal prolapse. In these individuals, the treatment differs from adult patients.

The patients with rectal prolapse often complaint of a mass protruding out of the anus, initially with straining then progressing to exteriorization with any increase in abdominal pressure and finally prolapse even at rest. Chronic prolapse is often associated with inflamed, irregular, and edematous mucosa, mucous discharge, ulceration, and bleeding. Nonetheless, incarceration and gangrene are rare phenomena. Fecal incontinence is reported in 50% to 75% of patients with rectal prolapse. This may be a consequence of sphincteric disruption or pudendal stretching. However, the association of rectal prolapse with chronic constipation is less well understood. A history of constipation is found in 25% to 50% of patients with rectal prolapse.[1,2]

The clinical assessment of a patient with rectal prolapse often includes the reproduction of the prolapse; thus, the patient may need to strain in the sitting position (toilet) in order to evaluate the extent of the prolapse. The use of enemas or suppositories (especially in children) may induce straining, facilitating the evaluation and extent of the prolapse. Full-thickness rectal prolapse presents with concentric rectal folds and double rectal wall on palpation, whereas mucosal prolapse is visualized as radial folds that rarely protrude greater than 5 cm.

Digital rectal examination often reveals weakened anal sphincters and may not always be required. However, colonoscopy or barium enema should be performed to exclude associated lesions such as tumors and rectal ulcer. Biopsies may also be obtained to exclude colitis cystica profunda or solitary rectal ulcer.

More than 100 procedures have been described for the management of rectal prolapse. We will describe the most frequently performed operations and their outcomes. Basically, operations for rectal prolapse are divided into perineal and abdominal approaches.

Perineal Operations

Perineal Rectosigmoidectomy

The Altemeier operation has been successfully used to treat rectal prolapse by many authors. However, controversies exist regarding recurrence rate. As described by William Altemeier, "In principle, the operation obliterates the pelvic pouch, plicates the levators, and resects the redundant bowel. However, the rectum is not fixated to the sacrum."[1]

The University of Minnesota's group reported their experience with the surgical treatment of rectal prolapse over an 18-year period.[3] Perineal rectosigmoidectomy was

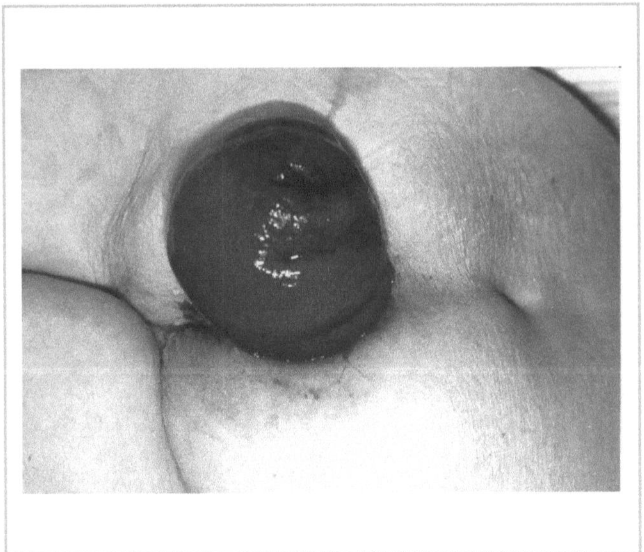

Figure 8-6.1. Full-thickness rectal prolapse.

performed in 183 of 372 patients treated for rectal prolapse; follow-up ranged from 12 to 165 months. It was noted that perineal rectosigmoidectomy was more frequently performed in patients with higher comorbidity and older age, compared with the abdominal approach. However, despite the higher recurrence rate (16% vs 5% for abdominal operations) in patients who underwent the perineal operation, this procedure was progressively more often performed over the last phase of the study. This finding may be attributed to relatively low morbidity, especially in high-risk patients, absence of an abdominal incision, therefore less adhesion formation, no intraabdominal anastomosis, and a relatively short hospitalization. Moreover, repeat perineal procedure can be safely performed.[4] Table 8-6.1 summarizes the most recent results of perineal rectosigmoidectomy in the treatment of rectal prolapse.

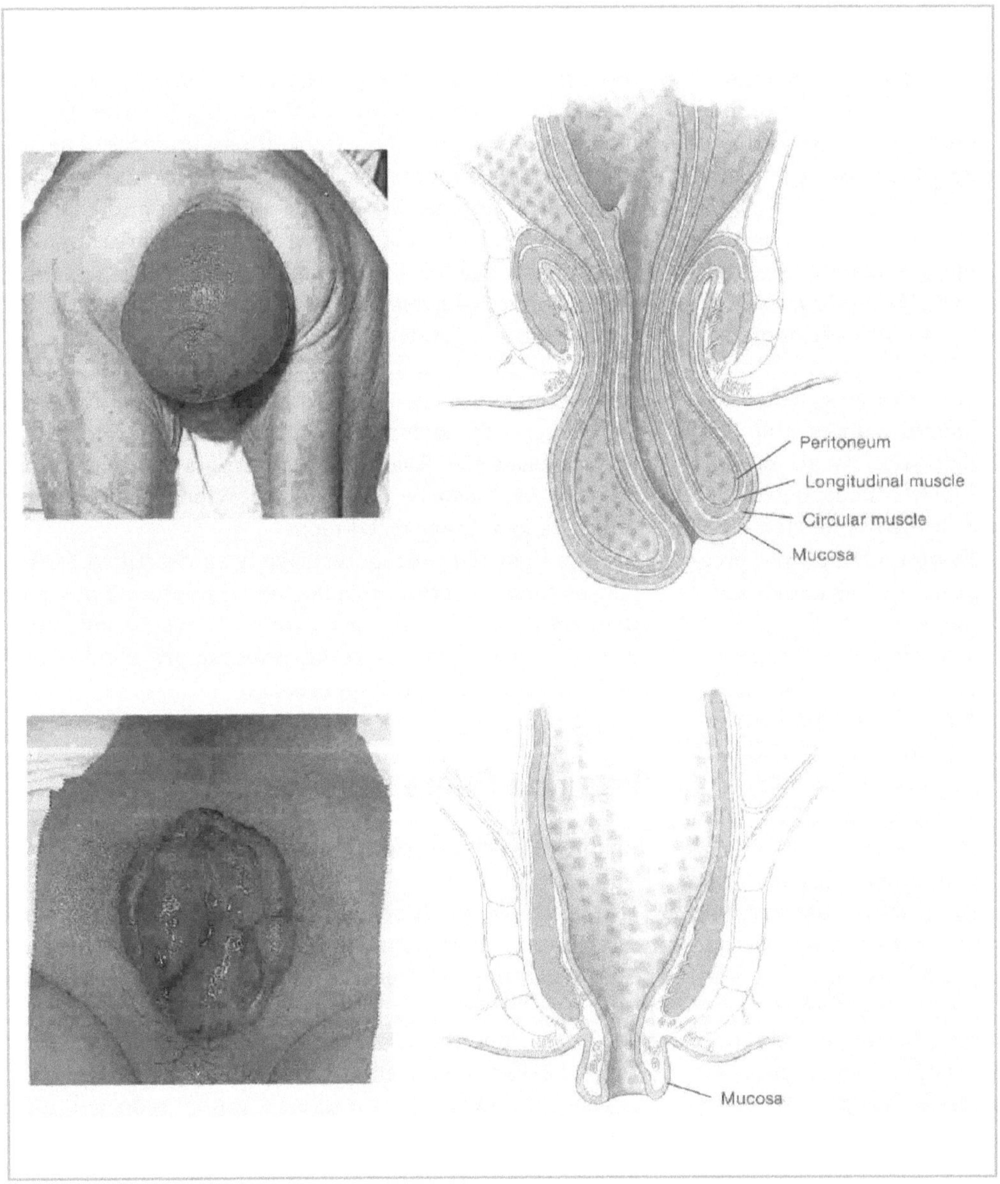

Figure 8-6.2. Difference between mucosal and full-thickness rectal prolapse. (Reprinted from Gordon PH, Nivatvongs S. Principles and Practice of Surgery for the Colon, Rectum and Anus. 2nd ed. Marcel Dekker Inc 2002, 509.)

Table 8-6.1. Outcome of perineal rectosigmoidectomy

Reference	Year	Patients	Follow-up (mo) (range)	Age (y)	Recurrence (%)	Leak (%)	Mortality (%)
Finlay[6]	1991	17	9–40	60–92	10	NR	1
Williams[7]	1992	114	3–90	21–100	10	NR	0
Thorne[8]	1992	16	3–37	72–96	12	NR	0
Agachan[9]	1997	32	6–72	48–101	12.5	2 (6.2)	3.1
Kim	1999	183	12–165	14–100	16	NR	0
Kimmins[10]	2001	63	6–88	39–96	6.3	2 (3.1)	0
Zbar[11]	2002	80	6–132	16–91	3.8	NR	0

NR, not recorded.

Delorme's Operation

The Delorme's procedure involves the removal of the rectal mucosa up to the apex of the prolapse, associated with pleating of the muscular layer of the rectum and anastomosis of the mucosa to the anal canal. This operation has mostly been used to treat mucosal prolapse. Therefore, it is difficult to evaluate the true outcome of the Delorme's operation in the treatment of full-thickness rectal prolapse. Moreover, it has been associated with a very high recurrence rate varying from 17% to 37%. In addition, improvement in fecal incontinence has been marginal.[1,2] We rarely recommend the Delorme's operation for the treatment of rectal prolapse, except in those few instances in which mucosal prolapse seems to be the main problem.

Abdominal Operations

Rectopexy

Several techniques, such as the Ripstein procedure and posterior rectopexy with synthetic mesh, have been described to perform the fixation of the rectum to the presacral fascia with the objective of preventing recurrence of the rectal prolapse. However, the use of synthetic material to fix the rectum has been associated with a multitude of complications including infection, fistulization of the prosthesis into the rectum, presacral bleeding, and obstruction.

The suture rectopexy using nonabsorbable sutures is an effective alternative to the use of synthetic mesh. This technique also avoids most of the problems encountered with the use of mesh. However, difficult evacuation, constipation, bleeding, and recurrence are still significant concerns. Although the recurrence rate is less than 10%, the onset of mucosal prolapse is a common form of recurrence.

Anterior Resection

Anterior resection for the treatment of rectal prolapse removes the redundant sigmoid colon, therefore avoiding problems such as constipation, torsion, and kinking when rectopexy is performed. Moreover, this technique can be safely performed laparoscopically. However, one should remember the risk of anastomotic complications. When this operation is performed for the treatment of prolapse, the rectum should be mobilized to the level of the lateral ligaments and the anastomosis should be fashioned at or just below the sacral promontory.

Anterior Resection with Rectopexy

The addition of rectal fixation to the sacrum may provide better results regarding the decrease in recurrence rate. This operation is safe and presents the lowest recurrence rate. The reported recurrence rate after sigmoid resection with proctopexy varies from 2% to 10%.[1] Initial procedures used the circumferential wrap. However, postoperative complications, including constipation and obstruction have led surgeons to modify the technique leaving the anterior rectum free. Others avoid the placement of the mesh altogether, adopting the use of nonabsorbable sutures to fixate the rectum to the presacral fascia. We currently prefer the latter approach to treat surgically fitted individuals with rectal prolapse.

The management of recurrent rectal prolapse depends on the previous procedure. An abdominal approach must be avoided if the patient had previous perineal rectosigmoidectomy, especially if sigmoid resection is contemplated, because of the potential ischemia and necrosis of the intervening segment between anastomoses. However, a repeat perineal rectosigmoidectomy can be safely performed in this patient population. Moreover, if the patient's previous operation was sigmoidectomy with proctopexy, a repeat procedure to treat the recurrence is also feasible. Nonetheless, one must perform a complete presacral rectal dissection despite the difficulties caused by the prior dissection.

In summary, one must match the operation with the patient. Healthy individuals and/or patients with low comorbidity should undergo the abdominal approach (laparoscopic[5] or open) for the treatment of rectal prolapse. Sigmoid resection with proctopexy seems to offer lower recurrence rates of the options described. However, compared with the perineal procedure, the abdominal operations are associated with a higher morbidity and mortality. Conversely, high surgical risk patients should generally be offered a perineal procedure, the best option of which is the perineal rectosigmoidectomy with levatorplasty. Compared with abdominal procedures, this operation has a significantly lower incidence of postoperative complications at the expense of a higher recurrence rate.

References

1. Corman ML. Colon and Rectal Surgery. 4th ed. Philadelphia: Lippincott-Raven; 1998:401–448.
2. Poritz LS. Rectal Prolapse. Available at: http://www.emedicine.com/med/topic3533.htm.
3. Kim DS, Tsang CBS, Wong WD, Lowry AC, Goldberg SM, Madoff RD. Complete rectal prolapse: evolution of management and results. Dis Colon Rectum 1999;42:460–469.
4. Pikarsky AJ, Joo JS, Wexner SD, et al. Recurrent rectal prolapse: what is the next good option? Dis Colon Rectum 2000;43:1273–1276.
5. Solomon MJ, Young CJ, Eyers AA, Roberts RA. Randomized clinical trial of laparoscopic versus open abdominal rectopexy for rectal prolapse. Br J Surg 2002;89:35–39.
6. Finlay IG, Altchison M. Perineal exclsion of the rectum for prolapse in the elderly. Br J Surg 1991;78(6):687–689.
7. Williams JG, Rothenberger DA, Madoff RD, Goldberg SM. Treatment of rectal prolapse in the elderly by perineal rectosigmoldectomy. Dis Colon Rectum 1992;35(9):830–834.
8. Thorne MC, Polglase AL. Perineal proctectomy for rectal prolapse in elderly and debilitated patients. Aust N Z J Surg 1992;62(10):791–794.
9. Agachan F, Reissman P, Pfeifer J, Nogueras JJ, Weiss EG, Wexner SD. Comparison of three perineal procedures for the treatment of rectal prolapse. South Med J 1997;90(9):925–932.
10. Kimmins MH, Evetts BK, Isler J, Billingham R. The Altemeier repair: outpatient treatment of rectal prolapse. Dis Colon Rectum 2001;44(4):565–570.
11. Zbar AP, Takashima S, Hasegawa T, Kitabayashi K. Perineal rectosigmoldectomy (Altemeier's procedure): a review of physiology, technique and outcome. Tech Coloproctol 2003;6(2):109–116. Review.

8-7

Rectoanal Intussusception, Solitary Rectal Ulcer, and Sigmoidoceles

Juan J. Nogueras and Susan M. Cera

Successful defecation requires the smooth coordination of neurologic and muscular events by the pelvic floor and rectum. Deterioration of a single component of this process may lead to symptomatic rectal dysfunction manifested as constipation from the inability to evacuate. Prolonged untreated dysfunction with straining may ultimately lead to occult or overt rectal prolapse with concomitant rectal ulceration or fecal incontinence.[1–3] The etiology of rectal dysfunction remains obscure but is believed to be multifactorial in nature, involving electromyogenic, psychologic, aging, and hormonal mechanisms. In addition, rectal dysfunction is usually one component of a pathophysiologic process that involves the entire pelvic floor. As a result, several clinical manifestations may occur together or separately as part of the spectrum of this disease process and include nonrelaxation of the puborectalis muscle, rectoanal intussusception, rectal prolapse, perineal descent, solitary rectal ulcer syndrome (SRUS), rectocele, sigmoidocele, and hemorrhoids. Consequently, the therapeutic approach to rectal dysfunction is multifactorial, with the primary focus on bowel and pelvic floor retraining, behavioral modifications, and, less frequently, surgical interventions. This chapter will primarily address rectoanal intussusception, solitary rectal ulcer, and sigmoidoceles.

Rectoanal Intussusception

Rectoanal intussusception and SRUS have many common features. Dysfunction of the rectum with the development of abnormal defecatory patterns is usually the underlying pathogenesis. Chronic straining may induce intussusception of the rectal mucosa, which in turn may develop into full-thickness rectal prolapse. The occult intussusception may also predispose to a persistent feeling of incomplete evacuation with the chronic urge to strain. Straining may cause excessive tensile forces on the anterior wall of the rectum resulting in ulceration.

Rectoanal intussusception, also known as occult rectal prolapse or internal procidentia, is an intussusception of the rectal wall that does not protrude through the anus. It may be asymptomatic or associated with incontinence or constipation. Although it is strongly associated with mucosal prolapse, full-thickness rectal prolapse, and perineal descent, the finding of occult rectal prolapse is not necessarily pathologic. Diagnosis is made by defecography and is seen in up to 50% to 60% of the defegrams of asymptomatic healthy volunteers.[4] In addition, the presence of intussusception does not correlate with rectal emptying,[5] and it seldom leads to rectal prolapse.[6] Care must be taken when associating symptoms with rectal intussusception because it is usually not a cause of symptoms but a marker for underlying rectal dysfunction in symptomatic individuals and is frequently found with other abnormalities of the pelvic floor.

For patients with internal intussusception and constipation from obstructive or dyssynergic defecation, the first line of therapy is conservative treatment aimed at restoring normal defecatory habits. Behavioral modifications include suppressing the urge to strain, minimizing toileting time, and decreasing the number of toilet visits. Additional behavioral modifications can be accomplished with biofeedback therapy, particularly in patients with obstructed defecation related to paradoxical, or nonrelaxing puborectalis syndrome. Dietary modifications include a fiber-enriched diet, fiber supplementation, and eight glasses of noncaffeinated beverages per day, all of which promote regularity and restoration of normal rectal motility. Laxatives and scheduled enemas may also be beneficial in rectal evacuation and suppressing the urge to strain.

Solitary Rectal Ulcer Syndrome

Solitary rectal ulcer syndrome is a rare syndrome associated with evacuation dysfunction.[7] The underlying pathophysiology involves paradoxical contraction, instead of relaxation, of the muscles of the pelvic floor during defecation. Consequently, patients strain against immobile, contracted muscles of the pelvic floor impeding the passage of stool. Chronic straining may lead to ischemia and ulceration of the anterior rectal wall. Classic symptoms include significant time straining, feeling of incomplete emptying, rectal bleeding, passage of mucous, anorectal pain, tenesmus, and the need for digital assistance for evacuation.

Clinical history and sigmoidoscopic findings, including single or multiple ulcers or hyperemia without ulceration, establish a diagnosis of SRUS. Ulcers are often shallow with gray-white base and zone of hyperemia. In addition, polypoid lesions may develop and must be differentiated from adenomatous polyps on histology. The lesions can be confused with benign and malignant neoplasms, or localized areas of inflammatory bowel disease, radiation proctitis, and pseudomembranous colitis. Biopsy is mandatory and confirms diagnosis. Histologic changes include mucosal thickening, edema of the lamina propria, fibrosis, and architectural derangement of the muscularis propria with extension of the smooth muscle fibers into the glandular crypts. Colitis cystica profunda is a related disorder in which the symptoms are indistinguishable from SRUS. Instead of ulcerations, examination reveals firm nodules on the anterior rectal wall. Histologically, these nodules are composed of normal colonic glands located submucosally and filled with pools of mucus. The displacement of glands into the submucosa can be misdiagnosed as invasive carcinoma if note is not made of the histologically benign mucosa overlying the glands.

Anorectal physiology studies may reveal paradoxical puborectalis contraction but this finding is not necessary to make the diagnosis and may be found independent of the signs and symptoms of SRUS. Internal intussusception is found in 80% of patients with SRUS but is also found in normal volunteers undergoing defecography.

Solitary rectal ulcer syndrome is notorious for its chronicity and refractory nature to treatment. No single treatment is entirely satisfactory and therefore the main goal is to adjust patient symptoms and resolve bleeding. Once the diagnosis is established, all patients should undergo a trial of medical management. Initial treatment consists of fiber and hydration maintenance to aid in retraining the bowel and promoting normal motility; topical agents such as steroid or Carafate enemas have a limited role because they do not address the underlying pathophysiology. Judicious use of laxatives and enemas facilitate evacuation with minimal straining.

The mainstay of treatment of SRUS is behavioral techniques (biofeedback) directed at retraining pelvic floor coordination. In addition, the patient is taught posturing and appropriate use of abdominal musculature to minimize straining. Biofeedback also teaches patients to restrict the number of visits to the toilet, the duration of these visits, digitations, laxative use, and, most importantly, psychological support. The psychological effects are most evident in the short-term improvement of symptoms with the ability to resume normal employment despite nonhealing of the ulcer on sigmoidoscopic examination.[8] Unlike surgery, it is noninvasive and free of side effects. Beneficial results deteriorate over time with only half of the patients maintaining benefits more than 36 months.[9] However, repeat courses of biofeedback therapy reestablish success with behavioral modification.

Surgical therapies that have been described for SRUS include local excision, rectopexy, rectal mucosectomy (Delorme's procedure), proctectomy, colostomy, and argon beam coagulation. Local excision is not recommended because it does not address the underlying etiology and the lesions tend to recur. However, it may be useful to excise the lesion if malignancy cannot be excluded. Results of the other surgical modalities are variable and recurrence rates are high because dysfunction of the pelvic floor remains problematic.

In one study of 13 patients followed for 57 months, Marchal et al.[10] found that simple resection did not improve symptoms, colostomy resolved the symptoms and healed the ulcer, and rectopexy and modified Delorme's procedures were prone to relapse of symptoms and ulcers. The authors concluded that considering the high failure rate of surgery, operative management should only be performed in patients with total rectal prolapse or intractable symptoms not amenable to behavioral therapy. In another study of 66 patients who underwent surgery for rectal prolapse and followed for 90 months, 22 of 49 patients who underwent rectopexy failed.[11] Four of these patients underwent subsequent proctectomy with coloanal anastomosis of which three also failed, signifying the refractory nature of this disease process to surgical intervention. Ultimately, 14 patients required a stoma. Four of nine patients who underwent Delorme's procedure for the initial operation also failed. Four of seven patients who underwent anterior resection as the initial procedure underwent stoma creation. Overall, the stoma rate was 30%. Anterior resection used as a salvage procedure was not successful. Anti-prolapse operations only result in satisfactory long-term outcome in approximately 60% of patients with SRUS. These conclusions are not surprising because surgery does not tend to address the underlying dysfunction or abnormal motility of the rectum and pelvic floor. Consequently, symptom recurrence often persists and ulcer recurrence remains problematic. In those patients who undergo surgery, biofeedback is an important adjunct to retrain the muscles of the pelvic floor to prevent recurrence of the symptoms and the ulcer.

Sigmoidoceles

A sigmoidocele, similar to an enterocele, is a true hernia with a break in the fascial supports of the upper vagina-uterosacral-cardinal ligament complex and the rectovaginal septum. With increased abdominal pressure, the sigmoid herniates into the cul-de-sac, through the defect, and abuts the vaginal mucosa. This requires a significant redundancy in the sigmoid and a deep cul-de-sac to allow herniation to occur.

Sigmoidoceles are believed to cause constipation by obstructing defecation. Diagnosis is made by defecography. The herniating sigmoid colon full of firm barium paste or stool is observed compressing or obstructing the rectum. The obstructed rectum does not completely empty and may lead to prolonged straining and a sense of incomplete emptying. Increased straining may cause further damage to the pelvic floor supports. Patients complain of pelvic or lower abdominal fullness and frequently need to apply manual pressure to the perineum, posterior vaginal wall, or lower pelvis to defecate. In addition, associated abnormalities such as rectoanal intussusception, rectocele, and paradoxical contraction of the puborectalis frequently contribute to the symptomatology.

Sigmoidoceles are rare and occur in 4% to 5% of all patients undergoing defecograms.[12,13] Clinical examination may reveal pelvic floor prolapse and apical or posterior support defects but the presence of a sigmoid hernia is indiscernible. Fenner[12] indicated that none of the nine patients with vaginal vault or uterine prolapse with posterior compartment defects diagnosed as rectocele were diagnosed on physical examination alone. She concluded that the possibility of a sigmoidocele should be considered in all patients with a posterior compartment defect and symptoms of constipation.

Jorge and colleagues[13] suggested a sigmoidocele classification according to the position of the lower loop please see Chapter 3.4, Figure 3-4). Although sigmoidoceles are found on defegrams of asymptomatic patients, a nonemptying sigmoidocele can be the cause of a sensation of incomplete evacuation.

Optimal treatment of sigmoidoceles is controversial. Studies are few, retrospective, and involve small numbers. In addition, the clinical presentation is prolapse and constipation, which is not unique to sigmoidoceles particularly because they are often found in association with a variety of other clinical entities involving defects in the pelvic floor. Consequently, outcomes of differing therapy are difficult to assess. Choice of repair is based on the few published studies and clinical experience. Little is known about the functional relationship between rectoceles, enteroceles, sigmoidoceles, and defecation. A large, multicenter, randomized controlled trial is needed to evaluate the various surgical options and their long-term outcome of sigmoidocele repair.

At Cleveland Clinic Florida, patients with constipation and a first or second degree sigmoidocele are approached with conservative measures or biofeedback therapy with the expectation of success in approximately 50%. Third degree sigmoidoceles are usually managed by sigmoid resection with a successful outcome achieved in the majority. This procedure is ideally suited to the laparoscopic approach with the expectation of shorter hospital stay and less disability. Coexisting intussusception can be treated with rectopexy at the time of surgery. If nonrelaxing puborectalis syndrome is coexistent, this entity should be addressed preoperatively with biofeedback therapy.

If defects in the posterior vaginal wall and pelvic organ prolapse coexist with the sigmoidocele, repair may require a multidisciplinary approach. Fenner[12] reported on seven patients with vaginal or uterine prolapse or rectocele in addition to sigmoidocele. All patients underwent pelvic reconstructive surgery to address the prolapse and posterior wall defects followed by one of three procedures to address the redundant sigmoid based on the severity of the constipation, degree of prolapse, and degree of sigmoid redundancy. Two patients underwent sigmoid resection with sigmoidopexy, one patient had sigmoidopexy alone, and four had Halban obliteration of the cul-de-sac. Follow-up at 2 years showed no recurrence of the posterior vaginal defect and all patients had resolution of constipation except two in whom the Halban procedure was performed. In summary, the choice of the surgical procedure should primarily address repair of the posterior compartment hernia. In addition, resection or sigmoidopexy performed for large hernias in addition to the pelvic reconstructive surgery.

Conclusions

Intussusception is an epiphenomenon rather than a cause for evacuatory dysfunction and should be conservatively managed. Solitary rectal ulcer syndrome is a consequence of excessive chronic straining, and therapy should be geared toward restoration of normal defecatory habit. Surgical therapies described for SRUS are reserved for patients with severely refractory ulcers, particularly those with persistent hemorrhage. Optimal treatment of sigmoidoceles depends largely on symptomatology and associated conditions. Repair involves sigmoid resection with additional pelvic floor reconstruction for coexisting defects.

References

1. Felt-Bersma RJ, Cuesta MA. Rectal prolapse, rectal intussusception, rectocele, and solitary rectal ulcer syndrome. Gastroenterol Clin North Am 2001;30(1):199–222.
2. Madoff RD. Rectal prolapse and intussusception. In: Beck DE, Wexner SD, eds. Fundamentals of Anorectal Surgery. 2nd ed. Philadelphia: WB Saunders; 1998:99–114.
3. Timmcke AE. Functional anorectal disorders. In: Beck DE, Wexner SD, ed. Fundamentals of Anorectal Surgery. 2nd ed. Philadelphia: WB Saunders; 1998:91–98.
4. Shorvon PJ, McHugh S, Diamant NE, Somers S, Stevenson GW. Defecography in normal volunteers: results and implications. Gut 1989;30:1737–1749.
5. Nielsen MB, Buron B, Christiansen J, Hegedus V. Defecographic findings in patients with anal continence and constipation and their relation to rectal emptying. Dis Colon Rectum 1993;36:806–809.

6. Mellgren A, Schultz I, Johansson C, Dolk A. Internal rectal intussusception seldom develops into rectal prolapse. Dis Colon Rectum 1997;40:817–820.
7. Vaizey CJ, van den Bogaerde JB, Emmanuel AV, Talbot IC, Nicholls RJ, Kamm MA. Solitary rectal ulcer syndrome. Br J Surg 1998;85(12): 1617–1623.
8. Vaizey CJ, Roy AJ, Kamm MA. Prospective evaluation of the treatment of solitary rectal ulcer syndrome with biofeedback. Gut 1997; 41:817–820.
9. Malouf A, Vaizey C, Kamm M. Results of behavioral treatment (biofeedback) for solitary rectal ulcer syndrome. Dis Colon Rectum 2001;44(1):72–76.
10. Marchal F, Bresler L, Brunaud L, et al. Solitary rectal ulcer syndrome: a series of 13 patients operated with a mean follow-up of 4.5 years. Int J Colorectal Dis 2001;16(4):228–233.
11. Sitzler PJ, Kamm MA, Nicholls RJ, McKee RF. Long-term clinical outcome of surgery for solitary rectal ulcer syndrome. Br J Surg 1998;85(9):1246–1250.
12. Fenner DE. Diagnosis and assessment of sigmoidoceles. Am J Obstet Gynecol 1996;175(6):1438–1442.
13. Jorge JM, Yang YK, Wexner SD. Incidence and clinical significance of sigmoidoceles as determined by a new classification system. Dis Colon Rectum 1994;37:1112–1117.

8-8

Case Presentation: Prolapse

G. Willy Davila and Daniel Biller

History and Physical Examination

The patient is a 38-year-old G4P2 woman, with a 3-year history of increasing bladder dysfunction and a 6-month history of exteriorized vaginal and rectal prolapse. She stated that the exteriorized portion is quite large, and she brought two Polaroid photos of what appeared to be at least a fourth degree cystocele, as well as a rectal prolapse of a large amount of rectal mucosa. She also complains of urge urinary incontinence for 2 years. She voids 7 to 10 times per day wearing 1 to 2 pads per day, reporting 1 to 2 episodes of incontinence per day. She reported four or more episodes of nocturia per night with no enuresis or recurrent urinary tract infections, and voids with a normal flow. Her obstructive defecation symptoms have been present for 3 years with no fecal incontinence. She is sexually active. She had been fit with a vaginal ring pessary by her primary physician, which she wears without significant difficulties.

Medical History

1. Vaginal delivery ×2, largest was 6 pounds 6 ounces
2. Multiple sclerosis
3. Hypertension
4. Irritable bowel syndrome

Medications

1. Escitalopram (Lexapro)
2. Acetaminophen/hydrocodone (Lorcet)
3. Interferon beta 1a (Avonex)

Physical Examination

Pelvic Examination

Normal external genitalia. Urethra, bladder, and vagina were normal. Empty bladder stress test was negative. Perineal sensation and reflexes were normal. Prolapse examination was remarkable for no evident prolapse externally with associated marked levator hypertonicity on digital vaginal examination. Vaginal caliber was decreased at the introitus with Valsalva efforts in the supine, sitting, and standing position. Third degree cystocele, and third degree rectocele elicited on examination. In addition, third degree uterine prolapse was noted. Bimanual examination was normal, with no masses. Excellent Kegel contraction was elicited.

Rectal Examination

Large full-thickness rectal prolapse, approximately 5 to 6 cm.

Work-up

- Urinalysis: negative
- Urine culture: negative

Defecography

Rectal emptying partial. Opening of the anal canal adequate. Relaxation of the puborectalis adequate. Anorectal angle normal. Straightening of the anorectal angle adequate. Perineal descent increased and fixed. Large anterior rectocele which was nonemptying. There was rectoanal intussusception.

Colonic Transit Study

Normal

Urodynamics

Postvoid residual of 40 mL. Multichannel cystometrogram to capacity of 403 mL revealed uninhibited detrusor contractions resulting in a large amount of urine loss beginning at about 350 mL. No stress incontinence was noted. Urethral pressures were high-normal at 195 cm H_2O. Leak-point pressure testing revealed no leakage up to a maximal Valsalva effort of 166 cm H_2O at capacity. Cough profile was negative and on uroflowmetry she had a peak flow of 14.7 mL per second. Q-tip angle was positive between 35 to 50 degrees. Results indicated detrusor instability without any stress incontinence.

Cystoscopy

Normal bladder with no significant trabeculations.

Assessment

Complex pelvic floor dysfunction including:

1. Remarkable levator hypertonicity
2. Reported exteriorized vaginal prolapse
3. Third degree cystocele and rectocele
4. Third degree uterine prolapse
5. Full-thickness rectal prolapse
6. Detrusor overactivity
7. Constipation

Procedures Performed

1. Sigmoid resection and rectopexy
2. Uterocolposacropexy
3. Abdominal enterocele repair
4. Anterior and posterior repair
5. Cystoscopy and suprapubic catheter placement

Commentary

Coexistence of advanced genital and rectal prolapse can be quite challenging to evaluate and treat. This is especially true in the younger, reproductive-age woman. As compared with the elderly woman presenting with very evident exteriorized vaginal and rectal prolapse who may be readily treated with a perineal proctosigmoidectomy and vaginal obliterative colpocleisis, the reproductive-age woman frequently presents with a complaint of rectal prolapse that is intermittent. This type of patient may now typically show up to the clinic with "Polaroid in hand." The advent of digital photography has facilitated demonstration of the maximum extent of rectal prolapse.

Although this patient did not demonstrate paradoxical contraction of the puborectalis muscle during defecography, she did demonstrate significant levator hypertonicity on vaginal and rectal examination. This likely contributed to her rectal prolapse because of the need to perform intense Valsalva efforts for bowel evacuation. Providing further evidence of increased pelvic floor tone is the rather elevated urethral closure pressure at 195 cm H_2O. This may have provided her protection from developing stress urinary incontinence.

This patient carried a diagnosis of multiple sclerosis. The detrusor overactivity demonstrated on multichannel cystometrogram is likely related to her underlying neurologic problem, because it is otherwise uncommon to find idiopathic detrusor overactivity in a young woman. She only takes anticholinergic medications on a p.r.n. basis.

Planning a combined reconstructive pelvic surgery in a patient such as this requires excellent communication among the involved surgeons (Figure 8-8.1). For example, the abdominal resection rectopexy was performed first, with care being taken to not contaminate the operative field. Appropriate sharing of the sacral promontory to achieve both rectal and vaginal elevation requires a methodical surgical approach. We have found that performing the rectopexy to the mid sacral region before using the upper sacral segments for the vaginal elevation allows for a smooth surgical flow. If there is no operative field contamination, we will use bone anchors for suture attachment to the sacral promontory, as we normally do during a non-

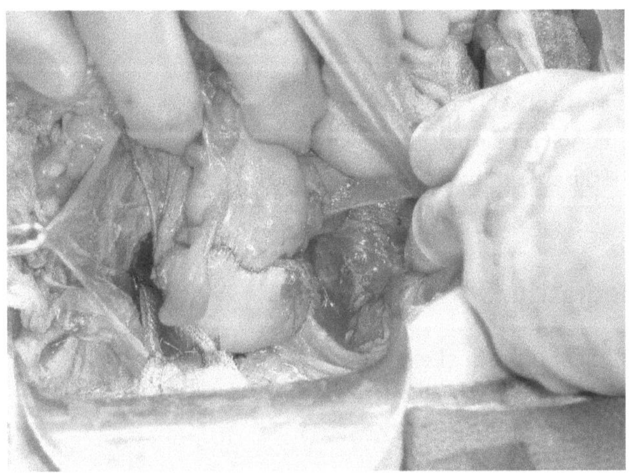

Figure 8-8.1. Image of combined sacrocolpopexy and resection rectopexy.

combined procedure. If there is contamination, copious irrigation should be performed and monofilament sutures used on the sacral promontory. We use polypropylene mesh for the colpopexy. This mesh tends to not get infected or cause tissue reaction. We have not had graft-related problems when performing combined procedures such as this one. We opted to leave the uterus in place because of patient preference, and also because the uterus provides a strong structure for graft attachment, and the likelihood of graft erosion at the vaginal apex is reduced. The rectopexy is performed without a graft, using bilaterally placed monofilament sutures. We have not had any cases of life-threatening periosteal bleeding.

This patient has done very well over the last 3 years of follow-up, with no evidence of prolapse recurrence. She has performed pelvic floor exercises regularly, as we recommend to any patient who undergoes pelvic reconstructive surgery.

Section IX

Pain and Irritative Syndromes Therapy

Section IX

Pain and Irritative Syndromes Therapy

Gamal M. Ghoniem

More than 9 million women have pelvic pain, and their management entails more than $2.8 billion in direct and greater than $555 million in indirect costs.[1] The prevalence of chronic pelvic pain (CPP) in women is approximately 3.8%. Chronic pelvic pain is a symptom, not a disease, and it rarely reflects a single pathologic process. Various pelvic floor structures and disorders may be the cause of CPP. The origin of pain may be urologic, genital, gastrointestinal, musculoskeletal, or psychological. Therefore, systemic evaluation, and different and combined therapies are required for patients with CPP. Therapies include invasive and noninvasive modalities. Noninvasive therapies include behavioral therapy, nutrition, physical therapy, and acupuncture. Dietary modification might help, because certain types of food, such as acidic foods, caffeine, and alcohol are common triggers for interstitial cystitis and irritable bowel syndrome. Pain in the patients with these diseases might respond to an appropriate change in diet. Pharmacologic therapies include nonsteroidal antiinflammatory analgesics and opioids, or both. Patients using pain medication should be assessed and evaluated for side effects, response, and development of tolerance or addiction. Polymedication is not uncommon in pelvic pain therapy. Pharmacotherapy of pelvic pain may include tricyclic antidepressants, anticonvulsants, and antihistamines, particularly when it is associated with a neuropathic condition or interstitial cystitis. Invasive therapies include nerve blocks, neurostimulation, and surgical procedures.

Reference

1. Mathias SD, Kuppermann M, Liberman RF, Lipschutz RC, Steege JF. Chronic pelvic pain: prevalence, health-related quality of life, and economic correlates. Obstet Gynecol 1996;87:321–327.

9-1

Interstitial Cystitis-Painful Bladder Syndrome

Gamal M. Ghoniem and Usama M. Khater

Interstitial cystitis (IC) is a chronic, debilitating disease of the urinary bladder characterized by urinary frequency, nocturia, urgency, and frequently pain. It affects more females than males by a ratio of approximately 10:1.[1] Recently, the International Continence Society has developed a somewhat broader term for IC described as "IC-painful bladder syndrome." This new term is defined as the complaint of suprapubic pain related to bladder filling, and is accompanied by other symptoms, such as increased daytime and nighttime frequency in the absence of proven urinary infection or other obvious pathology.[2] The true prevalence of IC is not determined and it may be underestimated. In 1997, Jones and Nyberg[3] estimated that up to one million patients had IC, many of them unable to cope with day-to-day activities. In Finland, Oravisto[4] estimated the incidence as 18.6 out of 100000 in 1975. Another Finnish study in 2002 used a wider definition and found 450 out of 100000 had IC.[5] There is little international agreement on the epidemiology of IC; it varies as the diagnostic tools vary.

Etiology and Pathogenesis

Interstitial cystitis is an indolent bladder disorder that has continued to be a challenging concern in urology. Despite aggressive investigation in the past two decades, the cause and pathophysiology of the disease remain elusive. Several theories of its pathogenesis have been proposed, but none fully account for the manifestation of the disease. Although the specific etiology of IC is unknown, many mechanisms may be involved.

Occult Infection

Attempts to show an infectious etiology go back to the dawn of the disease, but the case has never been a strong one. Bacterial, viral, and fungal studies were performed on IC patients, and they failed to substantiate an infectious etiology. Infection with "atypical" or fastidious organisms has been proposed by numerous investigators. Some studies showed isolation of fastidious bacteria and Ureaplasma urealyticum.[6–8] Although absence of bacterial DNA was reported, presence of bacterial 16Sr RNA was found in the bladder biopsies of some IC patients.[9,10] Domingue and Ghoniem[11] suggested that even if the organisms are not causative agents, their presence might lead to immune and host-cell responses that could initiate or exacerbate an inflammatory state.

Defective Mucosal Layer (Epithelial Dysfunction)

The healthy bladder surface is coated by a thin mucinous substance, termed bladder surface mucin (BSM), which is composed of numerous sulfonated glycosaminoglycans (GAGs) and glycoprotein. This mucus lining prevents urine and its contents from leaking through the urothelium and damaging the underlying nerves and muscles. In IC patients, this layer is defective and the epithelium is abnormally permeable. As a result, potentially toxic substances in urine are permitted to permeate the bladder muscle, depolarizing sensory nerves and causing the symptoms of IC. One of the urine constituents is potassium (K^+), which is highly toxic to the bladder muscularis. Therefore, investigators have developed the potassium sensitivity test to support this theory.[12,13] Based on this, GAG agents such as heparin and pentosan polysulfate have been used to treat IC.[14,15]

Mast Cell Involvement

Simmons[16] (1961) was the first to suggest mast cells as a cause of IC. Mast cells contain cytoplasmic granules, which in turn contain substances such as histamine, leukotrienes, prostaglandins, and tryptases. All these substances are capable of stimulating inflammation. Degranulation or activation of mast cells can occur as a part of an immunoglobulin E-mediated hypersensitivity reaction or in response to multiple other stimuli including neurotransmitters substance P, cytokines, anaphylatoxins (complement: C3a, 4a, 5a), bacterial toxins, and stress.[17,18] Mastocytosis has been reported in the bladders of 30% to 65% of patients with IC.[19,20] Increased substance P-containing fibers were found adjacent to mast cells in the bladder biopsies of IC patients.[21] Elevated levels of histamine and its metabolites, in the urine of IC patients, were reported by some investigators. Others found overlap or no difference in urinary histamine excretions in IC patients and controls.[22,23] Whether mast cells have a primary or secondary role in the etiology of IC is not exactly known.

Neurogenic Mechanism

Neurogenic inflammation is a process by which nerves may secrete inflammatory mediators, resulting in local inflammation and/or hyperalgesia. This pathogenesis is described in IC as well as in other painful syndromes. One central component of this mechanism is substance P, a short chain peptide that functions as a nociceptive neurotransmitter in the central and peripheral nervous system and as an inflammatory mediator. When released by peripheral nerves (C fibers or fibers associated with pain transmission), an inflammatory cascade occurs that results in processes such as mast cell degranulation and activation of nearby nerves. Supporting the role of neurogenic inflammation in IC is the finding of increased numbers of substance P-containing nerve fibers in the bladders of IC patients.[24] Likewise, an increased concentration of substance P has been found in the urine of patients with IC, and the concentration of substance P is affected by the patient's degree of pain.[25]

Autoimmunity and Inflammation

The exact role of autoimmunity in IC remains controversial, with no clear indication of a primary role for autoimmunity as the cause of IC. Urothelial activation in IC may result in aberrant immune responses and immune activation within the bladder wall that could relate to the pathogenesis of the disease.

Numerous inflammatory mediators have been studied with regard to their relation to IC. Bladder inflammation is categorized by elevated urinary interleukin-6 and activation of the kallikrein-kinin system.[26] Abdel-Mageed and Ghoniem[27] were the first to find activated nuclear factor-kappa B in the bladder biopsies of IC patients. This nuclear factor was also found in other inflammatory diseases including rheumatoid arthritis, inflammatory bowel disease, and bronchial asthma. Activation of this nuclear factor was found to be responsible for production of proinflammatory cytokines.[28]

Toxic Substances in Urine

The idea that the urine of IC patients carries a pathologic substance accounting for the disorder has been suggested. The initial observation that the urine of IC patients may contain pathologic substances was suggested when it was found that it inhibits the proliferation of cultured human transitional cells. Keay et al.[29] determined that the urine of IC patients specifically contains a low-molecular-weight protein factor that inhibits bladder epithelium proliferation, an antiproliferative factor (APF).

Clinical Picture

Patients can present with many symptoms. These symptoms include urgency, frequency, pelvic pain, pelvic pressure, bladder spasm, dyspareunia, dysuria, awakening at night with pain, and pain that persists for many days after intercourse. The location of pain includes the vaginal area, the lower abdomen, suprapubic area, groin, or low back. Many symptoms are aggravated by menstruation and most of the patients believe that sexual intercourse exacerbates their symptoms.

To confirm bladder origin of pain, the patient is asked whether pain worsens if the bladder is full and if it improves with voiding. Bladder pain of IC is experienced suprapubically, in the perineum, vulva, vagina, or in the back or medial thigh. Most patients will also have nocturia, at least one to two times per night.

Symptoms Scores

In 1997, O'Leary et al.[30] developed a questionnaire specifically to assess IC patients. The questionnaire is composed of two sections, including symptoms and problem indices. The maximum scores are 20 and 16, respectively. A second questionnaire, the University of Wisconsin IC Scale (UW-ICS), includes 7 points, with a 0 (not at all) to 6 (a lot) rating scale. The summated scale ranges from 0 to 42.[31] Both questionnaires are validated and either can be given to the IC patient for quantitative evaluation of their symptoms during the course of treatment. We evaluate our patients with the O'Leary questionnaire.

Voiding Diary

The number of daily voidings and average volume can be determined from a voiding diary, whereby each voiding is recorded and measured by the patient. At our institution, a 3-day voiding diary is used. Patients with IC void an average of 16 times per day. A baseline voiding diary is obtained at the initial patient visit. Subsequent voiding diaries are then used during and after the treatment for comparison and to determine the progress in therapy.

Diagnosis

The diagnosis of IC is suspected by the clinician, because no objective test exists. In August 1987, the National Institute of Diabetes and Digestive and Kidney Diseases (NIDDKD) established criteria to diagnose IC for research purposes.[32] However, the criteria were strict and patients meeting these criteria had advanced disease. Application of these criteria in routine clinical practice will miss 60% of patients with IC.[33] The NIDDKD in association with the IC Association is currently developing a broader definition for IC/painful bladder syndrome.

Physical Examination

The main goal of examination is to exclude identifiable causes that may be responsible for the patient's symptoms. Abdominal examination is usually normal in IC patients except for occasional suprapubic tenderness. On pelvic examination, identifiable diseases can be recognized or excluded. In female patients, lack of estrogen identified by inspection of atrophic mucosa, may contribute to vaginal pain, dyspareunia, and dysuria. Pelvic floor hypertonicity is suspected if the levator ani muscles are tight or tender to palpation. The urethra should be palpated to check for a mass, tenderness, or expression of pus, because this finding may indicate presence of urethral diverticulum.

Genital prolapse by itself is not usually relevant to the symptoms of IC but may coexist. A rectal examination should assess for other sources of perineal pain such as anal fissure, and the presence of masses. Rectovaginal and bimanual examinations may reveal masses or implants suggestive of endometriosis. In the classic IC patient, palpation of the anterior wall reveals a tender bladder base; pelvic floor muscle spasm and tenderness are also usually found. Occasionally trigger points may be found along the levator ani muscles.

Urinalysis and cultures are required. Urine cytology should also be obtained to rule out the possibility of carcinoma. If hematuria is found, a full urinary tract work-up should be performed to exclude malignancy.

Cystoscopy

Cystoscopy is performed to exclude other bladder pathology including carcinoma in situ and other malignancy. In the office, cystoscopy under local anesthesia is usually normal in IC. However, cystoscopy and hydrodistention under anesthesia have more of a therapeutic than diagnostic role, because the presence of glomerulations is not specific and their absence does not exclude the diagnosis of IC. A full cystoscopic examination of the bladder is performed first. We use water as a cystoscopic irrigant. Water is infused at a pressure of 80 cm of water weight into the bladder until filling stops. Diminished bladder capacity under anesthesia can be seen in patients with advanced forms of IC. The bladder is distended for 2 to 5 minutes before all irrigant is released from the bladder. Terminal bloody efflux of irrigant suggests the diagnosis of IC. The bladder is then reexamined cystoscopically. The appearance of glomerulation (petechiae) and/or Hunner's ulcers, which appear as fissures or cracks in the epithelium, are consistent with IC. The appearance of Hunner's ulcers is not common in IC (11%), although it has been suggested that its presence is a more specific sign than glomerulation.[34] The diagnostic specificity of appearance of post-distention glomerulation is in question, because it is seen in some normal controls after hydrodistention. Hunner's ulcers are significantly associated with reduced anesthetic bladder capacity, pain, urgency, and frequency, whereas glomerulations are not.

Hydrodistention has a therapeutic role, because it can relieve the IC symptoms in 30% to 50% of patients.[35] A dramatic reduction in pain, frequency, and urgency can occur after hydrodistention in these patients. It was found that the mechanism underlying this reduction may be attributed to a placebo effect or it may deplete substance P from C nerve fibers. However, in one of the studies, a normalization of urine growth factor abnormalities and APF was seen after hydrodistention.[36]

Urodynamics

Urodynamic studies are not routinely indicated in evaluation of IC patients. At our institution, urodynamic investigation is performed if the patient's symptoms include incontinence, severe urgency, or obstructive voiding complaints. Cystometry may demonstrate an intense urge to void with as little as 150 mL of water filling and a capacity of up to 350 mL, limited by discomfort. Compliance may be normal or decreased. Detrusor instability may coexist in IC patients.

Potassium Sensitivity Test

The potassium sensitivity (KCl) test was developed by Parsons et al.[12,15] as a method to diagnose IC. This test is based on the assertion that bladder epithelium is leaky in IC patients. Potassium in the urine crosses the urothelium to the submucosa and muscle layer, causing sensory nerve irritation and inflammation. The IC committee in 2003 did not recommend KCl test because neither high sensitivity nor specificity has been established,[2] and at our institution, KCl test is also not routinely used in evaluation of IC patients.

Bladder Biopsy

Bladder biopsy is not diagnostic for IC, but it is indicated to exclude other possible diseases, particularly carcinoma. Biopsy is taken during cystoscopy, after hydrodistention to avoid bladder perforation or rupture.

Urinary Markers

Clearly, a convenient, objective, and noninvasive diagnostic test of adequate sensitivity and specificity for IC is needed. Researchers have continued to seek objective tests that could be used in diagnosis. Many urinary substances have been described as increased or decreased in IC compared with controls. These substances, including histamine, interleukins, GAGs, hyaluronic acid, epithelial growth factors, nerve growth factor, and others, were selected based on theorized etiologies for IC. One of the major problems of using many of these substances as diagnostic markers is that, although the levels may be statistically significantly higher or lower in IC, most of the individual values significantly overlap among control and IC subjects. Extensive reviews of urine markers have been published.[37] Glycoprotein-51 and APF did not show overlap between controls and IC patients who met NIDDK criteria.[38,39] The level of APF is detected with indirect assay based on inhibition of H-thymidine incorporation. The sensitivity and specificity of APF were 94% and 95%, respectively.[40] Antiproliferative factor is more likely to be a diagnostic marker for IC.

Treatment

Dietary Modification

Avoidance of certain foods and beverages seems to improve the symptoms in many IC patients. To identify the type of food, the patient is encouraged to begin with a bland diet and to add each suspected food one by one. If the patient is sensitive to a certain type of food, the symptoms will worsen within 30 minutes to 6 hours. Fluid restriction, although it will decrease the frequency, is not advised, as pain often worsens because of a higher concentration of offending agents in the urine. Urine alkalinization may help.

Oral Therapy

Oral medications are used to enhance the GAG layer (pentosan polysulfate), stabilize mast cells in the bladder wall (hydroxyzine), modulate neural input (amitriptyline, gabapentin), control pain (narcotics), control inflammation (antiinflammatory), improve frequency (anticholinergics), control dysuria (phenazopyridine), and to improve sleep (zolpidem).

Pentosan Polysulfate Sodium (Elmiron)

Pentosan polysulfate is a highly sulfated synthetic polysaccharide similar in structure to heparin sulfate, which is one of the GAG constitutes of the BSM. When taken orally, 2% to 6% of pentosan polysulfate sodium is excreted unchanged in the urine. The response to treatment varies from 28% to 32%.[41,42] The dosage is 100 mg three times a day. The medication is generally well tolerated with adverse reaction rates varying from 1% to 4%.[43] The most common reported adverse effect is dyspepsia. Other uncommon reactions include reversible alopecia and increased bruising.

Amitriptyline

Amitriptyline is a tricyclic antidepressant, which can be used for pain management. Pain modulation seems to be the result of decreased serotonin and norepinephrine reuptake in the central nervous system. It stabilizes mast cells and has a moderate anticholinergic effect. Amitriptyline may improve the IC symptoms in 64% to 90% of patients.[44] Amitriptyline works better for patients with a substantial pain component to their symptoms and anesthetic bladder capacity greater than 600 mL.[45] The starting dose is 10 to 25 mg at suppertime, and may be increased on a weekly basis to 75 to 100 mg, but the lowest dosage often results in satisfactory improvement. The side effects of amitriptyline may include weight gain, fatigue, decreased libido and ability to achieve orgasm, palpitations, and anticholinergic side effects. It may also affect thyroid function. Hepatotoxicity and bone marrow suppression can rarely occur.

Hydroxyzine

Hydroxyzine is an antihistamine, H-1 receptor antagonist. It reduces the symptoms of IC in 40% of patients and in 55% of patients with a history of allergy.[46] The mechanism of action may include inhibition of mast cell degranulation. Hydroxyzine has a relaxation effect to the skeletal muscles, which may be beneficial to IC patients. The dose is 10 to 75 mg each evening. Clinical response may take up to 3 months.

Anticonvulsants

The GABA agonist, gabapentin, and the sodium channel blocker, carbamazepine, both antiseizure medications, are useful in treatment of IC patients. They are reserved as second line agents. Gabapentin is usually preferred, because it has an excellent safety profile. The dose can be increased slowly from 100 mg every day to 3600 mg per day in three divided doses. Its most common side effect is fatigue.[47]

Selective Serotonin Reuptake Inhibitors

Selective serotonin reuptake inhibitors (SSRIs) are used in depression and pain management. Frequently used SSRIs are paroxetine 10 to 20 mg/day and venlafaxine 37.5 to 150 mg/day.

Narcotics

Narcotics can be used in acute exacerbation of pain. They were not accepted for the treatment of chronic nonmalignant pain until the 1980s because of the fear of addiction. They are now used but not as first line of therapy. Patients with IC requiring narcotics on a chronic basis are usually referred to pain clinic.

Constipation produced from opioids is often associated with worsening symptoms of pelvic pressure. The adverse effects may also include sedation, respiratory depression, and nausea.

Intravesical Therapy

Intravesical therapy can be used alone or in combination with oral therapy. Its advantages include high local drug concentration in the bladder, minimal systemic absorption, and a lack of significant systemic side effects.

Dimethylsulfoxide

Dimethylsulfoxide (DMSO) was first synthesized in 1867 as a byproduct of the wood pulp industry. The antiinflammatory, analgesic, and bacteriostatic effects were first described by Jacob et al.[48] Since the 1960s, DMSO has been used clinically for treatment of IC. The beneficial effect of DMSO appears in its ability to release and ultimately deplete substance P from the bladder wall. Initially, DMSO may exacerbate the symptoms because it may stimulate mast cell degranulation and stimulate bladder efferent pathways to cause nitric oxide release.

Dimethylsulfoxide is excreted through the alveoli and causes a garlic-like odor of the breath. Dimethylsulfoxide is administered as a 50% solution or as a DMSO cocktail.[49] The component of the solution consists of 50 mL of 50% DMSO (Rimso-50), 10 mL of sodium bicarbonate, steroid (40 mg of triamcinolone), 10 000 U of heparin sulfate, and 80 mg of gentamicin. The mixture is administered and the patient is asked to hold it for 30 to 60 minutes. If the patient cannot hold the solution, a decreasing volume can be used in each visit or preceded with belladonna and opium suppository. Instillations are usually performed on a weekly basis for 6 weeks. Patients may experience an initial flare up of symptoms caused by increased substance P. These symptoms usually improve after 2 weeks of treatment. Therapy is ultimately discontinued when symptoms have been reduced and stabilized.

Intravesical Heparin

Heparin sulfate is a GAG constituent of the basement membrane (BSM). Similar to pentosan polysulfate, intravesical heparin is thought to enhance the protective feature of BSM. Heparin also has antiinflammatory properties, inhibiting angiogenesis and proliferation of fibroblasts and smooth muscle fibers. Self-instilled 20 000 IU of heparin can be used and no changes in coagulation studies can occur.[15] The solution is held for 30 to 45 minutes, and then voided. Four to 12 months is often needed to attain a beneficial response.

Anesthetic Cocktail

Because the bladder is suspected to be the source of pain in the IC patient, direct treatment of its surface may produce improvement of symptoms. The solution consists of bupivacaine and lidocaine. The patient is instructed to hold the solution for 30 minutes, and the instillation is repeated on a weekly basis for 8 to 12 weeks. Patients usually improve for a number of days, increasing in number of days as the instillations progress. An anesthetic cocktail can be helpful when symptoms flare. Worsening of symptoms and urine retention rarely occur as side effects.

Silver Nitrate and Clorpactin WCS-90

Instillation of caustic agents, including silver nitrate and oxychlorosene are used in treatment of IC. Success rates may vary from 50% to 80%.[50] Because these agents are caustics, voiding cystogram should be performed to exclude any vesicoureteral reflux and avoid upper urinary tract damage. Bladder biopsy is contraindicated along with instillation of these agents to avoid perivesical extravasations.

Bacillus Calmette-Guerin

Bacillus Calmette-Guerin (BCG) is an attenuated strain of Mycobacterium bovis. Bacillus Calmette-Guerin can provide a 60% improvement rate compared with 27% with placebo.[51] The mechanism of action is unclear, but it is suggested that BCG may stimulate type 1 T helper cell response, leading to a decrease in inflammatory cells and decreased type 2 T helper cell-mediated allergic response. Nitric oxide levels are found decreased in IC patients and increases are found with symptom improvement. Bacillus Calmette-Guerin seems to increase urinary nitric oxide level in bladder cancer patients. In oncology literature, BCG adverse effects may include irritative voiding symptoms, gross hematuria, and development of bladder inflammatory granuloma.

Sodium Hyaluronate

Sodium hydronate is a highly purified salt of hyaluronic acid. Hyaluronic acid is a GAG component of BSM. Intravesical administration of hyaluronic acid may improve the bladder lining and further protect the bladder wall from the irritating effect of urine. It is available commercially in Europe and Canada and, although under investigation in the United States, it can be used on compassionate basis.

Capsaicin and Resiniferatoxin

Capsaicin is the pungent component of the hot pepper. Chronic application is known to be associated with C fiber (unmyelinated nerve fibers known for transmitting painful stimuli) desensitization. Reports on the success vary because of the use of different concentrations. The main drawback of capsaicin is a severe burning sensation, which is described by the patients despite the preinstillation of anesthetic. Resiniferatoxin, which is an analog of capsaicin has no irritative or burning symptom, but is 100 to 10 000 times more potent than capsaicin.

Surgical Intervention

Hydrodistention

As previously mentioned, hydrodistention has therapeutic and diagnostic roles in IC. Hydrodistention may offer relief of symptoms in 30% to 50% of patients. The symptoms may worsen 2 to 3 weeks after the procedure and improve after that period. Patients with small bladder capacity under anesthesia tend to have a better response to hydrodistention than those with large capacity. The mean duration of improvement is approximately 3 months. While repeating hydrodistention, the improvement may progressively decrease.[52]

Neuromodulation

Neuromodulation is a unilateral sacral nerve (S3) stimulation (InterStim device; Medtronic, Inc., Minneapolis, MN). Neuromodulation has been approved by the Food and Drug Administration for treatment of urge incontinence, urinary retention, idiopathic urgency-frequency syndrome, and recently for IC.[53] Application of the InterStim device to IC patients has recently been shown to improve symptoms including urinary frequency, urgency, and pain.[54] In addition, abnormally high urine APF and abnormally low heparin-binding epidermal growth factor concentrations have been normalized after neuromodulation.[55] The procedure is performed in two stages, the first stage involving the percutaneous placement of a temporary electrode within the S3 foramen. Appropriate stimulation is then achieved through an external stimulator. If satisfactory symptom improvement occurs, the patient will be a candidate for permanent implantation of a small stimulator (InterStim) unit that rests in the upper buttocks.

Laser Resection of Ulcers

These procedures are reserved for those patients with gross inflammatory lesions of the bladder wall (Hunner's patches). In this procedure, neodymium (Nd):YAG laser has been used for ablation of Hunner's patch ablation.[56] The procedure has the advantage of deep bladder penetration and less scatter to adjacent normal tissue. A disadvantage of this procedure is that bladder perforation can occur. Excellent symptom improvement is usually achieved; however, recurrence is common within 1 year.

Cystectomy and Urinary Diversion

Cystectomy with urinary diversion is the last resort procedure being reserved for patients who have failed all other modalities of treatment. Pain may be persistent in some patients after this aggressive therapeutic approach, and before we consider surgery, patients are evaluated by the pain clinic to localize the pain. If there is evidence of central up-regulation of pain, surgery will not be performed because it will not relieve the pain.

Pain Severity and Localization

Before recommending surgical therapy, each patient should undergo tests for localization of pain, in cooperation with the pain clinic. In the urology clinic, intravesical instillation of lidocaine 2%, or subtrigonal Marcaine 0.5% block, may relieve bladder and trigonal pain, indicating its end organ origin. At our institution, nerve blocks are usually performed by an anesthesiologist at the pain clinic.

References

1. Hanno PM, Levin RM, Monson FC, et al. Diagnosis of interstitial cystitis [see comments]. J Urol 1990;143:278–281.
2. Allen P. Committee moves closer to revised definition of IC. Urol Times 2004;32:1–11.
3. Jones CA, Nyberg L. Epidemiology of interstitial cystitis. Urology 1997;49:2–9.
4. Oravisto KJ. Epidemiology of interstitial cystitis. Ann Chir Gynaecol Fenn 1975;64:75–77.
5. Leppilahti M, Tammela TL, Huhtala H, Auvinen A. Prevalence of symptoms related to interstitial cystitis in women: a population based study in Finland. J Urol 2002;168(1):139–143.
6. Hedelin HH, Mardh PA, Brorson JE, Fall M, Moller BR, Pettersson KG. Mycoplasma hominis and interstitial cystitis. Sex Transm Dis 1980;10S:327–330.
7. Potts JM, Ward AM, Rackley RR. Association of chronic urinary symptoms in women and Ureaplasma urealyticum. Urology 2000;55: 486–489.

8. Wilkins EGL, Payne SR, Pead PJ, Moss ST, Maskell RM. Interstitial cystitis and urethral syndrome: a possible answer. Br J Urol 1989;64: 39–44.
9. Domingue G, Ghoniem G, Bost K, Fermin C, Human LG. Dormant microbes in interstitial cystitis. J Urol 1995;153:1321–1326.
10. Haarala M, Jalava J, Laato M, Kiilholma P, Nurmi M, Alanen A. Absence of bacterial DNA in the bladder of patients with interstitial cystitis. J Urol 1996;156:1843–1845.
11. Domingue GJ, Ghoniem G. Occult infection in interstitial cystitis. In: Sant GR, ed. Interstitial Cystitis. Philadelphia: Lippincott-Raven; 1997:77–86.
12. Parsons CL, Greenberger M, Gabal L, Bidair M, Barme G. The role of urinary potassium in the pathogenesis and diagnosis of interstitial cystitis. J Urol 1998;159:1862–1867.
13. Parsons CL, Lilly JD, Stein PC. Epithelial dysfunction in non bacterial cystitis (interstitial cystitis). J Urol 1991;145:732–735.
14. Parsons CL, Mulholland SG. Successful therapy of interstitial cystitis with pentosan polysulfate. J Urol 1987;138:513–516.
15. Parsons CL, Housley T, Schmidt JD, Lebow D. Treatment of interstitial cystitis with intravesical heparin. Br J Urol 1994;73:504–507.
16. Simmons JL. Interstitial cystitis: an explanation for the beneficial effect of an antihistamine. J Urol 1961;85:149–155.
17. Lagunoff D, Martin TW, Read G. Agents that release histamine from mast cells. Ann Rev Pharmacol Toxicol 1983;23:331–351.
18. Theoharides TC, Sant GR. Bladder mast cell activation in interstitial cystitis. Semin Urol 1991;9:74–87.
19. Feltis JT, Perez-Marrero R, Emerson LE. Increased mast cells of bladder in suspected cases of interstitial cystitis: possible disease marker. J Urol 1987;138:746–752.
20. Lynes WL, Flynn SD, Shortliffe LD, et al. Mast cell involvement in interstitial cystitis. J Urol 1987;138:746–752.
21. Pang X, Marchand J, Sant G, Kream RM, Theoharides TC. Increased number of substance P positive nerve fibers in interstitial cystitis. Br J Urol 1995;75:744–750.
22. el-Mansoury M, Boucher W, Sant GR, Theoharides TC. Increased urine histamine and methylhistamine in interstitial cystitis. J Urol 1994;152:350–353.
23. Erickson D, Xie S, Bhavanandan VP, et al. A comparison of multiple urine markers for interstitial cystitis. J Urol 2002;167(6):2461–2469.
24. Hohenfellner M, Nunes L, Schmidt RA, Lampel A, Thuroff JW, Tanagho EA. Interstitial cystitis: increased sympathetic innervation and related neuropeptide synthesis. J Urol 1992;147:587–591.
25. Chen Y, Varghese R, Chiu P, et al. Urinary substance P is elevated in women with interstitial cystitis. J Urol 1999;161:26.
26. Lotz M, Villiger P, Hugli T, Koziol J, Zuraw BL. Interleukin-6 and interstitial cystitis. J Urol 1994;152:869–873.
27. Abdel-Mageed A, Ghoniem G. Potential role of Rel/nuclear factor-kappa B in the pathogenesis of interstitial cystitis. J Urol 1998;160: 2000–2003.
28. Bajwa A, Shenassa BB, Human L, Ghoniem GM, Abdel-Mageed AB. NF-kappa B-dependent gene expression of proinflammatory cytokines in T24 cells: possible role in interstitial cystitis. Urol Res 2003;31:300–305.
29. Keay S, Zhang CO, Trifillis AL, et al. Decreased 3H-thymidine incorporation by human bladder epithelial cells following exposure to urine from interstitial patients. J Urol 1996;156:2073–2078.
30. O'Leary MP, Sant GR, Fowler F Jr, Whitmore KE, Spolarich-Kroll J. The interstitial cystitis symptom index and problem index. Urology 1997;49:58–63.
31. Goin JF, Olaleye D, Peter KM, et al. Psychometric analysis of the University of Wisconsin Interstitial Cystitis Scale: implications for use in randomized clinical trials. J Urol 1998;159:1085–1090.
32. Gillenwater JY, Wein AJ. Summary of the National Institute of Arthritis, Diabetes, Digestive and Kidney Diseases Workshop on Interstitial Cystitis, Bethesda, Maryland, August 28–29, 1987. J Urol 1988;140:203–206.
33. Hanno PM, Landis JR, Matthews-Cook Y, Kusek J, Nyberg L Jr. The diagnosis of interstitial cystitis revisited: lessons learned from the National Institutes of Health interstitial cystitis data base study. J Urol 1999;161:553–557.
34. Messing E, Pauk D, Schaeffer A, et al. Association among cystoscopic findings and symptoms and physical examination findings in women enrolled in the Interstitial Cystitis Data Base Study. Urology 1997;49(suppl 5A):81–85.
35. Hanno PM, Buehler J, Wein AJ. Conservative treatment of interstitial cystitis. Semin Urol 1991;9:143–147.
36. Chi TX, Zhang CO, Shoenfelt JL, Johnson HW Jr, Warren JW, Keay S. Bladder stretch alters urinary heparin-binding epidermal growth factor and antiproliferative factor in patients with interstitial cystitis. J Urol 2000;163:1440–1444.
37. Erickson DR. Urine markers in interstitial cystitis. Urology 2001;57: 15–21.
38. Byrne DS, Sedor JF, Estojak J, Fitzpatrick KJ, Chiura AN, Mulholland SG. The urinary glycoprotein GP-51 as a clinical marker for interstitial cystitis. J Urol 1999;161:1786–1790.
39. Erickson DR, Ordille SD, Zhang CO, et al. Anti-proliferative factor (APF), heparin binding epidermal growth factor-like growth factor (HB-EGF) and epidermal growth factor (EGF) alterations in interstitial cystitis (IC) conformed [abstract]. J Urol 2000;163(suppl):62.
40. Keay SK, Zhang C, Shoenfelt J, et al. Sensitivity and specificity of antiproliferative factor, heparin-binding epidermal growth factor-like growth factor and epidermal growth factor as urine markers for interstitial cystitis. Urology 2001;57:9–14.
41. Mulholland SG, Hanno PM, Parsons CL. Pentosan polysulfate sodium therapy for interstitial cystitis. Urology 1990;35:552–558.
42. Parsons CL, Benson G, Childs SJ. A quantitatively controlled method to study prospectively interstitial cystitis and demonstrate the efficacy of pentosan polysulfate. J Urol 1993;150:845–848.
43. Nickel JC, Barkin J, Forrest J. Safety and efficacy of up to 900 mg/day polysulfate sodium (Elmiron) in patients with interstitial cystitis. Urology 2001;57(S1):122–123.
44. Hanno PM. Amitriptyline in the treatment of interstitial cystitis. Urol Clin North Am 1994;21:89–91.
45. Erickson DR. Interstitial cystitis: update on etiologies and therapeutic options. J Womens Health Gend Based Med 1999;8:745–758.
46. Theoharides TC, Sant GC. Hydroxyzine therapy for interstitial cystitis. Urology 1997;49:108–110.
47. Wasner G, Backonja MM, Baron R. Neuropathic pain syndromes. Neurol Clin North Am 1998;16:851–868.
48. Jacob SW, Bischel M, Herscheler RJ. Dimethyl sulfoxide (DMSO): a new concept in pharmacotherapy. Curr Ther Res 1964;6:134–135.
49. Ghoniem G, McBride D, Sood OP, Lewis V. Clinical experience with multiagent intravesical therapy in interstitial cystitis patients unresponsive to single-agent therapy. World J Urol 1993;11:178–182.
50. Moldwin RM, Sant GR. Interstitial cystitis: a pathophysiology and treatment update. Clin Obstet Gynecol 2002;45(1):259–272.
51. Peters KM, Diokno AC, Steinert BW, Gonzalez JA. The efficacy of intravesical bacillus Calmette-Guerin in the treatment of interstitial cystitis: long-term followup. J Urol 1998;159:1483–1487.
52. Rofeim O, Shupp-Byrne D, Mulholand SG, et al. Increased production of bladder surface mucin due to bladder trauma. J Urol 2000; 163:40.
53. Shaker HS, Hassouna M. Sacral nerve root neuromodulation: an effective treatment for refractory urge incontinence. J Urol 1998; 159:1516–1519.
54. Paszkiewicz EJ, Siegel SW, Kirkpatrick C, Hinkel B, Keeisha J, Kirkemo A. Sacral nerve stimulation in patients with chronic intractable pelvic pain. Urology 2001;57(S6A):124.
55. Chai TC, Zhang C, Warren JW, Keay S. Percutaneous sacral third nerve root neurostimulation improves symptoms and normalized urinary HB-EGF levels and antiproliferative activity in patients with interstitial cystitis. Urology 2002;55:643–646.
56. Rofeim O, Hom D, Freid R, Moldwin RM. The use of the neodymium: YAG laser in management of interstitial cystitis. J Urol 2001;166: 134–136.

9-2

Painful Conditions of the Urogenital Sinus

Nathan Guerette and G. Willy Davila

Diagnosis and treatment of lower urogenital pain conditions (vulvar vestibulitis, urethral pain syndrome, and trigonitis) can be challenging and frustrating for both the patient and clinician. However, a methodical approach to evaluation and management can lead to remarkable improvements.

Embryology

In early female gestation, the cloaca separates into anterior and posterior divisions. The anterior division develops into the urogenital sinus. This endodermal structure then morphs into the bladder, urethra, and vaginal bulb. The bulb forms a portion of the superior vagina and inferiorly fuses with the genital ectoderm. This fusion point forms the vulvar vestibule.

The common embryologic origin of the vestibule, urethra, and bladder has suggested an etiologic association among chronic painful conditions involving these tissues. This may help explain the similarity in physical and histologic findings, prevalence rates, and treatment strategies for vulvar vestibulitis, trigonitis, urethritis, and interstitial cystitis.[1,2] Interstitial cystitis is considered in a separate chapter.

Vulvar Vestibulitis

Anatomy and Definition

The vulvar vestibule is the most inferior portion of the urogenital sinus. It is a ringed structure delineated superiorly by the hymen, inferiorly by the more pigmented skin at the base of the labium minus (Hart's line), anteriorly by the clitoris, and posteriorly by the fourchette and fossa navicularis. It is composed of thinly keratinized stratified squamous epithelium. The urethra, Bartholin's glands, Skene's glands, and numerous minor mucous glands drain into the vestibule.

Friedrich[3] characterized vulvar vestibulitis as a persistent syndrome lasting more than 6 months consisting of a triad of findings: 1) severe pain with vestibular touch or attempted vaginal penetration, 2) tenderness in response to pressure within the vulvar vestibule, and 3) physical findings of erythema confined to the vestibule of varying degrees.

Clinical Presentation

Typically, patients will present with a history of vaginal pain or burning and dyspareunia. Upon further questioning, the pain is localized to the introitus and usually described as a burning or cutting that initiates with entry. Often tampon insertion or other activities increasing pressure in the vestibule will elicit similar symptoms. Many patients will present with "recurrent cystitis," caused by episodic burning with urination. Upon detailed questioning, it is determined that the burning is external, typically after voiding and frequently with negative urine cultures, and it resolves whether or not antibiotics are taken.

Other vulvar dysesthesias can be difficult to distinguish from vestibulitis. The key difference is that vulvar vestibulitis has provoked symptoms, with pain initiated by direct physical contact. Women with vestibulitis will often have great apprehension about sexual intercourse and largely cease activity secondary to the degree of pain.

Distinctions have been made between women with primary vestibulitis, dyspareunia at the initiation of their sexual experience, and secondary vestibulitis (dyspareunia after a time of asymptomatic intercourse) because patients with primary vestibulitis do not tend to respond as well to treatment.[4]

Epidemiology and Demographics

No large studies have been performed to accurately estimate the prevalence of vestibulitis in the population. In

1995, Baggish and Miklos[5] performed a meta-analysis of 15 studies with 450 patients in an attempt to more accurately characterize women with vulvar vestibulitis. The typical vulvar vestibulitis patient is white, nulliparous, well-educated, and young. The women have frequently seen multiple clinicians, and have sought treatment for 2 to 3 years before the diagnosis is made. Many are diagnosed with recurrent urinary tract infections, candidal infections, human papillomavirus (HPV), and allergies. An increased rate of prior sexual abuse in these patients has been reported, but is controversial.[6,7]

Psychological testing on women with vulvar vestibulitis has found higher rates of introversion, somatization, and interference with sexual function. Overall levels of psychological distress, negative feelings toward sex, self-esteem, and marital satisfaction are conflicted in the literature with some studies showing no difference from matched controls and others finding opposite results. Furthermore, the order of causation remains unclear, with many authors suggesting a circular relationship of the physical condition and psychological changes.[8–10]

Histopathology

Historic terms for vulvar vestibulitis (vulvar adenitis, erythematous vulvitis, burning vulva syndrome) all implicate inflammatory changes, but it was not until the late 1970s and 1980s that the histologic changes associated with vestibulitis were better characterized.

A 1988 study analyzed 41 surgical specimens from patients diagnosed with vulvar vestibulitis. Chronic interstitial inflammation was seen in all cases and the glands were not involved. The etiology of this inflammation was undetermined.[7] More recently, controlled studies have produced more mixed results with similar patterns of inflammation but often no differences from control specimens.[6,11,12]

When investigating hypotheses that vascular injury, immunologic alterations, or changes in nociception may be contributing factors, additional histologic findings have been noted. Inflammatory cytokines interleukin-1 and tumor necrosis factor were elevated in vestibulitis specimens compared with perineoplasty specimens from asymptomatic women.[13] Increased levels of perivascular immunoglobulin M, complement, and fibrinogen, as well as heightened angiogenesis have been noted in vestibulitis specimens compared with controls.[14] Studies have noted increased neural hyperplasia in vestibulitis subjects and found the degree of vestibular nerve formation significantly correlated with the degree of inflammation.[15,16] Whether these changes are causative or secondary remains unclear.

Immune dysfunction has also been postulated as an underlying factor in vulvar vestibulitis with decreased natural killer cell activity in the vestibular tissues of symptomatic women. It has also been observed that natural killer cell activity increases after application of interferon.[17]

Etiology

Numerous theories have been investigated to identify the etiology of vulvar vestibulitis. Chronic bacterial and viral infections have been the strongest focus of study. Candida, chlamydia, gonorrhea, mycoplasma, contact dermatitis and other allergic responses, hormonal reaction, and psychosomatic causes have also been considered. Despite these extensive investigations, the etiology remains unclear and may be multifactorial.

Human papillomavirus has been the most thoroughly studied virus as an etiology. Numerous studies have tried to correlate the presence of HPV with vulvar vestibulitis symptoms, but the results have been highly variable, reporting 0% to 85% rates of identifiable HPV in vestibulitis patients.[18–22] In addition, no difference has been noted in HPV-positive and -negative women in response to treatment.[23] Human papillomavirus does not seem to be causative or, if present, affect treatment outcomes.

During the initiation of urination, urine separates the labia minora and thus wets the entire vestibule, thus, contact with urinary irritants, particularly oxalates, has been researched as a possible etiology. Oxalates are organic acids excreted in the urine that can cause pruritus and burning. Many studies have shown higher rates of hyperoxaluria in women with vestibulitis. These findings, however, have been inconsistent.[24–26]

Evaluation

A careful evaluation is necessary to design an effective treatment regimen. We obtain a thorough history, including previous treatments and any contributing psychological factors, including sexual abuse, when timing is appropriate. A broad differential should be used to rule out other vulvar or vaginal conditions such as candidal vaginitis, lichen planus, or a chronic abscess of Bartholin's glands. Any potential allergens or irritants should be identified. Particular attention should be given to patterns of exacerbation of vestibulitis symptoms in conjunction with ingestion of foods with high oxalate concentrations. For example, we have found a strong relationship between consumption of various types of berries and worsening symptoms.

When vestibulitis is suspected, a detailed vulvar and vaginal examination must be performed. Any cutaneous or mucosal abnormalities as well as atrophic changes in the vaginal mucosa should be noted. The vestibule must be inspected for erythema and other signs of chronic inflammation. Pain mapping with a cotton-tipped swab should be done, before any vaginal instrumentation, to delineate the area of sensitivity (Figure 9-2.1). This

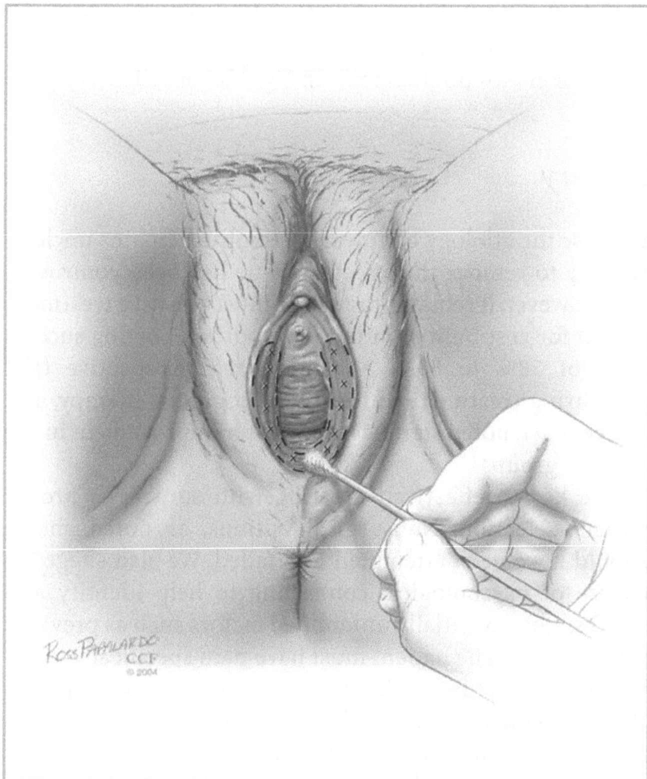

Figure 9-2.1. Vulvar vestibule and pain map. (Reprinted with the permission of The Cleveland Clinic Foundation.)

should be recorded in the medical record for future reference.

Pain along the posterior vestibule and inner aspect of the labia minora is typical. If the painful area is difficult to discern, diluted acetic acid on the swab helps define the area with acetowhite changes, but may be quite uncomfortable for the patient and is rarely necessary.

The urethra and trigone should be palpated transvaginally to identify other potential areas of tenderness. Purulent urethral discharge indicates the need for urethral cultures and possibly antibiotic therapy before further work-up. Isolated trigonal sensitivity should be further evaluated with cystourethroscopy to look for evidence of chronic infection. If localized trigonal pain is identified, we will frequently initiate empiric tetracycline therapy for 3 to 4 weeks.

The vaginal speculum examination is then performed, and is important in order to identify inflammatory or infectious vaginitis. Vaginal lichen planus characteristically presents as well-delineated marked mucosal erythema and desquamation with a mucopurulent discharge. Microscopy should be used to rule out bacterial vaginosis or yeast infections.

The vaginal mucosa of a woman with atrophic vaginitis usually will be pale, thin, and have loss of rugations, but milder degrees of atrophy may have a more subtle presentation. A maturation index or pH may be helpful in quantifying the degree of atrophy and therapeutic response. Upper genital tract pain, elicited on bimanual examination, may be suggestive of pelvic inflammatory disease, endometriosis, or a pelvic mass. A gentle, methodical approach to the vulvovaginal examination in patients with a history suggestive of urogenital sinus pain is imperative.

Treatment

Education

Women diagnosed with vulvar vestibulitis, particularly at specialty centers, have usually seen multiple clinicians, received numerous courses of medications for infections, and have often been told the problem is psychological. Giving patients a firm diagnosis, reassuring them the problem is real and not "in their head," and using the examination to educate them on their genital anatomy can assuage many of their anxieties and increase compliance with recommended therapy.

Elimination of Potential Irritants

Removing all possible irritants is a critical component of successful treatment. Women with vulvar vestibulitis should discontinue use of all hygiene products, creams, and lubricants. Minimizing the irritative effect of urine on the vulvar vestibule is also important. Women are instructed to spread their labia before initiation of voiding to reduce contact with urine. We place all patients on a low oxalate diet (Table 9-2.1), and recommend taking calcium citrate (Citracal) as a binding agent with each meal to reduce urinary excretion of oxalates. The currently available diet is by no means comprehensive, and oxalates may not be the only irritating dietary factor in the urine. Patients are informed to be diet conscious and look for temporal relationships between certain foods and worsening of their symptoms.

Medications

Topical corticosteroids and interferon have been used in an attempt to reduce the chronic inflammatory changes, but significant response rates are disappointing. Topical estrogen cream has been shown to be beneficial. Women are instructed to use the cream at a dose of 1 g intravaginally two times a week and to digitally apply a small amount daily to the vestibule. The cream is soothing and over time thickens the vestibular epithelium, which may distance the nociceptors from the surface of the skin. Alone, local estrogen has produced, generally, disappointing results. There may also be a direct effect of estrogen on vestibular nerve fibers. We routinely use estrogen warn in one multi-modality treatment of vulvar vestibulitis.

Table 9-2.1. Oxalate content of common foods

Low Oxalate	Medium Oxalate	High Oxalate
Cola	Coffee	Cocoa
Apple juice	Fruit juices	Most berry juices
Milk	Grape	Blackberries
Butter	Orange	Blueberries
Yogurt	Tomato	Raspberries
Cheese	Apples	Strawberries
Mayonnaise	Pineapples	Whole wheat bread
Red wine	White bread	Peanuts
Grapes	Pasta	Soy products
Lemons	Asparagus	Celery
Melons	Broccoli	Collards
Raisins	Carrots	Green peppers
Rice	Corn	Popcorn
Coconut	Cucumber	Spinach
Bacon	Garlic	Tomato sauce
Beef	Lettuce	Yams
Chicken	Mushrooms	
Eggs	Onions	
Fish	Peppers	
Squash	Potato and potato chips	
Zucchini	Tomato	
Peas		

Source: Adapted from Rowan's Resources. More dietary selections and additional information available at www.vulvarpainfoundation.org.

Biofeedback

Pelvic floor muscle hypertonicity is not infrequently a contributing factor to the dyspareunia associated with vestibulitis. Biofeedback with therapists experienced in managing the pelvic floor can notably reduce the degree of spasm in these muscles, providing a useful adjunct to treatment.

Alteration of Pain Sensation

Decreasing the sensitivity of the vestibular nociceptors is another valuable step in successful treatment. Locally, a brief course of 2% Xylocaine gel applied regularly (two to three times per day) may provide relief while additional treatment avenues are initiated. Patients should be warned, however, that the Xylocaine may burn with initial application.

Centrally, pain sensation can be altered with tricyclic antidepressants. These agents are widely used for management of chronic pain and are useful in the improvement of vulvar vestibulitis symptoms. Amitriptyline (Elavil), imipramine (Tofranil), nortriptyline (Pamelor, Aventyl), and desipramine (Norpramin) can all have significant side effects (dry mouth, dry eyes, constipation, sedation, palpitations, etc.) and should be slowly increased from a low dose to the usual therapeutic dose. We typically add Elavil (10 mg at bedtime) if a low oxalate, local Xylocaine, and estrogen cream do not result in significant relief.

Surgery

Because the etiology of vulvar vestibulitis remains unclear, surgery to remove the affected tissue has been controversial. However, it remains the single most effective treatment option for vestibulitis, with most studies reporting success rates of 70% to 80%. A randomized prospective trial comparing treatment with surgery, cognitive therapy, and biofeedback noted significantly greater improvement in the surgery group.[23]

The key to effectively treating with surgery is proper patient selection. Nonsurgical options, as noted above, should have been attempted and failed. We also suggest a course of psychological counseling to help identify and address any potentially unfounded factors such as previous sexual abuse. The patient must have pain specifically localized to the vestibule. This requires careful pain mapping. Women with more generalized vaginal, vulvar, or pelvic pain are not candidates for surgery.

Numerous procedures have been described to treat vulvar vestibulitis including laser, highly localized excision, and undercutting for denervation. Most of these have had variable success. At our center, because the inflammatory process involves the entire vestibule, we believe a vestibulectomy with vaginal advancement should be performed. This will connect thick vaginal mucosa directly to vulvar skin.

The vestibulectomy incision should be made in a horseshoe manner and include the hymen, periurethral mucosa, and posterior fourchette to the perineal skin pigmented border (Figures 9-2.2–9-2.4). The vaginal mucosa is then undermined and advanced to close the defect without tension. Our success rates with this technique have exceeded 80% in allowing reinitiation of sexual intercourse.

Urethral Pain Syndrome

Urethral pain syndrome, previously known as chronic urethritis or urethral syndrome is a rare sensory disorder of the urethra recently defined by the International Continence Society as persistent or recurrent urethral pain, with frequency and nocturia, lasting at least 6 months, in the absence of proven infection or other obvious pathology.[27]

Similar to vestibulitis, the etiology of this disorder remains unknown. Evaluation must be thorough to rule out infection, anatomic abnormalities (urethral diverticulum), atrophy, and other more generalized conditions of the bladder and/or vagina. Cultures, voiding diaries, urodynamics, and cystoscopy should all be included in the work-up.

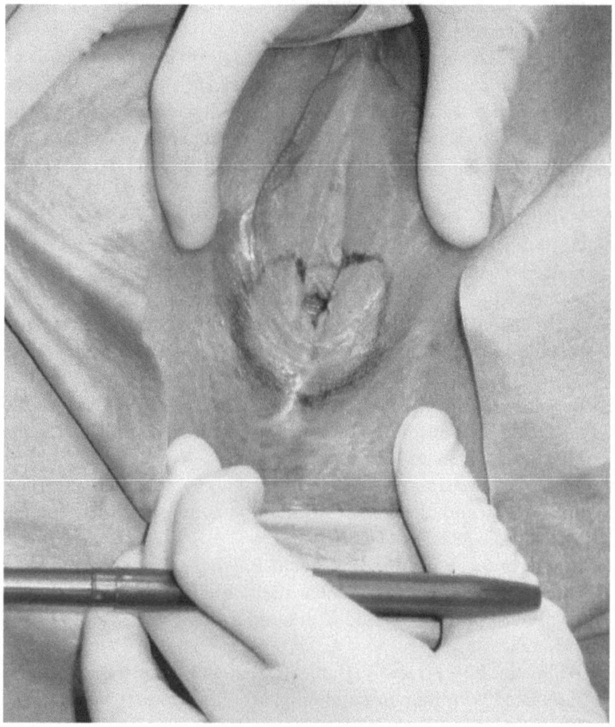

Figure 9-2.2. Vestibulectomy technique: A horseshoe area of vestibule from hymen to vulva is excised.

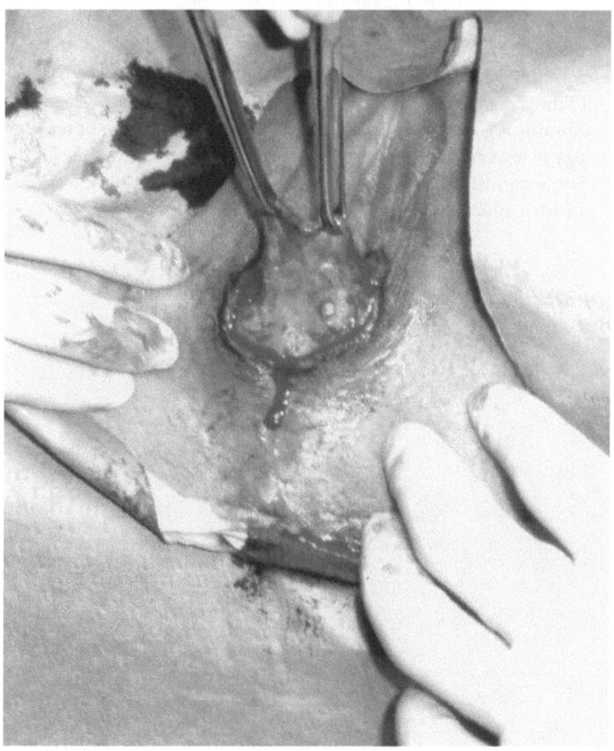

Figure 9-2.3. Vestibulectomy technique: The vaginal mucosa is dissected free.

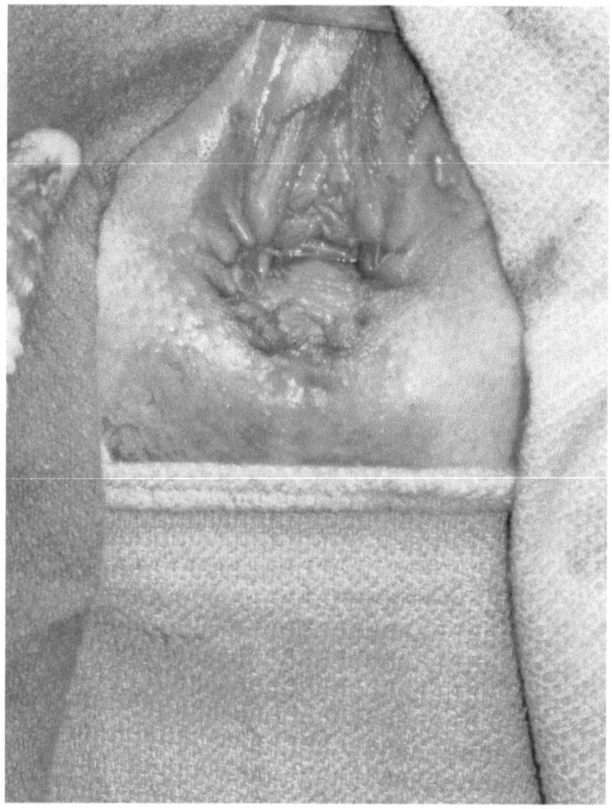

Figure 9-2.4. Vestibulectomy technique: The defect is closed in a running manner.

Treatment is focused on symptomatic relief. Urethral massage can be therapeutic and aids in making the diagnosis. Removing local and dietary irritants in the same manner as with vestibulitis is necessary. Biofeedback, including functional electrical stimulation, and physical therapy may be helpful. Pain sensation can be reduced with local anesthetics, analgesics, and tricyclics. A bladder-training regimen may decrease associated frequency. On occasion, multichannel urodynamics may demonstrate urethral instability, which may typically respond to functional electrical stimulation. Urethral dilation is no longer recommended as a treatment option, but has historically yielded good results.

Trigonitis

Trigonitis is inflammation localized to the bladder trigone. Symptoms manifest as bladder and urethral pain, as well as frequency and nocturia. Digital palpation will reveal pain localized to the trigone.

Acute trigonitis is usually obvious on cystoscopy and may be treated with a routine course of antibiotics. Chronic trigonitis is typically more subtle with mild inflammatory surface changes. Abundant squamous metaplasia, common to this tissue, is often mistaken for inflammation by the inexperienced cystoscopist.

Chronic trigonitis is often preceded by recurrent, documented urinary tract infections and is thought to represent a deep tissue bacterial infection. A long course of antibiotics with good tissue penetration is indicated as first-line therapy. We use Vibramycin 100 mg twice a day for 21 days. If antimicrobial therapy is unsuccessful, the patient should be evaluated and treated for interstitial cystitis.

References

1. Fitzpatrick CC, DeLancey JO, Elkins TE, McGuire EJ. Vulvar vestibulitis and interstitial cystitis: a disorder of urogenital sinus-derived epithelium? Obstet Gynecol 1993;81:860–862.
2. Gunter J, Clark M, Weigel J. Is there an association between vulvodynia and interstitial cystitis? Obstet Gynecol 2000;95(4 suppl 1):S4.
3. Friedrich EJ. Vulvar vestibulitis syndrome. J Reprod Med 1987;32(2): 110–114.
4. Bornstein J, Shapiro S, Rahat M, et al. Polymerase chain reaction search for viral etiology of vulvar vestibulitis syndrome. Am J Obstet Gynecol 1996;175:139–144.
5. Baggish MS, Miklos JR. Vulvar pain syndrome: a review. OG Survey 1995;50(8):618–627.
6. Chadha S, Gianotten WL, Drogendijk AC, Weijmar Schultz WC, Blindeman LA, van der Meijden WI. Histopathologic features of vulvar vestibulitis. Int J Gynecol Pathol 1998;17:7–11.
7. Pyka RE, Wilkinson EJ, Friedrich EG Jr, Croker BP. The histopathology of vulvar vestibulitis syndrome. Int J Gynecol Pathol 1988;7: 249–257.
8. Nunns D, Mandal D. Psychological and psychosexual aspects of vulvar vestibulitis. Genitourin Med 1997;73(6):541–544.
9. Stewart DE, Reicher AE, Gerulath AH, Boydell KM. Vulvodynia and psychological distress. Obstet Gynecol 1994;84:587–590.
10. Van Lankveld J, Weijenborg P, Ter Kuile M. Psychologic profiles of and sexual function in women with vulvar vestibulitis and their partners. Obstet Gynecol 1996;88(1):65–70.
11. Lundqvist EN, Hofer PA, Olofsson JI, et al. Is vulvar vestibulitis an inflammatory condition? A comparison of histological findings in affected and healthy women. Acta Derm Venereol 1997;77:319–322.
12. Slone S, Reynolds L, Gall S, et al. Localization of chromogranin, synaptophysin, serotonin, and CXCR2 in neuroendocrine cells of the minor vestibular glands: an immunohistochemical study. Int J Gynecol Pathol 1999;18:360–365.
13. Foster DC, Hasday JD. Elevated tissue levels of interleukin-1B and tumor necrosis. Obstet Gynecol 1997;89:291–296.
14. Stewart EG, Berger BM. Parallel pathologies? Vulvar vestibulitis and interstitial cystitis. J Reprod Med 1997;42:131–134.
15. Bohm-Starke N, Hilliges M, Falconer C, Rylander E. Neurochemical characterization of the vestibular nerves in women with vulvar vestibulitis syndrome. Gynecol Obstet Invest 1999;48:270–275.
16. Westrom LV, Willen R. Vestibular nerve fiber proliferation in vulvar vestibulitis syndrome. Obstet Gynecol 1998;91:572–576.
17. Masterson BJ, Galask RP, Ballas ZK. Natural killer cell function in women with vestibulitis. J Reprod Med 1996;41:562–568.
18. Bazin S, Bouchard C, Brisson J, Morin C, Meisels A, Fortier M. Vulvar vestibulitis syndrome: an exploratory case-control study. Obstet Gynecol 1994;83:47–50.
19. Bergeron C, Moyal-Barracco M, Pelisse M, Lewin P. Vulva vestibulitis: lack of evidence for a human papillomavirus etiology. J Reprod Med 1994;39:936–938.
20. Turner ML, Marinoff SC. Association of human papillomavirus with vulvodynia and the vulvar vestibulitis syndrome. J Reprod Med 1988;22:533–537.
21. Umpierre SA, Kaufman RH, Adam E, Woods KV, Adler-Storthz K. Human papillomavirus DNA in tissue biopsy specimens of vulvar vestibulitis patients treated with interferon. Obstet Gynecol 1991;78: 693–695.
22. Wilkinson EJ, Guerrero E, Daniel R, et al. Vulvar vestibulitis is rarely associated with human papillomavirus infection types 6, 11, 16, or 18. Int J Gynecol Pathol 1993;12:344–349.
23. Bergeron S, Binik YM, Khalife S, et al. A randomized comparison of group cognitive-behavioral therapy, surface electromyographic feedback, and vestibulectomy in the treatment of dyspareunia resulting from vulvar vestibulitis. Pain 2001;91:297–306.
24. Solomons CC. Calcium citrate for vulvar vestibulitis. Reprod Med 1991;36:879–882.
25. Snyder DS, Hatfield GM, Lampe KF. Examination of the itch response from the raphides of the fishtail palm Caryota mitis. Toxicol Appl Pharmacol 1979;48:287–292.
26. Schmidt RJ, Moult SP. The dermatitic properties of black bryony (Tamus communis L.). Contact Dermatitis 1983;9:390.
27. Abrams P, Cardozo L, Fall M, et al. The standardisation of terminology of lower urinary tract function: report from the Standardisation Sub-committee of the International Continence Society. Neurourol Urodyn 2002;21:167–178.

9-3

Anal Pain

Tracy L. Hull

Chronic idiopathic anal pain can be a frustrating and challenging problem for the clinician and patient.[1–4] This encompasses pain in the anal region without definable anatomic abnormalities.[5] To add to the situation, nomenclature in the literature is confusing, and terms are often used incorrectly. From intensive review of the literature, essentially three different patterns emerge: coccygodynia, levator ani syndrome, and proctalgia fugax. Levator ani syndrome and proctalgia fugax will be discussed in this chapter.

Levator Ani Syndrome

Levator ani syndrome consists of a dull aching or pressure discomfort high in the rectum.[6] Patients may have a sensation of sitting on a ball or other intrarectal object. The pain may be precipitated by prolonged sitting or defecation, and seems to occur more often on the left side. The diagnostic criteria stipulated by an international committee consist of chronic or recurrent anorectal pain for 3 months in the preceding year (they do not have to be consecutive weeks or months). The episodes must last 20 minutes or longer. This problem seems to primarily affect women under 45 years of age. It is postulated that spasm of the levator muscles leads to these symptoms. Physical examination may be normal or demonstrate a tight band like levator muscle. Palpation of the levator may precipitate the pain.

Proctalgia Fugax

Proctalgia fugax is recurrent pain lasting from seconds to minutes.[6] The pain can be severe, sharp, and awaken the patient from sleep. Initially it was thought that young men more often experience this problem; however, in some series, it was seen more often in women. Proctalgia starts in adult life and usually stops by middle age. Studies have suggested that anal smooth muscle dysfunction is the causative problem. Physical examination is normal.

Approach and Treatment

When evaluating these patients, a comprehensive history and physical examination are needed to rule out other pathology. In addition to a physical examination, which includes endoscopy of the rectum, a computed tomography or magnetic resonance imaging of the pelvis may be indicated if the diagnosis is in doubt. When embarking on treatment for levator syndrome or proctalgia, the literature is not straightforward, because authors frequently refer to all anal pain by the same name without deference to there being two distinct problems that can occur. Patient should be reassured that they do not have a serious condition such as cancer. They should also be encouraged to improve their fiber and fluid intake if deficient, as well as sitz baths and perineal strengthening exercises. Even without scientific evidence to corroborate this approach, this has proven helpful in the majority of patients. For those who are unresponsive to these measures, more intensive therapies are indicated, using individualized approaches.

Electrogalvanic Stimulation

Some patients believe they are dramatically helped by this treatment. It is best suited for levator syndrome. High-voltage, low-frequency oscillating electrical current given with a probe placed in the rectum causes fasciculation and fatigue in the levator muscle. Usually the treatments are given for 1 hour every other day for three sessions. There are also home units available. At best, only approximately 40% of patients find this treatment successful. However, the patients included in this treatment and reported in the literature may contain those with proctalgia.

Levator Massage

Deep, transanal digital massage to the puborectalis muscle in the operating room may benefit those with levator syndrome. Steroid injection into the muscle after the procedure may also be added.

Biofeedback

For patients with levator syndrome, teaching relaxation through electromyographic-based biofeedback can be therapeutic.[7] The theory is that feedback to patients concerning the tone in the pelvic muscles will allow alteration of the contraction and initiate relaxation.

Drug Treatments

The only randomized controlled study for anal pain used salbutamol (Ventolin) inhaler to treat proctalgia. This drug inhibits smooth muscle and two puffs at the onset of pain allowed rapid onset of action. Patients treated had significant reduction in pain over controls. Successful case reports using diltiazem (80 mg twice daily) and clonidine (150 μg twice daily for 3 days, then tapered to 75 μg twice daily for 2 days, then 75 μg daily for 2 days) for treatment of proctalgia-type pain have been reported. Both agents work by relaxing smooth muscle. Topical 0.2% nitroglycerin to the anal verge has also been reported to treat proctalgia.[8] This medication also produces smooth muscle relaxation.

Other Treatments

Since patients may be desperate for pain relief, nonconventional treatments such as acupuncture and hypnosis may help some. Referral to a pain management program may be useful, where differential blocks can be performed to evaluate for a supratentorial component of pain. Similarly, psychiatric evaluation may be indicated for select patients.

Summary

Anal pain in patients can be extremely frustrating to treat. Individualized treatment should be initiated beyond this based on available resources. Outlining an approach and moving on to another therapy when one does not work is acceptable because the reported success rates of most treatments are approximately 50% at best.

References

1. Hull TL, Milsom JW. Pelvic floor disorders. Surg Clin North Am 1994;74:1399–1413.
2. Wald A. Anorectal and pelvic pain in Women. Diagnostic considerations and treatment. J Clin Gastroenterol 2001;33:283–288.
3. Wald A. Functional anorectal and pelvic pain. Gastroenterol Clin North Am 2001;30:243–251.
4. Whitehead WE, Wald A, Diamant E, Enck P, Pemberton JH, Rao SSC. Functional disorders of the anus and rectum. Gut 1999;45(suppl II): II-55–II-59.
5. Hull TL. Unexplained anal/rectal pain. In: Cameron J, ed. Current Surgical Therapy. 7th ed. St. Louis: Mosby; 2001:307–308.
6. Vincent C. Anorectal pain and irritation: anal fissure, levator syndrome, proctalgia fugax, and pruritus ani. Prim Care 1999;26:53–68.
7. Gilliland R, Heymen JS, Altomare DF, Vickers D, Wexner SD. Biofeedback for intractable rectal pain: outcome and predictors of success. Dis Colon Rectum 1997;40:190–196.
8. Lowenstein B, Cataldo PA. Treatment of proctalgia fugax with topical nitroglycerin. Dis Colon Rectum 1998;41:667–668.

9-4

Pain Localization and Control

Wagih W. Gobrial

The female pelvis contains diverse, multiple, and intricately innervated structures that are potential sources of pain. As an example, when the etiologic process is gynecologic cancer, which tends to spread locally either by direct invasion or by spread of metastases to regional lymph notes, pain can be present at multiple sites simultaneously.

Pelvic pain is particularly difficult to manage because it is often vague, poorly localized, and tends to be bilateral or to cross the midline. Thus, a systematic approach to pelvic pain is the best approach. A careful review of history and physical examination may give a clue about the source and type of pain. Cooperation with a specialist in gynecology, urology, or colorectal surgery is helpful in identifying the most likely location of the painful stimulus during a directed pelvic examination.

Two types of diagnostic blocks are typically performed to try to ascertain the source of the pain and hence devise a treatment plan:

1. Differential spinal block – to differentiate between psychological, central, sympathetically medicated, or somatic pain
2. Sympathetic block – to impact nerves directly supplying the pelvic structures
   A. Superior hypogastric plexus block: Specifically useful for pelvic pain arising from the uterus and upper vagina, bladder, prostate, urethra, seminal vesicles, testes, and ovaries; pelvic pain secondary to radiation; sympathetically maintained pain (e.g., after rectal anastomosis, abdominoperineal resection, etc.); and chronic pelvic inflammatory processes.[1]
   B. Inferior hypogastric block: Primarily useful for perineal pain either malignant or sympathetically mediated, and superficial hyperesthesia including sensation of severe burning and urgency.

Both blocks are done initially with a local anesthetic, as diagnostic/therapeutic blocks. Local anesthetic blocks are therapeutic if done multiple times in patients with sympathetically mediated pain. Patients with malignancy, who get good relief although short-lived, might benefit from neurolytic blocks.

Neurolytic blocks are done using the same technique. However, instead of using a local anesthetic, a neurolytic agent is used, such as alcohol (because it is hypobaric and the patient is in a prone position) or 6% phenol (occasionally used, if a hyperbaric solution is needed). The patient must be made aware of the risks and side effects of the neurolytic block before proceeding.

Differential Spinal Block

There are two ways to accomplish this block:

1. Standard differential block. This is done by intrathecally injecting several solutions. Injections are spaced 5 to 10 minutes apart and the patient is evaluated after each injection.
   A. The first solution is 5 mL of preservative-free normal saline. If the patient gets pain relief from the saline injection, this might be attributed to:
      1. Placebo effect (reported in up to 30% of patients), which is usually short-lived and the pain usually comes back in a few minutes.
      2. Psychological pain, in which case the patient gets an extended pain relief, which may last for days or even permanently.
   B. If no relief is achieved, the second solution will be 4 to 5 mL of 0.5% procaine, with 5 mL of normal saline. This will block the sympathetic fibers without sensory or motor effects. If the patient gets pain relief with this injection, the pain is probably sympathetically mediated and the patient will benefit from a sympathetic block.
   C. The third solution is 1 mL of 5% procaine added to 9 mL of normal saline – somatic blockade. If the patient gets pain relief, the pain is somatic in origin and treatment should be focused on this direction.

D. The fourth solution is 2 mL of 5% procaine added to 2 mL of saline. This would cause a complete motor block. If the patient continues to have the pain after complete motor and sensory spinal block, the patient's pain is considered central. This might be caused by a true organic lesion above the level of the spinal block (that is why spinal differential block is done above the suspected pain level), encephalization of the pain because of the intensity and direction, psychologically mediated pain, or the patient may be malingering.

2. Retrograde differential spinal block (more frequently used). This is done by using two solutions.
   A. The first solution is 2 mL of normal saline, same interpretations as in the standard differential spinal block.
   B. The second solution is 1 mL of 10% procaine and 1 mL of cerebral spinal fluid. This will give complete motor and sensory spinal blockade. If the patient's pain continues, it is central. If the patient's pain resolves, pain assessment should be performed every 10 minutes until there is a return of motor and then sensory function. If the pain returns with the return of the sensory function, the pain is somatic in origin. If the pain returns a few hours after the return of the sensory function, it is sympathetically mediated pain.

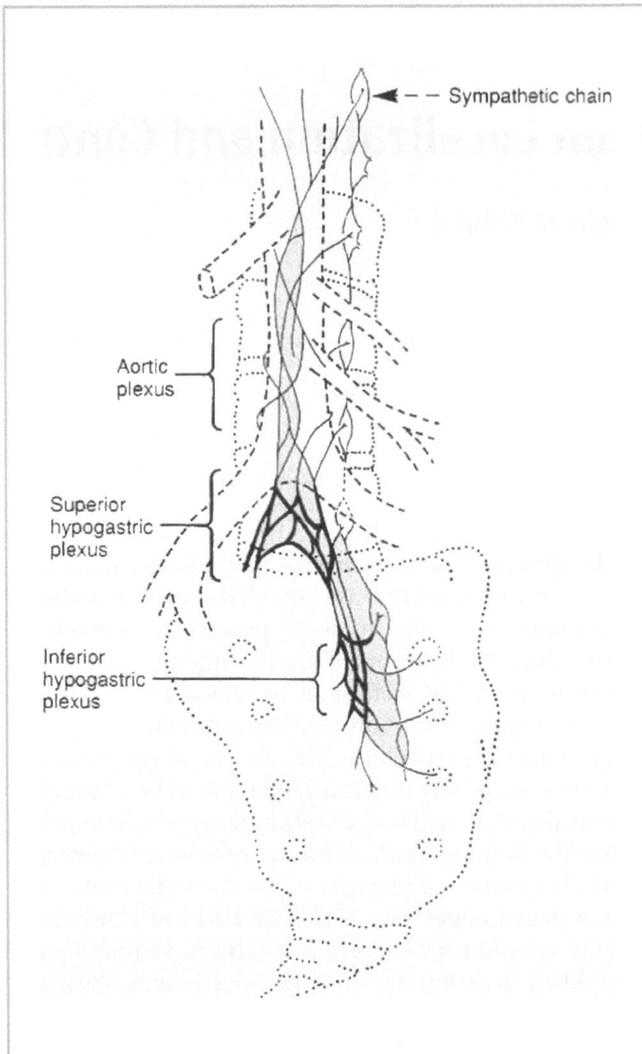

Figure 9-4.1. Anatomy of superior and inferior hypogastric plexuses. (Reprinted from Raj PP, Lou L, Erdine S, et al. Radiographic Imaging for Regional Anesthesia and Pain Management, p 231, Copyright 2003, with permission from Elsevier.)

Sympathetic Blocks

Anatomic Considerations

Sometimes referred to as presacral nerve, the superior hypogastric plexus is formed by the confluence of the lumbar sympathetic chains and branches of the aortic plexus, which contain fibers that traverse the celiac and inferior mesenteric plexuses (Figure 9-4.1). In addition, it usually contains parasympathetic fibers that originate in the ventral roots of S2-S4 and travel along pelvic splanchnic nerves through the inferior hypogastric plexus. The plexus is located in the retroperitoneum, anterior to the body of the lower part of L5, and upper part of S1. It is posterior to the bifurcation of the aorta and both common iliac arteries, and left common iliac vein. The superior hypogastric plexus divides into right and left hypogastric nerves, which descend lateral to the sigmoid colon and rectosigmoid to reach the two inferior hypogastric plexuses. The superior hypogastric plexus gives off branches to the ureteric and testicular (or ovarian) plexus, the sigmoid colon, and the internal iliac artery. In addition to the pathways that traverse the superior hypogastric plexus, sympathetic fibers also reach the plexus through perivascular pathways that include the inferior mesenteric plexus (sigmoid, colon, and rectum) and renal plexus (ureteric and ovarian or testicular plexuses).

Technique of Superior Hypogastric Plexus Block

The patient is positioned on the fluoroscopy table in prone position. Using fluoroscopic guidance, the L4-5 spinal process is identified. Going laterally, a 7-cm skin marker is made and this will be the point of needle entry. The lumbarsacral area is prepped and draped in a sterile manner. Multiple approaches have been described, including a lateral approach, medial approach, and the intradiskal approach – the approach that is most often used is the lateral approach (Figure 9-4.2). At 7-cm lateral to the L4-5 interspace, the skin and deeper tissue and muscles are infiltrated with lidocaine 0.5% using a 20-gauge 6-in needle. The needle is directed 45 degrees medially and caudally to miss the transverse process of L5 and the sacral ala on anteroposterior (AP) fluoroscopy view (Figure 9-4.3). The needle must be more than 1 cm from the bony outline. On the lateral view, the needle tip should be at the anterior

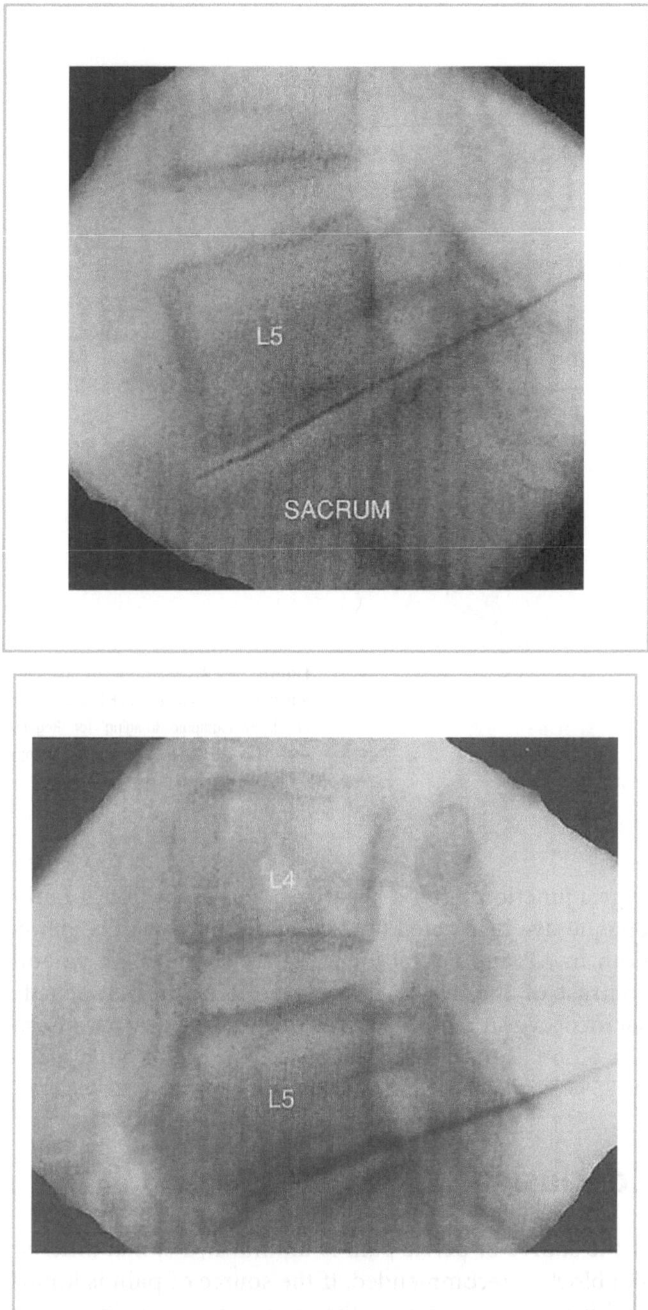

Figure 9-4.2. Lateral view of superior hypogastric block.

surface of the junction of L5-S1. It is further advanced, and loss of resistance usually occurs at this point. Confirmation of the location of the needle is done with injection of radio-opaque dye and the spread is followed both in AP and lateral views. After confirmation of the position of the needle and negative blood aspiration, a test dose of bupivacaine 0.375% (2–3 mL) is injected. Patient evaluation is done a few minutes after the injection, to confirm there is no sensory or motor blockade. A total of 12 to 15 mL of bupivacaine 0.375% is injected with intermittent aspiration. The block is performed unilaterally or bilaterally depending on the patient's symptoms. Patients usually get pain relief in 15 to 20 minutes. The patient is subsequently monitored in the postanesthesia care unit for the duration of pain relief, which should be for several hours. If it is short-lived (15–30 minutes), it is usually a placebo effect. CT guidance can be used for this block.[2]

Inferior Hypogastric Plexus

The inferior hypogastric plexus is a bilateral abdominal structure situated on each side of the rectum, lower part of the bladder and (in men) prostate and seminal vesicles or (in women) cervix, uterus, and vaginal fornices (Figure 9-4.4). The inferior hypogastric plexus supplies branches to the pelvic viscera directly, as well as from subsidiary plexuses (e.g., the superior, middle rectal, vesical, prostate, and uterovaginal plexuses). The sacral sympathetic trunk lies in the parietal pelvic fascia behind the parietal peritoneum and on the ventral surface of the rectum, just medial to its anterior foramina and the existing sacral nerves. Below they converge and unite to form a solitary small "ganglion impar" which is located anterior to the sacrococcygeal junction.

Technique of Inferior Hypogastric Block

Multiple approaches, such as lateral, prone, and lithotomy have been described (Figure 9-4.5).[3] The most frequently used approach is the prone position. The sacrococcygeal area is prepped and draped in a sterile manner. The entry site, just under the tip of the coccyx, is anesthetized with lidocaine 0.5% using a 25-gauge needle. A 22-gauge, 3.5-in spinal needle is bent into a C shape (commercial C shape 22-gauge, 3.5-in spinal needles are also available). At the site of entry, just under the coccyx, the needle is advanced in a semicircular manner under fluoroscopy (lateral view) until the top of the needle is just anterior to the sacrococ-

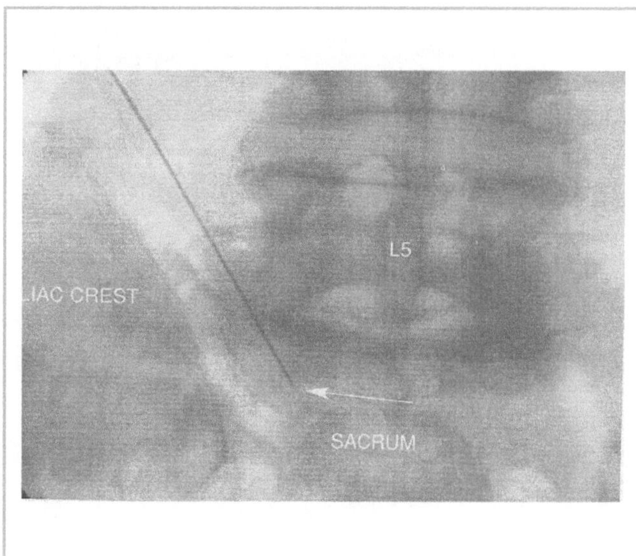

Figure 9-4.3. AP view of superior hypogastric block.

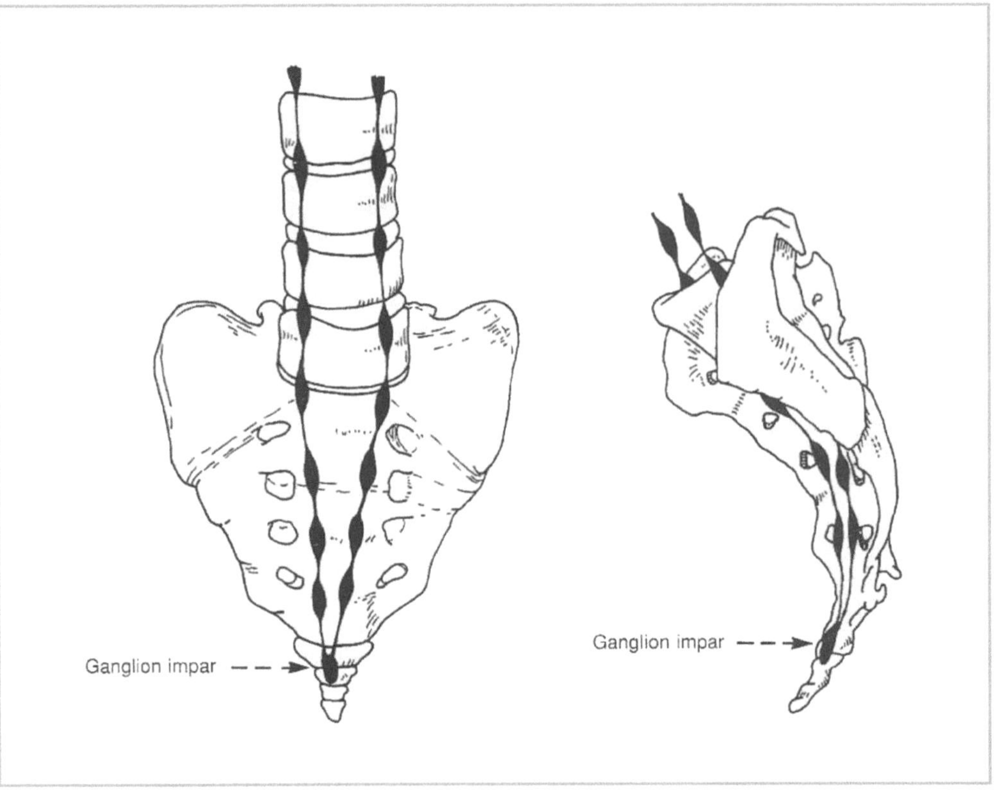

Figure 9-4.4. Anatomy of ganglion impar. (Reprinted from Raj PP, Lou L, Erdine S, et al. Radiographic Imaging for Regional Anesthesia and Pain Management, p 238, Copyright 2003, with permission from Elsevier.)

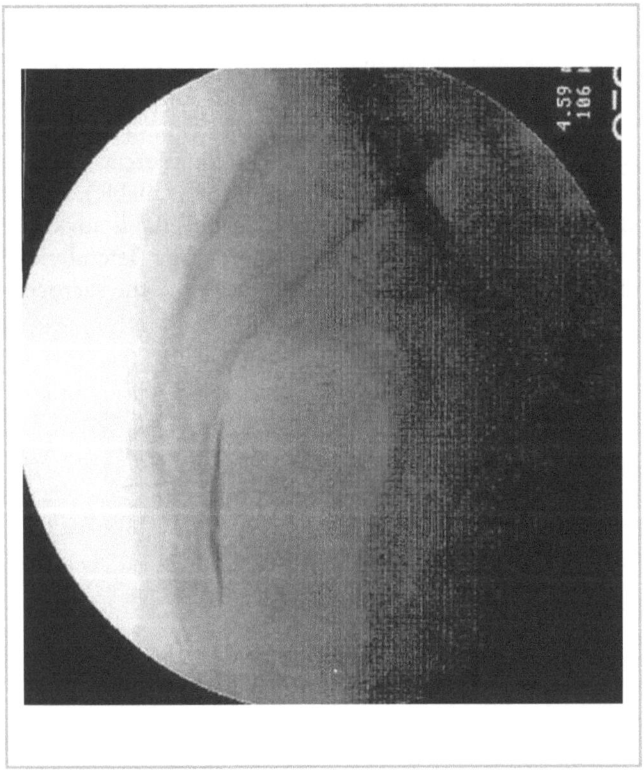

Figure 9-4.5. Lateral view of inferior hypogastric block.

cygeal junction. After a negative blood aspiration, a radioopaque dye is injected (1–2 mL) and the spread is viewed both in AP and lateral views. There should be a smooth contrast of the dye in the retroperitoneum between the sacrococcygeal region and the rectal bubble. After a negative blood aspiration, a total of 12 to 15 mL of bupivacaine 0.375% is injected with intermittent aspirations.

Conclusions

If the source of pelvic pain is unknown, a spinal differential block is recommended. If the source of pain is known and it is of pelvic origin, the treatment is usually a superior hypogastric or inferior hypogastric block depending on the organ innervations. Neurolytic blocks should be considered for pain related to malignancy or intractable pelvic pain.

References

1. de Leon-Casaola OA, Kent E, Lema MJ. Neurolytic superior hypogastric plexus block for chronic pelvic pain associated with cancer. Pain 1993;54:145–151.
2. Waldman SD, Wilson WL, Kreps RD. Superior hypogastric plexus block using a single needle and CT guidance. Reg Anesth 1991;16:286.
3. Plancarte R, Amescua C, Patt RB. Presacral blockade of the ganglion impar. Anesthesiology 1990;73:A751.

9-5

Acupuncture for Pelvic Floor Dysfunction

Lawrence P. Frank

Acupuncture is an important mainstay of traditional Chinese medicine (TCM). Empirically derived by thousands of years of observation, Chinese medical doctors recorded their treatments almost 2000 years ago in the then modern text, "Yellow Emperor's Book of Internal Medicine." Although the physiologic reasoning may seem bizarre to modern allopathic physicians, accepting and understanding the basic tenets does then provide a logical choice of therapy that has withstood the test of time, longer than any current Western therapeutics. Every patient, both responsive and unresponsive to conventional medical therapy, should be given the choice of undergoing a course of acupuncture.

Pelvic floor dysfunction and, in particular, pelvic pain, represents as much a dilemma for TCM acupuncture as it does for Western medicine. Precise diagnoses are difficult to differentiate. Indeed, every medical condition mentioned in this book can cause pelvic dysfunction in male and females. Acupuncture does have a role in the treatment of these patients. Fortunately, the TCM acupuncture approach is a much broader diagnostic system and therefore the available treatments will work theoretically for most conditions in the general pelvic region. Symptoms are just as important as a diagnosis, as is the patient's overall health and well-being to form a holistic picture of the disease or distress.

Modern randomized clinical trial studies have been done investigating acupuncture for a variety of pelvic dysfunctions. However, there is still only a limited number of such studies. The largest number of trials concern the syndrome of interstitial cystitis.[1–6] Then, there are generalized pelvic pain,[7,8] pelvic myofascial trigger points,[9,10] vulvodynia,[11] and review articles of pelvic pain listing acupuncture as a treatment modality.[12–14]

Regarding interstitial cystitis, significant reductions in frequency and incontinence have been demonstrated using acupuncture.[1] Others have found a significant increase in cystometric capacity and symptomatic improvement in frequency, urgency, and dysuria.[6]

Treatment Philosophies

A fundamental concept of acupuncture is the balanced flow of life energy, Qi, through the body. The Qi flows in well-defined, but as yet undetectable "vessels" called "meridians." When the flow of Qi is blocked, by disease or injury, symptoms of illness, e.g., pain, swelling, and tenderness, manifest themselves. Acupuncture points are well-demarcated areas along these "vessels," which are known to "unblock" the flow of Qi. This is known as the Principle Meridian System, with which Western medical doctors are most familiar. This is also the simplest, most straightforward system in acupuncture for resolving the blockages.

Patients with pelvic floor problems are anything but straightforward. Their treatments will involve other, more complex meridian systems in addition to the previously discussed Principle Meridian System. The next deeper system is known as the Distinct Meridian System. According to traditional acupuncture theory, this system can provide direct access to an organ and help to release this obstruction. Another deeper, denser meridian system is the "Curious" or "Irregular" system. The Curious system helps to balance and coordinate energy flows in the principle meridians, which affect all the systems. These two systems, the Curious and the Distinct, are difficult to use and activate, but can offer profound relief of symptoms if they work.

Normally, an acupuncturist would use only one meridian system per visit. Because of the profound nature of these problems, I combine two of three of the above treatment systems together at any one session rather than just using one, and I find a combination that works. That is, the patient either immediately reports improvement in symptoms, if they are present at the time, or lets me know at the following treatment session of the improvement in their chronic symptoms. I then use that acupuncture program for at least four sessions. Acupuncture promotes a subtle response, which increases over time. Generally, three or

Table 9-5.1. Acupuncture protocols for pelvic floor dysfunction

System	Command Points	Notes
TY-SY	KI3→+ .at 4 Hz BL60	Use KI10 instead of KI3 if tender
KI-BL	KI10 crossed to BL40 BL10 return	EA to focusing needles placed when tender and crossed as +: BL27, BL28, and/or BL30, or if more tender BL31,32,33, and/or 34 all + at15–60 Hz
Chong-Mo	SP4 crossed KI3 MH6 couplet	EA + to focusing needles used if tender: KI11, ST30, CV2,3,4 at 2–4 Hz
Dai-Mo	GB41, GB26 TH5 couplet	Focusing needle used if tender: GB27,28

four treatments are required to obtain significant reduction in symptoms for these chronic conditions.

Actual treatments consist of the Tai Yang-Shao Yin (TY-SY) Principle Meridian, the Kidney Bladder (KI-BL) Distinct Meridian, and the Chong Mo-Dai Mo Curious Meridian System. The important points for these circuits and their focusing points are listed in Table 9-5.1. Points used for focusing are selected if they are tender on palpation. Electroacupuncture (EA) is used at all sessions. Generally, I use lower frequency (2–4 Hz) with the Principle and Curious Meridian Systems. This has been shown to amplify the release of endorphins and presumably other active neurohumoral substances. I use higher frequency (EA) with the KI-BL Distinct Meridian System. This helps to deactivate the intense musculoskeletal trigger points so often present in these patients, similar to percutaneous nerve stimulation.

I often begin my female patients with a TY-SY Principle Meridian program and a Chong Mo-Dai Mo program. This has proven to be more effective in the female. Conversely, the KI-BL Distinct Meridian System seems to be more effective in the male, combined with a TY-SY Principle Meridian circuit with additional abdominal focusing points that are tender (Table 9-5.1). These abdominal points are needled at a very acute angle so the patient can be turned prone and not have the needle penetrate further or be uncomfortable. Incidentally, these treatments also help the irritable bowel symptoms that many of these patients exhibit concurrently.

Discussion

In practice, these protocols are quite effective. My typical patient has had symptoms for 2 to 5 years. They are generally referred to me when fellow medical doctor colleagues can no longer offer effective treatment. I can improve their symptoms in two or three once-a-week sessions. This encourages them to maintain their therapy, which can easily span several months. After six to eight weekly sessions, we then go to biweekly and eventually taper to a once-a-month regimen. With this particular symptom complex, it is unrealistic to reduce treatment frequency to less than once a month. The acupuncture needle is akin to a drug, and the brain needs periodic reeducating to keep the desired effect apparent.

Patient Examples

Patient S.B. is a 56-year-old professional male with a 3-year history of irritable urinary bladder symptoms. He is taking multiple medications including Ditropan, Elavil, Elmiron, Atarax, Flomax, Cysta-Q, Celebrex, and Saw Palmetto. Confounding the history of irritable bladder symptoms is a concurrent history of mild inflammatory bowel disease. Every time he moves his bowels, he has bladder spasms and pain, and this sets off a cycle of frequency and urgency. He wishes to be able to control his urinating and reduce the episodes of bladder spasm. He would also like to taper his medications.

Physical examination reveals a delicate, thin, balding man. Palpation of the lower abdomen acupuncture points (CV2, 3, 4, ST25, 30, SP12, KI11) detects tenderness. Turning the patient prone and palpating the buttock area (BL27–34) also reveals tender acupuncture points. I elect to give the patient a KI-BL circuit treatment. I have provided a diagram of the KI-BL Distinct Meridian Channel and the organ coverages (Figures 9-5.1 and 9-5.2). Needles are placed at the abdominal points first, and then the patient is turned prone. When I place the needles in the buttock, frank muscle twitches are noted. This would indicate correct assessment of pathology amenable to an acupuncture approach.

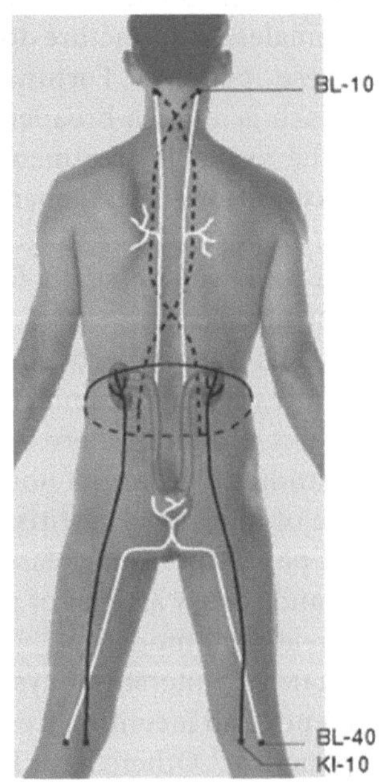

Figure 9-5.1. Kidney-Bladder Distinct Meridians. (Courtesy of Acupuncture Energetics, by J Helms, Medical Acupuncture Publishers, Berkeley, CA, 1995.)

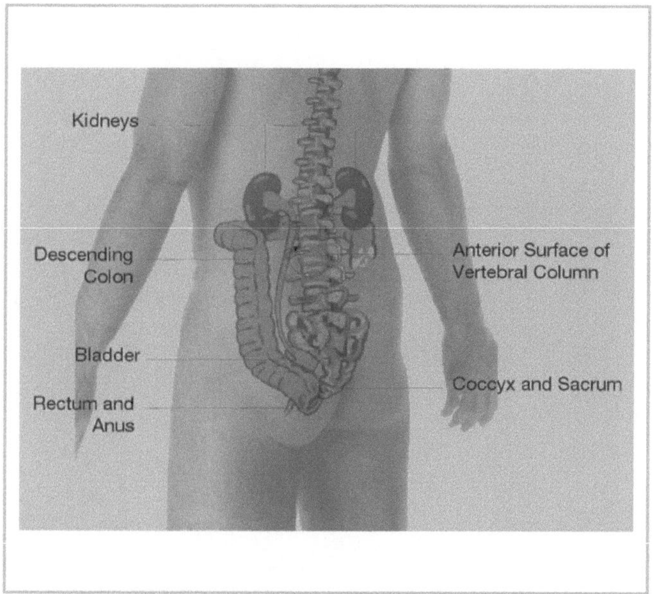

Figure 9-5.2. Target organs/zones. (Courtesy of Acupuncture Energetics, by J Helms, Medical Acupuncture Publishers, Berkeley, CA, 1995.)

The patient had a good response after the first treatment for 1 day. This was enough encouragement to continue weekly. After each session, he reported longer and longer responses. He continued weekly sessions for the next 12 weeks before tapering gradually to once a month. S.B. continues to receive regular treatment and is quite satisfied. Frequency and urgency are under good control and he has almost no bladder-type pain symptoms presently. He has also noted that his inflammatory bowel symptoms are under the best control ever.

Patient N.B. is a 51-year-old woman with chronic pelvic pain since the onset of menopause 3 years earlier. Her chief complaint, however, is dyspareunia, resulting she believes, from her pelvic pain. This has caused, needless to say, a great deal of stress between her and her husband. She is otherwise generally healthy, even athletic, and would like to get her pain under control and enjoy her life more.

Physical examination reveals a petite, well-groomed, middle-aged woman. Palpation of abdominal acupuncture points elicits only minimal response. No tender areas at all are found in the posterior buttock region. I decide to give this patient a Chong Mo circuit with all the focusing points described in Table 9-5.1. The change in pain pattern is only slight immediately after the first treatment. However, she reports at the following appointment that the symptoms continued to resolve for several days after. After another two treatments, for a total of three, she claims total resolution of the dyspareunia and a return to normal relations with her husband. She has not felt the need to return for treatment for several months now.

In most of these patients, there is a large myofascial component to their pain, which acupuncture can help break up using electrical simulation. I have painted the male example as a more difficult case requiring longer, more intense treatment. This has generally been the case.

Conclusion

There is no doubt that acupuncture works; the National Institutes of Health 1997 consensus statement[15] confirmed this. It is a time-tested, albeit empirical treatment we can offer our patients. Because acupuncture can only help, and never hurt our patients, we should always consider it among our armamentarium of therapies.

References

1. Chang PL. Urodynamic studies in acupuncture for women with frequency, urgency and dysuria. J Urol 1988;140:563–566.
2. Gerisson G, Wang Y, Lindstrom S, Fall M. Traditional acupuncture and electrical stimulation of the posterior tibial nerve. J Urol 1993; 27:67–70.
3. Lyons P. Acupuncture treatment for interstitial cystitis: a case report. Med Acupunct 2003;14:43–45.
4. Maher CF, Carey MP, Dwyer PL, Schlutter PL. Percutaneous sacral nerve root neuromodulation for intractable interstitial cystitis. J Urol 2001;165:884–886.
5. Phillip T, Shah PJR, Worth PHL. Acupuncture in the treatment of bladder instability. Br J Urol 1988;61:490–493.
6. Pigne A, De Goursac C, Barrat J. Acupuncture and unstable bladder. In: Proceedings of 15th Annual Meeting of International Continence Society, London, 1985:186–187.
7. Shaoguang W. Electroacupuncture treatment for constipation due to spasmodic syndrome of the pelvic floor. J Tradit Chin Med 2001; 21(3):205–206.
8. Wedenberg K, Moen B, Norling A. A prospective randomized study comparing acupuncture with physiotherapy for low-back pain in pregnancy. Acta Obstet Gynecol Scand 2000;79: 331–335.
9. Slocumb J. Neurological factors in chronic pelvic pain: trigger points and the abdominal pelvic pain syndrome. Am J Obstet Gynecol 1984;149(5):536–543.
10. Weiss J. Pelvic floor myofascial trigger points: manual therapy for interstitial cystitis and the urgency-frequency for interstitial cystitis and the urgency-frequency syndrome. J Urol 2001;166:2226–2231.
11. Powell J, Wojnarowska F. Acupuncture for vulvodynia. J R Soc Med 1992;92:579–591.
12. Martin DC, Ling FW. Endometriosis and pain. Clin Obstet Gynecol 1999;42:664–686.
13. Steege JF, Stout AL, Somkuti SG. Chronic pelvic pain women: toward an integrative model. Obstet Gynecol 1991:3–30.
14. Wesselmann U, Czakanski P. Pelvic pain: a chronic visceral pain syndrome. Curr Pain Headache Rep 2001;5:13–19.
15. National Center for Complementary and Alternative Medicine. Accessed June 2004. http://nccam.nih.gov/health/acupuncture.

Section X

Evacuation Disorders

Section X

Evacuation Disorders

Steven D. Wexner

Evacuatory disorders are very difficult to both evaluate and treat. Both urinary and fecal retention are troubling symptoms to the patients if they cannot completely void or evacuate and yet have the persistent urgency to repeatedly attempt evacuation. Patients can often spend significant parts of their lives in the bathroom seated on the commode and, between such episodes, contemplate the need to return to the bathroom. In many cases, they have such severe life-compromise that they would prefer major surgery. Because of the lack of uniform success with any of the procedures, these patients are often managed by psychologists and psychiatrists for depression. Prior psychological assessments of these patients have noted significant elevations in the "neurotic triad" including somatic sensation, hypochondriasis, and depression. As can be seen in the five chapters within this section, urinary and fecal retention is a vexing problem to which no adequate solution is universally applicable. These chapters do offer a review of the currently available therapies and discuss their anticipated rates of success and complications.

10-1

Conservative Management of Urinary Retention in Women

Tara L. Frenkl and Firouz Daneshgari

Prevalence and Etiology

The exact prevalence of urinary retention (UR) in women remains unknown, because many of the cases of acute retention are not reported. However, the prevalence of impaired bladder emptying as a cause of chronic UR in women who were referred to a urodynamic clinic was reported to range from 13% to 26%.[1,2]

In general, women with UR can present in acute or chronic status. The most common causes of acute UR are postsurgical, either after a pelvic (anti-incontinence surgery, hysterectomy, pelvic organ prolapse repair), abdominal (anterior pelvic resection, colectomy) or a non-pelvic organ–related surgery. The latter causes are usually attributed to the effects of anesthesia on the bladder function.

Causes of chronic UR can be broadly classified into five categories: neurologic, anatomic, inflammatory, functional, and idiopathic. Table 10-1.1 provides a list of these causes. Details are discussed in another chapter.

Evaluation

Our evaluation of patients who present with UR begins with a thorough history, physical examination, and urinalysis. At each visit, patients are asked to respond to two self-administered questionnaires, a urogenital distress inventory-short form (UDI-6) and an incontinence impact questionnaire-short form (IIQ-7), so that subjective improvement or progression with treatment can be monitored. Salient features of the history and physical are listed in Table 10-1.2.

A complete history and physical examination will reveal the nature (acute vs chronic) and possible cause (neurogenic, anatomic, postsurgical, functional, inflammatory, idiopathic) of the UR. If the clinical scenario warrants further assessment of the bladder function, urodynamic studies (UDS) including cystometrogram, pressure-flow studies, and electromyography are performed. Electromyography is strongly recommended in suspected cases of neurogenic, detrusor sphincter dyssynergia or Fowler's syndrome.[3]

It is important to remember that the characteristics of neurogenic bladders, as seen in patients with multiple sclerosis and spinal cord injury, can change with time and disease progression. Therefore, annual reevaluation with urodynamics and assessment of the upper urinary tracts is recommended.

Interpretation of the competent bladder requires a careful analysis of both the filling and voiding phases of the bladder function. Evidence of impaired bladder function in UDS may include decreased sensation (delayed first sensation) or increased bladder capacity during the filling phase of UDS, and a low or absent detrusor voiding pressure (P_{det}) with prolonged voiding time and slow flow rate. Many patients may demonstrate nonsynchronized or intermittent voiding pattern. In many of these cases, voiding is accentuated by abdominal straining.

In some patients with UR, UDS will show evidence of bladder outlet obstruction (BOO). Discussion of BOO in women requires a separate discussion independent of UR. In summary, however, detrusor pressure at maximum flow (P_{det}/Q_{max}) of greater than 20 cm H_2O and maximum flow rate of less than 15 mL/s should be suspected for BOO.[4,5]

Traditionally, postvoid residual (PVR) is considered abnormal if greater than 50 mL remains after voiding.[6] Often heard is the question of what is an acceptable PVR after pelvic or anti-incontinence surgery. Kleeman et al.[7] attempted to determine postoperative PVR that predicts a voiding efficiency after operation for incontinence and/or pelvic organ prolapse. Before removing the catheter, the bladder was filled to 300 mL or to subjective fullness. A catheter was left in the bladder if voided volume was less than 50% of the instilled amount. Repeat voiding trials were conducted on a weekly basis until the criteria were met. They concluded that patients with PVR less than 32% of bladder capacity never required a catheter in their postoperative course. Although this study had several limitations, it coincides clinically with the arbitrary value

Table 10-1.1. Etiologies of urinary retention in women

Neurologic	Inflammatory	Mechanical	Functional and Other
Multiple sclerosis	Herpes zoster	Pelvic organ prolapse	Functional bladder outlet obstruction
Spina bifida occulta	Herpes simplex II	Urethral stricture	Pseudomyotonia (Fowler's syndrome)
Lumbosacral disc disease	Cytomegalovirus	Urethral diverticulum	Psychogenic
Diabetes mellitus	Epstein-Barr virus	Prior Anti-incontinence	Pharmacologic
Reflex sympathetic dystrophy	Human immunodeficiency virus	Surgery	Fecal impaction
Hypothyroidism	Lyme's disease	Foreign body, tumor, or calculus	Bladder overdistension
Radical pelvic surgery	Acute cystitis	Pelvic/uterine mass Ectopic ureterocele	

Table 10-1.2. Salient features of the initial evaluation

History	Physical
Sequential progression of urinary symptoms	Back – skin or spinal cord anomalies
UDI-6 and IIQ-7	Abdomen – scars, mass, tenderness
Neurologic symptoms (vision, gait, coordination, paresthesia, etc.)	Pelvis and vagina – pelvic organ prolapse, benign or malignant uterus, or vaginal mass
Sensation of pelvic pressure or heaviness	Urethra – mass, tenderness
Bowel habits/constipation	Anus and rectum
Sexual history	Neurologic examination
Prior urologic disease	PVR (bladder scan or catheter)
Urogenital trauma	
Diabetes	
Thyroid disease	
Herpes	
Back pain or disc disease	
Pelvic, anorectal, or spinal surgery	
Medications	

of 100 mL, which many urologists use as an acceptable postoperative residual.

Cystourethroscopy may yield information helpful in making a diagnosis. Anatomic lesions such as urethral stricture, bladder neck fibrosis, trabeculation, and bladder lesions have been found in women with BOO.[5] Baseline upper tract imaging is performed in patients with neurologic disease, or if indicated by physical or baseline studies or a patient's history.

Management

Management of Acute Urinary Retention in Women

A majority of cases of acute UR in women could be alleviated by reversing the recognized cause (such as BOO after an anti-incontinence surgery) and thereby the UR. Discussion of these management options are presented in other chapters of this book. In cases in which permanent neurologic damage is suspected (such as after a colectomy or abdominoperineal resection), UR should be considered a chronic condition. In a small portion of cases, the neurologic damage could be of transient nature, and watchful waiting with conservative management would result in the return of bladder function. In such cases, a minimum of 3 months of conservative management is recommended. During this time, any of the noninvasive options discussed below can be used.

Management of Chronic Urinary Retention in Women

Chronic UR is defined as a condition in which short-term return of the bladder function is not anticipated. In such cases, a management plan that is most suitable to the patient's lifestyle, physical abilities, and psychological approach should be selected. Any combination of the following options could have a role in the management of chronic UR in women.

Medications

Bethanechol is a synthetic parasympathetic agonist designed to selectively act upon postganglionic muscarinic cells to elicit smooth muscle contraction. In vivo and in vitro studies have demonstrated pharmacologic activity on the detrusor muscles and bladder outlet; however, its clinical effectiveness in facilitating bladder emptying has not been demonstrated.[8]

Benzodiazepines have also been advocated in the past as effective in preventing postoperative UR. A recent small randomized placebo-controlled trial revealed no reduction in the incidence of retention after ambulatory gynecologic procedures.[9]

Catheterization

Intermittent Self-catheterization

The acute management of any type of UR requires bladder drainage. This can be accomplished by insertion of an indwelling urethral catheter, a suprapubic catheter, or intermittent self-catheterization (ISC). The primary goals of management are to empty the bladder and prevent renal deterioration. The ideal method would mimic the physiologic cycle, provide discretion and convenience for the patient, and have no risk of infection or injury to the urinary tract. Intermittent self-catheterization is able to fulfill many of these requirements.

There are numerous factors to consider when deciding on the best method of catheterization for a patient. Intermittent self-catheterization may not be possible for patients who lack the manual dexterity or mobility to reach the urethra by themselves. Often, the primary caregiver can perform the procedure; however, it places a large burden of responsibility on the caregiver, and forces complete dependence of the patient on the caregiver for bladder emptying.

At first, patients are often reluctant to accept the concept of self-catheterization. With teaching and reassurance, most patients quickly master the technique and enjoy the freedom from the indwelling catheter. Intermittent self-catheterization does not require sterile technique or special cleansing products and can be performed over a private or public toilet. A short female straight catheter and packet of lubrication can be kept conveniently in a purse or pocket. We instruct patients to perform catheterization as frequently as needed to achieve volumes between 300 to 400 mL, mimicking the normal physiologic bladder cycle and reducing the risk of damage to the upper tracts from overdistension of the bladder.

Urethral and Suprapubic Catheters

Indwelling urethral catheters are generally well tolerated for short intervals of time. However, because of chronic downward pressure of the catheter, urethral erosion eventually occurs. As the urethra widens, the patient either develops leakage around the catheter or is unable to retain the catheter balloon. Other complications of urethral catheters include pain, bladder spasms, and limitation of daily activities. Complications of chronic indwelling catheters may include the following:

- Infection
- Urethral erosion
- Bladder spasm
- Urine leakage
- Pain
- Bladder stones
- Urethral stricture
- Decreased quality of life

A suprapubic catheter offers several advantages over the urethral catheter. A suprapubic catheter is usually easier to care for and easier to change. As new treatments are tried, voiding trials with assessment of PVRs can be performed easily by capping the catheter. In addition, less pain and discomfort are reported with suprapubic catheters than with urethral catheters. Sexual relations may also be easier with suprapubic catheters rather than a urethral catheter.

We recommend yearly cystoscopy in patients with any indwelling catheter and patients performing ISC, because squamous cell carcinoma has been associated with catheterization and recurrent urinary tract infections.[10–12] Although the data regarding this matter remain incomplete, the low morbidity and efficiency of flexible cystoscopy contrasted with the aggressiveness and high mortality of squamous cell carcinoma of the bladder persuade us to perform annual cystoscopy in these patients.

Botulinum A

Botulinum A toxin inhibits the release of acetylcholine from the presynaptic neuromuscular junction that leads to muscle relaxation. The actions are temporary because axonal regeneration occurs in approximately 3 to 6 months. Botulinum A toxin has been used in patients with detrusor sphincter dyssynergia and refractory overactive bladder. Two studies have reported the use of botulinum A in patients with voiding dysfunction. Phelan et al.[13] performed a prospective evaluation of the injection of 80 to 100 U of botulinum A into the external urethral sphincter in 8 men and 13 women with voiding dysfunction secondary to neurogenic detrusor sphincter dyssynergia, pelvic floor spasticity, or an acontractile bladder. All patients except one were able to void spontaneously, and all but two were able to discontinue the use of catheterization.

Kuo[14] evaluated the effect of 50 U of botulinum A toxin injection in 20 patients (16 women) with voiding dysfunction. Fifteen patients had an areflexic bladder and five had dysfunctional voiding. After injection, 11 patients could void by abdominal straining without the use of a catheter and 7 patients urinated with less difficulty than they had previously. Voiding pressure, maximal urethral closing pressure, and PVR decreased in 90% of the patients. These patients showed improvement in symptoms scores and quality of life at 3 months.

One double-blind controlled study evaluated the efficacy of botulinum A toxin versus lidocaine in 13 patients with detrusor sphincter dyssynergia secondary to spinal cord disease.[15] One transperineal injection of botulinum A toxin in the external urethral sphincter was found to be superior to lidocaine injection as evidenced by decreased PVR and maximal urethral pressure – and increased patient satisfaction scores.

Larger randomized controlled trials are needed to determine long-term safety, efficacy, and durability of treatment. Potential side effects include exacerbation of autonomic dysreflexia, infection, urinary incontinence, and allergic reaction. None of these side effects have been reported in the literature to date. Long-term studies after repeated injections are necessary to determine safety, efficacy, and optimal dose in this clinical population.

Urinary Diversion

In a few cases in which none of the conservative or minimally invasive choices are available, a urinary diversion using a portion of bowel could be used for long-term management of the bladder drainage. The details of this option are discussed elsewhere in this book.

Functional Bladder Outlet Obstruction

Functional BOO is a condition, that deserves further attention. Terms such as female aseptic dysuria, female prostatitis, abacterial cystitis, and, most often, the "urethral syndrome" have been used to describe this condition in the literature.[16,17] Such variation in terms reflects both the common symptoms among these patients (recurrent episodes of urinary frequency, urgency, and dysuria without pelvic pain) and the lack of consensus over the etiology and pathophysiology of this condition. Some authors suggest that "the urethral syndrome is probably the most frequent reason for urological consultation among women."[15]

The pathophysiology of this syndrome has been addressed by several different theories including infection and inflammation of the paraurethral glands, spasticity of the urethral musculature, and increased periurethral scar formation. Animal studies have shown that partial outflow obstruction is first evidenced by increased frequency of voids and increased voiding detrusor pressure.[18] Over time, the detrusor muscle decompensates, and manifestations of the disease change. Bladder capacity, volume, and compliance increase as bladder contractility and voiding pressures decrease.[19] This can eventually lead to elevated PVRs, retention, and ultimately overflow incontinence.

In the past, treatments have included: a) urethral dilation,[20,21] b) urethrotomy,[22,23] c) combined use of antibiotics and muscle relaxant,[24] and d) cryosurgery[25] or diathermy.[26] Clearly, there remains a lack of an effective and generally accepted treatment modality for these patients.

Based on a review of the literature and our clinical observations, we have theorized that these patients have inflammation/infection of the urethra (urethritis) with a subsequent functional hyperactivity of the smooth muscle components of the urethra. Such spasm of the smooth muscle of the bladder outlet may manifest in symptoms of frequency, urgency, and frequent cystitis. The spasm may create a functional BOO that can be demonstrated in a pressure flow study by elevated detrusor pressure at maximum flow and decreased rate of urine flow.

We treat our patients with functional BOO with a combination of an alpha-adrenergic blocking agent and a prophylactic antibiotic before activities that may cause urethral irritation. This combined therapy may decrease the recurrence of infection and the tension of the urethral spasms – thereby breaking the vicious cycle between spasm and infection. This treatment is experimental, and, currently, a randomized placebo-controlled trial is underway at our institution to assess the efficacy of this treatment.

Summary

Women with UR can present in acute or chronic status. A careful history and physical examination will lead clinicians to the possible cause of the UR and, thereby, the appropriate management. In selected cases, use of UDS and cystourethroscopy will help to distinguish cases of UR with a bladder etiology from those of outlet/urethral obstruction.

A collection of conservative and surgical options are available for the management of UR in women. In the majority of the acute cases, a combination of medication, watchful waiting, and ISC will allow the recovery of the bladder function. In a case of chronic retention, the patient's medical, physical, social, and psychological status should be considered when selecting from the available management options that range from minimally invasive choices such as ISC and indwelling catheters to more invasive choices such as urinary diversions or sacral neuromodulation implantations. In all instances, with the variety of available choices, bladder drainage can be achieved successfully with minimal risk to the patient's overall health.

References

1. Dwyer PL, Desmedt E. Impaired bladder emptying in women. Aust N Z J Obstet Gynaecol 1994;34:73–78.
2. Groutz A, Gordon D, Lessing JB, et al. Prevalence and characteristics of voiding difficulties in women: are subjective symptoms substantiated by objective urodynamic data? Urology 1999;54:268–272.
3. Fowler CJ, Christmas TJ, Chapple CR, et al. Abnormal electromyographic activity of the urethral sphincter, voiding dysfunction, and polycystic ovaries: a new syndrome? BMJ 1988;297:1436.
4. Chassagne S, Bernier PA, Haab F, Roehrborn CG, Reisch JS, Zimmern P. Proposed cutoff values to define bladder outlet obstruction in women. Urology 1998;51(3):408–411.
5. Goutz A, Blaivas JG, Chaikin DC. Bladder outlet obstruction in women: definition and characteristics. Neurourol Urodyn 2000;19: 213–220.
6. Hughes PN, Abrahms P. Cystometry. In: Cardozo L, Staskin D, eds. Textbook of Female Urology and Urogynaecology. 1st ed. London: ISIS Medical Media; 2001:203.
7. Kleeman S, Goldwasses S, Vassallo B, Karram M. Predicting postoperative voiding efficiency after operation for incontinence and prolapse. Am J Obstet Gynecol 2002;187(1):49–52.
8. Finkbeiner AE. Is bethanechol chloride clinically effective in promoting bladder emptying? A literature review. J Urol 1985;134: 443–449.
9. Hershberger JM, Milad MP. A randomized clinical trial of lorazepam for the reduction of postoperative urinary retention. Obstet Gynecol 2003;102(2):311–316.
10. Denlay KM, Stonehill WH, Goldman H, Jukkola A, Domochowski RR. Bladder histological changes associated with chronic indwelling urinary catheter. J Urol 1999;161:1106–1109.

11. Pattison S, Choong C, Corbishley CM, Bailey MJ. Squamous cell carcinoma of the bladder, intermittent self-catheterization, and urinary tract infection – is there an association? BJU Int 2001;(88):441.
12. Zaidi SZ, Thaeker JM, Smart CJ. Squamous cell carcinoma in a patient on clean intermittent self-catheterization. Br J Urol 1997;80: 352–353.
13. Phelan MW, Franks M, Somogyi GT, et al. Botulinum toxin urethral sphincter injection to restore bladder emptying in men and women with voiding dysfunction. J Urol 2001;165:1107–1110.
14. Kuo HC. Effect of botulinum A toxin in the treatment of voiding dysfunction due to detrusor underactivity. Urology 2003;61(3):550–554.
15. de Seze M, Petit H, Gallien P, et al. Botulinum A toxin and detrusor sphincter dyssynergia: a double blind lidocaine-controlled study in 13 patients with spinal cord disease. Eur Urol 2002;42:56–62.
16. Bashi SA. The urethral syndrome. Int Urol Nephrol 1988;20(4): 367–374.
17. Susset J. Female urethral syndrome. Ann Urol (Paris) 1993;27(6–7): 329–330.
18. Mostwin JL, Karim OMA, Van Koeveringe G. The guinea pig model of gradual urethral obstruction. J Urol 1991;145:859–863.
19. Saito M, Ohmura M, Kondo A. Effects of long term partial outflow obstruction on bladder function in the rat. Neurourol Urodyn 1996;15:157–165.
20. Bergman A, Karram M, Bhatia NN. Urethral syndrome. A comparison of different treatment modalities. J Reprod Med 1989;34(2): 157–160.
21. Lemack GE, Foster B, Zimmern PE. Urethral dilation in women: a questionnaire-based analysis of practice patterns. Urology 1999; 54(1):37–43.
22. Choa RG, Abrams PH, Pynsent PB, Ashken MH. A controlled trial of Otis urethrotomy. Br J Urol 1983;55(6):694–697.
23. Hart RD Jr, Murphy BJ. Female urethral syndrome: treatment by internal urethrotomy. J Am Osteopath Assoc 1983;82(8):609–610.
24. Schmidt RA. The urethral syndrome. Urol Clin North Am 1985;12(2): 349–354.
25. Boreham P. Cryosurgery for the urethral syndrome. J R Soc Med 1984;77(2):111–113.
26. Taylor JS. Diathermy to the trigone and urethra in the management of the female urethral syndrome. Br J Urol 1977;49(5):407–409.

10-2

Surgical Management of Urinary Retention

Raymond R. Rackley and Tara L. Frenkl

Clean intermittent catheterization (CIC) is one of the major advances in the history of urologic practice. Developed in 1972 by Jack Lapides and colleagues,[1] this nonsterile technique has been adopted worldwide and few advances in urology have had the same impact on clinical practice and patient management as CIC for bladder evacuation or emptying dysfunction. It has implications for overactive bladder caused by elevated postvoid residuals (PVR), bladder outlet obstruction resulting in high voiding pressures or various degrees of urinary retention, as well as infectious urinary tract conditions because it actually results in improved urologic care and fewer complications than previously considered management options. Clean intermittent catheterization has improved the quality of life for countless people with transient or permanent voiding dysfunction who might otherwise have been treated with chronic indwelling catheters or surgery for urinary diversion such as an ileal conduit. This technique has allowed for the evolution of numerous surgical techniques for pelvic organ reconstruction such as anti-incontinence procedures, as well as orthotopic and nonorthotopic urinary diversions.

Clean intermittent catheterization is indicated if it represents the best option for urinary drainage in a patient with bladder-emptying dysfunction. By far, the most common indication for CIC is chronic bladder dysfunction as a result of non-neurogenic, as well as neurogenic conditions. Medical management of severe neurogenic and non-neurogenic detrusor instability can convert a bladder-storage problem into a bladder-emptying problem that, when managed successfully with the potential use of CIC, provides for a condition of continence. Likewise, people with overflow incontinence from hypocontractile conditions and chronic retention can avoid indwelling catheters with CIC. In patients with retention secondary to outlet obstruction from conditions such as pelvic organ prolapse, CIC is helpful when surgery must be delayed or avoided because of medical conditions. The entire field of reconstructive urology and continent urinary reservoir creation depends on CIC to ensure urinary drainage and is the reason for including this concept at the beginning of this chapter.

Urethral Dilatation, Incision, or Reconstruction

True urethral strictures in females are rare, but can occur as a result of chronic urethral infections, trauma from iatrogenic urethral dilations, and postreconstructive urethral surgery. In these cases, the narrowing of the urethra is easily recognized endoscopically. Although urethral dilations have been used in the past, idiopathic and postsurgical strictures can be treated with periodic self-catheterization or permanent CIC, transurethral incision, or urethral reconstruction as reported in males with similar conditions. As reported in the management of male urethral strictures, CIC can be used as a modality to self-obturate strictures and prevent stricture recurrences in females. Adjuvant topical estrogen replacement may complement this manipulative therapy.

Urethral dilatation has been used for many years as a treatment for recurrent cystitis, pelvic or bladder and urethral pain, and for nonspecific voiding dysfunction including bladder emptying dysfunction in women. Although frequently used to treat urethral syndrome or primary bladder neck obstruction in the female, this procedure often results in only temporary symptomatic relief and over time it may result in bladder outlet obstruction secondary to transmural urethral stenosis or stricture formation. The obstruction is caused by periurethral fibrosis and scarring of the urethral wall that results from multiple episodes of postdilatation bleeding or extravasation of urine into the periurethral tissues. This may result in rigidity of the urethral wall and narrowing of the urethral lumen.

Management of the postdilatation obstruction and urethral stenosis in females is challenging in that the goal of providing low-resistance bladder emptying must be

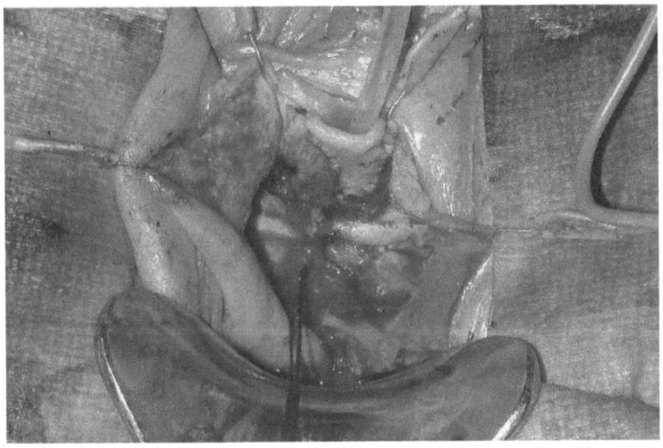

Figure 10-2.1. Buccal mucosa graft for urethral reconstruction.

weighed against the complication of developing urinary incontinence. Curative therapy ranges from transurethral incision of the urethra to various forms of urethral reconstruction depending on the degree (mucosal vs transmural) and the area involved. For urethral reconstructions without bladder neck involvement, there is usually sufficient tissue of the anterior vaginal wall to use as pedicle flaps for reconstruction. If the vaginal tissue is extensively scarred, ischemic, or atrophied, other potential donor sites should be considered. These may include labial and perineal pedicle flaps, and rarely the use of rectus and gracilis pedicle flaps. We prefer to use buccal mucosa grafts (Figure 10-2.1) for urethral reconstruction in women when the anterior vaginal wall tissue is not applicable for reconstruction or the urethral defect is too large. This obviates the more disfiguring and morbid complications of muscle flaps or anterior bladder flap repairs (Barnes' bladder flap urethroplasty) obtained through an abdominal incision.

Bladder Neck Incisions Versus Reconstructions

Surgical treatment of primary and secondary bladder neck obstruction after medical and conservative therapies such as CIC have been attempted, consists of transurethral bladder neck incision or bladder neck reconstructions. Whereas reported series of abdominal approaches for bladder neck reconstructions using anterior and posterior bladder flaps in women are restricted to small numbers from a few institutions because of the complexity of the technique, we prefer to use a transvaginal approach for performing the Tanagho anterior bladder neck reconstruction technique as described by Elkins and colleagues in 1992.[2–3] This unique approach provides simultaneous exposure of both the bladder neck and orthotopic urethral reconstructions using a Martius fat pad and a fascial sling placement in an effort to reduce the morbidity of the procedure for the patient and to reduce the complexity of the procedure for the reconstructive specialist (Figure 10-2.2). Our added modification to this technique has been the use of the incision for obtaining the rectus fascia sling harvest as a means for providing finger access into the retropubic space for the anterior bladder neck mobilization required for the transvaginal Tanagho bladder neck reconstruction.

We generally reserve bladder neck reconstructions for cases in which the bladder neck is obliterated or nonfunctional and prefer to perform transurethral bladder neck incisions as a first-line procedure for functional bladder neck conditions based on the simplicity and efficaciousness of this minimally invasive endoscopic approach. Most specialists perform two incisions at a depth of 0.5 to 1.0 cm at the 5 and 7 o'clock position of the bladder neck, beginning midway between the urethral orifice and the bladder neck, and ending 1 to 2 cm distally. To avoid potential incontinence, one may elect to stage the intervention by performing only one incision at a time and elect to make a second or more extensive incision at a different session depending on the results achieved. Using a staged approach for transurethral bladder neck incisions, Gronbaek et al.[4] reported a success rate of 76% at a mean follow-up of 55 months with only one case (3%) of developing interval incontinence.

As described above, the improvements in endoscopic technology and application of refinements in performing

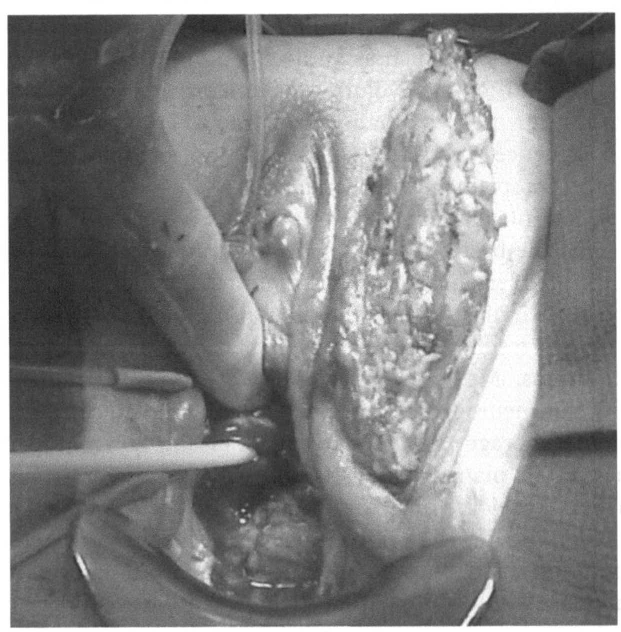

Figure 10-2.2. Transvaginal view of a combined transvaginal bladder neck and urethral reconstruction using placement of a circumferential Martius fat pad around the neourethra before securing the fascial sling.

bladder neck incisions, transurethral incisions for mucosal strictures, or urethral and bladder neck reconstructions have largely replaced the anecdotal reports and use of traumatic urethral sphincterotomy in women with the attendant risk of urethral bleeding, stricture, or fistula formation. Although the use and indications for urethral stents have generally been restricted to males, the development of a removable, shorter urethral stent for females may have an expanding role in cases of bladder outlet obstruction in the near future.

Neuromodulation

Botulinum Toxin Injections

Urethral sphincter botulinum toxin injections may be considered a treatment option for idiopathic and complex bladder outlet dysfunction in which fibrosis, stricture formation, or other anatomic distortions are not the cause of urinary retention. Although first reported in complex cases or neurogenic cases of detrusor sphincter dyssynergia, the encouraging clinical outcomes obtained with this safe, yet reversible treatment have promoted its use in people with various forms of bladder outlet obstruction and in people with an acontractile bladder in which a decrease in outlet resistance may lead to better bladder emptying by Valsalva maneuvers.

Botulinum toxin is a reversible inhibitor of acetylcholine release at the presynaptic neuromuscular junction and has proven clinical utility and safety in the treatment of striated, as well as smooth muscle spasticity disorders. The major disadvantage of using this neurotoxin as a neuromodulator of muscle function in lower urinary tract dysfunction is the repeated need for injections to maintain its efficacy. The chemical denervation produces muscle atrophy that results in a reversible process in 3 to 6 months as the primary axon regenerates the affected transport protein(s) involved with presynaptic acetylcholine exocytosis and secondary axonal sprouting occurs.

Since 1988, numerous reports exist on the expanding indications and variations in clinical outcomes, as well as methodologies and toxins used for performing urethral sphincteric botulinum toxin injections. Dosing ranges from 50 to 100 U of botulinum A toxin (Botox; Allergan Inc., Irvine, CA) and approximately 250 U of Dysport (Speywood, Portons Downs, UK) placed through a transurethral or transperineal approach have been used with or without electromyogram localization of the external sphincter. The toxin volumes injected and the location(s) within the sphincter have all varied in description, but most physicians report one to three injection sessions per case with improvement (clinical, myographic, or urodynamic outcomes) noted for 3 to 9 months (longer in non-neurogenic conditions) depending on the patient's underlying condition, initial toxin concentration used, and the reinjection schedule followed. Although no major complications of this therapy in women have been reported, the development of transient stress incontinence has been noted in some women with non-neurogenic conditions undergoing this therapy.

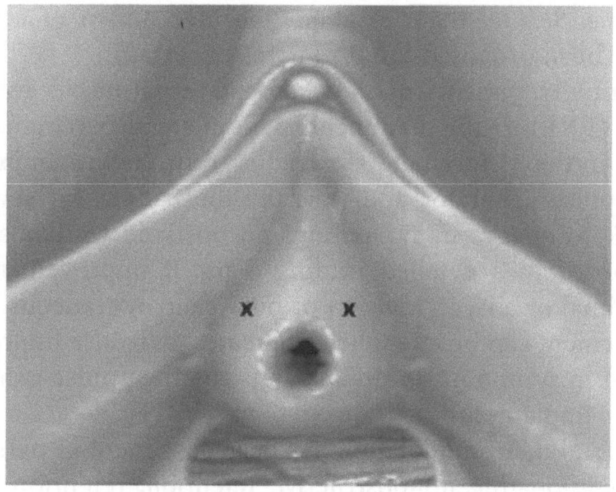

Figure 10-2.3. Periurethral access to the external sphincteric complex for botulinum toxin injections. (Reprinted with the permission of The Cleveland Clinic Foundation.)

Although favorable double-blinded and randomized trials of botulinum toxins used in males with neurogenic bladder dysfunction exist, botulinum toxin injection results in women with primary and secondary causes of urinary retention have been varied and limited to small series and anecdotal reports. In 1992, Fowler et al.[5] described the unsuccessful use of botulinum A toxin injections in six women with urinary retention secondary to abnormal activity in the striated muscle of the urethral sphincter. However, Chancellor's group in 2001[6] reported on a group 21 patients (8 men, 4 women with non-neurogenic and 9 women with neurogenic sphincteric issues) and noted an overall 67% significant subjective improvement in voiding. In a prospective ongoing series at our institution, 8 of 10 female patients with neurogenic bladder sphincter dyssynergia receiving 100 U of Botox injected periurethrally (Figure 10-2.3) into the external sphincter 3 times between 4-week intervals have had a cure or improvement in their validated outcomes at their 3- and 6-month follow-up evaluations. Novel use of botulinum toxin for urinary retention of various neuromuscular etiologies in women suggests that this reversible therapy may have a significant role in the treatment of bladder evacuation disorders.

Sacral Root Neurostimulation

Sacral root neurostimulation offers an alternative mode of treatment for patients with nonobstructive urinary retention. The device was approved by the United States Food and Drug Administration in 1999 for management of chronic urinary retention and frequency/urgency syndrome. Several theories about the mechanism of action have been proposed but it remains uncertain.

Initially, patients undergo percutaneous nerve evaluation (PNE) by the placement of a unilateral percutaneous lead in the S3 foramen. The lead is connected to an external pulse generator and worn by the patient for several weeks. Changes in symptoms and postvoid residuals are recorded in a detailed voiding diary. If improvement is minimal or absent, revision or bilateral percutaneous lead placement can be attempted. If greater than 50% improvement in symptoms is attained, a permanent pulse generator is implanted.

Sacral neuromodulation has been successful in patients with idiopathic nonobstructive retention, retention secondary to deafferentation of the bladder after hysterectomy, and in patients with Fowler's syndrome. Patient factors predicative of success have been sought. In 2003, Bross et al.[7] evaluated the predicative ability of the carbachol test and concomitant diseases in patients with an acontractile bladder. Eighteen patients were subcutaneously injected with 0.25 mg of carbachol. An increase in detrusor pressure to more than 20 cm within 20 to 30 minutes after injection was considered a positive test; however, only 33% of patients had a successful bilateral PNE, but a positive carbachol test was not predictive of success. Percutaneous nerve evaluation seemed particularly effective in patients who developed retention after hysterectomy.

A large prospective randomized multicenter trial to evaluate the efficacy of sacral nerve stimulation for urinary retention was performed by Jonas and colleagues[8] in 2001. After a PNE period of 3 to 7 days, 68 patients (38% of those evaluated) with chronic urinary retention qualified for permanent implantation. Patients were randomly assigned to the treatment or control group in which treatment was delayed for 6 months. Successful results were initially achieved in 83% of patients who received the implant with 69% able to discontinue intermittent catheterization completely. At 18 months, 71% of patients available for follow-up had sustained improvement.

In 2002, Aboseif[9] and colleagues evaluated the efficacy and change in quality of life in patients with idiopathic, chronic nonobstructive functional urinary retention. Thirty-two patients with idiopathic retention requiring intermittent catheterization underwent PNE. Permanent implants were placed in 20 patients (17 women) who showed greater than 50% improvement in symptoms. Eighteen patients were subsequently able to void and no longer required intermittent catheterization with one patient requiring bilateral implants. Average voided volumes increased from 48 to 198 mL and the PVR decreased from 315 to 60 mL. Eighteen patients reported a greater than 50% improvement in quality of life, although the questionnaire used in the study was not described. Significant score improvements in the Beck depression inventory and SF-36 after sacral root neuromodulation for retention have been demonstrated by Shaker and Hassouna in 1998.[10]

Overall success rates of the PNE range from 33.3% to 100%. Improvement in patients with retention may not be as rapid as in patients undergoing sacral root stimulation for other reasons. A PNE period of at least 2 to 3 weeks has been recommended. Reported complications of the procedure include pain at the implant or lead site, wound infection, seroma at the implantable pulse generator site, transient sensation of electrical shock, and skin erosion at the site of the implanted pulse generator as well as hardware problems such as insulation defects and lead fracture.

Urinary Diversion

In the last three decades, CIC proposed by Lapides has been accepted in the urologic community as an excellent method of management of the bladder that fails to empty. However, this option may not be applicable for some people, such as those without manual dexterity and limited assistance from caretakers. In these cases, options may include the use of chronic indwelling urethral or suprapubic catheter or urinary diversion. Chronic catheter placement is frequently associated with complications including stone formation, tissue erosion, frequent infection, and malignancy; therefore, it is uncommonly used as the first-line management. As discussed in a previous chapter, urinary diversion, which traditionally takes the form of ileal conduit involving ureteral reimplantation may lessen the risk for chronic catheter-associated problems. However, a different set of problems such as urinary reflux and ureteral obstruction may arise that are best avoided by performing urinary diversions that incorporate the native bladder and preserve the bladder trigone for drainage of the upper urinary tract.

Continent Catheterizable Subcutaneous Channel

For many people with urinary evacuation disorders in non-neurogenic and neurogenic conditions, the bladder capacity is often normal or increased and the major obstacle for improving the impact of their disorder is obtaining access to the bladder. In these complex cases, we have opted to perform a minimally invasive procedure, which we describe as a continent subcutaneous catheterizable diversion (Figure 10-2.4). This novel procedure uses a subcutaneous catheterizable umbilical channel that is attached to the bladder dome. The continence mechanism of the subcutaneous channel is provided by circumferential compression of the rectus muscle after we have crossed the medial fibers of its lower abdominal portion overlying the bladder dome. The channel is constructed of an in situ or free skin graft. In this way, we create a percutaneous subcutaneous tunnel that passes through augmented com-

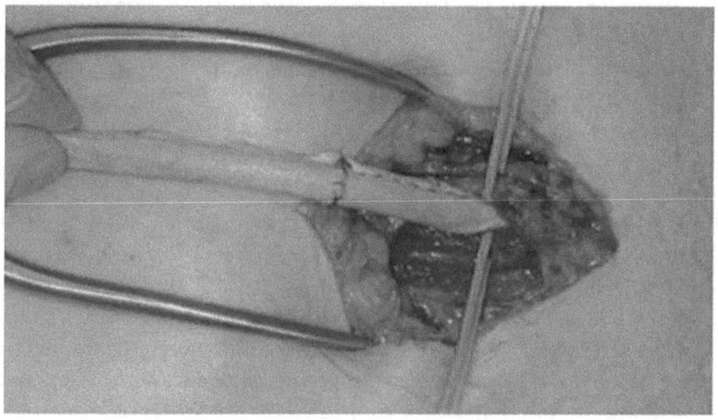

Figure 10-2.4. Formation of a continent catheterizable subcutaneous channel using a free dermal graft from the umbilicus to the bladder dome. Compression obtained from crossing the fibers of the rectus muscle serves to form the continence mechanism.

pression of the rectus muscle and is attached to the bladder dome for continent bladder access using CIC. This procedure is technically comparable to placement of a suprapubic tube and is performed in a same-day ambulatory surgical setting. Two weeks after the minimally invasive procedure, CIC is performed through the subcutaneous channel.

Prosthetic Access Device: Use of a Gastrostomy Button for the Bladder

An alternative to formation of a continent catheterizable subcutaneous channel is placement of a prosthetic device on the lower abdominal wall that allows CIC such as a gastrostomy button (CR Bard, Covington, GA). Bennett et al.[11] in 2003 described the outcomes of using such a device in 20 cases of evacuation disorders of the bladder that is placed through a matured suprapubic tube tract. Similar to catheter intubations for gastric feeding, this device serves as a continent abdominal portal for patients to perform CIC. Although symptomatic infections and encrustations of the portal device were low, the failure rate in this series was attributed to technical issues of skin erosions surrounding the device that were caused by the short length of the prosthetic in relationship to abdominal wall thickness for 60% of patients. Patients with less than 4 cm of abdominal wall thickness do well with this device, and development of a device specifically for the bladder (i.e., variation in device length) will expand the use and adoption of this minimally invasive technique.

Nonorthotopic Urinary Diversions

Incontinent ileovesicostomy providing low-pressure urinary storage and drainage was first introduced in 1994 to address the problems associated with conventional incontinent urinary diversions. Without ureteral mobilization or reimplantation, it decreases the operative time and avoids the risk for ureteral complications. Preserving the ureterovesical junction, it maintains the antireflux mechanism and lessens the risk for pyelonephritis. Furthermore, it does not require the use of any catheter or foreign material, either intermittently or chronically. These advantages, together with excellent renal function preservation and the low rate of complications, have been confirmed by several recent reports with long-term follow-up, and ileovesicostomy has been recommended as a better alternative to all other types of incontinent urinary diversion.

All ileovesicostomy procedures described to date have been performed in the conventional, open manner. The major surgical components of ileovesicostomy included the following: (a) bladder mobilization with cystotomy creation, (b) harvesting of a well-vascularized bowel segment, (c) establishment of bowel-to-bowel anastomosis with closure of mesenteric window, and (d) performance of full-thickness, mucosa-to-mucosa ileovesical anastomosis in a tension-free, watertight manner. The laparoscopic technique we have been using for this procedure has replaced the traditional open approach to accomplish these steps efficaciously [Figure 10-2.5(a)].

Augmentation cystoplasty with the right colon using a reinforced ileocecal valve catheterizable continence mechanism is another form of urinary diversion preferred for preservation of the bladder and upper urinary tract; a continent catheterizable access port to the bladder is needed as well [Figure 10-2.5(b)]. One hundred years have passed since the fundamental open technique for this procedure has become an established reconstructive technique, which we have been performing laparoscopically.[12]

We recently reported on a comparative study of nine cases of open versus nine cases of laparoscopic approaches

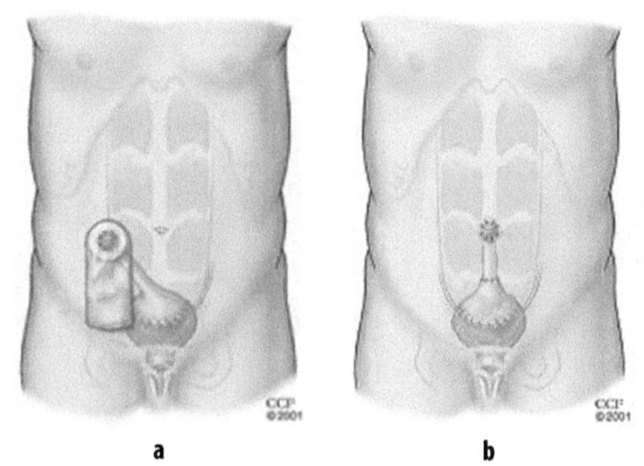

Figure 10-2.5. a, Laparoscopic ileovesicostomy formation. b, Laparoscopic augmentation cystoplasty with the right colon using a reinforced ileocecal valve continence mechanism that is catheterized through the umbilicus. (Reprinted with the permission of The Cleveland Clinic Foundation.)

to this complex reconstructive procedure. The mean operative time for the open approach was 278 versus 468 minutes for our initial cases using a laparoscopic approach ($P < .001$). There was no significant difference regarding the blood loss between the two approaches. Mean time necessary to meet discharge criteria was significantly shorter after the laparoscopic approach (2.8 days), than after the open approach (7.1 days). Mean hospital stay was significantly shorter with the laparoscopic approach (4.4 days) than with the open approach (8.2 days) ($P < 0.001$). There was significant improvement in validated outcomes of bladder function with no adverse effect on bowel function after both procedures.

For people with complex voiding dysfunction who need bladder drainage or access procedures as described above for evacuation disorders, using a laparoscopic approach for these traditional open procedures makes the selection of these options more of an attractive consideration. A clinically significant positive impact on their postoperative quality of life related to their bladder function compared with their preoperative status can be achieved using a laparoscopic approach. Furthermore, this benefit in their quality of life from improvement of their bladder function can be achieved without a negative impact on their bowel control.

References

1. Lapides J, Diokno AC, Silber SJ. Clean intermittent self-catheterization in the treatment of urinary tract disease. J Urol 1972; 107:458–461.
2. Tanagho EA. Bladder neck reconstruction for total urinary incontinence: 10 years of experience. J Urol 1981;125:321.
3. Elkins TE, Ghosh TS, Tagoe GA, Stocker R. Transvaginal mobilization and utilization of the anterior bladder wall to repair vesicovaginal fistulas involving the urethra. Obstet Gynecol 1992 Mar;79(3):455–460.
4. Gronbaek K, Struckman JR, Frimodt-Moller C. The treatment of female bladder neck dysfunction. Scand J Urol Nephrol 1992;26(2): 113–118.
5. Fowler CJ, Betts CD, Christmas TJ, Swash M, Fowler CG. Botulinum toxin in the treatment of chronic urinary retention in women. Br J Urol 1992 Oct;70(4):387–389.
6. Phelan MW, Franks M, Somogyi GT, Yokoyama T, Fraser MO, Lavelle JP, Yoshimura N, Chancellor MB. Botulinum toxin urethral sphincter injection to restore bladder emptying in men and women with voiding dysfunction. J Urol 2001 Apr;165(4):1107–1110.
7. Bross S, Braun PM, Weiss J, Martinez Portillo FJ, Knoll T, Seif C, Juenemann KP, Alken P. The role of the carbachol test and concomitant diseases in patients with nonobstructive urinary retention undergoing sacral neuromodulation. World J Urol 2003 May;20(6):346–349. Epub 2003 Jan 28.
8. Jonas U, Fowler CJ, Chancellor MB, et al. Efficacy of sacral nerve stimulation for urinary retention: results 18 months after implantation. J Urol 2001;165:15–19.
9. Aboseif S, Tamaddon K, Chalfin S, Freedman S, Mourad MS, Chang JH, Kaptein JS. Sacral neuromodulation in functional urinary retention: an effective way to restore voiding. BJU Int 2002 Nov;90(7): 662–665.
10. Shaker HS, Hassouna M. Sacral root neuromodulation in idiopathic nonobstructive chronic urinary retention. J Urol 1998 May;159(5): 1476–1478.
11. Bennett SG, Bennett S, Bell TE. The gastrostomy button as a catheterizable urinary stoma: a pilot study. J Urol 2003 Sep;170(3): 832–834.
12. Rackley RR, Abdelmalak JA. Laparoscopic Augmentation. Urological Clinics of North America. Philadelphia: WB Saunders; 2001.

10-3

Bladder Outlet Obstruction after Anti-Incontinence Surgery/Urethrolysis

Gamal M. Ghoniem and Usama M. Khater

Surgical correction of female stress urinary incontinence (SUI) generally results in success rates between 80% to 95%, depending on the procedure used.[1] Bladder outlet obstruction (BOO) after surgical correction of SUI is not infrequent. Reported rates of urethral obstruction range from 5% to 20% after Marshall-Marchetti-Krantz procedure,[2,3] 5% to 7% after needle suspension and retropubic urethropexy,[4] and 4% to 10% after pubovaginal sling.[5] Obstruction after these procedures may result from excessive suture tension, periurethral suture placement, or excessive sling tension. Placing a supportive rather than obstructive sling is the best strategy for prevention of BOO.

With more slings being performed because of their long-term efficacy in all types of SUI, complications including obstruction are more prevalent. Failure to empty the bladder is not the only symptom that indicates obstruction. Physicians should be aware of this condition, to ensure that early recognition and treatment can be offered to these frustrated patients.

Medical History

The type of anti-incontinence procedure, material used, and the temporal relationship between the procedure and the development of lower urinary tract symptoms should be identified from the patient's history. Women with BOO present with a variety of urinary symptoms. As a reaction to the obstructive mechanism, the detrusor muscle can become overactive and unstable, resulting in frequency, urgency, and urge incontinence, or even progress to the phase of detrusor decompensation heralded by urinary retention and recurrent urinary tract infections or overflow incontinence.

We evaluated 15 patients with obstructive symptoms after sling procedures who were referred to the incontinence center at Cleveland Clinic Florida.[6] Fourteen of the patients (93%) presented with irritative symptoms of urgency and frequency. Twelve patients (80%) presented with urine retention (complete and incomplete) and increased postvoid residual volume. Eight patients (53%) had urge incontinence, and four (27%) presented with recurrent urinary tract infections. Pain (suprapubic or during micturition) or dysuria and poor urine stream were also reported, but at a lower rate than irritative symptoms and retention.

Voiding Diary

Urinary symptoms in women with suspected BOO can be assessed primarily through two methods: patient recall during history taking, and self-monitoring of voiding behavior using a diary. Although both methods have pitfalls in terms of providing reproducible information, the voiding diary has been shown to be more accurate for demonstration of a patient's symptoms.

Physical Examination

The physical examination of patients with suspected BOO should include a general abdominal examination, a neurologic examination, and a detailed combined perineal/vaginal examination. Abdominal examination may reveal a distended bladder. The neurologic examination should include testing of the bulbocavernosus reflex, anal tone, and perineal sensation to exclude any neurologic cause of voiding dysfunction.

During vaginal examination, periurethral scarring may become apparent through observation and palpation. A lighted speculum is very helpful in the vaginal examination, especially in the detection of signs of erosion. The entire urethra should be palpated carefully for any tender area or mass. In cases of excessive scarring, the urethra will be hypersuspended in a high retropubic position and cemented to the anterior vaginal wall.

Q-tip Test

The Q-tip test is usually negative with minimal or no urethral hypermobility in patients with BOO. However, with mid urethral slings, hypermobility may be maintained.

Cystourethroscopy

Cystourethroscopy should be included in the evaluation of patients who develop BOO after anti-incontinence procedures. Urethroscopy with a rigid scope and lens at 0 or 30 degrees may demonstrate a high shelf, scarring, narrowing, occlusion, kinking, deviation, or erosion of the urethra. In cases of voiding obstruction caused by needle suspension, urethroscopy may show sutures. Cystoscopy is also indicated to exclude other causes of irritative symptoms, such as bladder tumors or stones.

Urodynamic Investigations

Urodynamic investigations have an essential role in the evaluation and appropriate diagnosis of patients with suspected BOO.

Uroflowmetry

Uroflowmetry is a noninvasive study for patients with suspected obstruction. Prolonged voiding time and decreased maximum (Q_{max}) and average flow rates (Q_{ave}) may indicate obstruction. The caveat is that uroflow interpretation cannot distinguish between obstruction with a high pressure detrusor contractility or impaired detrusor contractility as the cause of weak stream. After uroflowmetry, postvoid residual volume measurement is necessary. Presence of significant postvoid residual volume (>100 mL) indicates either an impaired detrusor contraction or urinary outlet obstruction. In patients who have recently undergone surgical correction of SUI in absence of preoperative voiding difficulties, large postvoid residual volume may be suggestive of urinary outlet obstruction. Otherwise, pressure-flow studies can differentiate between the causes of elevated postvoid volumes.

Pressure-flow Studies

The simultaneous measurement of detrusor pressure and urine flow during voiding offers valuable objective evidence for diagnosis of BOO in women. This is based on the physical principle that obstruction exists in a fluid-transporting system if an elevated pressure is required to transport the usual rate of flow through a relative narrowing. Several groups have attempted to study obstruction in women on the basis of pressure-flow analysis. Massey and Abrams[7] defined the BOO at Q_{max} less than 12 mL/s, $P_{det}Q_{max}$ more than 50 cm H_2O, urethral resistance ($P_{det}Q_{max}/Q_{max}$) more than 0.2, and "significant" residual urine. Farrar et al.[8] used an arbitrary definition based on Q_{max} of less than 15 mL/s and $P_{det}Q_{max}$ of more than 50 cm H_2O. Chassagne et al.[9] reported that Q_{max} of 15 mL/s or less and $P_{det}Q_{max}$ more than 20 cm H_2O are reasonable pressure-flow parameters to define female BOO. Blaivas and Groutz[10] constructed a normogram for women with lower urinary tract symptoms. We suspect obstruction when Q_{max} is less than 12 mL/s and $P_{det}Q_{max}$ is more than 30 cm H_2O.

Cystometry

Cystometry may demonstrate involuntary detrusor contractions secondary to the obstruction as well as reduced bladder capacity. Patients with long-standing BOO may have low bladder compliance.

Urethral Pressure Profile

Urethral pressure profile is technically challenging and is not currently recommended by most authors for investigation of patients with BOO. Its role in evaluation of patients with SUI continues to be debated.

Fluorourodynamics

At Cleveland Clinic Florida, we use fluorourodynamics in the evaluation of all patients with suspected urinary obstruction after anti-incontinence surgery. In addition to providing urodynamic measurements, fluorourodynamics demonstrates the location and mobility of the urethra and indicates whether there are any associated bladder trabeculations, diverticula, or cystoceles.

Nitti et al.[11] proposed fluorourodynamic criteria for diagnosis of BOO in women. According to these criteria, obstruction is defined as radiographic evidence of obstruction in the presence of a sustained detrusor contraction of any magnitude.

Treatment

The treatment options for voiding obstruction after anti-incontinence surgery in women depend primarily on the duration and severity of the obstruction. After anti-incontinence surgical procedures, many patients experience temporary voiding dysfunction; therefore, it is reasonable to wait at least 2 to 3 months before surgical treatment.

During this period, intermittent catheterization can be used. However, early intervention may be required for patients with an obstructive synthetic sling, because readjustment of some slings is possible (e.g., loosening of the sling using a urethral sound).

After the 2- to 3-month "waiting" period, surgical intervention is indicated. There are a variety of surgical approaches for the treatment, including simple sling incision, sling incision and free vaginal wall flap, urethrolysis, and Martius flap interposition.

Urethrolysis

Transvaginal urethrolysis was first described in 1984 by Leach and Raz.[12] The goal of this procedure is to release the urethra from the surrounding obstructing fibrous tissue. It is performed through an anterior midline or inverted U-shaped incision, and dissection of the vaginal wall off of the underlying periurethral fascia. The submucosa is infiltrated with lidocaine and Marcaine before the incision to facilitate the dissection. The dissection proceeds bilaterally to the urethra and up to the pelvic fascia. When the endopelvic fascia is reached it is perforated. Dissection is performed until adequate urethral mobility is obtained. If dissection between the urethra and pubic bone is excessively difficult, a suprameatal incision may be made and dissection performed anterior to the urethra. In some cases it may desirable to interpose a free vaginal wall or Martius flap between the urethra and the pubic bone to prevent readherence of these structures.

Retropubic urethrolysis has been described by Webster and Kreder.[13] It is performed through a Pfannenstiel or low midline incision. It is less often used than a transvaginal approach because it is associated with higher morbidity and more difficult dissection.

Infrapubic urethrolysis has been described by Petrou et al.[14] It involves an inverted U incision made around and 1 cm away from the urethral meatus, between the 3 and 9 o'clock positions. Using sharp dissection, a plane is developed above the urethra to the bone of the symphysis pubis. Sharp and blunt dissection frees the urethra, vesical neck, and bladder from the pubic and pelvic attachments anteriorly and laterally. The retropubic space is reached with the index finger, and further mobilization can be accomplished. With this exposure, the lateral wings of the sling or suspension sutures can be cut. As with a traditional transvaginal urethrolysis, a Martius flap may be harvested and placed between the urethra and pubis. Equivalent success rates have been demonstrated with all of the approaches to urethrolysis.[15]

Sling Incision

The technique of sling incision was originally described by Ghoniem and Elgamasy,[16] with use of a midline sling incision and free vaginal interposition graft. Later, midline sling incisions without interposition were used by Amundsen et al.[17] and Nitti et al.[18] With this technique, a midline or inverted U incision, with the apex halfway between the bladder and urethral meatus, is made to expose the area of the bladder neck and proximal urethra. As the vaginal flap is dissected off, the sling should be identified above the periurethral fascia (Figure 10-3.1). The sling is incised in the midline and its two edges are dissected laterally to free them from adhesions. The vaginal mucosal surface is cauterized slightly to prevent inclusion cyst formation. The most distal portion of the vaginal flap, marked previously, is cut free, inserted, and sutured between the two edges of the sling. The remainder of the vaginal wall flap is advanced without tension to cover the wound.

At Cleveland Clinic Florida, we use different treatment approaches for patients with obstructive symptoms after sling placement, and other anti-incontinence surgery. These approaches vary, and include simple sling release, sling release and interposition of a free graft of vaginal wall, urethrolysis, and urethrolysis and Martius flap. Simple sling release can be performed for patients who have undergone a single sling procedure within 3 to 12 months and have minimal scarring. Sling incision in these patients is easier and requires less patient recovery than does

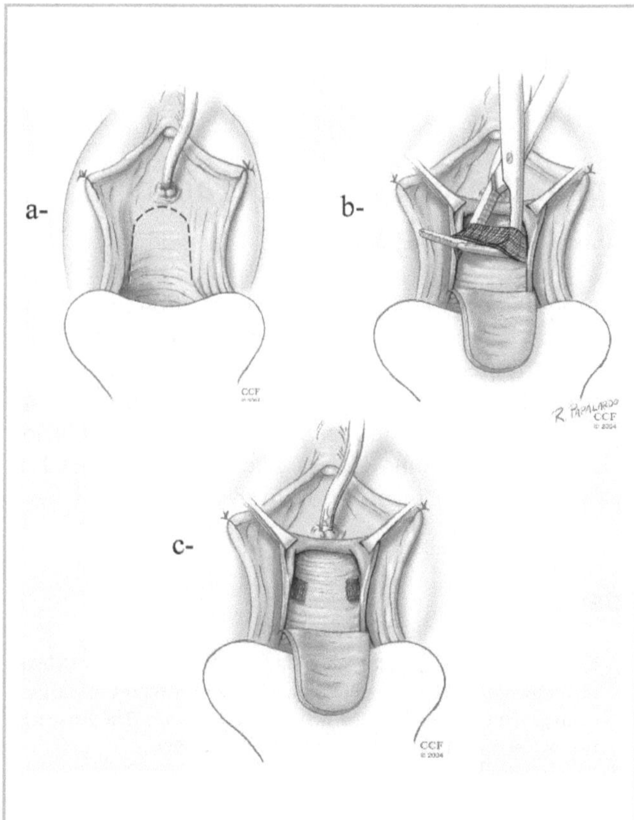

Figure 10-3.1. a, An inverted U-shaped incision in the anterior vaginal wall with the Apex halfway between the bladder neck and urethral meatus. b, Identification of the sling and separation of the sling from urethra by a right-angle clamp. c, Incision of the sling in the midline. (Reprinted with the permission of The Cleveland Clinic Foundation.)

formal urethrolysis, whereas the success rates are comparable.[18] The retropubic space is not entered, and the urethra is not freed from the surface of the pubic bone; thus, the lateral support is not disturbed. With previous multiple anti-incontinence surgeries performed more than 1 year previously and with significant scarring, complete urethrolysis is required, and Martius graft interpositions are performed in the presence of extensive fibrosis or dense scarring.[19,20]

Outcomes of Treatment

Our experience has shown that early intervention ensures simple corrective procedures. Urethrolysis and anterior vaginal wall or Martius flap interposition in patients who present after 1 year from the last anti-incontinence surgery and have extensive fibrosis decreases the rate of recurrence of obstructive symptoms. However, irritative symptoms may persist, even after successful resolution of the obstruction. When we evaluated 15 patients referred to us for management, 53% reported cure, 40% reported improvement of their symptoms, and 27% reported persistence of their irritative symptoms. Three had recurrence of incontinence and received collagen injections.

The recurrence rates of SUI after urethrolysis range from 0% to 19%.[21,22] Failure of urethrolysis may be attributed to persistence or recurrence of obstruction, detrusor instability, impaired detrusor contractility, or learned voiding dysfunction.

Repeat Urethrolysis after Failed Initial Urethrolysis

Urethrolysis successfully relieves obstruction in 65% to 93% of cases. Persistence of obstruction after urethrolysis may be attributed to incomplete urethrolysis or recurrence of fibrosis. In these cases, aggressive repeat urethrolysis can be successful for relieving retention. Complete resolution of irritative symptoms is less likely. Recurrent SUI is similar to that after primary urethrolysis.[23]

References

1. Leach GE, Dmochowski RR, Appell RA, et al. Female Stress Urinary Incontinence Clinical Guidelines Panel summary report on surgical management of female stress urinary incontinence. The American Urological Association. J Urol 1997;158(3):875–880.
2. McDuffie RW Jr, Litin RB, Blundon KE. Urethrovesical suspension (Marshall-Marchetti-Krantz). Experience with 204 cases. Am J Surg 1981;141:297–298.
3. Zimmern PE, Hadley HR, Leach GE, Raz S. Female urethral obstruction after Marshall-Marchetti-Krantz operation. J Urol 1987;138: 517–520.
4. Holschneider CH, Solh S, Lebherz TB, Montz FJ. The modified Pereyra procedure in recurrent stress urinary incontinence: a 15-year review. Obstet Gynecol 1994;83:573–578.
5. Horbach NS. Suburethral sling procedures. In: Ostegard DR, Bent AE, eds. Urogynecology and Urodynamics. Theory and Practice. 3rd ed. Baltimore: Williams & Wilkins; 1991:413–421.
6. Ghoniem GM, Kothandaraman P, Khater UM. Different techniques for the treatment of obstructive slings. J Pelvic Med Surg 2004; 10(suppl 1):S48.
7. Massey JA, Abrams PH. Obstructed voiding in the female. Br J Urol 1988;16(1):36–39.
8. Farrar DJ, Osborne JL, Stephenson TP, et al. A urodynamic view of bladder outflow obstruction in the female: factors influencing the results of treatment. Br J Urol 1976;47:815–822.
9. Chassagne S, Bernier PA, Haab F, Roehrborn CG, Reisch JS, Zimmern PE. Proposed cutoff values to define bladder outlet obstruction in women. Urology 1998;51:408–411.
10. Blaivas JG, Groutz A. Bladder outlet obstruction nomogram for women with lower urinary tract symptomatology. Neurourol Urodyn 2000;19:553–564.
11. Nitti VW, Tu LM, Gitlin J. Diagnosis of bladder outlet obstruction in women. J Urol 1999;161:1535–1540.
12. Leach GE, Raz S. Modified Pereyra bladder neck suspension after previously failed anti-incontinence surgery. Surgical technique and results with long-term follow-up. Urology 1984;23:359–362.
13. Webster GD, Kreder KJ. Voiding dysfunction following cystourethropexy: its evaluation and management. J Urol 1990;144: 670–673.
14. Petrou SP, Brown JA, Blaivas JG. Suprameatal transvaginal urethrolysis. J Urol 1999;166:1268–1271.
15. Scarpero HM, Nitti VW. Management of urinary retention following surgery for stress urinary incontinence. Curr Urol Rep 2002;3(5): 354–359.
16. Ghoniem GM, Elgamasy AN. Simplified surgical approach to bladder outlet obstruction following pubovaginal sling. J Urol 1995;154: 181–183.
17. Amundsen CL, Guralnick ML, Webster GD. Variations in strategy for treatment of urethral obstruction after a pubovaginal sling procedure. J Urol 2000;164:434–437.
18. Nitti VW, Carlson KV, Blaivas JG, Dmochowski RR. Early results of pubovaginal sling lysis by midline incision. Urology 2002;59:47–52.
19. Carey JM, Chon JK, Leach GE. Urethrolysis with Martius labial fat pad graft for iatrogenic bladder outlet obstruction. Urology 2003; 61(4 suppl 1):21–25.
20. Ghoniem GM, Monga M. Modified pubovaginal sling and Martius graft for repair of the recurrent vesicovaginal fistula involving the internal urinary sphincter. Eur Urol 1995;27(3):241–245.
21. Goldman HB, Rackley RR, Appell RA. The efficacy of urethrolysis without re-suspension for iatrogenic urethral obstruction. J Urol 1999;161:196–198.
22. Petrou SP, Young PR. Rate of recurrence of stress urinary incontinence after retropubic urethrolysis. J Urol 2002;167:613–615.
23. Scarpero HM, Dmochowski R, Nitti VW. Repeat urethrolysis after failed urethrolysis for iatrogenic obstruction. J Urol 2003;169: 1013–1016.

10-4

Conservative Management of Constipation

Gregory F. Bonner

Constipation is a common complaint in medical practice. Estimates of the frequency of constipation in the American population have varied from 2% to 28%.[1] This wide range in estimated frequency is in part related to differences in definitions. Not all patients who complain of constipation experience the same symptoms. Physicians typically define constipation as inadequate frequency of bowel movements. Normal frequency of bowel movements can vary from three times per day to three times per week. These parameters have led to one of the more commonly accepted definitions of constipation, i.e., a frequency of defecation of two times per week or less. However, a patient's definition of constipation may be quite different. Many patients believe that any frequency of less than a daily bowel movement is abnormal. Other patients will consider themselves constipated if their stools are too hard or too small, if defecation is associated with pain or excessive straining, or if they experience a feeling of incomplete evacuation. Given these variables, the clinician will need to use a combination of both subjective complaints and objective criteria when addressing constipation. Self-reported constipation tends to occur more often in women and the prevalence increases with age.

Evaluation

When assessing potential therapies for constipation, we initially consider potentially reversible factors external to the gastrointestinal tract. Hypothyroidism and hypercalcemia are metabolic causes of constipation, which occur with enough frequency that they require evaluation. Multiple drugs can cause constipation, with opiate analgesics the most classic example. Other potentially contributing medications include antispasmodics (often given for other gastrointestinal and genitourinary disorders), tricyclic antidepressants, antiparkinsonian agents, and a variety of antihypertensive agents (most classically calcium channel blockers).

Diagnostic studies, which can detect possible physiologic or neurologic abnormalities associated with constipation, have already been discussed. Most constipated patients do not require extensive evaluation. Conversely, most constipated patients who are evaluated are not found to have an obvious cause to explain their symptoms. We recommend a trial of empiric therapy before pursuing diagnostic testing unless alarm symptoms such as rectal bleeding, weight loss, or abdominal distention are present. For patients seeking evaluation for constipation from a specialist, colonoscopy is the single most expensive part of the evaluation.

For the practicing clinician, it is helpful to try to differentiate patients with constipation with predominant irritable bowel syndrome from those with significantly delayed colonic transit or other physiologic disorders. Slow transit constipation is exhibited by slower than normal movement of fecal contents from the proximal to the distal colon. Slow transit constipation can result from delayed transit through the entire colon (colonic inertia) or uncoordinated motor activity in the distal colon. A colonic transit study is useful in confirming a diagnosis of colonic inertia by demonstrating generalized slow movement of the markers throughout the colon. Patients with uncoordinated motor activity of the distal colon are referred to as having pelvic floor dysfunction or dyssynergia. For these patients, colonic transit studies show a preferential accumulation of markers in the rectosigmoid. Anorectal physiology studies, addressed elsewhere, can identify more specifically the disorder in patients with pelvic floor dysfunction.

Most patients with significant pain associated with constipation have irritable bowel syndrome. Additionally, symptoms of bloating and incomplete evacuation are also suggestive of irritable bowel syndrome. This entity is addressed in more detail elsewhere. Even among patients whose symptoms are most suggestive of colonic inertia, approximately 30% will have normal colonic transit studies. These patients with normal transit have a psychological profile similar to those with irritable bowel syndrome, and perhaps should be addressed as a variation of irritable bowel syndrome.

Treatment

Nonsurgical treatment of constipation can be divided into several broad categories. There is the dietary approach with use of a fiber-rich diet and fiber supplements. Use of laxatives is deeply rooted in our society, and they are often used inappropriately. Laxatives are classified based on their mode of action. Bulk-forming laxatives are simply fiber supplements, which have already been noted. They are a benign form of therapy generally recommended as the first-line therapy for constipation. They should be avoided if there is any clinical suspicion of obstruction. Behavioral approaches include the encouragement of good toileting habits or biofeedback. Lastly, there are pharmacologic interventions (Table 10-4.1).

Our first recommendation for all adults with a complaint of constipation is a fiber-rich diet and fiber supplements. Many organic polymers interact with water and have the ability to hold extra water in the stool. Commercially available supplements include psyllium, bran, polycarbophil, or methylcellulose. Food and Drug Administration recommendations are for 20 to 25 g of dietary fiber per day. The average Western diet is far short of this recommendation. Although fiber is almost uniformly recommended for treatment of constipation, the evidence supporting its use is largely anecdotal. We know that people in nonindustrial nations typically consume significantly more fiber and that this typically results in the passage of several daily bowel movements of considerable volume. Still, there is no evidence to support that in Western society, people with constipation consume less fiber than people without this complaint. It is helpful to increase the use of fiber supplements gradually, especially in patients with a significant component of irritable bowel symptoms who find the increased gaseousness troublesome. Patients need to be educated that dosages among the different fiber supplements are not equivalent. A dose of psyllium will contain 3.5 to 4 g of fiber whereas methylcellulose contains only 2 g, and a single pill of polycarbophil (which patients often consider equivalent to a full dose) is only 500 mg. Lastly, patients need to be advised not to expect the immediate response to a fiber supplement that they may have previously experienced with a purgative laxative.

Table 10-4.1. Classification and trade names of some frequently used laxatives

Class/Generic Name	Trade Names
Fiber supplements	
Psyllium	Metamucil, Konsyl
Methylcellulose	Citrucel
Polycarbophil	FiberCon, Equalactin
Emollients	
Mineral oil	Fleet Mineral Oil, Liqui-Doss
Docusate sodium	Colace
Docusate calcium	Surfak
Osmotic agents	
Magnesium citrate	Citroma, Citrate of Magnesia
Magnesium hydroxide	Phillips' Milk of Magnesia, Haley's MO
Magnesium sulfate	Epsom Salts
Sodium phosphate	Fleet's Phospho-Soda
Lactulose	Chronulac, Duphalac
Polyethylene glycol	Miralax
Stimulant laxatives	
Senna	Correctol Herbal Tea, Senokot, Ex-Lax Gentle Nature Pills
Danthron	Doxidan
Castor oil	Castor oil
Bisacodyl	Correctol, Dulcolax

Fiber supplements can be very effective in many patients with constipation, but not all patients. The patients who respond best are those with normal colonic transit and normal defecation dynamics.[2] Patients with significantly delayed transit constipation and those with functional outlet obstruction often do not respond well to fiber. Patients with irritable bowel can have difficulty with the increased distention caused by gas production.

Proper toilet training is routinely used for childhood constipation and probably should be more widely recommended for adults. This involves routinely scheduling adequate time to use the bathroom after breakfast or dinner to take advantage of meal-stimulated increases in colonic activity. The use of biofeedback for rectosphincteric dyssynergia, is addressed in the section on physiotherapeutic approaches.

Emollient laxatives are mineral oil and ducosate salts (Colace, Surfak). They penetrate and soften the stool to allow easier defecation. These agents also are generally thought to be benign and can be added if fiber alone is inadequate. The ducosate salts also act to a degree as stimulant laxatives. Aspiration of mineral oil has been associated with lipid pneumonia and is contraindicated for this reason in elderly debilitated patients.

Osmotic laxatives, as their name implies, act by increasing stool osmolarity. This will secondarily result in water being drawn into the colon and a subsequent increase in stool volume. If chronic therapy beyond fiber is required, osmotic laxatives are preferred over stimulant laxatives, because osmotic laxative use in the long term does not result in worsening of colonic motility. Osmotic laxatives include a variety of magnesium-based saline laxatives, which are available over the counter (Milk of Magnesia, Citrate of Magnesium, Haley's MO). Phosphate salts (Fleet Phospho-Soda) is available for use as a purgative for colonoscopy. It is markedly hypertonic and must be ingested with two to three glasses of water to prevent vomiting. Hyperphosphatemia may occur, especially in patients with renal insufficiency. Glycerin also is active as an osmotic laxative as it is not absorbed by the colon, and, when given as a suppository, it will draw water into the rectum. Glycerin is well absorbed in the small bowel and of no effect as a laxative when taken orally. Osmotic agents available by prescription include lactulose and polyethylene glycol (GoLYTELY, CoLyte, Miralax). Lactulose is a synthetic disaccharide that cannot be digested in the small intestine. It is then fermented by colonic bacteria, which in addition to the osmotic effect produces lactic acid as a byproduct. GoLYTELY and CoLyte have been designed as a large-volume purge in preparation for colonoscopy; once the salt and liquid have been mixed, it must be ingested within several hours. Miralax is packaged in such a way that a single dose of 17 g can be taken on a daily basis, if necessary.

Stimulant laxatives include castor oil, anthraquinones such as cascara and senna, and bisacodyl. They all result in increased in intestinal motility and either stimulate intestinal secretion or decrease absorption. The anthraquinones may cause melanosis coli with chronic use. There is concern that this class of drug may cause smooth muscle atrophy and damage to the myenteric plexus. The term "cathartic colon" originated as a radiologic term.[3] Stimulant laxatives have been associated with dilatation of the colonic lumen, colonic redundancy, and loss of haustral markings. Still, the risk of "cathartic colon" with use of stimulant laxatives remains a point of debate. Many senna-based laxatives are available in health food stores as "all natural" herbal laxatives. Phenolphthalein had been a widely used stimulant laxative but was withdrawn from the market several years ago over concern of a possible association with carcinogenesis based on animal studies.

Prokinetic agents have also been introduced in recent years. Metoclopramide affects gastric motility but is ineffective for constipation. Cisapride (Propulsid) was helpful for some cases of constipation but was withdrawn from the market because of concern of arrhythmias; it is currently only available for very restricted compassionate-use protocols. The newest prokinetic agent is tegaserod (Zelnorm), which is a serotonin agonist acting specifically on the $5HT_4$ receptor. Tegaserod enhances smooth muscle activation and peristalsis by facilitation of cholinergic transmission. A typical starting dose for tegaserod is 6 mg twice daily although lower doses are sometimes preferable. The Food and Drug Administration has approved tegaserod only for use by women, since the controlled trials did not demonstrate statistical significance in male patients.

Summary

A stepwise approach to medical therapy for constipation should start with a high-fiber diet, fiber supplement, and adequate fluid intake. An emollient such as ducosate would be suggested next. Although almost any laxative can be used on an infrequent basis, osmotic laxatives such as Milk of Magnesia or lactulose are preferred if chronic laxative use is needed or desired by the patient. For patients with constipation-predominant irritable bowel syndrome, tegaserod can be added for patients refractory to the simpler therapies.

References

1. Locke GR, Pemberton JH, Phillips SF. AGA technical review on constipation. Gastroenterology 2000;199:1766–1778.
2. Schiller LR. Review article: the therapy of constipation. Aliment Pharmacol Ther 2001;15:749–763.
3. Xing JH, Soffer EE. Adverse effects of laxatives. Dis Colon Rectum 2001;44:1201–1209.

10-5

Surgical Management of Constipation

Tracy L. Hull

Patients who seek medical consultation for constipation usually respond to conservative therapy.[1] However, some have such severe difficulty that surgical intervention is needed. When considering surgical intervention, it is important to verify which portion of the large intestine malfunctions. Usually problems originate either with the colon proper from abnormal transit or in the pelvis. Combined problems from both areas also can exist.

Initial Evaluation

Patients initially need a thorough history and physical examination. Examination of the colon will rule out problems from strictures or tumors. Laboratory work (full electrolytes, thyroid stimulating hormone, and calcium) is considered in those who do not respond to medical treatment. This will eliminate problems from an organic source.

Specific Investigations

After medical treatment fails and no abnormality is found on laboratory values, specific investigations are performed.[2] At our institution, this would consist of a multiple contrast defecating proctogram, a colonic transit study, and anal physiology testing. The multiple contrast defecating proctogram involves contrast given from the oral, bladder, vaginal, and rectal route. The aim is to identify enteroceles, rectoceles, sigmoidoceles, or internal intussusception, all of which can lead to debilitating constipation. Paradoxical puborectalis contraction may also be seen, but this is usually diagnosed with anal physiology testing.

Colonic Transit Study

The colonic transit study can be performed by many protocols. Uniformly, patients must stop all laxatives 48 hours before the study begins. They begin a high-fiber diet and do not use laxatives before or during the test. Commercially prepared capsules that contain radio-opaque markers are orally ingested. Abdominal X-rays are then taken daily or every other day and compared. Usually, 80% of the markers are passed by the fifth day and all by the seventh day.

Anal Physiology Testing

Anal physiology testing is used at our institution to verify that the puborectalis muscle relaxes when straining occurs. Paradoxical puborectalis contraction occurs when this does not happen and is treated with biofeedback. We also use anal physiology testing to look for an anorectal inhibitory reflex. Absence of this points to Hirschsprung's disease. A small number of adults with lifelong constipation are found annually in our institution with short segment Hirschsprung's. Verification that a patient can defecate a balloon transanally placed in the rectum is also done and is performed in the anal physiology laboratory.

Before undertaking a surgery that involves resection of bowel, and there is concern about the motility of the small bowel, additional investigations should be ordered. Hints of small bowel problems are early epigastric discomfort after eating, or enlarged stomach on abdominal flat plate X-ray. Various studies are used to examine the small bowel. At our institution, ingestion of a hamburger or egg mixture combined with a radionucleotide is given. The progress of the radionucleotide out of the stomach and through the small intestine is traced in the nuclear medicine laboratory.

Surgical Treatment

Colectomy with Ileorectal Anastomosis

Surgical treatment is guided by the results of the specific investigations.[3–5] Patients with slow transit constipation, as demonstrated by the failure to eliminate all markers by 7 days during the colonic transit study should be considered for a colectomy and ileorectal anastomosis (IRA). However, this may not be the best option for patients who are unable to evacuate a balloon in the anal physiology laboratory as this may signal rectal dysfunction. Nonetheless, these patients may still be candidates, because the stool produced after this surgery is liquid or very soft and evacuation of soft stool may not be a problem even if rectal dysfunction exists. Instead, if they are unsatisfied and request surgical intervention, an ileostomy is the procedure of choice. A loop ileostomy can usually be done laparoscopically is reversible. Many individuals experience great relief after an ileostomy in this situation.

Poor anal sphincters with lax sphincter tone may preclude perfuming a colectomy and IRA. The stool frequency after this procedure can be eight or more very soft stools daily. For compromised sphincters, this could lead to anal incontinence. Many patients would understandably avoid an ileostomy and opt for a colectomy and IRA despite the possibility of incontinence. They feel that a subsequent ileostomy is still possible if this fails.

A frequent dilemma is the order in which to manage patients who exhibit slow transit constipation and paradoxical puborectalis contraction. Traditionally, some surgeons thought that the paradox should be managed with biofeedback before the surgical intervention. However, many patients fall into the paradoxical muscle problem because they need to strain and bear down in any attempt to evacuate stool and gain relief. In our experience, until the need to excessively strain is eliminated, biofeedback is not as helpful, therefore, the colectomy should be performed first. If the patient still has difficulty after surgical recovery, they are reassessed with anal physiology. If paradox is found, biofeedback is then recommended.

Segmental Colectomy

As a general rule, a segmental colectomy is not performed for constipation, as it is very difficult to accurately determine what part of the colon is malfunctioning from the transit study. There are some who believe that using a scintigraphic transit study will delineate if a particular segment malfunctions and only resect that segment. However, because of the need to reoperate on individuals after a segmental resection, colorectal or ileosigmoid anastomosis, a full colectomy and IRA should be performed as the initial operation. Furthermore, the scintigraphic transit study can more precisely show a specific segment of colonic dysmotility.

Notable exceptions are individuals with rectal prolapse and constipation. It is important to realize that internal intussusception has been found to lead to constipation in some individuals. For those with severe constipation, a transit study should be performed. If this is prolonged, a colectomy with IRA and rectopexy is combined to surgically treat both problems. Some patients with rectal prolapse also will have an element of constipation, but this is not considered "severe." This group will benefit from a sigmoid (segmental) colectomy with colorectal anastomosis and rectopexy. Additionally, when performing a rectopexy for rectal prolapse and the sigmoid kinks over severely when the rectum is pulled up, consideration should be given to performing the sigmoid resection with anastomosis and rectopexy. This could eliminate potential post-operative constipation due to kinking.

A few patients will experience unusual constipation symptoms from sigmoidoceles; these are typically extremely difficult to diagnose. If this is suspected, communication with the radiologist helps to specifically look for this finding while performing the defecating proctogram. A segmental sigmoid colectomy and colorectal anastomosis with closure of the pelvic floor defect (if possible) is the surgical treatment of choice.

Pelvic Ileal Pouch Procedure

For the few patients who continue to have constipation after colectomy and IRA, completion proctectomy and ileal pelvic pouch is considered.[6] In a select few, this will eliminate the problem. The anal sphincters must function well, as the stool produced after a pouch is looser and more frequent than that after the IRA. Pelvic Floor Function should be assessed to avoid continued problems after this more involved procedure. This may be difficult to assess, but excessive perineal descent with strain may be one clue that warns against the pelvic pouch.

Stoma

As mentioned previously, when the colon, rectum, and pelvic floor do not correctly function, an ileostomy will treat the severe constipation. Specifically, patients with abnormal colonic transit studies and inability to defecate a balloon (or the inability to eliminate contrast on defecating proctogram) to consider an ileostomy.[7]

A colostomy is rarely offered. This would be considered in an individual with severe rectal or hindgut dysfunction and pelvic floor dysfunction. For instance, an individual with a colonic transit study in which the markers all come to the top of the rectum and no further and the inability to eliminate contrast on a defecating proctogram may benefit from with a colostomy.

Proctectomy with Coloanal Anastomosis

For adult patients diagnosed with Hirschsprung's disease, the colon usually needs decompression as the initial treatment. At laparotomy, a loop ileostomy is performed and the colon is purged of stool. The segment above the aganglionic portion is usually massively dilated and the stool is hard and must be broken up and irrigated. A tube is usually placed in the ileum at the area of the future ileostomy and irrigated with large amounts of warm saline while breaking up the stool downstream. An ileostomy is selectively chosen so as not to interfere with the next operation. It usually takes approximately 6 months for the sigmoid to regain tone and decrease to an acceptable size, and at this point, a proctectomy with coloanal anastomosis is performed. Ganglia in the distal colonic segment are verified by frozen section at the time of operation. After 3 months, the anastomosis is cleared for leaks with a Gastrografin enema, and the ileostomy is closed.

In very rare circumstances, a proctectomy with coloanal anastomosis is advocated for patients with rectal or hindgut dysfunction. These patients should be screened very carefully to rule out colonic dysmotility or pelvic floor dysfunction.

Rectocele

Many women have rectoceles, but not all of them need repair. Generally speaking, large rectoceles greater than 4 cm on defecating proctogram and/or the need to digitally support the back wall of the vagina or perineal body to expel stool are good indicators of those needing repair. Patients have usually been started on a high-fiber diet and have continued problems before considering repair. In my practice, I find working from the vaginal aspect is the most beneficial for the patient. This is usually the urologist, gynecologist, or urogynecologist. Many times, an unanticipated enterocele is found. The symptoms from the enterocele may be overlooked and become more evident when the rectocele is eliminated.

Antegrade Enema

A procedure used more frequently in children is the construction of a stoma in the proximal colon that allows a catheter to pass and instill irrigation fluid. This procedure also seems to be used more frequently outside of the United States for patients with slow transit constipation instead of a colectomy and IRA or stoma.

Laparoscopic Surgery

As experience is gained with laparoscopic bowel surgery, many operations performed via the open technique can also be done laparoscopically. For instance, some individuals are thin and ideally suited for laparoscopic colectomy and IRA.

Future Horizons

With some scattered discussions about sacral stimulation improving constipation in patients with pelvic etiology constipation, it will require further study to verify this finding and ascertain which patients will benefit. Perhaps even a surgically implantable colonic pacemaker will be available in the future for those with colonic dysfunction.

Summary

Most constipation can be managed medically. However, significant relief from agonizing abdominal symptoms can be accomplished by surgical treatment of constipation in selected individuals. Preoperative evaluation is critical to assess who these selected individuals are and what can be done to help them.

References

1. Eccersley AJ, Maw A, Williams NS. Comparative study of two sites of colonic conduit placement in the treatment of constipation due to rectal evacuatory disorders. Br J Surg 1999;86:647–650.
2. Hull TL. Constipation. In: Walters MD, Karram MM, eds. Urogynecology and Reconstructive Pelvic Surgery. 2nd ed. St. Louis: Mosby; 1999:269–276.
3. Lundin E, Karlbom U, Pahlman L, Graf W. Outcome of segmental colonic resection for slow-transit constipation. Br J Surg 2002;89: 1270–1274.
4. Mollen RM, Kuijpers HC, Claassen AT. Colectomy for slow-transit constipation: preoperative functional evaluation is important but not a guarantee for successful outcome. Dis Colon Rectum 2001;44: 577–580.
5. Pikarsky AJ, Singh JJ, Weiss EG, Nogueras JJ, Wexner SD. Long-term follow-up of patients undergoing colectomy for colonic inertia. Dis Colon Rectum 2001;44:179–183.
6. Rotholtz NA, Wexner SD. Surgical treatment of constipation and fecal incontinence. Gastroenterol Clin North Am 2001;30:131–166.
7. Thakur A, Fonkalsrud EW, Buchmiller T, French S. Surgical treatment of severe colonic inertia with restorative proctocolectomy. Am Surg 2001;67:36–40.

Section XI

Hormonal Influences on the Pelvic Floor

11-1

Hormonal Influences on the Pelvic Floor

G. Willy Davila

Reproductive hormones, especially estrogen, have a significant impact on pelvic floor function. Hormonal changes that occur during a woman's lifespan impact many aspects of female physiology. With the onset of menopause, it becomes evident that estrogen sensitivity is greatest in the central nervous system and the genital tissues. The most frequently occurring initial symptoms of estrogen deprivation include hot flushes, mood changes, and emotional irritability. These central nervous system symptoms are promptly reversible with systemic estrogen replacement. More prolonged duration of estrogen deprivation will lead to other known consequences such as osteoporosis and urogenital atrophy. It is estimated that 80% of postmenopausal women have hot flushes, and that not all women will develop osteoporosis. However, urogenital atrophy occurs universally. Interestingly, many women with urogenital atrophy are asymptomatic.[1] Many treatments are currently available for prevention of and therapy for loss of bone mineral content. However, local estrogen therapy is the only means of treating symptoms and signs of urogenital atrophy.

As related to the pelvic floor, urogenital atrophy is present at both ends of the reproductive cycle, and consequences such as labial fusion may be present both premenarchally as well as postmenopausally. After the onset of menopause, hypoestrogenism is expressed in the lower genitourinary tract by thinning of the vaginal and urethral mucosa, as well as other well-recognized changes (Table 11-1.1). Estrogen receptors have been found in most body tissues. The presence of estrogen receptors in the pelvic floor mucosa has been demonstrated, as has the importance of the estrogen-replete state in physiologic cellular proliferation.[2,3] The implications of the lack of estrogen on urogenital well-being will be discussed in this chapter.

Vaginal Atrophy

The urethral and vaginal mucosas are rich in estrogen receptors and share a common embryologic origin from the urogenital sinus. As such, these tissues are exquisitely sensitive to estrogen deprivation, and symptoms may appear promptly as soon as estrogen levels begin to decline. Vaginal atrophy will present with a variety of symptoms along a continuum of severity. Symptoms are typically vaginal dryness and associated dyspareunia, which may be initially presented as vaginal irritation. This will progress to loss of vaginal rugation and development of a progressively pale, hypovascular mucosa. Eventually, a thin inflammatory exudate can develop, sometimes in large quantities. This may result in a watery vaginal discharge that on microscopic examination is replete with inflammatory and basal squamous epithelial cells, but with no evidence of bacterial infection. As with other more obvious mucosal hypoestrogenic states, a chronic watery vaginal discharge without an infectious cause in a postmenopausal woman warrants a course of local estrogen therapy. Up to this stage of atrophy progression, resultant changes are readily reversible with local estrogen administration. Left untreated, chronic vaginal atrophy will progress to loss of vaginal caliber and length, changes that are not reversible with estrogen therapy alone. Progressive coalescence of the apical vaginal mucosa will result in vaginal foreshortening. Lack of sexual activity exacerbates the progression of urogenital atrophy.

Because the progression of urogenital atrophy is in large part attributed to a devascularization of the mucosal tissues, systemic hormone replacement may not be sufficient to alleviate symptoms of vaginal atrophy. Systemically administered estrogen will not reach a devascularized mucosa. As such, many women receiving low-dose systemic replacement therapy will still have urogenital atrophy and will require concomitant local hormonal replacement therapy.

Upon initiation of local hormone replacement therapy, there is a prompt revascularization of the vaginal and urethral mucosa. This can be clinically detected within 6 weeks of initiation of local estrogen therapy. As might be expected, vaginal irritative symptoms (i.e., tingling, itching) may occur during the initiation of the neovascularization process. These symptoms are typically short-lived.

Table 11-1.1. Signs and symptoms of urogenital atrophy

Vaginal
Dryness
Pallor
Decreased rugation
Mucosal thinning
Inflammation with/without discharge
Decreased caliber and depth
Increased pH
Vulvar
Thinned skin
Loss of hair volume
Labial fusion
Decreased introital diameter
Urethral
Decreased alpha receptors
Decreased vascularity
Decreased epithelial thickness
Formation of caruncle
Bladder
Increased bladder wall collagen
Decreased bladder epithelial thickness
Increased parasympathetic sensitivity
Decreased beta sympathetic activity

Vaginal atrophy can be easily detected on vaginal examination with identification of a thin, pale, and dry vaginal mucosa. In addition, vaginal pH will increase to greater than 5.0. This can be assessed during a pelvic examination, using pH paper. To quantify the degree of vaginal atrophy, a maturation index is most useful. This is a cytologic evaluation of the vaginal mucosa determining the percentage of basal, parabasal, and superficial squamous cells identified on a smear of the vaginal epithelium, much like a pap smear. A ratio distribution of cell types can be determined (% superficial/% parabasal/% basal). Women who are well-estrogenized will have a predominance of superficial cells, whereas atrophic women will have greater than 50% of the cells being basal or parabasal, with few superficial cells. A change in maturation index can be documented during treatment with local estrogen therapy, as the percentage of superficial cells increases. This is typically an inexpensive test, with which most pathologists are familiar.

Urethral Effects

Changes in the urethral mucosa occur in a near identical manner to those in the vaginal mucosa. There is a thinning and loss of rugation of the urethral mucosa associated with an increasingly hypovascular state. The urethral mucosa will become thin, pale, and there will be loss of mucosal coaptation along the length of the urethra. These changes are promptly reversible with intravaginal administration of local estrogen.

The changes in the vaginal and urethral mucosa occur in parallel. Therefore, a woman with vaginal atrophy can be assumed to have urethral atrophy. Attempts have been made to quantify urethral atrophy using urethral cytology. Our efforts to use this cytologic evaluation tool to quantify urethral atrophy have not been fruitful because pathologists are typically not familiar with urethral cytology, and thus we do not consider urethral cytology to be a reliable assessment tool.

Symptomatology of Urogenital Atrophy

Typical symptoms of vaginal atrophy include vaginal dryness, vaginal irritation, as well as loss of vaginal caliber and depth in more advanced degrees of atrophy. The initial symptoms of vaginal atrophy can be promptly reversed with local estrogen therapy. However, the more long-term effects of urogenital atrophy such as loss of vaginal caliber and depth are less likely to be readily reversible. This has recently become a clinically important issue, because male erectile dysfunction can be treated pharmacologically, whereas the female partners' advanced urogenital atrophy has not been taken into account. Any attempt at sexual intercourse after a long hiatus of inactivity in a postmenopausal woman can result in sexual dysfunction and even vaginal trauma including lacerations upon attempted penetration. It is thus critical that both partners be considered when initiating erectile dysfunction therapy in the male.

Vaginal atrophy is also associated with vulvar atrophic changes. These may include lichen sclerosis as well as hypertrophic dystrophies. As a consequence, external atrophic changes may also be seen, including reduction in pubic hair volume, labial fusion, loss of labial volume, irritative vulvar symptoms, and eventual reduction in introital diameter (Figure 11-1.1).

Symptoms of urethral atrophy are typically attributed to the loss of urethral tone associated with urethral devascu-

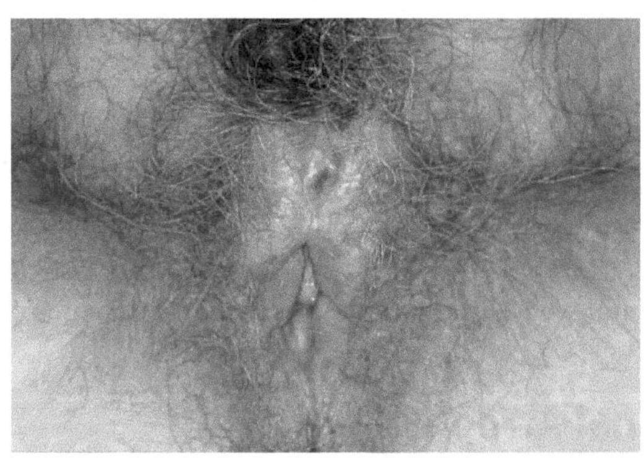

Figure 11-1.1. Advanced urogenital atrophy with marked mucosal paleness, vulvar atrophy, and labial fusion.

larization. It is estimated that the urethral submucosal vascular plexus is responsible for up to 30% of urethral closure pressure. As a consequence, devascularization will lead to decreased urethral tone. This frequently presents with urinary urgency and frequency, presumably because of wetting of the proximal urethra during bladder filling. Nocturia is an exceedingly common symptom of urethral atrophy, and one that is promptly reversible with local estrogen in a more efficient manner than with an anticholinergic medication. Although hypoestrogenism leads to devascularization of the urethral submucosa, stress incontinence typically does not develop in an isolated manner secondary to menopausal urethral atrophy. Local estrogen is thus an important cofactor in the treatment of overactive bladder and stress incontinence symptoms. When used in combination with alpha-agonist medications, stress urinary incontinence symptoms may be reduced (Chapter 6-2).

We require a greater understanding of the natural history and symptomatology of urogenital atrophy. Although the occurrence of urogenital atrophy is universal after menopause, most women are asymptomatic. In a recent trial, we attempted to correlate the symptoms of urogenital atrophy with objective assessment of urogenital atrophy severity.[1] We found no significant correlation between atrophy symptomatology and objective assessment tools such as maturation index, vaginal pH, or visual examination. It is thus unclear why some women become highly symptomatic of their urogenital atrophy, whereas others remain completely asymptomatic.

Impact on Nerve Function

One of the proposed reasons for differences in urogenital symptomatology seen in postmenopausal women may have to do with the effect of estrogen on nerve function. Treatment of postmenopausal sexual dysfunction has demonstrated that systemic and local estrogen therapy improves clitoral, perineal, and vaginal sensation. Thus, there is likely a neurotrophic effect of estrogen on nerve conduction and function. On a systemic basis, this is best demonstrated in the effect of hypoestrogenism on short-term memory. Systemic estrogen supplementation has a positive impact on memory, which is likely attributed to a direct neuronal effect.

Treatment of Urogenital Atrophy

Estrogen replacement therapy is the hallmark of treatment for urogenital atrophy. Because the underlying factor for atrophic changes is a devascularization process, systemic estrogen replacement may not impact the urogenital tissues. This was best described in the recent HERS trial, where stress incontinence symptoms actually worsened on low-dose systemic estrogen replacement therapy.[4]

Table 11-1.2. Urogynecologic aspects of local estrogen replacement

Primary indications
Atrophy with urgency, frequency, nocturia
Atrophy with recurrent cystitis
Along with pessary use
Preparation for reconstructive surgery
Other uses
With alpha-agonist for stress incontinence
Postoperative mucosal maintenance
Safe usage regimen
Intravaginal cream 1 g 2 nights per week
Vaginal tablets 25 μg 2 nights per week
If necessary, confirm lack of absorption by measuring estradiol level after 6–8 wk of therapy

Local estrogen therapy is the preferred therapy for symptoms and signs of urogenital atrophy (Table 11-1.2). There are various preparations available for this purpose:

1. **Vaginal estrogen cream.** Various preparations are available to supply estrogen to the urogenital tissues in a cream vehicle – conjugated estrogens (Premarin, Wyeth Pharmaceuticals), estradiol (Estrace, Mead Johnson), and estriol (not available in the United States). There do not seem to be any significant differences in bioavailability or potency between the different products available in the United States. The recommended dosage for local replacement therapy is 0.5 to 1.0 g, 2 to 3 nights per week.[5] We typically recommend 1.0 g 2 nights per week, placed deep within the vagina with an applicator for women who are symptomatic, being prepared for reconstructive surgery, and postoperatively in women who have undergone pelvic reconstructive therapy.

2. **Vaginal estrogen tablets.** Twenty-five-microgram estradiol tablets (Vagifem, Novo Nordisk) are available for intravaginal use. These tablets contain estradiol in a cellulose matrix such that as the tablet dissolves, the material coats the vaginal mucosa. It is considered by patients to be less messy, although it seems to not have as prompt an effect on vaginal atrophy symptoms as does cream. When we use Vagifem, we recommend 1 tablet, 2 to 3 nights per week.

3. **Estrogen ring.** The estradiol ring (Estring, Pfizer) is placed within the vaginal canal and left in place for 3 months at a time. Its popularity has been somewhat less than the other two formulations, but it is useful in providing local estrogen therapy. It is small and therefore does not function as a pessary. Vaginal rings are also available for systemic estrogen replacement therapy.

4. **Other formulations.** In Europe, Latin America, and other places around the world, vaginal suppositories and pessaries are available for local estrogen therapy. We do not have any experience with that form of therapy in the United States.

It must be emphasized to patients that low-dose local vaginal estrogen therapy, in the doses noted above, does not result in significant absorption into the systemic circulation.[5] As such, patients should not be concerned about systemic effects of local estrogen therapy. In patients who are concerned, we will obtain a serum estradiol level before initiating therapy, as well as 6 weeks after the patient has begun therapy, to demonstrate the lack of significant systemic absorption. This may be particularly important for young women who have undergone chemotherapy for breast cancer, and are severely atrophic.

In some women, the use of concomitant systemic and local estrogen replacement therapy will result in a revascularization of the vaginal mucosa and the possibility of discontinuation of local estrogen therapy once increased vaginal blood flow is documented.

Effect on Continence and Urethral Function

Our knowledge regarding the effect of local estrogen on urethral function and continence is limited and somewhat unclear. It is assumed that urethral mucosal changes mimic those occurring in the vagina during local estrogen therapy. Atrophic women with urinary urgency, frequency, and particularly nocturia – in the absence of cystitis or detrusor instability – will typically benefit from local estrogen therapy. Presumably, this is attributable to thickening of the urethral mucosa with resultant improved mucosal coaptation and thus improved sphincteric function. It has been demonstrated in vivo and in vitro that there is increased contraction of the periurethral smooth muscles with estrogen therapy. This is thought to be mediated through alpha-2-adrenoceptors. Both of these effects should positively impact continence. Some studies have shown improvement in urethral function on dynamic urethral profilometry during multichannel urodynamics, whereas others have not.[6,7] The overall consensus regarding the effects of local estrogen therapy on urethral function is that there is a neovascularization and increased sensitivity to alpha-adrenergic medications. Therefore, local estrogen replacement is an important cofactor in the treatment of a patient with stress urinary incontinence, especially when used in combination with an alpha-agonist medication, such as phenylpropanolamine or imipramine. In our practice, we routinely use local estrogen therapy in patients with any form of pelvic floor dysfunction who are atrophic. However, we rely on other pharmacologic, physiotherapeutic, or surgical therapies as our primary means of addressing underlying continence and anatomic support problems.

We do not believe systemic estrogen replacement therapy has a significant impact on continence status, as seen in multiple previously published studies.

Hormonal Impact on Colorectal Function

Very few data are available regarding the impact of estrogen deprivation after menopause on colorectal function. Many of the changes in colorectal function, such as increased incidence of hemorrhoids and weakening of the anal sphincter are thought to be age-related, rather than caused by hormonal changes. However, estrogen receptors have been found in the external anal sphincter, and there is some anecdotal evidence that local estrogen therapy may be useful in the treatment of hemorrhoids.[8]

A Few Words Regarding Women's Health Initiative

Recent reports regarding the adverse effects of combination hormone replacement therapy found during the Women's Health Initiative (WHI) trial have further fueled the controversy regarding the usage of estrogen therapy in postmenopausal women. The WHI is a National Institutes of Health–funded study beginning in 1993 evaluating the influence of postmenopausal management on cardiac disease, breast and colon cancer, and fractures. The study intervention groups included a low-fat diet group, calcium and vitamin D supplementation group, conjugated equine estrogens-alone group, combined conjugated equine estrogens and medroxyprogesterone acetate group, and a placebo group. The findings reported in 2002 demonstrated an increased risk of deep venous thrombosis, stroke, myocardial infarction, and breast cancer in women in the combination estrogen/progestin group. There was a reduction in colon cancer and risk of bone fractures. The adverse changes were only seen in the combination hormonal therapy group, and not in the estrogen-alone group. The estrogen-alone group comprised hysterectomized women, and no increase in breast cancer was noted. A mild increase in stroke was found. This study has been the subject of much debate, and its conclusions have been questioned. Criticism has focused on various factors, including the fact that most of the patients were already menopausal, most older than age 60. Thus, the results apply to the impact of hormone replacement therapy after the onset of menopause, and not with initiation of therapy at the onset of menopause. Approximately 35% of patients were hypertensive and only half of the patients had never smoked. Thus, multiple other cofactors are present, which could be responsible for increasing the risk of cardiac and vascular disease. Nevertheless, the wide publication of these results in lay magazines has led to massive discontinuation of hormone replacement therapy by women in the United States and elsewhere.

It must be remembered that the estrogen-alone group has not been found to have an increased incidence of breast cancer. This suggests that it is the progestin (medroxyprogesterone acetate), and not the estrogen component of replacement therapy, that may be responsible for these adverse effects. Once this study is fully completed in 2005, we will have a better idea of the safety characteristics of estrogen replacement therapy in postmenopausal women. Because most urogynecologic patients will benefit from low-dose local estrogen therapy, rather than systemic therapy, the findings of this large study should not significantly impact the care of women with pelvic floor dysfunction.

Summary

Hormonal replacement therapy has a prominent role in the treatment of pelvic floor dysfunction. The preferred route of administration is local with vaginal creams, rings, or tablets. This low-dose route of administration is not associated with any significant systemic absorption, thus minimizing any concerns regarding potential vascular, cardiac, or breast consequences of systemic hormonal replacement therapy. We routinely prescribe low-dose local estrogen therapy for our postmenopausal urogynecologic patients, combined with patient education materials in order to optimize compliance with therapy.

References

1. Davila GW, Singh A, Karapanagiotou I, et al. Are women with urogenital atrophy symptomatic? Am J Obstet Gynecol 2003;188:382–388.
2. Blakeman PJ, Hilton P, Bulmer JN. Cellular proliferation in the female lower urinary tract with reference to oestrogen status. Br J Obstet Gynecol 2001;108:813–816.
3. Smith P, Heimer G, Norgren A, Ulmsten U. Steroid hormone receptors in pelvic muscles and ligaments in women. Gynecol Obstet Invest 1990;30:27–30.
4. Hulley S, Furberg C, Barrett-Connor E, et al. Noncardiovascular disease outcomes during 6.8 years of hormone therapy: heart and estrogen/progestin replacement study follow-up (HERS II). JAMA 2002;288:58–66.
5. Handa VL, Bachus KE, Johnston WW, Robboy SJ, Hammond CB. Vaginal administration of low-dose conjugated estrogens: systemic absorption and effects on the endometrium. Obstet Gynecol 1994;84:215–218.
6. Bhatia NN, Bergman A, Karram MM. Effects of estrogen on urethral function in women with urinary incontinence. Am J Obstet Gynecol 1989;160:176–181.
7. Hilton P, Stanton SL. The use of intravaginal oestrogen cream in genuine stress incontinence. Br J Obstet Gynecol 1983;90:940–944.
8. Haadem K, Ling L, Ferno M, Graffner H. Estrogen receptors in the external anal sphincter. Am J Obstet Gynecol 1991;164:609–610.

Section XII

Physiotherapeutic Approaches

Section XII

Physiotherapeutic Approaches

G. Willy Davila

Neuromuscular alterations underlie most, if not all, pelvic floor symptomatic dysfunctions. Direct trauma to the pelvic floor muscles and the nerve supply to the pelvic organs occurring during pregnancy and the vaginal birth process is linked by increasing degrees of objective evidence to the development of pelvic floor problems in women. This also explains the greater propensity of these problems in women. In most women, anatomic and functional pelvic floor dysfunction symptoms tend to worsen over time, as a result of the increased trauma of further deliveries, chronic increases in intraabdominal pressure, bipedal ambulation, the aging process, and hormonal alterations such as menopause.

Similar to other medical conditions, which are attributable to neuromuscular dysfunction (such as chronic back pain), pelvic floor problems frequently respond to rehabilitative therapies. Behavioral approaches such as bladder retraining and directed muscle strengthening exercises, such as Kegel exercises, are effective in the treatment of most bladder, bowel, and pelvic support problems. Unfortunately, many clinicians are not aware of the utility of these conservative approaches for the treatment of their symptomatic patients. Kegel exercises were popularized during the 1950s, but it has not been until recently that their utilization has been supported in a more widespread manner. Because women cannot visualize or frequently feel their pelvic floor muscles, appropriate and correct performance of pelvic floor exercise can be quite challenging. The increased availability of biofeedback techniques has improved our ability to train women to perform pelvic floor exercises correctly. Directed functional electrical stimulation of the pelvic floor neuromuscular unit has demonstrated significant value in the treatment of specific forms of dysfunction. The use of functional electrical stimulation is limited in the United States. However, in Europe, where it has achieved a greater degree of acceptance, its usage is more extensive.

At our institution, we will use physiotherapy as first-line therapy for many conditions such as constipation, stress incontinence, and obstructive defecation syndromes. We will use it as concomitant therapy with pharmacotherapy in the treatment of overactive bladder, irritable bowel, and other irritative symptoms. We will use it sequentially with surgical therapy in patients with fecal incontinence, urinary stress incontinence, or genital prolapse. The use of physiotherapy in the peripartum phase is also very important and underutilized in the United States. We do not have an obstetrics unit at Cleveland Clinic Florida, and therefore will not be discussing this aspect of physiotherapy in great detail. However, the role of pelvic physiotherapy antepartum, and more importantly postpartum, should be given significant consideration, especially in patients at risk for pelvic floor dysfunction.

This section will review the various behavioral and rehabilitative techniques available for the treatment of symptomatic women. The importance of a well-trained and motivated therapist will be highlighted along with the importance of patient motivation on achieving success. The role of pelvic floor rehabilitative therapy by itself, before or after surgical therapy, and use in combination with pharmacotherapy, will be discussed in the hope that more clinicians will avail themselves and their patients of these effective and cost-efficient techniques.

12-1

Kegel Exercises and Biofeedback

Dawn Vickers and G. Willy Davila

Indications

Biofeedback therapy may be considered first-line treatment for stress, urge, and mixed urinary incontinence, fecal incontinence, paradoxical puborectalis contraction, pelvic pain, and other forms of pelvic floor dysfunction. In 1989, the National Institutes of Health Consensus Conference on Urinary Incontinence in adults concluded that less-invasive procedures such as biofeedback treatment should be the first-line treatment for patients with many types of incontinence. This was further promoted by the Agency for Health Care Policy and Research, which recommended behavioral interventions before other forms of treatment in the patient with a history of stress, urge, or mixed incontinence after a basic evaluation.

Biofeedback Defined

Biofeedback is defined as a "a group of therapeutic procedures that utilizes electronic instruments to accurately measure, process, and give feedback to individuals and their therapists, meaningful physiological information with educational and reinforcing properties regarding both normal and abnormal neuromuscular and autonomic activity, both normal and abnormal, in the form of analog, binary, auditory and/or visual feedback signals."[1] This process helps patients develop a greater awareness of, and confidence in, voluntary control over physiologic processes. Using biofeedback instruments without proper cognitive preparation, instruction, and guidance is not appropriate biofeedback therapy.[1]

Pelvic Muscle Exercise – Kegel Exercises

In the late 1940s, Arnold Kegel developed a vaginal balloon perineometer to teach pelvic muscle exercises for poor tone and function of the genital muscles. He was instrumental in developing a standardized program for treating urinary stress incontinence and recommended structured home practice with the perineometer along with symptom diaries. His clinical use of these techniques showed that muscle reeducation and resistive exercises guided by sight sense is a simple and practical means of restoring tone and function of the pelvic musculature and improves urinary incontinence.

Unfortunately, clinicians taught Kegel exercises without the use of instrumentation. It has been shown that verbal or written instructions alone are often inadequate and that up to 50% of patients perform Kegel exercises incorrectly. There is a strong tendency to substitute abdominal and gluteal contractions for weak pelvic floor muscles (PFMs). For patients with fecal or urinary incontinence, abdominal contractions increase intraabdominal pressure, thus increasing the probability of an accident or even further worsening pelvic floor weakness. For patients to begin performing isolated pelvic muscle contractions, they are instructed to contract the PFMs without contracting abdominal, gluteal or leg muscles, and to hold this contraction to the best of their ability. The patient must tighten the pelvic diaphragm (levator ani) in a manner similar to stopping the passage of gas or to stop the flow of urine. Patients should be advised that the initial aim of treatment is not to produce a contraction of maximum amplitude, but to contract the PFM in isolation from other muscles without undue effort. To build muscle endurance, training proceeds with gradual increases in the duration of each contraction along with gradual increases in the number of repetitions. Rhythmic breathing patterns during contractions should also be encouraged.

Recommended home practice is tailored according to the patient's ability and the degree of muscle fatigue observed during the session. At each stage of treatment, patients are encouraged to practice these exercises daily without instrumentation feedback.

The goal of PFM retraining includes attaining normal resting tone, quick recruitment of the PFMs, sustained isolated pelvic muscle contraction, quick release to a normalized resting tone, and appropriate relaxation during defecation or micturition.

Pelvic Muscle Exercise Training Principles

Important training principles for any exercise program include overload, specificity, and maintenance. Overload states that in order to strengthen pelvic muscles they must be pushed to the limit and just a little beyond. Specificity states that the pelvic muscles are composed of fast and slow twitch fibers in a ratio of approximately 35%:65%; some fibers have a combination of fast and slow twitch components. Fast twitch fibers improve in speed and strength with quick contractions, whereas slow twitch fibers strengthen and gain optimal resting length and tone with longer "hold" contractions. Fast twitch fibers fatigue quickly whereas slow twitch fibers are designed for endurance and postural tone; therefore, repetitions are low for fast twitch fibers and higher for slow twitch fibers. Maintenance describes exercising for continence as a lifelong endeavor. Pelvic muscle strength may be maintained by one to three daily 7- to 10-minute sessions. After exercising and symptomatic improvement, discontinuing exercises will typically result in symptom recurrence over time.

Quick Contract and Relax Exercises

This exercise improves the strength and function of the fast twitch muscle fibers primarily of the urogenital diaphragm and external sphincter muscles. These fast twitch muscle fibers are important for preventing accidents caused by increased intraabdominal pressure exerted during lifting, pulling, coughing, or sneezing. The use of quick flick contractions may also assist in inhibiting urgency. This not only interrupts the urinary stream but also inhibits a detrusor contraction once it is started. Once patients learn to perform isolated pelvic muscle exercises, they are instructed to perform quick contract and release repetitions 5 to 10 times at the beginning and end of each exercise session practiced at home.

Behavioral Strategies

Patient Education and Behavior Modification

Many misconceptions can be dispelled as patients gain a better understanding of their disorder. This begins with reviewing the anatomy of the pelvic floor musculature along with a review of normal bowel and bladder function with the use of visual aids. This is followed by reassurance that irregular bowel habits and other defecatory symptoms are common in the healthy general population. Another frequently observed behavioral pattern, common among elderly women with symptoms of urinary incontinence, is the restriction of fluid intake to avoid leakage; in fact, this may worsen symptoms of constipation as well as symptoms of urinary incontinence. A bowel and bladder diary can be used to accurately document fluid intake and voiding habits along with any symptoms of urgency, incontinence, or constipation.

Habit Training

A voiding diary is the starting point for bladder training by allowing the patient to see how often voiding and incontinence actually occur. This is then used to set a voiding interval. Patients who are infrequent voiders are instructed to do timed voiding every 2 to 2-1/2 hours during waking hours. Regular bladder emptying should result in less incontinent episodes. Patients are then instructed to increase the voiding interval by 15 to 30 minutes per week in order to achieve voiding every 3 to 4 hours while awake with less urgency and less incontinence. Timed voiding has been proven effective in patients with urgency, frequency, and urge and stress incontinence. The goal of retraining is increased functional bladder capacity. The initial prescribed voiding interval may be as little as every 30 minutes in patients whose baseline diaries show voiding occurs more often than every 30 minutes. Urge suppression strategies assist with maintaining bowel and bladder control by educating patients to respond adaptively to the sensation of urgency. Rather than rushing to the toilet, which increases intraabdominal pressure and exposes patients to visual cues that can trigger incontinence, patients are encouraged to pause, sit down if possible, relax the entire body, and contract PFMs repeatedly to diminish urgency, inhibit detrusor contractions, and prevent urine loss. When urgency subsides, patients are instructed to proceed to the toilet at a normal pace. Patients with mixed urinary incontinence are also taught stress loss prevention strategies, which consist of contracting the PFMs just before and during any physical activities such as coughing or sneezing that may trigger stress incontinence. These strategies, although intended for urinary incontinence, are quite helpful in maintaining control for patients with fecal incontinence. Norton and Kamm[2] reported that an enhanced ability to contract the anal sphincter is likely to diminish large bowel peristalsis, may even induce retrograde peristalsis, or may simply allow continence to be preserved until the urge (bowl contractions) ceases. This seems to relate to the ability of biofeedback treatment to modify urgency.[2]

Bowel habit training is recommended for patients with symptoms of incomplete, difficult, or infrequent evacuation. Patients are encouraged to set aside 10 to 15 minutes at approximately the same time each day for unhurried attempts to evacuate. The patient should not be overly concerned with any failure because another attempt later in the day is acceptable. This is best initiated after a meal, which stimulates the gastrocolic reflex.

Most commodes are approximately 35 to 40 cm in height; if a patient's feet or legs hang free or dangle above the floor while sitting, simulation of the squatting position will not be accomplished. Full flexion of the hips stretches the anal canal in an anteroposterior direction and tends to open the anorectal angle, which facilitates rectal emptying. This may be achieved by the use of a footstool to elevate the legs and flex the hips.

Dietary Modification

Dietary information is reviewed with all patients to assist in improving bowel and bladder function. Patients are provided with written informational handouts regarding foods that are high in fiber or foods that either stimulate or slow transit. Offering creative fiber alternatives that may be more appealing for the individual to easily incorporate in their daily diet regime assists with compliance. Such alternatives include unrefined wheat bran that can be easily mixed with a variety of foods, cereals, muffins, as well as over-the-counter bulking agents. Adequate fluid intake and limiting caffeine intake are essential for normal bowel and bladder function. It is essential to monitor the amounts and types of fluid intake. Urinary oxalates can be irritating to tissue; therefore, avoidance of oxalate-containing foods may improve irritative bladder and vulvar symptoms.

Practical Aspects of Biofeedback Therapy for Pelvic Floor Dysfunction

Practical aspects of biofeedback therapy for PFM dysfunction to treat symptoms of urinary incontinence, voiding dysfunction, constipation, and fecal incontinence include the technical, therapeutic, behavioral, and pelvic muscle rehabilitation (PMR) components. The technical component involves the instrumentation used to provide meaningful information or feedback to the user. Devices include surface electromyography (sEMG), water-perfused manometry systems, and the solid-state manometry systems with a latex balloon. Although each system has inherent advantages and disadvantages, most systems provide reproducible and useful measurements. A solid-state system is preferable to a water-perfused system because there is no distraction or embarrassment from leakage of fluid and the patient can be moved to a sitting position without adversely affecting calibration. Surface electromyography instrumentation is widely used and proven effective for biofeedback training. Although not suitable for either coordination training or sensory conditioning for fecal incontinence, sEMG, used at Cleveland Clinic Florida for the routine care of patients with pelvic floor dysfunction, is more cost effective and suitable for office use.[3] Patients are able to remain fully clothed during the session and position changes are easily accomplished to assist with functional maneuvers. The therapeutic component involves the clinician taking an active role by establishing a rapport with the patient, reviewing bowel and bladder habits, educating the patient, and interpreting data. Clinicians must have a complete understanding of bowel and bladder functioning, considering the coexistence of multifactorial concomitant PFM dysfunction. For example, increased PFM sEMG activity during voiding is indicative of functional outlet obstruction, which may inhibit a detrusor contraction, thus requiring excessive straining by increasing intraabdominal pressure to empty the bladder. This also consequently produces a dysfunctional defecation pattern and contributes to symptoms of constipation. Chronic straining at stool is another source of PFM denervation that contributes to PFM weakness and incontinence. Patients with fecal incontinence may complain of multiple daily bowel movements and a feeling of incomplete evacuation resulting in postdefecation seepage along with concomitant symptoms of urinary incontinence. For these reasons, it is difficult to offer a specific standard biofeedback therapy protocol that is beneficial for all patients. Therefore, the clinician must address all bowel and bladder symptoms and develop an individualized program for each patient with progressive realistic goals. The behavioral component is aimed toward systematic changes in the patient's behavior to influence bowel and bladder function. The PMR component involves designing an exercise program suitable for each patient to achieve the ultimate goal of efficient PFM function.

Surface Electromyography Instrumentation

Surface electrodes summate the electrical action potentials from the contracting muscle and establish electrical pathways from skin contact of the monitored muscle site (Figure 12-1.1). The sEMG instrument receives and processes this electrical correlate of a muscle activity measured in microvolts (μV) (Figure 12-1.2). Intracavitary or surface sensors may detect one or two channels of sEMG activity. The two-channel multiple electrode probe (MEP) anal sEMG sensor (Figure 12-1.3) allows discrimination between proximal and distal external anal sphincter activity, thereby allowing the clinician to target specific areas of external anal sphincter inactivity in the rehabilitation process.

Surface Electromyography Evaluation

Abdominal and pelvic floor channels are the two sEMG channels most frequently used and should be simultaneously monitored during sEMG evaluation and sEMG biofeedback-assisted pelvic muscle exercise training. Interpretive problems arise when monitoring only PFMs without controlling changes in the intraabdominal pressure. The transmission of abdominal artifact to perineal

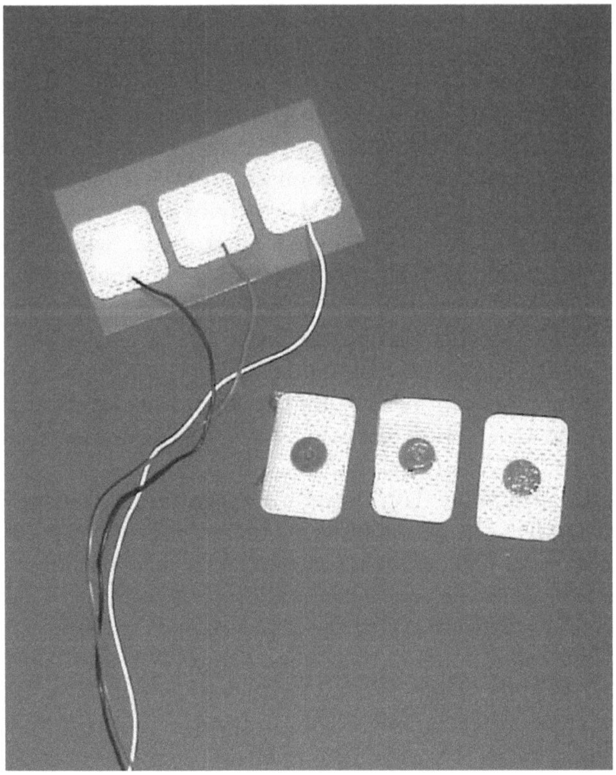

Figure 12-1.1. Surface EMG electrodes.

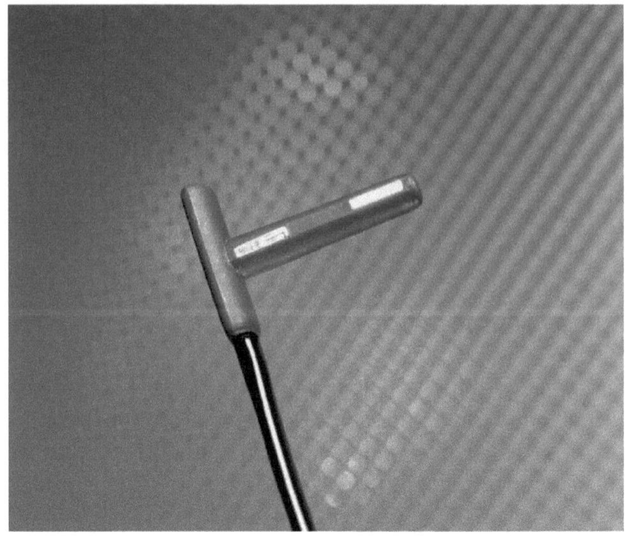

Figure 12-1.3. The multiple electrode probe (MEP) internal sensor. (Courtesy of SRS Medical, Redmond, WA.)

measurements invalidates changes in the PFM measurements and can inadvertently reinforce maladaptive abdominal contractions. The recommended surface electrode placement for monitoring abdominal muscle activity is along the lower quadrants of the abdominal wall. Intravaginal, intrarectal, or perianal placement of surface electrodes may be used to monitor the PFMs. To obtain an evaluation, patients are instructed to simply relax, and then perform an isolated pelvic muscle contraction over a 10-second period, followed by performing a Valsalva maneuver; this sequence is repeated 2 to 4 times for accuracy. During contraction, the abdominal muscle activity should remain relatively low and stable indicating the patient's ability to isolate PFM contraction from abdominal contraction (Figure 12-1.4). During the Valsalva maneuver,

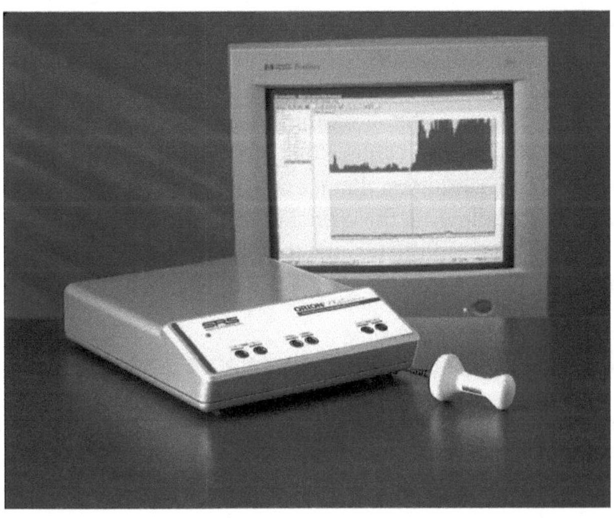

Figure 12-1.2. The Orion Platinum Multi-Modality Biofeedback System shows a typical display during a PFM contraction. (Courtesy of SRS Medical, Redmond, WA.)

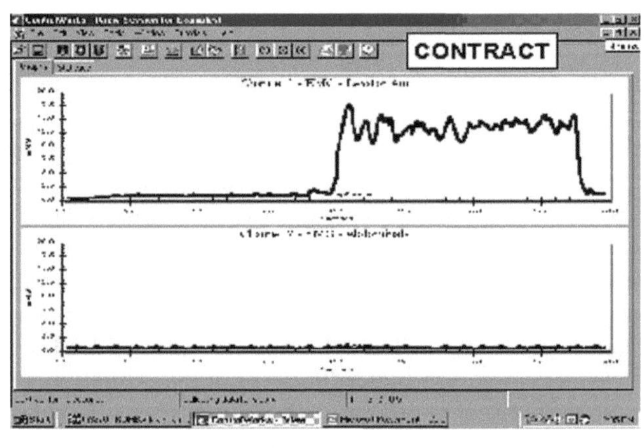

Figure 12-1.4. Channel 1 – PFM sEMG tracing during a contraction. Channel 2 – Abdominal sEMG muscle activity.

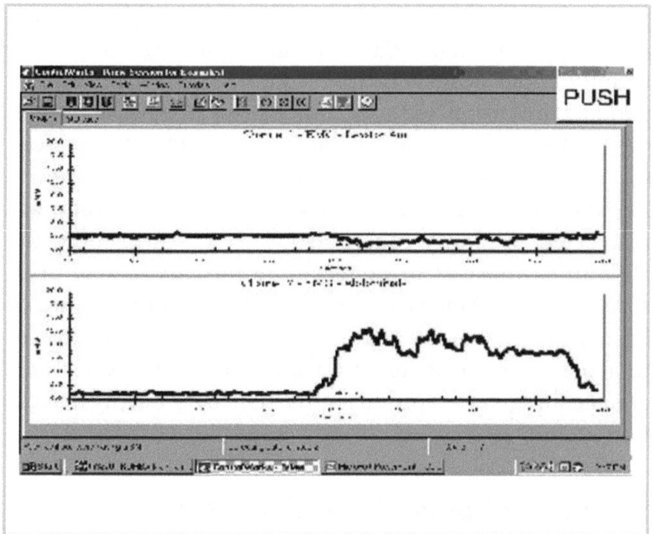

Figure 12-1.5. Channel 1 – sEMG tracing of the PFM during a Valsalva maneuver. Channel 2 – Abdominal sEMG muscle activity.

PFM muscle activity should decrease below the resting baseline (to <2 μV), while the abdominal sEMG activity increases with elevated intraabdominal pressure (Figure 12-1.5). These objective measurements are documented and reviewed with the patient. This also provides the clinician with initial objective measurements to gauge training and recommended at-home practice according to individual capabilities.

Biofeedback Sessions

At Cleveland Clinic Florida, all patients are requested to keep a 3- to 7-day diary of bladder and bowel habits, laxative, enema, or suppository use, fluid intake, number of home exercises completed, fiber intake, and any associated symptoms of constipation or incontinence.

The initial session begins with a history and description of the anatomy and physiology of the bowel, bladder, and pelvic muscle function using anatomic diagrams and visual aids. This is followed by a description of the biofeedback process, instrumentation, and PMR exercises. Results are not immediate; as with any exercise program, muscle improvement requires time and effort. The initial goals of isolated pelvic muscle contractions are established and an example of an sEMG tracing showing efficient muscle function is reviewed. Patients are given instructions on proper insertion of the internal sensor and remain fully clothed during the session. They are placed in a comfortable semi-recumbent position for training. Surface electrodes are then placed on the right abdominal quadrant along the long axis of the oblique muscles, used to monitor abdominal accessory muscle use. The cables are attached to the SRS Orion PC/12 (SRS Medical Systems, Inc., Redmond, WA) multimodality instrumentation that provides the ability to simultaneously monitor up to four muscle sites (Figure 12-1.2). Electromyography specifications include a bandwidth of 20 to 500 Hz and 50/60 Hz notch filter. Surface EMG evaluation is performed and reviewed with the patient.

Training for dyssynergia, incontinence, or pain begins with the systematic shaping of isolated pelvic muscle contractions. Observation of other accessory muscle use such as the gluteal or thighs during the session is discussed with the patient. Excessive pelvic muscle activity with an elevated resting tone more than 2 μV may be associated with dyssynergia, voiding dysfunction, and pelvic pain. Jacobson's progressive muscle relaxation strategy implicates that, after a muscle tenses, it automatically relaxes more deeply when released. This strategy is used to assist with hypertonia, placing emphasis on awareness of decreased muscle activity viewed on the screen as the PFM becomes more relaxed. This repetitive contract–relax sequence of isolated pelvic muscle contractions also facilitates discrimination between muscle tension and muscle relaxation. Some patients, usually women, have greater PFM descent with straining during defecation associated with difficulty in rectal expulsion. Pelvic floor weakness may result in intrarectal mucosal intussusception or rectal prolapse, which contributes to symptoms of constipation, along with vaginal prolapse. Multifactorial concomitant PFM dysfunction accounts for the rationale to initiate all patients with isolated symptoms on comprehensive pelvic muscle rehabilitative therapy. Home practice recommendations depend on the observed decay in the duration of the contraction accompanied by the abdominal muscle recruitment. The number of contractions the patient is able to perform before notable muscle fatigue occurs gauges the number of repetitions recommended at one time. Fatigue can be observed in as few as 3 to 4 contractions seen in patients with weak PFMs. An example of home practice may begin with the patient performing an isolated PFM contraction, holding for a 5-second duration, relaxing for 10 seconds, and repeating 3 to 10 times, which constitutes one set. One set is performed 3 to 5 times daily, at designated intervals, allowing for extended rest periods between sets. The lower the number of repetitions, the more frequent interval sets should be performed daily. Excessive repetitions may overly fatigue the muscles and exacerbate symptoms.

Subsequent sessions begin with a diary review and establishing further goals aimed toward individualized symptom improvement. This is followed by an sEMG evaluation, which may include the addition of quick contract and release repetitions and Valsalva maneuvers, depending on the patient's progress. These objective measurements gauge improvements in muscle activity that should be seen with each visit and occur before symptomatic improvement; this provides positive reinforcement for the patient to continue treatment. To assist with compliance, additional tasks should be limited to no more than three at any given time. These tasks, tailored to the individual needs, may include increasing the duration and number of PFM exercises, habit training, physiologic quieting, anorectal coordination maneuvers, altering fiber and fluid intake, increasing activity and/or modifying laxative use or other methods of evacuatory assistance. Although the ideal goal may be to abolish all symptoms, this may not always be accomplished because of underlying conditions. Therefore, patients should be directly involved in setting their treatment goals. Some patients may be satisfied simply with the ability to leave home without fear of a significant relative fecal accident. Improved quality of life and patient satisfaction should be considered a treatment success.

Session Duration and Frequency

At the onset of biofeedback therapy, it may be difficult to ascertain how many sessions are required for successful training. The number of biofeedback training sessions should be customized for each patient depending on the complexity of their functional disorder as well as the patient's ability to learn and master a new skill. They are typically scheduled from 1 to 1.5 hourly visits once or twice weekly. Additionally, periodic reinforcement is recommended to improve long-term outcome.

Adjunctive Treatment Methods

Various adjunctive biofeedback treatment methods have been used throughout the years. Patients with symptoms of difficult, infrequent, or incomplete micturition or evacuation often exhibit increased PFM activity while performing the Valsalva maneuver during the initial evaluation and are taught the anorectal coordination maneuver. It remains unclear which of the three components – sphincter training, sensory conditioning, or rectoanal coordination – is most useful in the treatment for fecal incontinence. However, most agree that additional treatment methods may be helpful with symptomatic improvement depending on the underlying condition.

Anorectal Coordination Maneuver for Constipation and Dysfunctional Voiding

The goal is to produce a coordinated movement that consists of increasing intraabdominal (intrarectal) pressure while simultaneously relaxing the pelvic muscles. During the initial sEMG evaluation of the Valsalva maneuver, patients are asked to bear down or strain as if attempting to void or evacuate, which may elicit an immediate pelvic muscle contraction and closure of the anorectal outlet. This correlates with symptoms of constipation including excessive straining and incomplete evacuation. The results of the sEMG activity observed on the screen display must first be explained and understood by the patient before awareness and change can occur. Change begins with educating the patient on diaphragmatic breathing, proper positioning, and habit training. Relaxation and quieting the muscle activity while observing the screen is reviewed. Initially patients are instructed to practice these behavioral strategies; however, some patients may continue to feel the need to " push" or strain to assist with expulsion. While observing the sEMG muscle activity on the screen, they are instructed to slowly inhale deeply while protruding the abdominal muscles to increase the intraabdominal pressure. They are then asked to exhale slowly through pursed lips. The degree of the abdominal and anal effort is titrated to achieve a coordinated relaxation of the PFMs. Patients are encouraged to reproduce this maneuver during defecation attempts.

Sensory Discrimination Training for Fecal Incontinence

The sensory discrimination training technique involves a series of brief balloon inflations, noting the volume that induces a sensation of the urge to defecate thereby establishing a current sensory threshold. The volume is subsequently reduced by 25% and a series of insufflations are repeated until the patient is able to promptly recognize the new stimuli. Once patients learn to associate the increase in intrarectal pressure with balloon inflation, the patient is encouraged to recognize sequentially smaller volumes of distention. Thus, after each session, new sensory thresholds are established. The mechanisms by which biofeedback training improves rectal perception are unclear. It is suggested that biofeedback training may recruit sensory neurons adjacent to damaged afferent pathways. However, the speed with which sensory thresholds improve during biofeedback suggests that patients use existing afferent pathways, but learn to pay more attention to weak sensations and to recognize their significance (discrimination training).

Coordination Training for Fecal Incontinence

The aim of coordination training is to achieve a maximum voluntary squeeze in less than 1 to 2 seconds after inflation of the rectal balloon and to control the reflex anal relaxation by consciously contracting the sphincter muscles. This maneuver mimics the arrival of stool in the rectum and prepares the patient to react appropriately by using the appropriate muscle group. With each balloon insufflation, patients are asked to signal when they have perceived rectal distention and react to this distention by promptly contracting the PFM and maintaining this contraction without increasing intraabdominal pressure; this maneuver is performed while the patient observes changes on the visual display. The key element is to condition the external anal sphincter responses to improve the squeeze profile and the ability to respond to small volumes of rectal distention.

Efficacy of Biofeedback: Literature Review

When interpreting the reported clinical outcomes, one should keep in mind that there are no established guidelines regarding the number of sessions, teaching methods, clinician qualifications, type of equipment used, or patient inclusion criteria, nor are there subjective or objective data used to establish success. Hyman et al.[4] reported in their critical review that perhaps most importantly, there is no identified standard for training biofeedback clinicians to treat pelvic floor disorders. As with any therapy, the competence of the clinician is likely to have a significant impact on the outcome of treatment.[4] Norton and Kamm[2] reported that many patients lack the motivation or are unconvinced about the possible value of what they perceive to be simple exercises; therefore, the results of treatment are largely patient dependent, unlike drug or surgical therapy. Gilliland et al.[5] reported that patient motivation and willingness to comply with treatment protocols was the most important predictor of success. We fully agree with that premise.

Constipation

The many variants in these clinical trials may account for the wide range of reported success rates ranging from 30% to 100%. The end-points for successful treatment have not been clearly defined, and the duration of follow-up has been quite variable.[6] Furthermore, outcome was assessed by various methods including diary cards, reviews, telephone interviews, and questionnaires. These evaluation techniques are unreliable when the recorded event, such as defecation, is infrequent in nature. Furthermore, diagnostic data from physiologic testing beyond confirmation of spastic pelvic floor syndrome are often not reported. Concomitant conditions such as the presence of rectoceles, rectal sensory thresholds, and previous surgery disclose a significant variance in inclusion criteria, which presumably contributes to the success of treatment. In a recent study of 49 patients with idiopathic constipation, pre- and post-biofeedback using objective measurements as well as patient symptom diaries, symptomatic improvement occurred in 59% of patients.[7] We believe these are typical expected outcomes.

Fecal Incontinence

Enck's critical review[8] summarized a total of 13 clinical studies published between 1974 and 1990 using biofeedback therapy for the treatment of fecal incontinence. He reported that weighing the number of patients included into each study yields an overall success rate of 79.8%. Despite the wide variety in almost all criteria used to compare these studies, the therapy outcome is homogenous, ranging between 50% to 90%. In a review of 14 biofeedback studies performed between 1988 and 1997, Rao et al.[9] reported that 40% to 100% of patients were improved. The mechanism by which training effects are achieved is controversial. Some have argued that the most important ingredient is sensory discrimination training in which patients are taught to recognize and respond to increased intrarectal pressure or to squeeze more quickly in response to rectal distention. Others believe that biofeedback works primarily by strengthening the external anal sphincter muscles. On one hand, sensations consistently improve with biofeedback. Because this improvement occurs rapidly, it is likely associated with relearning of neurophysiologic patterns that are essentially intact but not in use because of faulty sensation. When the muscles are weak but sensation is intact, symptom reduction would depend on changing muscle strength through an extended and well-designed exercise protocol. In general, most experts believe that all components are useful and that the treatment program should be customized to each patient depending on the underlying dysfunction.

Urinary Incontinence

Outcomes of studies that compare pelvic muscle exercises alone to pelvic muscle exercises with biofeedback emphasize the importance of appropriate PFM identification. Bergio et al.[10] concluded that both groups significantly reduced the frequency of incontinence. The biofeedback group averaged a 75.9% reduction in incontinence, significantly greater than the 51% reduction shown by the verbal feedback group. Furthermore, whereas the biofeedback group improved in the strength and selective control of the PFMs, the same improvement was not seen in the verbal group. In a review of randomized clinical trials of physical therapy for stress urinary incontinence (SUI) and based on levels of evidence criteria, strong evidence was found to suggest that PFM exercises are effective in reducing the symptoms of SUI.[11] They found no evidence that PFM exercises with biofeedback were more effective than PFM exercises alone. In contrast in, a meta-analysis of studies that examined the patients with SUI, PFM exercises with biofeedback were more effective than PFM exercises alone.[12] Patients who do undergo biofeedback have a more rapid reduction in leakage compared with those who perform pelvic floor exercises alone.[13] In clinical practice, behavioral therapy is often combined with drug therapy. A randomized controlled study of 197 women with urge urinary incontinence or mixed incontinence with urge as the predominant pattern compared the effectiveness of biofeedback-assisted behavioral treatment with drug treatment.[14] Subjects were randomized to four sessions of biofeedback-assisted behavioral treatment, drug treatment with oral immediate-release oxybutynin chloride, or a placebo control group. Behavioral treatment, which yielded a mean of 80.7% reduction of incontinent episodes, was significantly more effective than drug treatment (mean 39.4% reduction). Patient-perceived improvement was greatest for behavioral treatment. Only 14% of patients receiving behavioral treatment wanted to change to another treatment versus 75% in each of the other groups. Multimodality approaches, combining physiotherapy, drug therapy, and behavioral pattern modification, seem to have the highest success rate.[15]

Summary

Despite the many variants in the clinical trials for biofeedback, most experts agree that biofeedback is an attractive outpatient, conservative treatment option that is cost effective, relatively noninvasive, easy to tolerate, morbidity free, and does not interfere with any future treatment options that may be recommended by the physician. It is gratifying to note that this simple technique can ameliorate symptoms and improve the quality of life in many patients with functional bowel and bladder symptoms attributed to pelvic muscle dysfunction.

References

1. Schwartz MS and Associates. Biofeedback: A Practitioner's Guide. 2nd ed. New York: The Guilford Press; 1995.
2. Norton C, Kamm MA. Outcome of biofeedback for faecal incontinence. Br J Surg 1999;86:1159–1163.
3. Rao SS. The technical aspects of biofeedback therapy for defecation disorders. Gastroenterologist 1998;6:96–103.
4. Hyman S, Jones KR, Ringel Y, Scarlett Y, Whitehead WE. Biofeedback treatment of fecal incontinence. Dis Colon Rectum 2001;44:728–736.
5. Gilliland R, Heymen S, Altomare DF, Park UC, Vickers D, Wexner SD. Outcome and predictors of success of biofeedback for constipation. Br J Surg 1997;84:1123–1126.
6. Rao SS, Welcher KD, Leistikow JS. Obstructive defecation: a failure of rectoanal coordination. Am J Gastroenterol 1998:93:1042–1050.
7. Emmanuel AV, Kamm MA. Response to a behavioral treatment, biofeedback in constipated patients is associated with improved gut transit and autonomic innervation. Gut 2001;49:214–219.
8. Enck P. Biofeedback training in disordered defecation: a critical review. Dig Dis Sci 1993;38:1953–1959.
9. Rao SS, Welcher KD, Pelsang RE. Effects of biofeedback therapy on anorectal function in obstructive defecation. Dig Dis Sci 1997;42:2197–2205.
10. Burgio KL, Robinson JC, Engel BT. The role of biofeedback in Kegel exercise training for stress urinary incontinence. Am J Obstet Gynecol 1986;154:58–64.
11. Berghmans LC, Frederiks CM, de Bie RA, et al. Efficacy of biofeedback, when included with pelvic muscle exercise treatment, for genuine stress incontinence. Neurourol Urodyn 1996;15:37–52.
12. Weatherall M. Biofeedback or pelvic floor muscle exercises for female genuine stress incontinence: a meta-analysis of trial identified in a systemic review. BJU Int 1999;83:1015–1016.
13. Pages IH, Jahr S, Schaufele MK, Conradi E. Comparative analysis of biofeedback and physical therapy for treatment of urinary stress incontinence in women. Am J Phys Med Rehabil 2001;80:494–502.
14. Burgio KL, Locher JL, Goode PS, et al. Behavioral vs drug treatment for urge urinary incontinence in older women: a randomized controlled trial. JAMA 1998;280:1995–2000.
15. Davila GW, Bernier F. Multimodality pelvic physiotherapy treatment of urinary incontinence in adult women. Int Urogynecol J 1995;6:187–194.

12-2

Timed Voiding and Fluid Management

Marie Fidela R. Paraiso and George Abate

Timed voiding (also referred to as bladder retraining/training, bladder drill, bladder discipline, and bladder reeducation) is a behavioral exercise used to establish bladder control in adults. Evidence indicating that bladder training is effective in women with urodynamic urge incontinence with or without associated detrusor overactivity, sensory-urgency without incontinence,[1] and urodynamic stress incontinence,[2] has led to the 1996 Clinical Practice Guideline's recommendation of bladder training as a first-line therapy for urge, stress, and mixed incontinence.[3] Urodynamic studies are not required before initiation of behavioral therapy.

The three components of timed voiding are education regarding continence and incontinence mechanisms; scheduled voiding with systemic delay of voiding by implementing distraction and relaxation techniques; and positive reinforcement provided by a caregiver. The specific goals of bladder training include correcting inappropriate habits of frequent urination, improving control of bladder urgency, extending intervals between voids, increasing bladder capacity, reducing incontinence episodes, and building patient self-confidence in bladder control. There is very little understanding of how bladder retraining works. Several hypotheses exist, including improved cortical suppression of sensory stimuli from an uninhibited bladder, improved cortical inhibition of an overactive detrusor muscle, maintenance of bladder pressures lower than the urethral closure pressure during stress, increasing bladder "reserve" volume, and altered patient behavior to avoid incontinence triggers and to gain awareness of the lower urinary tract.

Two randomized trials showed significant improvements in incontinent patients when compared with an untreated control group. Jarvis and Millar[4] reported that 90% of the treatment group was continent and 83.3% was symptom free at 6 months; whereas, the control group reported 23.3% continence and lack of symptoms. Fantl et al.[2] demonstrated that 12% of the treatment group was continent and 76% were improved at 6 weeks and subsequently maintained at 6 months. They also reported a 55% improvement of quality of life. One multicentered randomized trial by Wyman et al.[5] indicated that bladder retraining and pelvic floor muscle exercises showed similar efficacy in women with urge, stress, and mixed incontinence.

The literature regarding dietary factors and fluid intake and their effect on incontinence is sparse, and there are few poorly powered randomized trials. Caffeine intake increases detrusor pressure on cystometry, and women with detrusor overactivity have a higher mean caffeine intake.[6,7] Decrease of caffeine intake decreases urine loss episodes.[8] In the only randomized trial by Bryant et al.,[9] reduction of daily caffeine intake to less than 100 mg along with bladder exercises, decreased incontinent episodes by 74% versus a 32% reduction in those who continued to take normal caffeine and performed the same bladder exercises. In a multivariable analysis, there was no association between coffee drinking or alcohol consumption and incontinence.[10] Griffiths et al.[11] showed a strong relationship between evening fluid intake, nocturia, and nocturnal voided volume. However, the relationship was not as strong for diurnal intake, voiding, and voided volume. A modest positive association has been noted between fluid intake and incontinence severity in women older than age 55, whereas no such correlation was seen in women with detrusor overactivity.[12] In a small randomized trial of 32 women who were assigned to three different groups of fluid intake, adherence to protocol was so poor that the results were difficult to interpret.[13] There is anecdotal evidence that supports elimination of certain dietary irritants. In summary, fluid intake has a minimal effect on incontinence and caffeine intake, and its role in the pathogenesis of incontinence is controversial.

Timed voiding and fluid management go hand in hand as first-line therapy for urinary incontinence. A thorough history and physical, urinalysis, postvoid residual, and urinary diary are the baseline assessment of all incontinent patients before this first-line treatment. Bladder diaries are essential in helping the clinician and patient identify habits that the patient may modify to reduce incontinence.

Voiding and fluid diaries can be used to determine whether decreasing fluids or increasing the voiding interval is appropriate. (Please refer to Chapter 14-1 on bladder diaries for more detail.) Women with excessive fluid intake can decrease incontinence by limiting intake. However, other causes for incontinence, such as diabetes mellitus, diabetes insipidus, and hypercalcemia should be ruled out as causes for polydipsia and polyuria. If a patient complains of urinary frequency and drinks large volumes of fluid per day, we will obtain a fluid diary, and if voided volume is more than 3 to 4 L, a urine osmolality (Uosm) is ordered. If the Uosm is 200 or less, a plasma osmolality (Posm) is obtained, if less than 280 mOsm/kg, psychogenic polydipsia is likely and if greater than 290 mOsm/kg, diabetes insipidus is the diagnosis (Figure 12-2.1).[14]

Patients often are not aware of how much or what types of fluids they drink. With fluid diaries, patients learn to associate bad habits, such as caffeine intake, with worsening incontinence. The recommended daily water intake is 0.5 ounce per pound of body weight, or six to eight glasses a day. However, few scientific data are available to support this recommendation.[15] Constipation, which is a common problem that can worsen incontinence, can be lessened by adequate fluid consumption. Caution should be taken when advising patients to decrease fluid intake because this may lead to constipation, urinary tract infections, and dehydration. Adequate fluid intake ensures nonirritating, dilute urine, which reduces incontinence. Only patients with abnormally high fluid intake should be counseled to decrease fluid intake.

Generally, acidic and caffeinated foods and beverages, as well as oxalate-containing foods, are bladder irritants. The following foods are common bladder irritants:

- Alcoholic beverages
- Apples and apple juice
- Aspartame and saccharin
- Cantaloupe
- Carbonated beverages
- Chili and spicy foods
- Chocolate

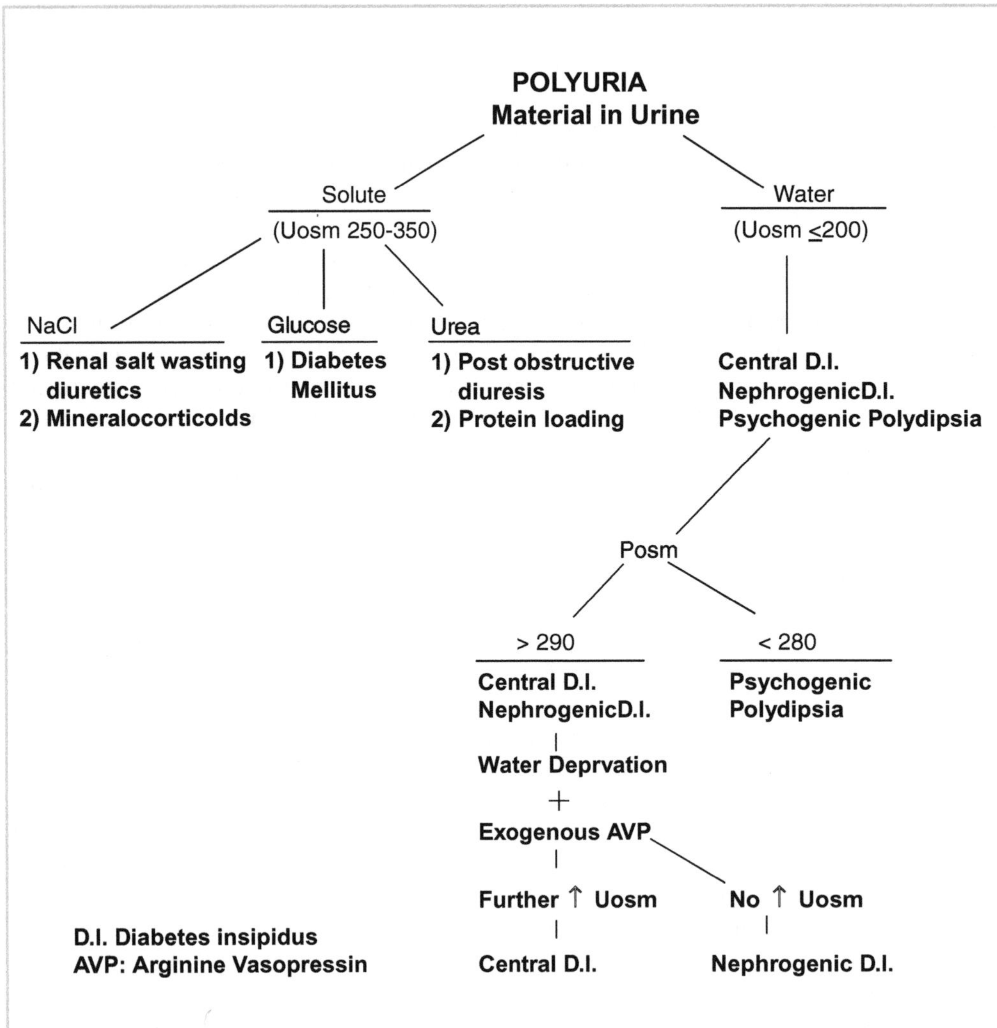

Figure 12-2.1. Evaluation of polyuria.

- Citrus fruits and juices
- Coffee (including decaffeinated)
- Cranberries and cranberry juice
- Grapes and grape juice
- Guava peaches
- Pineapple
- Plums
- Strawberries
- Sugar
- Tea
- Tomatoes
- Vitamin B complex
- Vinegar

Oxalate-containing foods include all berries, as well as spinach. The acid-restricted diet is most effective when 64 ounces of water are ingested daily. Low-acid substitutions include apricots, papaya, pears, and watermelon. Coffee drinkers may consume kava. Tea drinkers may substitute noncitrus herbal and sun-brewed teas. Citracal tablets, taken with meals, may reduce urinary excretion of dietary oxalates.

Strict adherence to a nonirritant diet often brings marked improvement in symptoms in just 10 to 15 days. Once a baseline is maintained, then irritants can be individually resumed. If symptoms return, then the most irritating sources are identified. This systematic approach allows patients to not eliminate all irritants, thereby maintaining quality of life. Patients with suspected interstitial cystitis should more closely follow an antiinflammatory diet. (Please refer to the chapter on interstitial cystitis.)

Voiding diaries can help to elucidate whether patients have a stress or urge-predominant incontinence pattern. The diaries document involuntary leaking episodes, precipitating factors, degree of urgency, timing, and volume of voids. The diaries also confirm that behavioral changes work to prevent incontinence. This allows for estimates of 24-hour urine volume, nocturia, voiding frequency, and functional bladder capacity. A voiding schedule is developed from the patient's voiding diary. Women with frequency symptoms undergo bladder retraining to expand the bladder's capacity. Patients use relaxation techniques (deep breathing and positive self-acknowledgments), distraction methods (hand-held computerized games, puzzles, problem solving), Kegel exercises, and/or perineal pressure to delay nonscheduled voids. The initial assigned voiding interval may vary, from 30 minutes to 60 minutes being the most common interval. The voiding interval may be increased by 15 or 30 minutes based on the patient's tolerance of the schedule. Self-charting of voiding patterns often leads to increased bladder awareness. Over 6 weeks, the patient will increase the interval between voiding with a goal of 2 to 3 hours between voids (Table 12-2.1). Women who rarely void are taught to schedule their voids at regular intervals. Follow-up visits to review the diaries are usually scheduled for between 1 to 2 weeks. Reassurance, enthusiastic support, and follow-up are important, because this treatment modality is driven by patient compliance.

Table 12-2.1. Bladder training instructions

Many of our bodily functions are influenced by habit, and if these habits are bad we can change them. This is certainly true of the bladder. The object is for you to reestablish your brain's control over your bladder function rather than the other way around.

To change your bladder control habit, start by urinating (voiding, peeing, emptying your bladder) every 45 minutes during the day. You **must** urinate by the clock, whether you need to urinate or not. At night, urinate only as the need arises. But, the next morning you **must** get back on a rigid schedule.

Remember, if you are scheduled to urinate at 6:30, but at 6:10 you get a strong urge to urinate, you **must** try to wait until 6:30 to urinate even if you leak urine before the scheduled time. Should you develop urgency in between the voiding intervals, immediately sit down in a comfortable position. Take slow deep breaths in and out through your mouth and try imagining yourself in a favorite vacation spot or use some other relaxation technique until the urge passes.

When you are able to follow this schedule for 7 days without losing control of your bladder, increase the interval between voids by 15 minutes. (That is, increase the time between urinating to 60 minutes, etc., until you can comfortably go 2 to 3 hours without urinating.)

This requires a lot of discipline on your part but you **can** do it, and you will be pleased with how well you and your bladder get along in the future.

Bladder Training Week	Timed Voiding Interval (min)
Week #1	
Week #2	
Week #3	
Week #4	
Week #5	
Week #6	

General bladder training recommendations include:

- Fluid restriction after 6 pm to decrease nocturia.
- Routine bladder emptying. Women with stress incontinence can significantly reduce their symptoms if a full bladder is avoided. Patients will make a concerted effort to set aside time to void, once this connection is made.
- Taper caffeinated and carbonated beverage usage to avoid caffeine withdrawal headaches. High caffeine intake is more than 3 cups of coffee per day.
- Do not rush to the bathroom. This increases abdominal pressure and contributes to poor muscle coordination.

Conclusion

Timed voiding and fluid management are first-line therapy options for the treatment of incontinence. Complex testing need not be performed before instituting this therapy. Patient compliance and clinician reinforcement are the basis for success of this intervention.

References

1. Frewen WK. A reassessment of bladder training in detrusor dysfunction in the female. Br J Urol 1982;54:372–373.

2. Fantl JA, Wyman JF, McClish DK, Bump RC. Efficacy of bladder training in older women with urinary incontinence. JAMA 1991;265: 609–613.
3. Fantl JA, Newman DK, Colling J, et al. Urinary Incontinence in Adults: Acute and Chronic Management. Rockville, MD: U.S. Department of Health and Human Services. Public Health Service. Agency for Health Care Policy and Research; 1996.
4. Jarvis GJ, Millar DR. Controlled trial of bladder drill for detrusor instability. Br Med J 1980;281:1322–1323.
5. Wyman JF, Fantl JA, McClish DK, Bump RC. Comparative efficacy of behavioral interventions in the management of female urinary incontinence. Continence program for Women Research Group. Am J Obstet Gyneol 1998;179(4):999–1007.
6. Arya LA, Myers DL, Jackson ND. Dietary caffeine intake and the risk for detrusor instability: a case-control study. Obstet Gynecol 2000; 96:85–89.
7. Creighton SM, Stanton SL. Caffeine: does it affect your bladder? Br J Urol 1990;66:613–614.
8. Tomlinson BU, Dougherty MC, Pendergast JF, Boyington AR, Coffman MA, Pickens SM. Dietary caffeine, fluid intake, and urinary incontinence in older rural women. Int Urogynecol J Pelvic Floor Dysfunct 1999;10:22–28.
9. Bryant CM, Dowell CJ, Fairbrother G. A randomized trial of the effects of caffeine upon frequency, urgency, and urge incontinence. Neurourol Urodyn 2000;19:501–502.
10. Brown JS, Seeley DG, Fong J, Black DM, Ensrud KE, Grady D. Urinary incontinence in older women: who is at risk? Study of Osteoporotic Fractures Research Group. Obstet Gyneol 1996;87(5):715–721.
11. Griffiths DJ, McCracken PN, Harrison GM, Gormley EA. Relationship of fluid intake to voluntary micturition and urinary incontinence in geriatric patients. Neurourol Urodyn 1993;12:1–7.
12. Wyman JF, Elswick RK, Wilson MS, Fantl JA. Relationship of fluid intake to voluntary micturitions and urinary incontinence in women. Neurourology 1991:463–473.
13. Dowd TT, Campbell JM, Jones JA. Fluid intake and urinary incontinence in older community-dwelling women. J Community Health Nurs 1996;13:179–186.
14. Pohl MA. Critical fluid and electrolytic abnormalities in clinical practice. In: Stoller JK, Ahmad M, Longworth DL, eds. The Cleveland Clinic Intensive Review of Internal Medicine. 2nd ed. Philadelphia: Lippincott Williams & Wilkins; 2000.
15. Valtin H. "Drink at least eight glasses of water a day." Really? Is there scientific evidence for 8 × 8? Am J Physiol Regul Integr Comp Physiol 2002;283:R993–R1004.

12-3

Bowel Retraining for Functional Disorders of the Colon, Rectum, and Anus

Susan M. Cera and Jonathan E. Efron

Bowel retraining refers to physiotherapeutic, or nonoperative, approaches to functional colorectal disorders. These nonsurgical or conservative measures may form the initial management for patients with both constipation and anal incontinence. They also provide means of effective therapy for mild forms of functional colonic disorders, high-risk surgical candidates, and for those patients who decline an operation. They may be as an adjunct to surgical therapy, or as part of a continuum once surgery has been completed. Bowel retraining involves diet manipulation, medical intervention, and biofeedback therapy.[1–3] In this chapter, bowel retraining will be discussed with respect to the two major classes of functional colorectal disorders, constipation and fecal incontinence.

Bowel Retraining for Constipation

The surgeon's first priority when faced with a patient complaining of constipation is to eliminate mechanical obstruction as the cause of the constipation. A complete history and physical examination should be performed. The history needs to focus on multiple factors, but specific attention should be given to the patient's medications as well as their diet and exercise regimen. If mechanical obstruction has not been ruled out before evaluation of the patient, this should be done either by colonoscopy or barium enema.

Diets low in fiber coupled with poor bowel habits are common causes of constipation. Management requires reassurance and simple guidance. Patients, especially those with fast-paced and stressful lifestyles, should be advised not to ignore the call to stool because it can lead to stasis and impaction. Regular exercise (e.g., a walk every morning) promotes regularity. Narcotics, diuretics, calcium channel blockers, antidepressants, and irritant laxatives may all result in constipation through different mechanisms. Fortunately, the effects of most of these agents are reversible, and discontinuing the medication usually results in cessation of the constipation. Meal patterns such as meal omission, fast foods, ingestion of large amounts of constipating foods, and inadequate intake of fluids should be recognized and modified. These changes, with the addition of fiber, aid to retrain the bowel to promote regularity and homeostasis as well as deter the onset of disease such as diverticulosis and cancer.

The addition of a fiber supplement to the diet is often therapeutic in the management of constipation. Inadequate fiber intake is a consequence of industrialization and results in small, hardened stools, poor peristalsis, and ultimately, constipation. The usual daily fiber intake by members of Western society is 10 to 14 g. The recommended daily fiber intake is 25 to 30 g of insoluble fiber with a maximum of 36 g in a 24-hour period. Insoluble fiber may be found in the form of psyllium or other vegetable fiber. Psyllium (3 g per os once or twice a day) is usually an adequate supplement to restore regularity or normalcy to most diets and bowel patterns. Patients are instructed to drink at least six, 8-oz glasses of water a day when taking a fiber supplement to avoid concretions. The majority of patients with complaints of constipation, inadequate rectal emptying, rectal pressure, straining, hard stools, and irregularity will respond to fiber supplementation with adequate fiber intake and need no further treatment. Thus, a fiber trial is not only a test but also a treatment and should be the initial therapy before embarking on an otherwise costly work-up yielding a diagnosis that may be amenable to fiber supplementation.

Despite counseling patients on eating properly, ingesting additional fiber, engaging in regular exercise, and not ignoring the urge to defecate, the addition of laxatives for the persistently constipated patient may be needed, especially in the elderly (Table 12-3.1). Laxatives are divided into two classes: stimulants and mechanical cleansers. Stimulants act to irritate the intestinal lining resulting in increased water, electrolyte, and mucus secretion and con-

Table 12-3.1. Frequently used bowel agents for constipation

Agent Type	Name	Dosage
Fiber		
Psyllium seed preparations	Metamucil, Konsyl	1–2 tsp q.d.–t.i.d.
Synthetic methylcellulose	Citrucel	1 tbsp PO q.d.–t.i.d.
Mechanical cleansers		
Saline laxatives		
Magnesium salt	Magnesium citrate	120–240 mL PO q.d.
Phosphate salt	Fleets Phospho-Soda	30–45 mL PO q.d.
Polyethylene glycol	GoLYTELY	12 oz PO q.d.
	Miralax	17 g PO q.d.
Mineral oil		15–45 mL PO q.d.
Lactulose		15–30 mL PO q.d.–t.i.d.
Stimulants		
Senna	Senokot	2–4 tabs PO q.d.–t.i.d.
Biscodyl	Dulcolax	10 mg PO or PR q.d.
Prokinetic agents		
Metoclopramide	Reglan	5–10 mg PO q.a.c. and q.h.s.
Cisapride	Propulsid	10–20 mg PO q.i.d. (restricted access in U.S.)
Erythromycin base		20 mg/kg/d PO divided t.i.d.–q.i.d.
Tegaserod	Zelnorm	6 mg PO b.i.d. for 4–8 wk

q.d., every day; t.i.d., three times a day; PO, per os; q.a.c., before each meal; q.h.s., before bedtime; q.i.d., four times a day; b.i.d., twice a day

sequent rapid evacuation. These agents are useful in the treatment of acute constipation, but long-term use should be discouraged because it can lead to a poorly functioning large intestine. Mechanical cleansers exert their effect through osmotic activity of poorly absorbed salts (magnesium or phosphate salts), disaccharides (lactulose), or electrolyte solutions (polyethylene glycol, also known as GoLYTELY or Miralax). The net result is increased peristalsis and evacuation of large amounts of watery stools. Liquid petrolatum (mineral oil) retards the absorption of water from the stool and thus softens and bulks fecal material. These substances can be safe adjuncts to a regular regimen for both short- and long-term use in bowel retraining for severe constipation; however, long-term use of mineral oil may lead to anal stenosis or stricture and should be avoided. A daily 12-oz glass of polyethylene glycol (GoLYTELY, Miralax) or lactulose may be used to propel colonic contents and is particularly effective in intractable constipation from colonic inertia. However, caution must be taken to avoid adverse effects that may result from over-consumption such as electrolyte abnormalities, vitamin deficiencies, dehydration, and malabsorption.

Regular use of suppositories and enemas are useful in evacuating the lower bowel. Various types of suppositories are available and primarily work through a reflex mechanism. Enemas work by causing distention, increasing stool volume through an osmotic effect, and direct irrigation of the lower bowel. Suppositories have the advantage of being easier to administer; however, both can safely be used as part of a regular bowel routine in those patients who require assistance with scheduled bowel evacuation. For cases of intractable constipation and the inability to evacuate, the performance of antegrade continent enemas (ACE) has been shown to be effective in bowel retraining in patients who have traditionally relied on laxatives, enemas, digital stimulation, and manual disimpaction.[4] Several minimal techniques have been developed (ACE procedures) to facilitate the administration of antegrade enemas. Most involve the creation of a continent colonic conduit consisting of either an appendicocecostomy or button cecostomy forming a channel that can be easily catheterized to perform regular colonic irrigations. More recently, the laparoscopic ACE (LACE) procedure has been described, which offers the potential advantages of a shorter hospital stay, faster recovery, less pain, and better cosmesis.

Current management of constipation has focused on the use of prokinetic drugs that enhance propulsive activity rather than acting as intestinal irritants. The function of these drugs is primarily to augment the intrinsic motor function of the gut. They are particularly efficacious in colonic inertia in which constipation is believed to result from the degeneration of colonic nerve fibers or abnormal terminal synapse function in the colon. Reglan, cisapride, and erythromycin may be used as part of a comprehensive bowel regimen in this condition to promote motility and stimulate peristaltic activity. The most recent promotility agent is tegaserod (Zelnorm), a 5-HT4 receptor partial agonist, which has been shown to successfully relieve abdominal pain and constipation associated with irritable bowel syndrome by aiding in gastrointestinal motility and modulating visceral sensation.

Along with dietary changes and pharmacotherapies, biofeedback is a viable option for the treatment of constipation. Biofeedback refers to therapy in which patients are trained to be more aware of and responsive to their bowels. It has been increasingly used in the management of functional pelvic floor disorders, such as constipation from obstructed defecation, fecal incontinence, and rectal pain. In patients with constipation secondary to obstructed defecation, biofeedback is used to heighten the patient's awareness of the sphincters and levator muscles to retrain these muscles to consciously relax during the act of defecation. One cause of pelvic outlet obstruction occurs as a result of nonrelaxation of the puborectalis muscles. This condition may be related to significant psychosocial stresses that may cause the patient to alter their normal defecatory patterns.

Biofeedback facilitates bowel retraining through counseling and audio or visual feedback that allows the patient to properly relax their puborectalis muscles resulting in defecation. Through the use of manometry, balloon defecography, surface electromyography (EMG), or intraanal EMG, the patient is shown the effects of his or her own squeeze efforts and the resultant muscular actions of the external sphincter and puborectalis complex. This reinforces to the patient appropriate behavior and modifies errors in the muscular activity. The goal is to retrain the pelvic musculature to elicit the appropriate response in performing the actions related to defecation. Success rates reported in the literature are variable although it seems to be more successful using EMG feedback as opposed to balloon expulsion exercises. Reports of success with biofeedback for constipation vary between 60% to 80%. At our institution, EMG-based biofeedback for patients with constipation is successful approximately 60% of the time.

Bowel retraining in constipation involves a multimodal regimen that can be tailored to the degree of the patient's condition. Patients who are unresponsive to simple dietary modifications, fiber therapy, or medical interventions may require scheduled laxatives or colonic irrigations. Patients with obstructed defecation may respond to a trial of biofeedback in which behavioral therapy is used to modify and retrain the muscles of the pelvic floor in coordinating defecation. Patients who do not respond to medical management of their constipation and who may be surgical candidates should undergo further work-up including colonic transit studies, defecography, possible EMG, and small bowel transit studies. These tests may then be interpreted with possible surgical intervention, if warranted.

Bowel Retraining for Anal Incontinence

Anal incontinence is the involuntary loss of control of rectal contents (solid, liquid, or gas) and can be secondary to a variety of causes. Conservative measures such as dietary manipulation, pharmacologic intervention, scheduled rectal emptying, perineal exercises, and biofeedback therapy are effective in many patients who are not surgical candidates or who do not desire surgical intervention for fecal incontinence. The bowel can then be trained to effectively reestablish continence using combinations of these techniques.

The approach to patients with fecal incontinence begins with the identification of the etiology of the incontinence. Stool consistency, colonic transit, rectal sensation, rectal compliance, pudendal nerve integrity, and sphincter complex function all have a role in continence. Increased stool consistency and colonic transit can result from a number of medical illnesses and alterations in colorectal function. Those patients who do not have concomitant dysfunction of the pelvic floor are best treated with dietary manipulation and constipating agents. All patients should be encouraged to adhere to a well-balanced, high-fiber diet with adequate fluid intake. The addition of fiber to the diet aids in bulking and solidifying the stool to facilitate control in defecation. Some patients may benefit from dietary restrictions such as lactose-free or gluten-free diets.

Multiple medications are available to help slow intestinal transit (Table 12-3.2). Loperamide (Imodium) inhibits intestinal motility through direct effects in the circular smooth muscle of the bowel. It also contributes to the continence by solidifying stool, increasing rectal compliance, and increasing the resting pressure of the sphincter complex. Diphenoxylate hydrochloride (Lomotil) can also be added to slow intestinal transit. Opium derivatives such as tincture of opium, paregoric, and codeine are very potent in slowing colonic transit via direct effects on the colonic musculature; however, the risk of addiction makes these agents less suitable for long-term use. Cholestyramine (Questran) is effective in slowing diarrhea in patients with known alterations in bile metabolism. It also helps bulk stool and may be of benefit in patients with chronic diarrhea.

Pseudo-incontinence, or overflow incontinence, may develop in patients with fecal impaction secondary to constipation. This cause of incontinence can be elicited from physical examination and is treated by removing the impaction and implementing a bowel regimen that prevents constipation and stool stasis.

Minor incontinence from either pudendal neuropathy or sphincter defects may also respond to conservative measures. Frequently, the addition of fiber and increased stool consistency are adequate to facilitate control in these patients.

For patients who are unresponsive to dietary and pharmacologic interventions, the institution of a regular bowel regimen can reeducate the bowel to evacuate at a predictable and convenient time. Colonic irrigations or suppositories administered at the same hour each day, preferably after meals when the gastrocolic reflex may contribute, serve to retrain the bowel to empty with regularity. Patients are taught abdominal massage and positioning exercises to reinforce patterns of defecation. The goal is to promote scheduled evacuations such that the majority of the time, the rectum is empty and episodes of incontinence are minimized. If a disorder in rectal evacuation accompanies the problem of fecal incontinence, ACE through a continent colonic conduit may facilitate regularity.

Evacuation retraining can also be accomplished with the help of a continence plug. The anal plug (the Procon device) (Figure 12-3.1) consists of a catheter that is inserted in the rectum and held in place by a balloon. The catheter has a sensor that detects flatus and stool. The catheter is attached to a beeper that signals when the rectum is full, preventing seepage and allowing adequate time to reach a bathroom and evacuate.

Table 12-3.2. Frequently used bowel agents for fecal incontinence

Agent Type	Name	Dosage
Loperamide	Imodium	2–4 mg PO b.i.d.–q.i.d.
Diphenoxylate hydrochloride	Lomotil	1–2 tabs PO t.i.d.–q.i.d.
Tincture of opium		0.3 mL PO t.i.d.–q.i.d.
Cholestyramine	Questran	2–8 g PO q.d.–b.i.d.

PO, per os; b.i.d., twice a day; q.i.d., four times a day; t.i.d., three times a day; q.d., every day

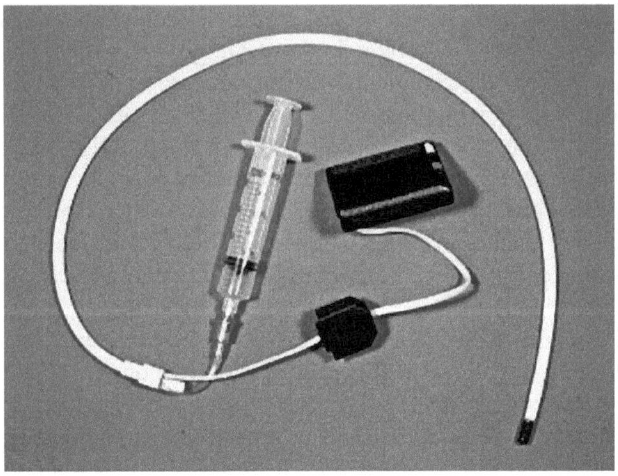

Figure 12-3.1. Procon device.

Perineal (Kegel) exercises have proven to be beneficial in the medical management of both urinary and fecal incontinence. Perineal exercises retrain the muscles of the pelvic floor by increasing muscle bulk and tone of the external anal sphincter, puborectalis, and levators. Exercises consist of contracting the muscles used to hold in bowel movements for 15 seconds at a time. This exercise is repeated 15 to 20 times a day with results seen after several weeks. In patients who are unable to perform such exercises because of physical disability, sphincter contractility and tone can be increased by electrical stimulation of the sphincter muscle using an anal canal electrode. The device applies an electrical stimulus to the anal and pelvic floor muscles causing tonic contraction for 30-minute time periods. The repetitive contractions strengthen the muscles in the same manner as perineal exercises.

Biofeedback, or operant conditioning, is an effective form of behavioral therapy for fecal incontinence. The use of biofeedback is most successful in patients with minor incontinence in whom a viable, functioning, innervated sphincter exists. Using anal EMG probes or manometry catheters, the subject is shown the results of the anal and rectal pressure changes generated during the squeeze. Exercises that contract and relax muscles of the pelvic floor are performed while the records of the muscle activity are monitored on a screen. The result is improved strengthening and efficiency of the anal sphincter, improved rectoanal coordination, and increased rectal compliance and sensation. The patient is often able to strengthen the muscles of continence and train alternate muscles to compensate for weak or defective muscles, when needed. Each session takes 1 to 2 hours and occurs at 4- to 8-week intervals for a total of four to eight sessions or until the biofeedback therapist thinks the patient has obtained maximum benefit. Short-term results include success rates of 65% to 85% in most series and are dependent on the degree of initial dysfunction and patient attendance of all sessions. Effectiveness tends to deteriorate with time requiring reinitiation of biofeedback in some patients.

Bowel retraining for anal incontinence involves pharmacologic treatment and dietary interventions to modulate diarrhea and reestablish normal colonic function. Scheduled rectal emptying minimizes the incidence of fecal incontinence. Perineal exercises in combination with biofeedback serve to retrain and strengthen the muscles of the pelvic floor to allow for anorectal coordination for controlled defecation. Patients with severe incontinence or large anatomic defects in the anal sphincters may benefit from surgery if conservative interventions do not improve fecal control.

References

1. Beck DE, Wexner SD, eds. Fundamentals of Anorectal Surgery. London: WB Saunders; 1998:127–132.
2. Corman ML, ed. Colon and Rectal Surgery. 4th ed. Philadelphia: Lippincott Williams & Wilkins; 1998:301–305, 380–385.
3. Gordon PH, Nivatvongs S, eds. Principles and Practice of Surgery for the Colon, Rectum, and Anus. 2nd ed. St. Louis: Quality Medical Publishing; 1999:376–378, 1199–1204.
4. Karpman E, Das S, Kurzrock EA. Laparoscopic antegrade continence enema (Malone) procedure: description and illustration of technique. J Endourol 2002;16(6):325–328.

Section XIII

Surgical Therapy: Mutual and Combined Aspects

Section XIII

Surgical Therapy: Mutual and Combined Aspects

G. Willy Davila

There is no one single clinician who can address all of the problems that may occur in the pelvic floor. In addition, it is not possible to address a problem in one of the compartments of the pelvic floor without potentially affecting its adjacent compartments. This is especially true with the performance of surgical therapy. It is thus of critical importance for the clinician addressing the patient's pelvic floor dysfunction to be keenly aware of the impact that specific treatment may have on adjacent organ systems.

Combined surgical therapy of pelvic floor problems affecting various organ systems at one setting has been demonstrated to be effective and not associated with increased morbidity. Clinicians should be able to operate together during one surgical setting to avoid sequential surgeries and their potential negative impact with prolonged recovery, increased risk of further denervation injury, and scar tissue formation.

Anatomic pelvic floor problems may directly involve two adjacent organ systems. The occurrence of a fistula between any of the three components of the pelvic floor requires an understanding of the impact of the fistula, and its repair, on both (if not all three) involved compartments. We will frequently operate as a team when addressing a complex fistula to prevent increasing dysfunction in one of the compartments during surgical correction of the fistula.

When addressing surgical complications, an understanding of potential dysfunction, which can result from correction of the underlying complication, is also important. This section will emphasize many of the clinical scenarios in which the expertise of multiple clinicians benefit the patient, applied concomitantly, will pelvic floor.

13-1
Vesico-Vaginal Fistula

Gamal M. Ghoniem and Usama M. Khater

History

A vesicovaginal fistula (VVF) is an epithelialized or fibrous communication between the bladder and vagina. It represents a distressful and difficult situation for both the patient and the physician. This debilitating condition has plagued women with obstructed labor and traumatic delivery for thousands of years. The first recorded references of VVFs were made in 1550 BC in ancient Egypt.[1] Avicenna, renowned Arabic physician, was the first person to document the relationship between VVFs and obstructed labor in 1037.[2] The first reference in European literature to urinary fistulas was made by Platter in 1597.[3] Many attempts at fistula closure followed. However, successful cases were rare. Over many decades, the principles of treatment slowly began to emerge. In 1852, James Marion Sims established the foundations of VVF repair after a series of experimental surgeries on slaves beginning in 1845 in Montgomery, Alabama.[4] These fundamentals included proper exposure with the knee-chest position, use of a weighted vaginal retractor, silver wire sutures, tension-free closure of the defect, and proper postoperative bladder drainage.

Iatrogenic Etiology of Adult Vesicovaginal Fistula – Postsurgical (81%–91% in developed countries)

Vesicovaginal fistulas occur as a result of injury to the vesicovaginal septum. Pelvic surgery is the most common cause of these fistulas in western countries, and hysterectomy is responsible for approximately 80% of all VVFs.[5] The incidence varies with the approach to repair [1:1000 transabdominal (TA), 0.2:1000 transvaginal (TV), and 2.2:1000 with laparoscopic].[6] The most common site is superior to the trigone, corresponding to the vaginal cuff. Other gynecologic procedures account for 11%, including caesarean delivery and dilation and curettage. Incontinence surgery, formalin injections, and laparoscopy are less reported causes of VVF.[7]

Pelvic radiation for malignancy can lead to VVF formation even after many years. The resulting fibrosis with arteritis nodosa, and decreased blood supply leads to tissue necrosis, sloughing, and fistula formation.

Noniatrogenic Etiology of Adult Vesicovaginal Fistula

Obstructed labor and prolonged second stage is the leading cause of VVF in underdeveloped countries, especially south of the Sahara in Africa. In Western countries it is encountered in only 5% of all VVFs.[8] Other noniatrogenic causes include pelvic tumor, pelvic trauma, foreign body, or abscess.

Presentation

The most common time before presentation of VVF is 7 to 10 days after surgery, but it can range from immediate up to 6 weeks. If associated with hysterectomy, a VVF can present in the immediate postoperative phase, upon catheter removal, if unrecognized direct trauma to the bladder occurred intraoperatively, or after 10 to 14 days as a result of a hematoma or stitch abscess at the vaginal cuff. Postradiation VVF takes months to years to develop. Usually the patient presents with continuous leakage of urine. This can be mistaken from early postoperative discharge, leading to more frustration of the patient and delayed diagnosis. If there is intraabdominal urine extravasation, the patient will present with abdominal pain and ileus.

Diagnosis

Complete history and physical examination are essential in evaluation of VVFs. History should include the details of gynecologic and past obstetric history, surgical history, pelvic malignancy, radiotherapy, or pessary usage. Abdominal examination may reveal abdominal or flank tenderness secondary to ureteral obstruction or retroperitoneal urinary extravasation. On pelvic examination, the vagina should be carefully inspected with use of a speculum. If this is difficult to perform in the office, an examination under anesthesia should be performed. In acute fistulas, the mucosa surrounding the fistula tract may appear inflamed and erythematous. For a more mature fistulous tract, a small opening may be visualized in the vaginal wall. Examination of the vaginal vault may reveal a collection of fluid at the apex of the vagina. To confirm urine leakage, the fluid is analyzed for urea and creatinine. Urine levels of urea and creatinine should be several folds higher than of serum. Rectal examination is also required for detection of any mass or tumor. A phenazopyridine test is an excellent test to be performed during pelvic examination to diagnose VVF and to differentiate between it and ureterovaginal fistula (UVF). At our institution, we use the double-dye technique to differentiate between VVF and UVF.[9] The patient intakes 400 mg of phenazopyridine hydrochloride (Pyridium). One hundred milliliters of diluted methylene blue solution is instilled into the bladder via a urethral catheter. The catheter is removed and a tampon is inserted vaginally, and the patient is asked to come back in 2 hours. The tampon is inspected. If it is stained blue, it is a VVF. If stained orange, it is a UVF, and further work-up is needed.

Radiographic examination by intravenous urography is useful to rule out UVFs. In patients with VVF, up to 25% will have hydroureteronephrosis with 10% having a concomitant UVF.[10] Other radiologic investigations include oblique and lateral cystograms, and vaginograms.

Cystoscopic examination can identify the presence and the size of fistula, and its relation to the trigone and ureteral orifices. Careful cystoscopy is required to identify the presence of other fistulas. Fistulas occurring after hysterectomy are typically found along the anterior vaginal vault and the interureteric ridge. Fistulas after radiation for pelvic cancer are often located in the caudal portion of the trigone, distal to the interureteric ridge. Biopsy is recommended if there is a history of malignancy.

Bilateral retrograde pyelography can confirm the diagnosis of a concomitant UVF. Vaginoscopy is performed simultaneously with cystoscopy to assess the quality of the vaginal aspect of the fistula. Additionally, adjacent tissue is inspected for prospective use as flaps. If intravenous urography and retrograde pyelograms are inconclusive, then fistulogram may be performed transvaginally.

Conservative Treatment

Only in small, clean, nonmalignant VVFs can conservative treatment be tried. Prolonged continuous catheter drainage for 3 to 5 weeks is suggested. Additionally, anticholinergic medication to relax the detrusor muscle and prevent spasms is used. Estrogen is used routinely in postmenopausal women to promote healing. In the last few years, we have been successful in managing small VVFs with cystoscopic fulguration with and without fibrin glue injection. The same principles of conservative management are followed after the procedure.

Surgical Repair Considerations

Each VVF case should be individualized. The first attempt is associated with the highest success rate. In cases in which the VVF is the result of complicated obstetric trauma, fistula repair can be delayed 3 to 6 months to allow inflammation to be resolved. If the VVF is secondary to uncomplicated hysterectomy, or a clean surgical wound is present, early repair (6–8 weeks) is indicated. The surgical approach for repair depends on surgeon preference and the location of the fistula. However, the TV approach is less morbid than the TA and is the most frequently used for repair. Transabdominal approach is indicated for reimplantation, augmentation, and if the vagina is narrow or the fistula is high. Transvaginal approach can be used for VVF repair when entering the peritoneal cavity is to be avoided.

Transvaginal Approach

The principals of TV repair are creation of anterior vaginal flap and a tension-free multilayered closure. The TV approach is the safest and less morbid for patients. It is ideal for fistulas located low in the vaginal wall, or in the bladder neck or proximal urethra. TV repairs do not require the excision of the fistulous tract, and therefore no need to perform ureteral reimplant. For simple, small, and mature fistulas, a transvaginal partial colpocleisis (Latzko procedure) is indicated.[11] If the fistulous tract is large, it can be excised, and closure is performed in overlapping layers once tissue is widely mobilized. The bladder is closed continuously with 4-0 chromic sutures, and the pubocervical fascia is closed in two layers (Figure 13-1.1, Figure 13-1.2). More complicated fistulas may require a labial fat or peritoneal pad (Figure 13-1.3).

Partial Colpocleisis (Latzko Procedure)

This technique was described by Latzko as an alternative repair of simple, deep vaginal vault fistulas. This procedure is convenient for fistulas less than 1.5 cm. It does not involve excision of the fistula nor does it place any intravesical sutures, thus eliminating the need for performing ureteral reimplantation. The procedure requires a short period of hospitalization. However, the disadvantage of this partial colpocleisis is shortening of the vagina.

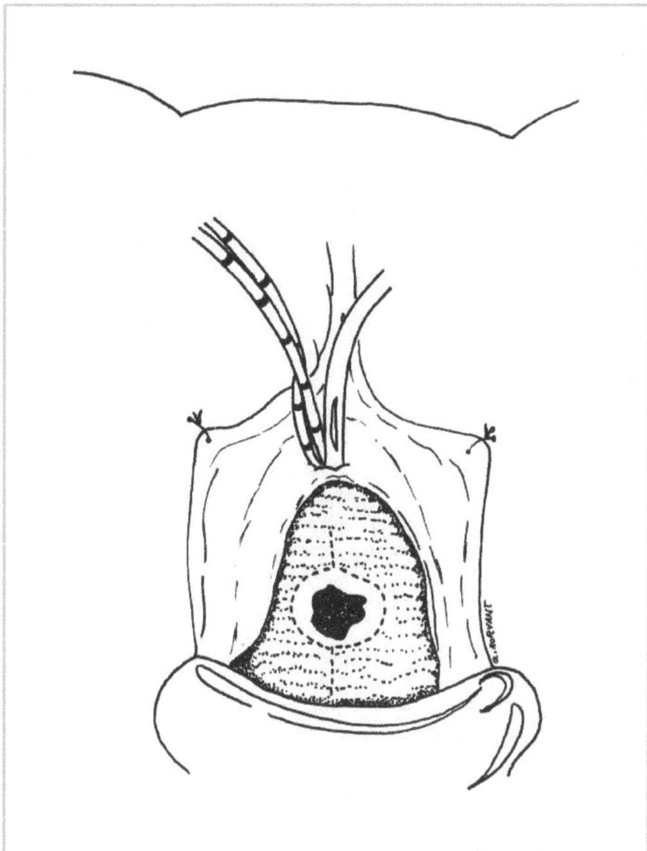

Figure 13-1.1. Circumferential incision around the fistula tract with anterior and posterior extension to allow wide dissection. (Reprinted from Ghoniem GM, Monga M. Modified pubovaginal sling and Martius graft for repair of the recurrent vesicovaginal fistula involving the internal urinary sphincter. Eur Urol. 1995;27(3):241–5, with permission from S. Karger AG, Basel.)

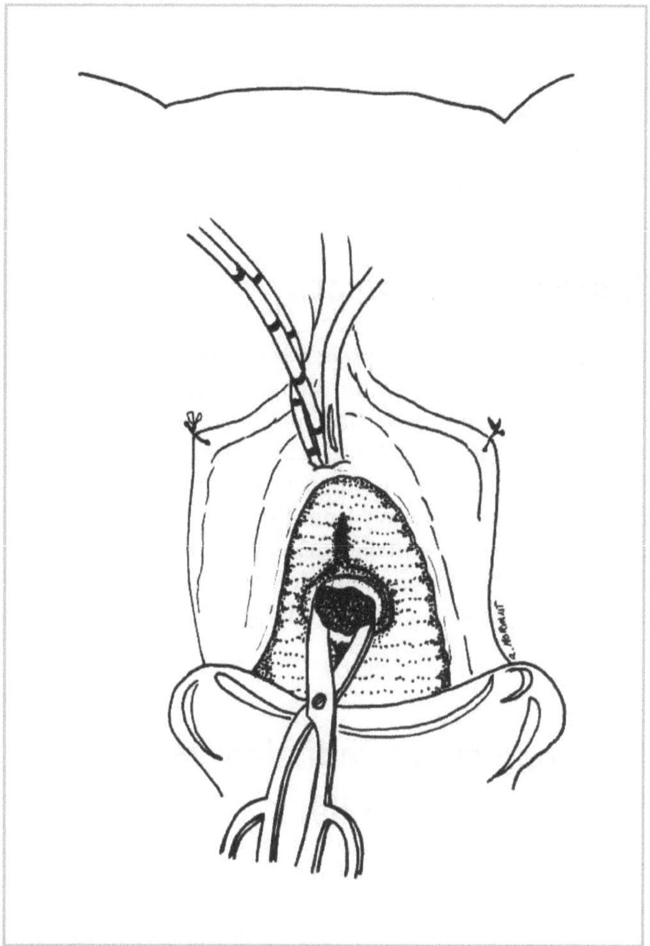

Figure 13-1.2. Excision of the fistula tract. (Reprinted from Ghoniem GM, Monga M. Modified pubovaginal sling and Martius graft for repair of the recurrent vesicovaginal fistula involving the internal urinary sphincter. Eur Urol. 1995;27(3):241–5, with permission from S. Karger AG, Basel.)

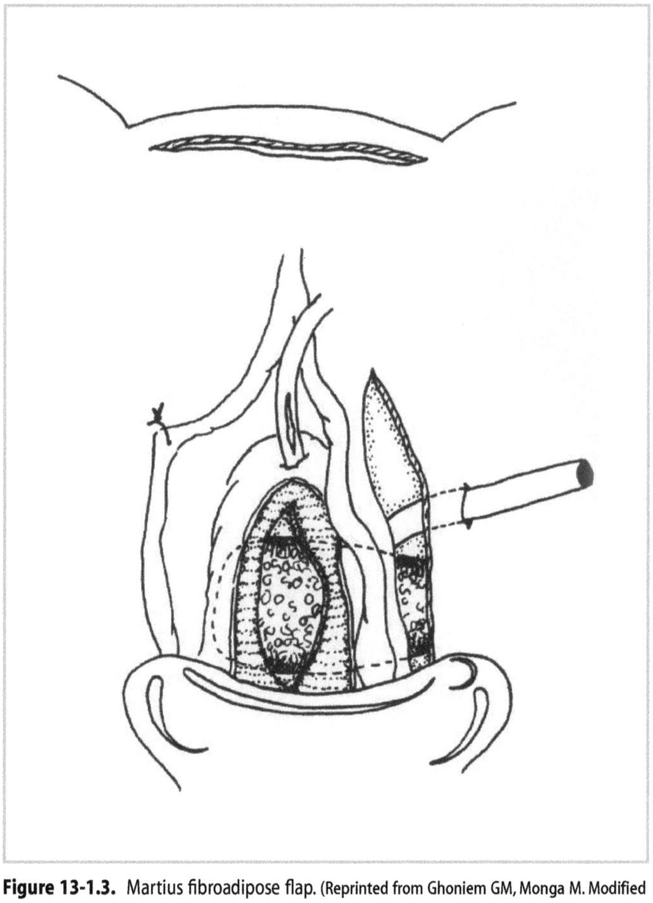

Figure 13-1.3. Martius fibroadipose flap. (Reprinted from Ghoniem GM, Monga M. Modified pubovaginal sling and Martius graft for repair of the recurrent vesicovaginal fistula involving the internal urinary sphincter. Eur Urol. 1995;27(3):241–5, with permission from S. Karger AG, Basel.)

Postoperative Care after Transvaginal Repair

The vaginal pack is removed on postoperative day one. The patient is given perioperative antibiotics and anticholinergics to prevent bladder spasm. Two weeks postoperatively, the urethral catheter and/or suprapubic tube are removed if the cystogram reveals no persistent fistula or extravasation. If the fistula is still present, bladder drainage should continue for another 2 to 3 weeks and the cystogram repeated.

Transabdominal Approach

The abdominal approach is recommended for complex VVFs associated with other pelvic organs such as uterus or bowel, those associated with malignancies, or as a result of radiation therapy. The vesicovaginal component of the fistula may be reached by a sagittal cystostomy to provide access to the site of fistula. The fistulous tract is excised and the vesicovaginal space is widely dissected. The opening into the vagina is closed in two layers and the other opening in the bladder is closed in three layers with delayed absorbable sutures. It is recommended that the omentum or peritoneum be placed to separate the vaginal and bladder repairs.

Transvesical Approach

The advantage of this approach is the avoidance of entering the peritoneal cavity. In this approach, the bladder is opened with a midline vertical incision. The VVF is located and the ureteral orifices are cannulated with stents. The fistulous tract is circumscribed carefully and excised. The bladder is mobilized away from the vagina. The bladder is closed in two layers and the omentum is brought down and fixed to the vagina. Suprapubic and Foley catheters are placed. Postoperative care is similar to that previously described.

Management of Radiation-induced Vesicovaginal Fistula

Vesicovaginal fistula caused by radiation therapy can be repaired as previously described, provided that the tissues are well-vascularized. The placement of nonirradiation tissues between the bladder and vaginal closure is required. When radiation-induced VVF is associated with radiation necrosis and fibrosis, mobilization of vascular pedicles is required. Vascular sources may include muscles such as the gluteus maximus or rectus abdominis, or a labial or omental flap. Before attempting repair, recurrent malignancy must be ruled out by biopsy of affected tissues.

Outcome

Reported success rates within the last two decades range from 92.5% to 96%, and 85% to 100% for TV and TA, respectively.[7,12,13] Complications include recurrence of the fistula, ureteral or bowel obstruction, development of stress urinary incontinence, and irritative voiding symptoms. With end-stage bladder and multiple failures, the only alternative would be diversion.

References

1. Derry DE. Note on five pelves of women of the eleventh dynasty in Egypt. Br J Obstet Gynecol 1935;42:490.
2. Bissada N. Vesicovaginal fistulas. In: O'Donnel PD, ed. Urinary Incontinence. St. Louis: Mosby Year Book; 1997:294–302.
3. Mahfouz N. Urinary fistulae in women. Br J Obstet Gynecol 1957; 64:23.
4. Sims JM. On the treatment of vesicovaginal fistula. Am J Med Sci 1852;60:100.
5. Tancer ML. Observations on prevention and management of vesicovaginal fistula. J Urol 1980;123:839–840.
6. Harkki-Siren P, Sjoberg J, Tiitinen A. Urinary tract injuries after hysterectomy. Obstet Gynecol 1998;92(1):113–118.
7. Lee RA, Symmonds RE, Williams TJ. Current status of genitourinary fistulas. Obstet Gynecol 1988;72:313–319.
8. Symmonds RE. Incontinence: vesical and urethral fistulas. Obstet Gynecol 1984;27:499–514.
9. Ghoniem G, Monga M. Modified pubovaginal sling and Martius graft for repair of the recurrent vesicovaginal fistula involving the internal urinary sphincter. Eur Urol 1995;27(3):241–245.
10. Goodwin WE, Scardino PT. Vesicovaginal and ureterovaginal fistulas: a summary of 25 years of experience. J Urol 1980;123: 370–374.
11. Kaser O. The Latzko operation for vesicovaginal fistulae. Acta Obstet Gynecol Scand 19977;56(4):427–429.
12. Eilber KS, Rosenblum N, Rodriguez LV, Raz S. Ten-year experience of transvaginal vesicovaginal fistula repair utilizing peritoneal flap. J Urol 2002;167(4 suppl). Abstract 814.
13. Hedlund H, Lindsted E. Urovaginal fistulas: 20 years of experience with 45 cases. J Urol 1987;137:926–928.

13-2

Rectovaginal Fistula

Susan M. Cera and Juan J. Nogueras

Rectovaginal fistulas are abnormal epithelial-lined communications between the vagina and rectum (Figure 13-2.1). Communications that occur below the dentate line are considered anovaginal fistulas; however, their approach and management are similar to rectovaginal fistulas and are classified as such. Fortunately, rectovaginal fistulas are an uncommon problem comprising approximately 5% of all anorectal fistulas.[1] However, surgical management remains a challenge and requires a methodical approach in choosing the appropriate corrective procedure to ensure operative success.

Etiology

Common causes of rectovaginal fistulas include obstetric trauma, inflammatory bowel disease (IBD), radiation, sepsis, iatrogenic injury from anal or vaginal surgery, forceful coitus or impalement, carcinoma, or congenital abnormalities. Etiology is important to the pathophysiology and the ultimate response to surgical therapy. Frequently associated or underlying conditions must be addressed either preoperatively or at the time of surgery to facilitate successful outcome.

Obstetric trauma is the most common cause of rectovaginal fistulas. Labor and delivery result in rectovaginal fistulas in a number of ways. Prolonged labor with pressure necrosis of the rectovaginal septum is an uncommon but possible event that results in a mid-vaginal fistula and occurs more often in underdeveloped countries. More frequently, fistulas occur low in the vaginal fourchette after either an episiotomy or a third/fourth degree laceration that goes unrecognized, is inadequately repaired, or develops an infection with subsequent failure of repair. Fortunately, rectovaginal fistulas after vaginal delivery are uncommon, occurring in less than 0.1% of cases.[2] Most of these injuries are associated with trauma to the sphincter mechanism. Recognition of a concomitant sphincter defect becomes an important part of the surgical approach and requires an appropriate anorectal physiologic evaluation before intervention. Trauma that is not obstetric related and can result in rectovaginal fistulas include forceful coitus, impalement, introduction of foreign body, and violence.

Rectovaginal fistulas can occur in both ulcerative colitis and Crohn's disease, although they tend to occur in the latter more frequently. In Crohn's disease, rectovaginal fistulas are included in the spectra of perianal disease and may precede the development of intestinal symptoms. In addition, recurrent rectovaginal fistulas and fistulas that develop in patients with a previous diagnosis of ulcerative colitis should raise the suspicion of the possibility of Crohn's disease. These patients should undergo an appropriate evaluation, including inquiry into intestinal symptomatology and imaging that may include radiologic assessment or endoscopic procedures.

A variety of vaginal or anorectal surgeries can lead to the development of rectovaginal fistulas including hysterectomy, especially if associated with dense pelvic adhesions; posterior colporrhaphy; transanal resections of anterior tumors such as villous adenomas or leiomyomas; fulguration of anterior lesions such as condyloma or recurrent adenocarcinoma not amenable to re-resection; sphincter repairs; low anterior resections in which the vaginal wall is partially incorporated in the anastomosis; or after leaks from low anastomoses involving colonic or ileal pouches.

Any perineal infection can lead to the formation of a rectovaginal fistula. Perirectal abscesses, venereal diseases, or Bartholin's cysts may induce inflammation leading to low rectovaginal fistulas. High fistulas are most often caused by diverticulitis, although tuberculosis and lymphogranuloma venereum have been reported.

Cancer of the anus, rectum, vagina, cervix, or uterus can give rise to rectovaginal fistulas. Leukemia, aplastic anemia, agranulocytosis, and endometriosis have also been implicated. Approach to these types of fistulas is dependent on the extent and stage of the disease. Malignant fistulas carry a poor prognosis. When surgical removal is not practical, they are treated palliatively with fecal diversion or endoluminal stenting. Therapy for

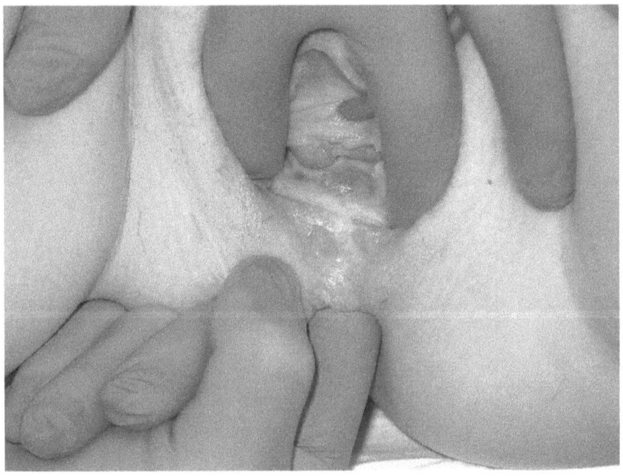

Figure 13-2.1. Rectovaginal fistula.

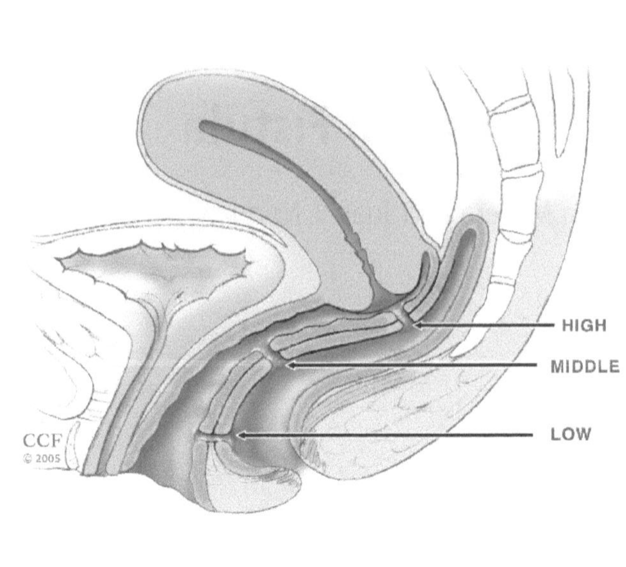

Figure 13-2.2. Classification of rectovaginal fistulas based on location. (Reprinted with the permission of The Cleveland Clinic Foundation.)

these types of fistulas is secondary to that of the cancer treatment.

Pelvic radiation after gynecologic and anorectal malignancies leads to fistulas that are particularly difficult to manage especially if accompanied by diabetes mellitus and hypertension. Radiation causes small vessel obliteration and scarring in localized tissues that subsequently exhibit poor wound-healing capability. Irradiation of the rectovaginal septum may lead to ulceration and eventual breakdown, with fistula formation, of this thin wall. Because of the inherent poor vascularity in the surrounding irradiated tissue, these types of fistulas are not amenable to simple local repairs and often require delayed intervention and fecal diversion for successful repairs. High doses of radiation predispose to fistula formation. Consideration must be made at the time of fistula evaluation, of the possibility of cancer recurrence with procurement of appropriate biopsies.

Classification

Rectovaginal fistulas are classified as either simple or complex according to etiology, size, and location with respect to the rectovaginal septum (Table 13-2.1). Approximately 9 cm of the anterior rectal wall lies adjacent to the vaginal wall below the peritoneal reflection. Low fistulas occur from any point distal to just above the dentate line and low in the vaginal fourchette (Figure 13-2.2), high fistulas occur behind or near the cervix in the apex of the vagina, middle fistulas occur somewhere between high and low, small fistulas are considered less than 2.5 cm.

Table 13-2.1. Classification of rectovaginal fistulas

Simple	Complex
Low	High
<2.5 cm	>2.5 cm
Trauma or infection	Underlying disease (Crohn's, radiation, malignancy)
No previous repairs	Recurrent

Simple fistulas are low and small. They usually result from infection or trauma, have healthy, well-vascularized surrounding tissue, and are amenable to local repairs. Complex fistulas are large, high, and frequently recur because the scar has less vascularity. They may also be caused by IBD, cancer, or radiation, all of which are conditions involving unhealthy localized tissues.

Evaluation

Symptoms depend on size, location, and etiology and may vary from asymptomatic to distressful and intolerable. Most complain of passing stool or flatus per vagina with foul vaginal discharge and chronic vaginitis. Specific history should be elicited regarding incontinence with concern for concomitant sphincter injury and intestinal symptoms, to evaluate for IBD. On physical examination, the goal is to identify the fistula, etiology, extent of underlying disease (Crohn's, carcinoma, irradiation, infection are cause), condition of surrounding tissues, and assessment of the sphincter. The fistula is often found at the base of an anterior midline pit on rectal examination, and the dark red rectal mucosa can be seen on the background of lighter pink vaginal mucosa during vaginal examination. In addition, stool may be seen in the vaginal vault along with signs

Table 13-2.2. Key points of evaluation

History	Examination	Evaluation	Ancillary Tests
Determine etiology from history	Location	Rigid proctoscopy and vaginal speculum examination	Tampon and methylene blue test
Assess degree of incontinence	Thickness of perineal body	Biopsy and culture abnormal tissue	Vaginogram
Severity of symptoms	Sphincter function	Examination under anesthesia	Barium enema
		Endoanal ultrasound	Computed tomography scan
			Anorectal physiology

of chronic vaginitis. The key points to the evaluation of rectovaginal fistulas are listed in Table 13-2.2.

Endorectal ultrasound is routinely performed to elucidate the presence of a sphincter defect.[4] At our institution, the ultrasound examination is enhanced with hydrogen peroxide, which is injected into the vaginal opening to aid in the location of the rectal opening. Examination under anesthesia with patient relaxation, anesthesia, lighting, and probes may facilitate localization that cannot be achieved in the office setting. In the medically fit patient, this form of evaluation should be undertaken first because direct assessment of the vaginal and rectal mucosa can be performed at the time of fistula identification. Appropriate biopsies should be procured if diseased tissue is present and seton placement is liberally performed to control inflammation and sepsis. Alternatively, colpocolonoscopy allows simultaneous endoscopic visualization of the rectum and vagina and can be performed in the endoscopy suite with sedation.

Ancillary studies may be useful in the identification of the fistula if physical examination and ultrasound are unrewarding. Contrast studies such as barium enema or vaginogram may be helpful for high fistulas if localization in the gastrointestinal tract is unclear. Computed tomographic scanning may be helpful in defining inflammatory processes or the staging of malignancies. If the fistula remains elusive or the appropriate tests are not available, instilling an enema of fluid containing methylene blue after inserting a vaginal tampon can identify the fistula. After the enema is retained for 15 to 20 minutes, the tampon is removed. Absence of blue staining indicates that the diagnosis of rectovaginal fistula is probably incorrect.

Regardless of the methodology used in locating the fistula, proctoscopy should be included in the evaluation to assess the rectal mucosa. Signs of proctitis should be followed by an evaluation for infectious etiology and inflammation, with cultures and biopsies obtained. If a history of radiation is present, recurrence of cancer should be suspected and local tissues biopsied.

Finally, assessment of the anal sphincter should be a part of every rectovaginal fistula evaluation, particularly those in which the etiology is obstetric trauma because of the high risk of associated sphincter damage. Fecal incontinence may be caused by the fistula, underlying disease, or sphincter damage. In addition, the symptoms of fecal incontinence may be masked by the fistula. Consequently, physical examination and anorectal physiologic testing, including endorectal ultrasound, manometry, and pudendal nerve terminal motor latencies, are important in unveiling the presence of an occult injury and the planning of surgical repair.

Operative Principles

Surgical options are categorized by approach: perineal or abdominal. The perineal approaches are divided into transanal, transperineal, and transvaginal. The route chosen depends on the severity of symptoms, patient medical status, location of the fistula, underlying etiology, the surgeon's expertise, the presence of a sphincter defect, and the number of failed prior attempts. Patients with minimal symptoms may elect not to undergo any intervention. Elderly patients or those with multiple comorbid conditions should undergo minimally invasive procedures such as placement of setons to control sepsis or fibrin glue instillation before embarking on complex interventions. Although success rates with fibrin glue in the treatment of rectovaginal fistulas have not been reported, it is used by our faculty as a minimally invasive approach in those who are unfit or do not desire to undergo surgical treatment. Location is particularly important for high and complex fistulas that may require an intraabdominal approach to facilitate repair.

Underlying etiology has an important role in determining approach to surgical treatment. Rectovaginal fistulas resulting from infection are amenable to successful repair after the appropriate treatment and resolution of the underling infection. Operative repair of rectovaginal fistula in the setting of medically controlled Crohn's proctitis may be possible; however, at a higher risk of recurrence. Rectovaginal fistulas in patients with medically refractory Crohn's proctitis may be controlled with seton placement and possibly infliximab. However, definitive repair usually requires fecal diversion or proctectomy. Repair of rectovaginal fistulas secondary to radiation proctitis demands the introduction of new and well-vascularized tissue into the area of radiation. These techniques involve mobilization of flaps from either the abdomen or leg often in conjunction with fecal diversion.

The surgeon's expertise has a role in surgical decision-making. Gynecologists approach repairs through the vagina or perineum whereas colorectal surgeons approach through the rectum. If an associated sphincter defect is present, a concomitant sphincteroplasty is performed. A history of prior failed attempts often requires fecal diversion and muscle flap transpositions that can be performed in a single or staged procedure.

Table 13-2.3. Types of rectovaginal fistula repairs

Repairs for Simple Fistulas		Repairs for Complex Fistulas	
Transvaginal	Transperineal	Perineal	Abdominal
Inversion technique	Fistulotomy	Tissue interposition (i.e., gracilis flaps)	Low anterior resection
Layered closure	Cutting seton		Onlay patch anastomosis
	Conversion to perineal laceration with layered closure		Proctectomy with coloanal anastomosis
	Sphincteroplasty		Abdominoperineal resection
			Fecal diversion

The most important factor in obtaining good outcome is ensuring the health of the involved tissue by minimizing inflammation and infection and optimizing the underlying disease process. A waiting period is advocated, because some fistulas will heal spontaneously with treatment of localized sepsis. In those fistulas secondary to obstetric trauma, the patient need not wait until childbearing is completed, but a period of 3 to 6 months may be prudent. In those that do not heal spontaneously, this time period allows resolution of inflammation and normalization of surrounding tissues to facilitate a successful, local repair. Repair of recurrent fistulas is facilitated by delaying repair.[3] In contrast, large or high fistulas, or those secondary to irradiation, IBD, or cancer, have a low likelihood of spontaneous healing.[5] However, a period of waiting may be necessary to optimize medical management in the case of IBD or to allow resolution of inflammation in the case of radiation proctitis after fecal diversion has been performed. As with management of all fistulas, optimizing nutrition is paramount before embarking on surgical intervention.

Despite choice of surgical repair, several intraoperative principles are important to a successful outcome. Appropriate mobilization of flaps or rectum to facilitate a tension-free repair, gentle dissection to minimize tissue trauma, apposition of healthy tissue edges with resection of diseased bowel if appropriate, and transposition of healthy well-vascularized tissue if localized tissues are insufficient, are the guiding principles during the operative procedure that should be used to maximize the likelihood of success. A description of various procedures and their indications is outlined in Table 13-2.3.

Treatment of Simple Rectovaginal Fistulas

Simple fistulas may be approached through the vagina, perineum, or rectum. Vaginal approaches are most often used by gynecologists and include the inversion technique and the layered closure. With the patient in the lithotomy position, the inversion technique involves exposure of the fistula by pressure on the rectal side. The vaginal mucosa is incised circumferentially around the fistula. A pursestring suture is placed in the fistula tract and the needle is passed to the rectal side. Tying of the suture at this point inverts the fistula to the rectal side. Muscle layers and the vaginal mucosa are reapproximated over the repair. The layered closure is the more common vaginal approach and involves a longitudinal elliptical incision around the fistula. A mediolateral episiotomy may be performed for better exposure. The vaginal mucosa is circumferentially dissected for 2 to 3 cm and the fistulous tract excised. The rectal mucosa is closed followed by reapproximation of the rectal muscle, rectovaginal septum, and vaginal mucosa in separate layers. The success rate for both is 84% to 100% and complications include bleeding, hematomas, cellulitis, urinary retention, and fistula at the episiotomy site.[6]

Perineal approaches include fistulotomy, cutting setons, conversion to perineal laceration with layered closure, and sphincteroplasty. The use of fistulotomy and cutting setons is mentioned only to be condemned because of the high risk of postoperative incontinence from sphincter damage. Conversion of the fistula to a perineal laceration with layered closure is more often used although not widespread because of the inherent division of the sphincter and the unknown rate of subsequent incontinence (Figure 12-2.3). For this technique, the patient is placed in the prone position after antibiotic and mechanical bowel preparation. The bridge of skin and muscle distal to the fistula are divided and the fistula tract is excised. The rectal mucosa is closed and the levators and the external anal sphincter are mobilized and reapproximated. The perineal body is reconstructed and the vaginal mucosa closed. This technique has been evaluated in small studies and demonstrates close to a 100% success rate.[6] The complications include hematoma, fever, and urinary tract infection. The disadvantage with this approach is the division of the sphincter mechanism and unknown long-term incidence of fecal incontinence.

Sphincteroplasty is considered a perineal approach to the correction of low rectovaginal fistulas, particularly those that are transsphincteric and result from obstetric trauma. The symptoms of rectovaginal fistula may mask true fecal incontinence that may manifest once the fistula is repaired. In addition, the sphincter muscle is the most vascular tissue between the rectum and vagina such that its absence diminishes success of local repair and its use in local repair predisposes to a higher success rate and functional outcome.[7] Consequently, patients with either an occult or symptomatic sphincter defect are candidates for this technique. The technique of overlapping sphincteroplasty is explained in detail in the chapter on anal sphincter repairs. Briefly, it involves placing the patient in the prone position under general anesthesia after mechanical and antibiotic bowel preparation. A curvilinear incision is made between the anus and vagina. A flap of anoderm and

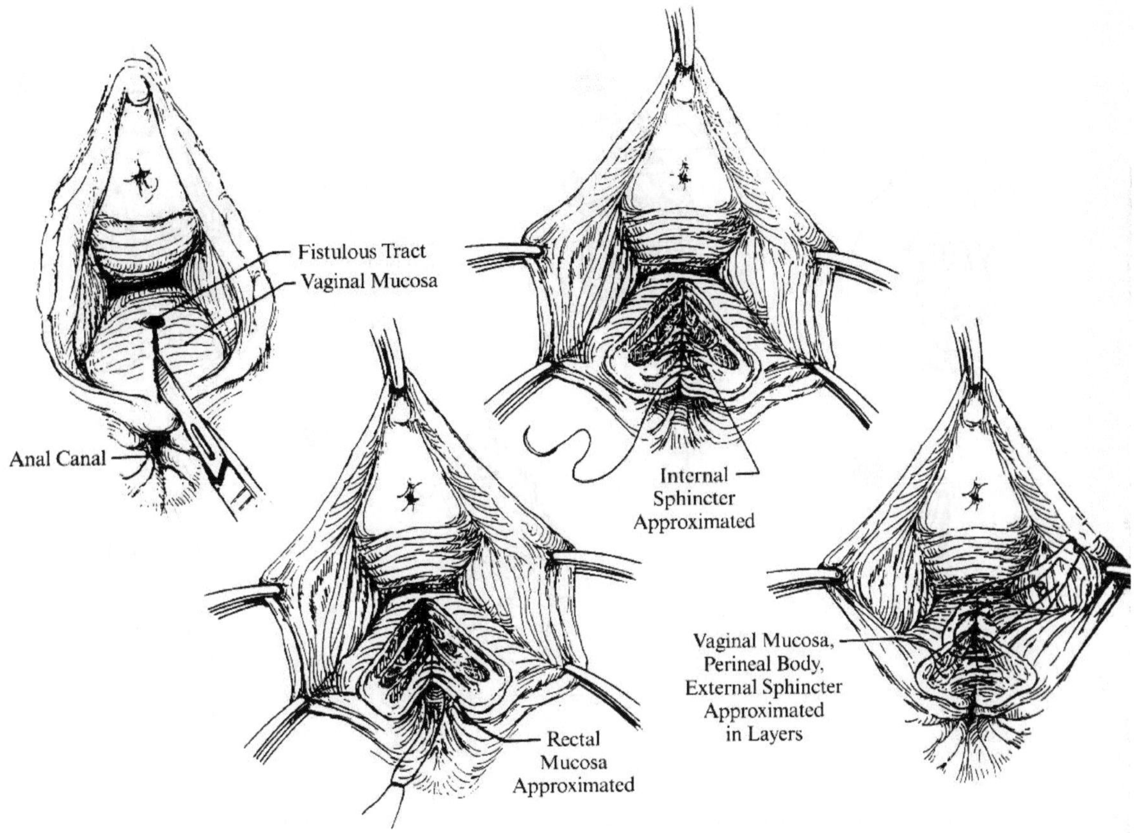

Figure 13-2.3. Perineal laceration with layered closure. (Reprinted from Beck DE, Wexner SD, eds. Fundamentals of Anorectal Surgery. 2nd ed., p 178, Copyright 1996, with permission from Elsevier.)

rectal mucosa and submucosa is raised off the internal anal sphincter. The external and internal anal sphincters are dissected free. The levators and the internal anal sphincter are separately plicated. The external anal sphincter is divided in the midline and the ends are overlapped and sutured in place. The endorectal flap is sutured to the reconstructed sphincter. The vaginal wall is reefed and the perineal body reconstituted before closure of the wound. The success rate with this technique is 78% to 100%.[6]

The most popular approach and the procedure advocated at our institution is the sliding endorectal advancement flap (Figure 13-2.4). The patient undergoes mechanical and antibiotic bowel preparation and is placed in the prone position with either local or general anesthesia after placement of a urinary catheter. A rectal flap is created with a base that is twice the length of the apex to ensure vascularity. The flap is composed of mucosa, submucosa, and circular muscle and is mobilized at least 4 cm to minimize the risk of tension. The fistulous tract is curetted of granulation and unhealthy tissue. The vaginal opening is closed with interrupted Vicryl sutures. The internal anal sphincter is mobilized bilaterally and reapproximated midline. The portion of the tract containing the fistulous opening is excised and the new edge of the flap brought distally and sutured in place using Vicryl sutures. Perioperative antibiotics are administered for 24 hours. Bulking agents are advocated to prevent trauma from stool to the repair. Tampons and sexual intercourse are prohibited for 6 weeks postoperatively. Success rates are reported between 71% to 100%.[6]

In knowledgeable hands, simple rectovaginal fistulas can be successfully repaired in any one of several ways. The advantage of the endorectal advancement flap is that the repair is directed at the high-pressure side of the fistula and an associated defect in the external anal sphincter can be concomitantly undertaken. In general, the advantage of the transanal and vaginal approaches is the lack of sphincter mechanism disruption, whereas perineal repairs involve transection of the sphincters and the unnecessary risk of permanent damage.

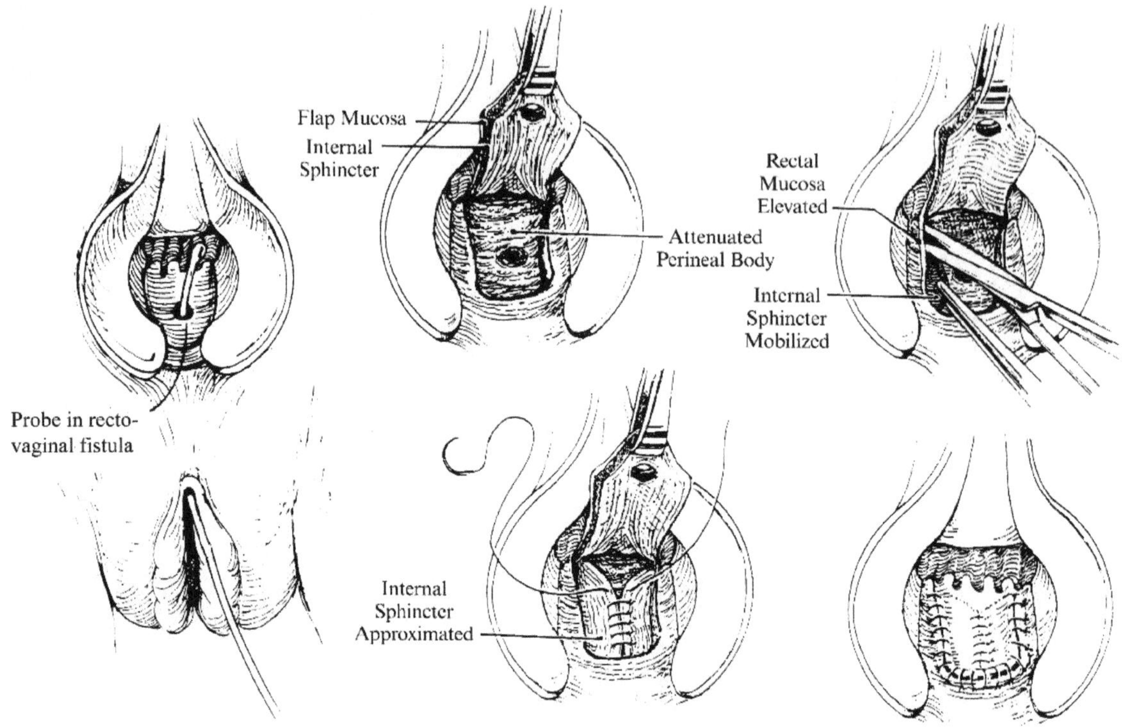

Figure 13-2.4. Sliding endorectal advancement flap. (Reprinted from Beck DE, Wexner SD, eds. Fundamentals of Anorectal Surgery. 2nd ed., p 178, Copyright 1996, with permission from Elsevier.)

Treatment of Complex Rectovaginal Fistulas

Complex fistulas are high, large, or recurrent and are often secondary to cancer, IBD, or radiation. Consequently, the condition of surrounding tissues (involved with cancer, IBD, scar, or irradiation) may preclude repair with local procedure such as the advancement flap.[3] These types of fistulas occur more often in older patients with numerous comorbid conditions, and therefore require more careful preoperative evaluation. In addition, the fistulas themselves require more intense evaluation by biopsy to ensure no recurrent cancer (in the case of irradiation) and to exclude the presence of IBD (in the case of recurrent fistulas). Radiologic and endoscopic assessment should be used judiciously to determine the extent of disease and associated involvement of adjacent organs. The presence of incontinence should be determined and addressed. Repairs for complex rectovaginal fistulas can be divided into two groups: perineal approaches with muscle transposition or abdominal procedures with bowel resection or fecal diversion. Fecal diversion is often used before repair of the fistula from radiation injury.

Repair of rectovaginal fistulas using tissue interposition techniques serves to introduce well-vascularized tissue to the area and to separate the two suture lines. Grafts of omentum, gracilis, sartorius, gluteus maximus, rectus, and bulbocavernosus muscle have been described. The bulbocavernosus muscle transposition is most frequently reported (Figure 13-2.5). The procedure is performed with the patient in the lithotomy position and a transperineal incision is made between the anus and vagina. The rectum is dissected from the vaginal wall, and both defects are sutured closed. A vertical incision is made over one of the labia majora, and skin flaps are created. The bulbocavernosus muscle and associated fat pad are mobilized and tunneled through a subcutaneous space to lie between the two closures. The vascular supply to this flap is the perineal branch of the pudendal artery.

The gracilis transposition is advocated at our institution because of the larger size, length, and bulk of this muscle and the ease of its mobilization (Figure 13-2.6). After undergoing mechanical and antibiotic bowel preparation, the patient is placed in the lithotomy position under general anesthesia. A transperineal incision is made, the walls of the rectum and vagina are separated, and the fistulous tract is divided [Figure 13-2.6(a)]. The fistulous openings in the rectal and vaginal walls are sutured closed with absorbable sutures. In the proximal leg, a small horizontal incision is made just below the edge of the palpable adductor muscles. The gracilis is identified and circumfer-

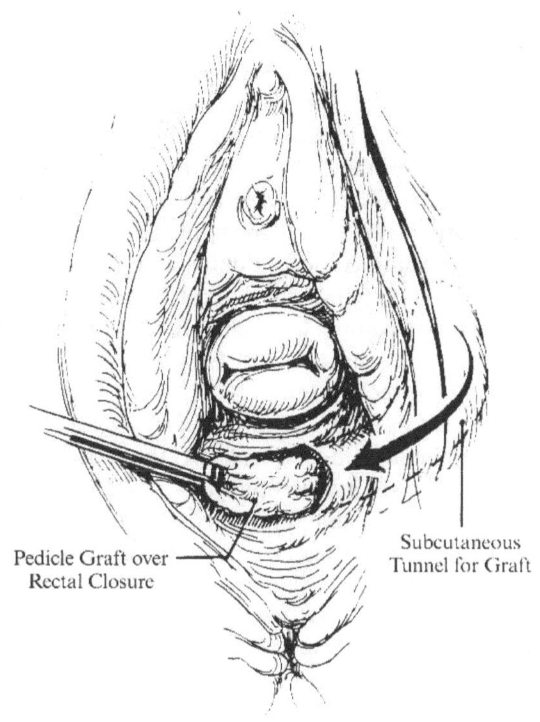

Figure 13-2.5. Bulbocavenosus muscle transposition. (Reprinted from Beck DE, Wexner SD, eds. Fundamentals of Anorectal Surgery. 2nd ed., p 178, Copyright 1996, with permission from Elsevier.)

entially dissected free, taking care to preserve the neurovascular bundle located in the proximal portion of the muscle. A second incision is used in the distal aspect of the leg to complete the circumferential dissection [Figure 13-2.6(b)]. The muscle is divided just proximal to its insertion at the knee and mobilized through a subcutaneous tunnel to the perineal incision. It is sutured into place on the rectal wall to prevent retraction after ensuring proper placement between the two fistulous closures [Figure 13-2.6(c)]. The perineal incision is closed from lateral to medial on each side leaving the central portion open for drainage. A closed external drainage system is left in the space previously occupied by the gracilis muscle and brought out through a separate stab wound. The leg wounds are then closed in layers. Adduction straps are placed around the patient's legs postoperatively to prevent any abduction and subsequent retraction of the flap. The patient is confined to bed rest with subcutaneous heparin and a urinary catheter for 3 days, after which the patient is allowed to ambulate and is discharged home.

The abdominal approach for the treatment of rectovaginal fistulas involves either resection of diseased bowel or associated organs, or fecal diversion. Fecal diversion is almost always used for radiation fistulas but may be used during repair of recurrent fistulas.

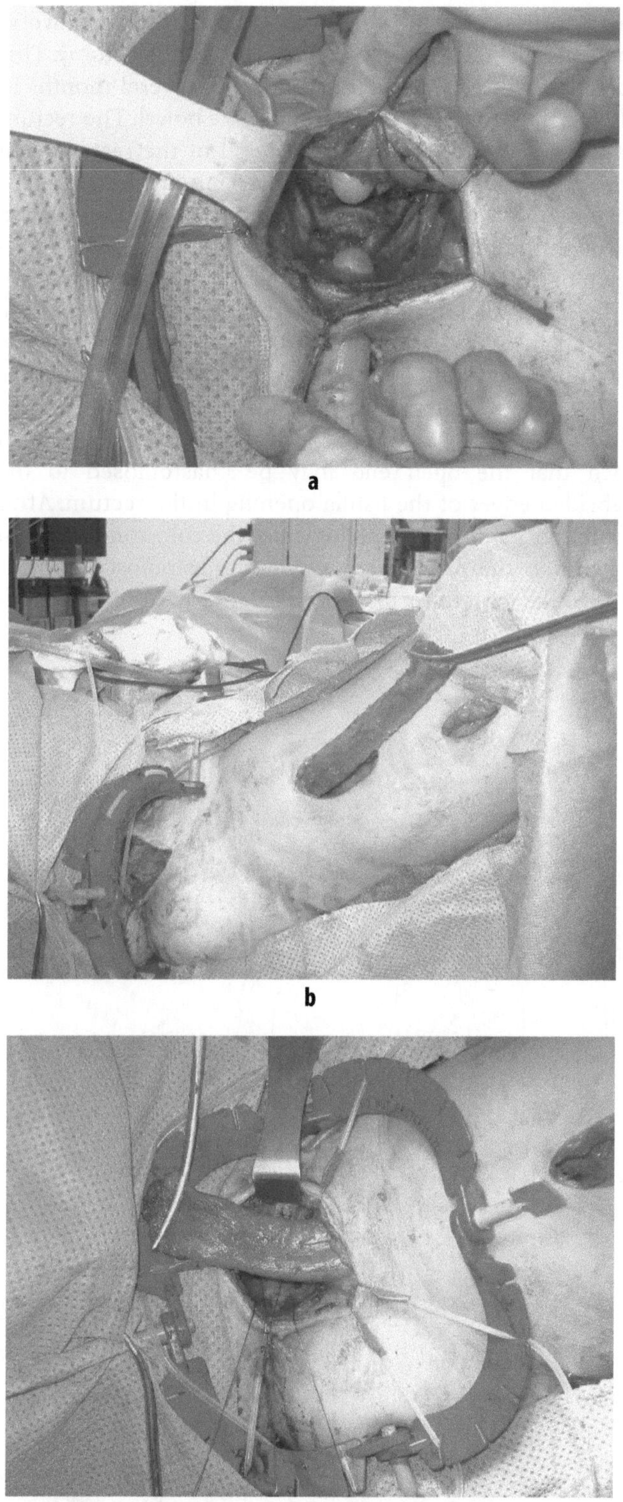

Figure 13-2.6. a, Through a perineal incision, the plane between the rectum and vagina is dissected to reveal both sides of the fistulous tract. b, The gracilis is mobilized through two incisions in the medial aspect of the leg. c, The gracilis is placed in the plane between the rectum and vagina separating the two ends of the fistula and reconstructing the perineal body.

Parks and associates[8] devised the sleeve (coloanal) anastomosis to treat postirradiation rectovaginal fistulas. The procedure is performed in stages with the first stage involving fecal diversion via a colostomy or loop ileostomy. This stage is followed by a waiting period of several months to allow resolution of inflammation of the bowel. The rectum is mobilized and divided at the point of the rectovaginal fistula. Through a perineal approach, an anorectal mucosectomy is performed and the healthy bowel is threaded through the muscular sleeve covering the fistula. A handsewn or double-stapled anastomosis is then performed.

Bricker and Johnston[9] devised an onlay patch for the treatment of radiation-induced fistulas. The rectum is mobilized and the fistula is exposed. The rectosigmoid is transected, with the proximal bowel made into an end colostomy. The distal line of transection is folded onto itself such that the open end may be anastomosed to the debrided edges of the fistula opening in the rectum. After healing is confirmed radiologically, the colostomy is taken down and the proximal sigmoid is anastomosed end-to-side to the loop of rectosigmoid (Figure 13-2.7). This technique is much easier to perform because mobilization of the rectum in an irradiated pelvis is not undertaken. However, irradiated bowel is left in place, with a potential for bleeding, pain, and malignant degeneration.

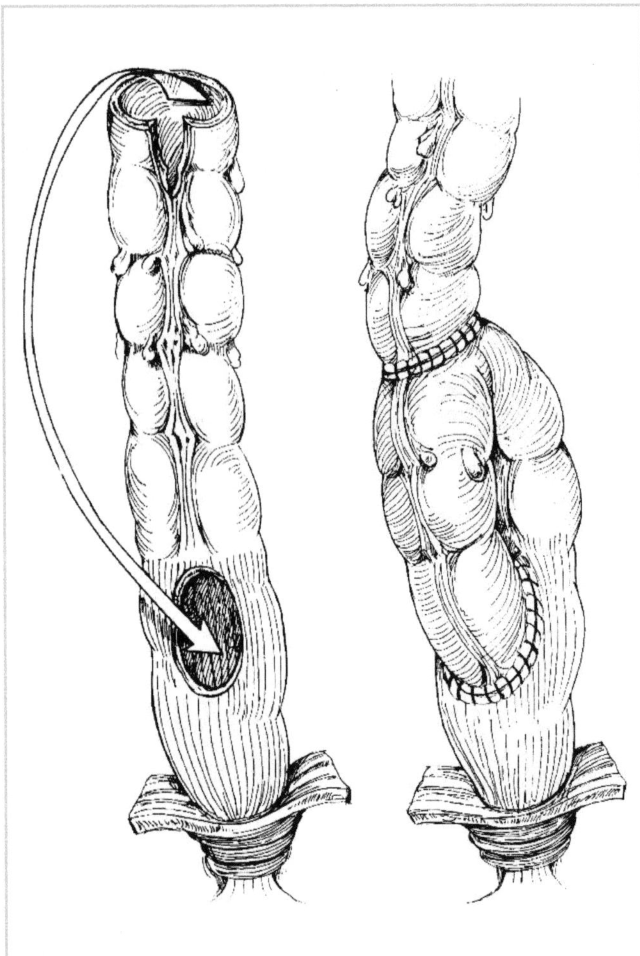

Figure 13-2.7. Bricker and Johnston's onlay patch for the treatment of radiation-induced fistulas. (Reprinted from Beck DE, Wexner SD, eds. Fundamentals of Anorectal Surgery. 2nd ed., p 178, Copyright 1996, with permission from Elsevier.)

At our institution, the procedure of choice for rectovaginal fistulas secondary to pelvic irradiation is a proctectomy with colonic J-pouch and double-stapled coloanal anastomosis with temporary diverting loop ileostomy. The rectum is mobilized circumferentially down to the level of the levators and resected. A 6-cm stapled J-pouch is created with the descending colon. A double-stapled anastomosis is performed involving the circular stapler inserted transanally to the colonic J-pouch. A diverting ileostomy is created until the anastomosis is healed and reversed approximately 6 weeks later.

The choice of operative repair for complex rectovaginal fistulas is primarily dependent on the etiology. High fistulas secondary to hysterectomy, diverticulitis, or previous surgery are amenable to proctectomy with anastomosis. Low, recurrent fistulas are most appropriately treated with the gracilis transposition or Martius bulbocavernosus graft. If the fistula is associated with cancer, resection appropriate for that specific cancer is undertaken. Abdominoperineal resection or pelvic exenteration may be needed for extensive disease, whereas diversion via a colostomy may be needed for palliation in unresectable disease. If the fistula is secondary to radiation, highly symptomatic patients that are otherwise healthy are candidates for repair once recurrent neoplasm has been excluded by multiple biopsies. Temporary diversion is usually undertaken in association with repair that may be accomplished with muscle transposition if low, and with low anterior resection with anastomosis if high. At Cleveland Clinic Florida, the preference is proctectomy with colonic J-pouch with double-stapled coloanal anastomosis and temporary loop ileostomy. An alternative is the Bricker onlay patch anastomosis, although this is not performed at our institution. Finally, if the complex fistula is secondary to IBD, assessment must be made of the type of IBD and the presence of proctitis. Medical management rarely results in closure. Proximal diversion may palliate symptoms but rarely results in closure. If proctitis is present, success is most likely with proctectomy. If no active proctitis is present, endorectal advancement flap may be attempted.

References

1. Laird DR. Procedures used in treatment of complicated fistulas. Am J Surg 1948;76:701–708.
2. Venkatesh KS, Ramanyam PS, Larson DM, Haywood MA. Anorectal complications of vaginal delivery. Dis Colon Rectum 1989;32:1039–1041.
3. Halverson AL, Hull TL, Fazio VW, Church J, Hammel J, Floruta C. Repair of recurrent rectovaginal fistulas. Surgery 2001;130(4):753–757.

4. Sudol-Szopinska I, Jakubowski W, Szczepkowski M. Contrast-enhanced endosonography for the diagnosis of anal and anovaginal fistulas. J Clin Ultrasound 2002;30(3):145–150.
5. Bahadursingh AM, Longo WE. Colovaginal fistulas, etiology and management. J Reprod Med 2003;48(7):489–495.
6. Lowry AC. Rectovaginal fistulas. In: Beck DE, Wexner SD, eds. Fundamentals of Anorectal Surgery. 2nd ed. Philadelphia: WB Saunders; 1992:175–186.
7. Tsang CB, Madoff RD, Wong WD, et al. Anal sphincter integrity and function influences outcome in rectovaginal fistula repair. Dis Colon Rectum 1998;41(9):1141–1146.
8. Parks AG, Allen CL, Frank JD, McPartlin JF. A method of treating post irradiation rectovaginal fistulas. Br J Surg 1978;65:417–421.
9. Bricker EM, Johnston WD. Repair of post irradiation rectovaginal fistula and stricture. Surg Gynecol Obstet 1979;148:499–506.

13-3

Enterourinary Fistula

T. Cristina Sardinha, Samir M. Yebara, and Steven D. Wexner

Enterourinary fistulas are usually the result of underlying pathology involving the gastrointestinal or genitourinary tract. Pathologies such as diverticulitis, Crohn's disease, radiation enteritis, trauma, iatrogenic injury, bladder cancer, appendicitis, colon carcinoma, and gynecologic tumors are causes of enterourinary fistulas. The true incidence of enterourinary fistula is unknown, although the most frequent site of fistulization is between the bowel and the bladder.

Most enterourinary communications involve the bladder and the colon. Although these colovesical fistulas often result from sigmoid diverticulitis (60% of enterovesical fistulas), only 2% of patients with diverticular disease develop colovesical fistula. Malignancy accounts for 20% of enterovesical fistulas. Colorectal adenocarcinoma may adhere and directly invade the bladder, leading to fistulization in 0.6% of patients.[1,2] Moreover, cancer of the cervix, prostate, ovary, and lymphoma can also occasionally result in enterourinary fistulas. However, primary bladder cancer has rarely been found to fistulize to the bowel, perhaps as a result of its relatively early detection.

Colovesical fistulas are more common in men, probably because of the male anatomy maintaining close proximity between the sigmoid colon and the bladder. This is contrary to the female anatomy, where the uterus forms a barrier between the bladder and the sigmoid colon, except in posthysterectomy patients. Colovesical fistulas are not a usual complication of diverticulitis, because most fistulas are at the dome of the bladder. Colovesical fistulas often present with symptoms of bladder irritability, dysuria, pneumaturia, fecaluria, and recurrent urinary tract infections.[3] In addition to clinical signs and symptoms of an abnormal colovesical communication, complementary tests may be required to confirm the diagnosis (Table 13-3.1). Computerized tomography findings include air in the bladder, focal bladder wall and adjacent bowel wall thickening, and surrounding inflammation. A water-soluble enema may also identify the fistulous tract in addition to diverticular disease or tumor. Moreover, a colonoscopy with biopsies should be performed to exclude invasive carcinoma. Cystography is more accurate than intravenous pyelography, and if bladder cancer is suspected, a cystoscopy with biopsies should also be performed. However, cystoscopy rarely reveals the fistula; common findings include bullous edema and erythema typically at the dome of the bladder.

Diverticulitis represents 65% of enterovesical fistulae.[4] Diverticular fistulae are almost entirely colovesical. The principal cause of a fistula in diverticular disease is a pericolic abscess. The abscess is walled off by omentum and small bowel and penetrates the dome of the bladder, and may eventually produce perforation into the bladder, causing a fistula.[4]

Crohn's disease accounts for approximately 10% of vesicoenteric fistulae and is the most frequent cause of an ileovesical fistula because of the anatomic proximity of the ileum to the dome of the bladder.[5] Pneumaturia in a patient with Crohn's disease is a strong indicator of the presence of an enterovesical fistula.[6] The transmural nature of the inflammation characteristic of Crohn's colitis often results in adherence to other organs. Consequent erosion into adjacent organs can then give rise to a fistula. The mean duration of Crohn's disease at the time of first symptom of fistula formation is 10 years, and the average patient age is 30 years.

The treatment of colovesical fistulae involves colon resection and drainage of the bladder with a Foley catheter. Typically, a laparoscopic sigmoid colectomy is the ideal option. Primary repair of the bladder opening is only necessary when easily identified with relatively healthy surrounding tissues. In the former case, a few sutures can be placed, whereas in the latter, a resection of the inflamed area may be necessary. The operation should be performed electively after adequate bowel preparation. Emergency surgery is frequently unjustified, because a colostomy will likely be required in this situation. In case the patient presents with urinary sepsis, this should be controlled with adequate broad-spectrum antibiotics and fluid resuscitation before surgery.

Nonoperative therapy of colovesical fistula with prolonged antibiotic therapy has been reported as a therapeu-

Table 13-3.1. Clinical features and diagnostic tools.[2,3]

Signs and Symptoms	Diagnostic Investigations
Cystitis	Cystoscopy
Pneumaturia	Contrast enema
Fecaluria	Computed tomography scan
Hematuria	Oral contrast: 1. barium, 2. charcoal, 3. methylene blue (followed the next day by a centrifuged urine sample)
Fever	
Abdominal tenderness	Cystography

tic option in high-risk diverticulitis patients. However, fecal diversion is often required for recurrent urinary tract infections. Nonetheless, some patients will present complete healing of the fistulous tract, and requirement for resection will depend on the nature of the colonic pathology. Iatrogenic enterourinary fistulas are often a consequence of surgical procedures such as prostatectomies, resection of rectal lesions, hysterectomy, laparoscopic inguinal hernia repair, and other pelvic operations. Radiation therapy for gynecologic or urologic malignancies may lead to progressive bowel injury culminating with perforation and often fistulization. A common example is the development of rectourethral fistula postradiation therapy of prostate cancer. This complex problem presents as urethral elimination of gas and stool as well as recurrent urinary sepsis. Diagnostic modalities include retrograde urethrogram and proctoscopy. The rectum is often inflamed and the mucosa friable with easy bleeding. This problem is more evident in the anterior rectal wall directly under the prostate or prostatectomy site. Fecal diversion and/or urinary diversion alone do not completely eliminate recurrent urinary infections or healing of the rectourethral fistula. Moreover, patients prefer to avoid permanent stomas. The use of rectal advancement flaps has been reported, but the success rate of this approach is unclear.[7] We advocate local procedures with interposition of healthy muscle flap. As well we prefer to use the gracilis muscle because of its low morbidity and easy harvesting (see Chapter 13-2, Figure 13-2.6). We have successfully performed the gracilis muscle transposition with very gratifying results in more than 30 patients with rectourethral fistula posttreatment of prostate cancer. Temporary fecal diversion and suprapubic cystostomy are also performed either at the time of or before muscle harvest and transposition. Reversal of the fecal diversion and removal of the cystostomy tube was performed after documentation of complete fistula healing.

Enterourinary fistula involving the small bowel and bladder is often diagnosed with a small bowel series or computerized tomography and frequently treated with segmental small bowel resection. The bladder can be primarily repaired and drained with a bladder catheter or just drained if the bladder opening is not clearly identified. In both instances, a follow-up cystogram should be performed approximately 6 weeks after the operation, before removal of the bladder catheter. The surgical treatment of radiation-induced fistula can be challenging, and a diverting proximal colostomy or ileostomy is advisable. Moreover, enterourinary fistulas in patients with Crohn's disease may respond to medical management.[5] Therefore, drugs such as tumor necrosis factor inhibitors should be considered in patients without evidence of significant infection. However, these agents may only delay the need for surgery by transient healing. The transmural inflammatory process that occurs in the bowel is not seen in the bladder, which allows for safe primary closure with absorbable suture.

In summary, enterourinary fistulas often require a multidisciplinary approach. We advocate elective laparoscopic surgery to treat most fistulas involving the small bowel or colon and the bladder. Nonetheless, all patients should be consented for and advised of the possibility of a laparotomy. Fecal diversion must be kept in mind, particularly in patients with significant radiation enteritis. However, rectourethral fistulas are better managed with local procedures, with our preference being the gracilis interposition.

References

1. Najjar SF, Jamal MK, Savas JF, Miller TA. The spectrum of colovesical fistula and diagnostic paradigm. Am J Surg 2004;188(5):617–621.
2. Larsen A, Bjerklund Johansen TE, Solheim BM, Urnes T. Diagnosis and treatment of enterovesical fistula. Eur Urol 1996;29(3):318–321.
3. Kirsh GM, Hampel N, Shuck JM, Resnick MI. Diagnosis and management of vesicoenteric fistulas. Surg Gynecol Obstet 1991;173(2):91–97.
4. Woods RJ, Lavery IC, Fazio VW, Jagelman DG, Weakley FL. Internal fistulas in diverticular disease. Dis Colon Rectum 1988;31(8):591–596.
5. Manganiotis AN, Banner MP, Malkowicz SB. Urologic complications of Crohn's disease. Surg Clin North Am 2001;81(1):197–215, x.
6. Gruner JS, Sehon JK, Johnson LW. Diagnosis and management of enterovesical fistulas in patients with Crohn's disease. Am Surg 2002;68(8):714–719.
7. Dreznik Z, Alper D, Vishne TH, Ramadan E. Rectal flap advancement: a simple and effective approach for the treatment of rectourethral fistula. Colorectal Dis 2003;5(1):53–55.

13-4

Management of Urinary Tract Injuries

Gamal M. Ghoniem and John C. Hairston

Iatrogenic injury to the urinary tract is caused by surgery in or around the pelvis and the retroperitoneal space. Injuries to the ureter are the most common, followed by injuries to the bladder and urethra. This is because of the similarity of the ureters to vascular structures, difficulty in identifying them as a result of their close adherence to the posterior peritoneum, and their long course in the abdomen and pelvis. Add to this the inherent difficulties of reoperation or operating in a pelvic malignancy, or the occasional unexpected congenital anomaly such as ureteral duplication or retrocaval ureter. Furthermore, the recent use of laparoscopy and endoscopy carries with it additional risks. Lack of recognition at the time of injury often mandates further surgical procedures with their associated morbidity and may result in loss of renal function or, worse, loss of a renal unit.

Injuries to the Ureter

The incidence of intraoperative ureteral injuries is 0.3% to 1.5%.[1,2] Historically, gynecologic procedures have accounted for the majority of ureteral injuries, greater than 50% in some series.[3] Traditionally, this had been followed by general surgical operations (5%–15%). More recently, with increased endoscopy usage, there is an increase in the incidence of ureteral injuries as a result of urologic procedures such that it contributes to 30% of all iatrogenic injuries.[4]

Diagnosis

It cannot be stressed enough that *intraoperative* recognition is the most important factor in achieving a successful outcome after ureteral injury. Types of injury can vary from crush injuries to complete transection. Urine in the operative field indicates obvious injury, and immediate surgical repair should be undertaken once the site is identified. If the site of injury cannot be readily identified, but transection is suspected, intravenous indigo carmine and Lasix can be administered to verify injury, identify the site, and allow immediate repair. Alternatively, one can perform an on-table intravenous pyelogram (IVP) by administering 2 mL/kg of contrast and taking a plain film of the abdomen after 10 minutes.[5] If extravasation is noted, surgical repair is performed. If there is no extravasation, then a significant ureteral injury is ruled out and the surgery can continue. If ligation with either a suture or clip is suspected, the area in question should be dissected and the ureter inspected. If one is unsure of the level of ligation, but suspicion is high, then a cystotomy can be made with retrograde passage of a ureteral catheter. If the catheter passes freely, then there is no injury. Inability to pass the catheter confirms ureteral obstruction and provides the surgeon with an idea of the level of injury. In the case of inadvertent ligation, the clip or offending suture should simply be removed and the ureter inspected for viability. Determination of viability is somewhat subjective, but active peristalsis of the segment in question indicates an intact smooth muscle layer and can usually be managed with observation. If there is any question of viability, then a ureteral stent should be left in place. If the ureter has been partially transected, then primary repair can be performed with a spatulated anastomosis over a ureteral stent as described later.

Quite often, however, the injury is not recognized until the postoperative period. The patient may present with hematuria, fever, flank or abdominal pain, ileus, or even fistula with drainage of urine. With unilateral injury, urine output will be of little or no value in making the diagnosis. Bilateral ureteral injury is rare but does occur, usually at the time of hysterectomy, and is easily identified by the presence of anuria in the recovery room.

Postoperative diagnosis is best made with an IVP, which may show extravasation, delayed renal function, or obstruction with hydroureteronephrosis. Computed tomography may show any of these as well as the presence of a urine collection in the retroperitoneal space; however, an IVP will usually give much better anatomic information when planning reconstruction. In the presence of any of these findings, retrograde pyelography (RPG) should be performed. The RPG will allow assessment of the level and, possibly, the length of injury. If a stent can be placed safely at the time of RPG, then this may be all that is necessary. If a stent cannot be safely manipulated past the injury because of complete obstruction or transection, then immediate decompression with percutaneous nephrostomy should be performed. Antegrade stenting can then be attempted, either immediately, or sometimes more successfully after a period of 48 hours. If stenting fails, then exploration with surgical repair is the only option. Nuclear renograms are usually only necessary if the time delay from injury to recognition is long (>6 weeks). In these cases, measurement of residual renal function may help to determine whether a reconstructive salvage procedure is warranted versus a nephrectomy.

Surgical Repair

Of utmost importance is accurate determination of the length and level of defect to be repaired, which can be done by a combination of the aforementioned radiologic studies. The upper ureter and ureteropelvic junction can be accessed via either a flank approach or through a dorsal lumbotomy incision. Extraperitoneal access to the middle and distal ureter is best obtained with a Gibson incision. Although the incision can be extended medially to expose the bladder, we prefer to use a midline incision.

Ureteroureterostomy

If the defect is short (<3 cm), then primary end-to-end anastomosis can be considered. Success of primary repair depends almost exclusively on obtaining a watertight tension-free anastomosis. After debridement of all nonviable tissue, the ends should be spatulated and the anastomosis performed using fine absorbable suture over an appropriately sized stent. If possible, retroperitoneal fat or omentum, if available, should be placed over the anastomosis. We use closed suction drainage for 24 to 48 hours or until drainage is minimal. A Foley catheter should be left in place for a few days because urine will reflux with a stent in place and impair wound healing. Most literature supports leaving the stent in place after ureteral reconstruction for 4 to 6 weeks and this can easily be removed in the office. High success rates are expected. Most failures and recurrent strictures are the result of excessive tension on the anastomosis, injudicious devascularization, or prolonged urinary leak.

Ureteroneocystostomy (Ureteral Reimplantation)

If enough mobilization cannot be achieved and the injury is in the lower middle or distal third of the ureter, then ureteral reimplantation becomes necessary. Direct ureteral reimplantation such as that performed by urologists for congenital reflux is uncommonly performed to repair defects from injury except for those in the most distal 3 to 4 cm, as the ureter enters the bladder.[6] We prefer to perform a nonrefluxing anastomosis. There are no definitive studies to show that nonrefluxing anastomosis decreases the risk of postoperative infection or increases the risk of renal deterioration in adults with normal bladders.

Psoas Hitch

In 1969, Turner-Warwick and Worth[7] described a means to replace length lost in the lower third of the ureter by bringing the bladder up to the level of the defect and affixing it to the psoas muscle and/or fascia. The bladder is mobilized from its lateral attachments as much as possible. The vas deferens or round ligament can also be divided to gain additional mobility. With these maneuvers, the bladder dome may be brought as high as the iliac vessels. A small, contracted bladder is a relative contraindication to a psoas hitch because of the difficulty in attaining adequate mobility. It is then preferably fixed to the psoas minor tendon with several absorbable sutures. Alternatively, the psoas muscle will often suffice with care taken to place the sutures longitudinally so as to not injure the genitofemoral nerve. A ureteroneocystostomy is then performed and a Foley catheter and pelvic drain are left in place. A double-J stent should be left in place as well.

Boari Flap

If the ureteral defect necessitates mobilization of the bladder above the level of the pelvic brim, then a Boari flap can be performed.[8] This procedure is also particularly useful when dealing with a long mid ureteral stricture/injury where a primary anastomosis may not be possible. This technique was first described by Boari in 1894 and involves constructing a bladder flap based on the ipsilateral superior vesical artery posteriorly and connecting the tubularized flap to the ureter (Figure 13-4.1). The length of flap should be equal to the defect plus a few centimeters. The contralateral superior vesical artery can also be divided for more length. The flap is brought up to the ureter and anastomosed. Fine absorbable suture is used and, again, a stent and drain should be placed. This procedure should not be attempted when the bladder is small or contracted because there will be insufficient remaining bladder capacity and postoperative voiding dysfunction will be significant.

Transureteroureterostomy

Transureteroureterostomy (TUU) is mainly used for the reconstruction of congenital anomalies and is not a fre-

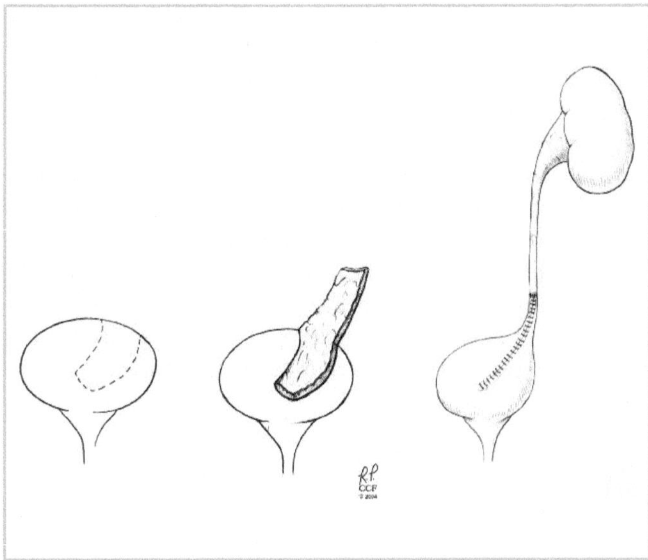

Figure 13-4.1. Boari flap. (Reprinted with the permission of The Cleveland Clinic Foundation.)

quently used procedure in any other situation. The only absolute contraindication to TUU is either a ureter of insufficient length to reach the contralateral side or a diseased recipient ureter. Repair of a ureteral injury secondary to aortic aneurysm surgery in this manner would not be recommended because of the fibrosis likely to be encountered. A TUU may, however, be the only viable option in a patient with significant pathology precluding dissection and exposure deeper in the pelvis.

Renal Descensus

For proximal ureteral defects, or to provide less tension on more distal repairs, the kidney may be mobilized inferiorly and affixed to the retroperitoneal muscle.

Ureteral Substitution

First described in 1909, the ileal ureter has become the most accepted and studied form of ureteral substitution.[9] Complications include early extravasation with fistula or urinoma formation and obstruction from edema, mucous plugging, or kinking of the loop. Ischemic necrosis of the loop has also been described and should be suspected in a patient with significant, unexplained pain postoperatively. Although the majority of patients have no metabolic consequences, hyperchloremic metabolic acidosis can develop, especially in patients with preexisting renal insufficiency (serum creatinine >2 mg/dL). It also should not be performed in a patient with a history of inflammatory bowel disease or radiation enteritis.

Autotransplantation

In patients with a solitary functioning kidney or poorly functioning contralateral kidney, renal autotransplantation can be performed when other methods of ureteral substitution are not feasible.

Injuries to the Bladder

Iatrogenic injuries to the bladder can occur as a result of any pelvic surgery, most often during vaginal or abdominal hysterectomy, colorectal oncologic resections, and endoscopic resections of bladder tumors. In a large review of the complications of laparoscopic procedures, the incidence of bladder injuries ranged from 0.02% to 8.3%.[10] This was usually the result of ill-advised placement of a trocar in addition to dissection injuries. Injury is more likely if the bladder is left full during trocar placement. Meanwhile, having the bladder slightly full can sometimes facilitate dissection of the tissue planes surrounding it during pelvic dissection.

Diagnosis

Injuries to the bladder are more easily recognized than injuries to the ureter because of immediate extravasation of urine into the operative field when operating near the organ. For this reason, most bladder injuries are repaired at the time of original injury. Unrecognized bladder injury may be clinically suspected with unexplained gross hematuria, decreased urine output, or postoperative fever, abdominal pain and/or suprapubic tenderness. Urine in the peritoneum is quite caustic and causes a significant inflammatory response. The diagnosis should be confirmed with a cystogram. A good-quality cystogram should include oblique views in addition to anterior/posterior views as well as a postdrainage film to completely rule out injury. Extraperitoneal injury is heralded by contrast extravasation limited or confined to the lateral pelvic sidewalls or perivesicular deep pelvic space whereas free extravasation of contrast with outlined bowel loops or a "ground glass" appearance of the abdomen indicates intraperitoneal injury.

Management

When discovered intraoperatively, extraperitoneal injuries should be immediately repaired. Most postoperatively discovered extraperitoneal injuries may be managed conservatively with indwelling Foley catheter drainage and will heal spontaneously. Whether discovered intraoperatively or postoperatively, most intraperitoneal bladder injuries require exploration and primary closure. A two-layer closure is generally recommended using absorbable suture in a running watertight manner, at least in the mucosal layer. Care should be taken not to extend the injury to the trigone or bladder neck or incorporate these into the repair, or significant bladder dysfunction and/or incontinence may ensue. If the ureterovesical junction is compromised or if there is simultaneous injury to the distal ureter, then any of the aforementioned techniques may be used with bladder repair. Postoperative drainage via a suprapubic cystotomy, urethral catheter, or both, should be typically used for a period of 7 to 10 days and only removed after a negative repeat cystogram. Pelvic drains should also be placed and removed when there is no more evidence of urinary leakage.

Urethral Injuries

Direct injuries to the urethra are fortunately extremely rare but can occur, especially during surgery for rectal or vaginal malignancies because of the close proximity of these structures.

Diagnosis

Urine extravasation in the field denotes lower urinary tract injury, and localization of the site becomes imperative. Injection of methylene blue through a feeding tube inserted into the urethral meatus, next to the urethral catheter, will show the site of injury. Urethroscopy will help to diagnose and test the integrity of the repair. Unrecognized injuries may have different presentations postoperatively. Fistula formation is the most common presentation. A retrograde urethrogram and, if possible, a voiding cystourethrogram will show the site and the extent of injury (complete vs incomplete). Intravenous pyelogram with a voiding film should be obtained. Urine analysis and culture should be obtained.

Management

Urethral injury during difficult dissection will be directly detected by seeing the catheter exposed in the field. Palpation of the catheter during dissection will help to avoid injury. Primary repair of iatrogenic urethral injuries is the method of choice. Mobilization of the urethra and periurethral tissue is necessary to obtain tension-free repair. Watertight anastomosis with absorbable suture in overlapping layers ensures good results. In case of loss of urethral wall, reconstruction can be performed using a vaginal wall flap in females or by creating an anteriorly based tubularized bladder flap. In case of compromised blood supply and presence of scar, a modified Martius graft is useful.[11] In males, many grafts have been used, but our preference is penile skin or foreskin pedicled grafts. These can usually reach up to the bulbomembranous urethra and cover up to 8-cm gaps.

References

1. Liapis A, Bakas P, Giannopoulos V, Creatsas G. Ureteral injuries during gynecological surgery. Int Urogynecol J 2001;12:391–393.
2. St. Lezin MA, Stoller ML. Surgical ureteral injuries. Urology 1991; 38(6):497–506.
3. Gangai MP, Agee RE, Spence CR. Surgical injury to the ureter. Urology 1976;8(1):22–27.
4. Assimos DG, Patterson LC, Taylor CL. Changing incidence and etiology of iatrogenic ureteral injuries. J Urol 1994;152(6 Pt 2): 2240–2246.
5. Morey AF, McAninch JW, Tiller BK, Duckett CP, Carroll PR. Single shot intraoperative excretory urography for the immediate evaluation of renal trauma. J Urol 1999;161(4):1088–1092.
6. Stefanovic KB, Bukurov NS, Marinkovic JM. Non-antireflux versus antireflux ureteroneocystostomy in adults. Br J Urol 1991;67(3): 263–266.
7. Turner-Warwick RT, Worth PHL. The psoas bladder-hitch procedure for the replacement of the lower third of the ureter. Br J Urol 1969; 41(6):701–709.
8. Franke JL, Smith JA Jr. Surgery of the ureter. In: Walsh PC, Retik AB, Vaughan ED, Wein AJ, eds. Campbell's Urology. Philadelphia: WB Saunders; 1998:3062–3084.
9. Verduyckt F, Heesakkers J, Debruyne F. Long-term results of ileum interposition for ureteral obstruction. Eur Urol 2002;42(2):181–187.
10. Ostrzenski A, Ostrzenska KM. Bladder injury during laparoscopic surgery. Obstet Gynecol Surv 1998;53(3):175–180.
11. Ghoniem GM, Monga M. Modified pubovaginal sling and Martius graft for repair of the recurrent vesicovaginal fistula involving the internal urinary sphincter. Eur Urol 1995;27(3):241–245.

13-5

Combined Versus Sequential Reconstructive Surgeries

G. Willy Davila

Surveys have demonstrated that incontinence and prolapse syndromes affecting the various compartments of the pelvic floor (PF) have a high rate of coexistence (Chapter 1-1). This frequent coexistence of multiple PF problems in a given patient can complicate the making of treatment decisions. This is particularly true if surgical therapy involving multiple surgeons is indicated. Once all PF problems have been identified, evaluated, and treatment options determined, it is not an uncommon finding that separate surgical procedures will be necessary. The decision then needs to be made to either perform sequential procedures at remote times or repair all defects at one setting. Multiple factors will influence this decision-making process. This chapter will discuss the implications of performing joint surgical procedures and review our own observations and experiences.

Barriers to Performing Combined Surgical Procedures

Traditional approaches to treating PF problems involved either one surgeon performing all of the indicated repairs at one setting, or scheduling sequential surgeries at different times by different surgeons. In most communities, a clinician will evaluate and treat the symptoms within his area of expertise. It is still rare that a team of clinicians is available for multidisciplinary evaluation. Identified barriers to multidisciplinary care include:

1. Expertise availability. Lack of availability of subspecialists capable of addressing specific PF problems is a significant barrier in many communities or hospital systems. This may be particularly true regarding urogynecologists, who are currently present in relatively small numbers.
2. Training issues. Most postgraduate training programs do not foster the concept of a team approach to evaluation or therapy. As such, combining surgical procedures performed by different specialty surgeons at one setting is not the standard in most universities or communities.
3. Turf issues. Certain PF problems are within the surgical armamentarium and can be surgically treated by more than one subspecialist. Examples include anal sphincter repairs, rectocoele and cystocele repairs, and stress incontinence procedures. These procedures can be performed by various specialty surgeons, each using different approaches. With the goal of optimizing patient outcomes, we believe that the approach associated with the highest reported success rates should be the one the patient undergoes. As such, clinicians should be aware of these outcome differences when planning surgical therapy. Based on these principles, at Cleveland Clinic Florida, the colorectal team performs all anal sphincteroplasties and the urogynecology team performs all rectocoele repairs. We do not believe that one surgeon can satisfactorily address all PF problems. Turf issues should be put aside in the patient's best interest.
4. Procedure scheduling challenges. A requirement for a multidisciplinary surgical intervention is the availability of different specialty surgeons. This can lead to inefficient use of time. Thus, preoperative planning is crucial. Ideally, two patients are operated on simultaneously, such that each team performs their designated procedure on a patient in one room, while the other operates in the other room. The teams then switch, after completing their respective operations. We have found this to work rather efficiently, as long as the procedures are planned and scheduled appropriately.[1]

Benefits of Combined Surgical Procedures

The multifactorial nature of PF dysfunction, especially as related to the impact on quality of life, must be kept in mind when planning surgical therapy. In most cases, an optimal outcome will not be achieved until all PF problems are addressed. As such, the concept of treating all PF problems at one surgical setting makes sense. There are very limited data as to outcomes and risks of concomitant surgeries. Most of the available data come from our reported experiences, here in Florida and in Cleveland.[1-2] We have identified the following benefits:

1. Multiple PF defects are corrected at one time. In our experience, treatment outcomes from concomitantly performed procedures are not inferior to those of individually performed procedures.
2. Reduce overall recovery phase. The recovery phase from combined surgeries is not longer than that from single procedures. Thus, the sequential recovery phases from sequential surgeries are avoided.
3. Optimize ability to perform multiple procedures. Because many reconstructive procedures use the same support structures, subsequent procedures could be jeopardized because of scarring around the used structure. A good example of this is the usage of the sacral promontory for both sacrocolpopexy as well as rectopexy. There is typically no difficulty in sharing the promontory when performing a combined procedure (see Prolapse Syndromes Case Presentation, Chapter 8-8). Once scarring has developed from one procedure, achieving access to the sacral promontory for another procedure at a remote time can be impossible.
4. Reduce potential surgical complications. This relates primarily to anesthetic, thrombotic, and other surgery-related complications. We have not noted complications such as graft infection or a higher failure rate when combined procedures are performed.
5. Reimbursement issues. In the United States, insurance companies reimburse for multiple surgeries performed by one surgeon at one setting in a proportionate manner - 100% for the main procedure, 50% for the secondary, and 25% for the tertiary. This can be seen as a disincentive to performing multiple repairs at one setting. However, this does not apply to combined procedures performed by multiple surgeons of different specialties.

When a bowel resection procedure is performed, care should be taken to avoid fecal material spillage. This is particularly important if a synthetic or biologic graft will be used for urogynecologic reconstruction. We routinely perform abdominal sacrocolpopexy procedures at the same setting as resection rectopexy and have not found a higher graft infection rate. However, we are fastidious about avoiding fecal spillage and irrigate the operative field extensively.

Procedures Amenable to Combined Surgical Intervention

Most reconstructive procedures seem to be amenable to being combined. Some procedures, especially those performed through different incisions and approaches, may be seen as more readily able to be combined. Those in which tissue is shared (i.e., those on either side of the rectovaginal septum) may require more consideration by one surgeon for the other surgeon's needs, but rarely have we found ourselves in a situation of jeopardizing a repair. Thus, there are some combinations of procedures that are better served as combined procedures (Table 13-5.1). Care must be taken when sharing the fascia of the rectovaginal septum. This thin layer of tissue can be jeopardized by direct trauma from either approach. Thus, care being taken to remain within tissue planes during dissection is important. When performing a combined anal sphincteroplasty and rectocoele repair, a preoperative decision should be made as to who will perform the levator plication, if it is to be performed, because it may affect the second procedure. If performed by both surgeons, restriction of rectal caliber may result.

Table 13-5.1. Combining reconstructive surgical procedures

	Amenability
Sphincteroplasty	
ASCP/PVDR/Burch	+++
AR/sling/SSF	+++
PR	+
Abdominal rectopexy	
ASCP	+/–
PVDR/Burch	+++
AR/PR/sling/SSF	+++
Perineal resection	
ASCP/PVDR/Burch	+++
AR/sling	+++
PR/SSF	+
Colpocleisis	+/–

ASCP, abdominal sacrocolpopexy; PVDR, paravaginal defect repair; AR, anterior repair/colporrhaphy; SSF, sacrospinous fixation; PR, posterior repair/colporrhaphy.

Requirements for Performing Safe and Effective Combined Surgical Procedures

Various routine surgical precautions should be taken as a combined surgical procedure is performed. We routinely obtain preoperative medical clearance, administer intravenous wide-spectrum prophylactic antibiotics (e.g., Ancef or Cefotan), and use pulsatile antiembolism stockings. If the procedure is going to exceed 4 hours, we consider low-dose heparin therapy as well. Almost all surgeries are performed under general anesthesia. This is in variance to most urogynecologic procedures, which we routinely perform under spinal anesthesia. Many procedures will require a patient to be changed from lithotomy to prone, jack-knife position (or vice versa). Great caution must be taken during the "flip" to avoid soft tissue damage or joint dislocations. This procedure requires at least four persons.

We have not had any complications related to this position change.

Results of Combined Surgical Procedures

Overall, surgical outcomes have not been jeopardized by the performance of combined surgical procedures. To date, we have not had reason to reconsider our concept that all necessary reconstructive procedures should be performed at one surgical setting. Our secretaries have become quite adept at scheduling procedures such that time is used efficiently. Time to resumption of normal activities has not been delayed and no complications more than what is expected with individual surgical procedures have been encountered.

Summary

Patients seem to benefit significantly from combined surgical procedures. Morbidity is not increased, and outcomes are not jeopardized. Traditional barriers to this innovative approach can be overcome by a cooperative spirit between surgical teams in order to facilitate procedure scheduling and pre- and postoperative care.

References

1. Sun JH, Aguirre OA, Davila GW. Team approach to pelvic floor dysfunction [abstract]. Dis Colon Rectum 1999;42(4):A15.
2. Halverson AL, Hull TL, Paraiso MF, Floruta C. Outcome of sphincteroplasty combined with surgery for urinary incontinence and pelvic organ prolapse. Dis Colon Rectum 2001;44(10):1421–1426.

Section XIV

Severity Assessment

Section XIV

Severity Assessment

Steven D. Wexner

Before treating any pelvic floor problem, albeit one of inability to evacuate or one of inability to prevent evacuation, an assessment of the severity of illness is imperative. A physician or surgeon would certainly not embark upon a course of therapy for cancer without adequate knowledge of the stage, and functional pelvic floor disorders are certainly no different. The only problem lies in the fact that whereas cancer and other organic pathologies can be objectively staged by clinical and histopathologic criteria, the staging of functional pelvic floor disorders relies on the conversion of subjective to objective information. The three chapters within this section provide an excellent up-to-date review of the most widely used currently available severity assessment scoring systems for these functional disorders. Use of these systems is imperative not only to decide upon appropriate therapies for individual patients but in fact to decide upon whether or not to even embark upon potentially invasive and/or expensive testing. An example is that if a patient has a mild score with limited impact on quality of life, one might elect not to pursue investigation, knowing that extensive treatment would not be offered. Conversely, a patient with a high score or a severe compromise in quality of life, would by all means undergo extensive evaluation. Therefore, it is incumbent upon every patient to have a score obtained during initial evaluation so as to determine the necessity for any more extensive preliminary investigations. Such scores should again be obtained at every stage after conservative and more aggressive managements. Certainly, such scores are also useful in the longer term for data collection in terms of providing both prognostic information to future generations of patients, allowing comparison of results in individual patients to the preoperative baseline and allowing research-gathering for data assessment and scientific presentation and publication. The ability to share such data allows not only a comparison within and among patients in a given department but also among groups of patients throughout the world. Although no single severity assessment score is universally used for either bladder or bowel dysfunction, and no single quality-of-life instrument is always applicable, providers should endeavor to, at the very least, be internally consistent with use of the same scoring system(s) for all patients at all stages of treatment.

14-1

Voiding Diary

Gamal M. Ghoniem and Usama M. Khater

The voiding diary is a record of micturition behavior completed by the patient. Although not frequently used, it is among the best possible means of obtaining objective data on subjective symptoms.[1] The International Continence Society (ICS) recommends the inclusion of voiding diaries in the clinical assessment of patients with lower urinary tract symptoms (LUTS).[2] There are various terms or types of voiding diaries such as micturition time or frequency charts, frequency volume charts, and bladder or urinary diaries. Frequency charts were introduced in 1972 to evaluate the effects of sacral nerve root section on LUTS in a group of patients with detrusor instability.[3] The patient was asked to record each daytime or nighttime void with an "X" in the appropriate area of the chart, for 7 days (Figure 14-1.1). Versions of the early chart required the patient to record the volumes voided, estimated fluid intake in cups or mugs, and the number of incontinence pads used per day.[4] Recently, computerized voiding diaries have been developed.

Frequency Chart

The frequency chart is the simplest type of voiding diary, because the patient is asked to record only micturition and incontinence episodes, as either ticks or ticks within a circle (Figure 14-1.1).

Frequency Volume Chart

The frequency volume chart was introduced in 1975 by Abrams and Klevmark,[4] and required the patient to record the amount of urine at each micturition, as well as the time of each void. Each leak was recorded with a "W" (wet). Parameters added later included the number of pads used and estimated fluid intake in cups or mugs.

Urinary Diaries

The use of the term "diary" implies the recording of more extensive details than are captured by the frequency volume chart. In addition to recording frequency, volume, and leakage episodes, the voiding diary captures pad use and degrees of incontinence and urgency. Furthermore, the voiding diary includes patient comments on the situations associated with or related to their LUTS. Some voiding diaries also capture fluid intake. Urgency may be recorded as 0, +, or ++, or on a scale from 0 to 10, depending on the diary used. Urgency can also be evaluated in minutes,[5] by asking the patient to estimate how long he or she could wait before voiding (Figures 14-1.2 and 14-1.3).

Computerized Voiding Diary

Computerized voiding diaries have been developed in an attempt to simplify, augment, and automate bladder symptomatology recording. They have been introduced as alternatives to the standard written diaries, which capture limited amounts of information. A drawback of the computerized diary may be a fear of technology, which is common in certain patient populations, such as the elderly.[6]

In our clinic, we use a voiding diary that captures urinary frequency, leakage, assessment of leakage, comment on physical activity at the time of leakage, fluid intake, and fluid output (Figure 14-1.4). We use a 3-day diary, which is long enough to be reliable, and short enough to decrease patient burden and increase compliance.[7] Patients are given the form at the first clinic visit, with verbal and written instructions on how to complete it.

Frequency Chart recording micturition and leakage episodes - Each time you pass your water, mark the chart with an X. If you leak or wet your bed, mark the chart with V. Keep the record for seven consecutive days (or as long as possible). Day is from the time you get up to the time you go to bed

	Day	Night
Day 1	XXX	
Day 2	XXX	O
Day 3	XXXX	
Day 4	XXX	X
Day 5	XXXXX	O
Day 6	XXXX	X
Day 7	XXXXX	

Figure 14-1.1. Simple 7 days' frequency chart recording micturition and leakage episodes.

Date ____________ Name ____________________

Record TIME, VOLUME, URGENCY (0= non, + = moderate, ++ = severe) and LEAKAGE (+ drops/ wet pants, ++ soaked clothes or pad)

TIME	VOL	URGENCY	LEAK	DRINKS (cup/mug)	ACTIVITY (note pad changes)

Figure 14-1.2. Urinary diary showing a page for single day use of 7-day form.

Name__________________ Date_____________

For URGENCY, circle the most appropriate number, 0, 2, 5, 15+

For LEAKAGE grade O, W, WW

TIME	VOL	URGENCY	LEAK	DRINKS (cup/mug)	ACTIVITY COMMENT
		0, 2, 5, 15+			
		0, 2, 5, 15+			
		0, 2, 5, 15+			
		0, 2, 5, 15+			
		0, 2, 5, 15+			
		0, 2, 5, 15+			
		0, 2, 5, 15+			

Figure 14-1.3. Seven-day urinary diary introducing the concept of evaluation of the severity of urgency in minutes.

VOIDING DIARY – DAY 1

DATE:

(1) TIME VOIDED	(2) Y/N	(3) LEAK (D/W/S)	(4) PAD OR CLOTHES (P / C)	(5) ACTIVITY AT TIME OF LEAKAGE	(6) FLUID INTAKE (OZ)*	(7) FLUID OUTPUT (ML)*
12:00 - 1:00 A						
1:00 - 2:00 A						
2:00 - 3:00 A						
3:00 - 4:00 A						
4:00 - 5:00 A						
5:00 - 6:00 A						
6:00 - 7:00 A						
7:00 - 8:00 A						
8:00 - 9:00 A						
9:00 - 10:00 A						
10:00 - 11:00 A						
11:00 A - 12:00 P						
12:00 - 1:00 P						
1:00 - 2:00 P						
2:00 - 3:00 P						
3:00 - 4:00 P						
4:00 - 5:00 P						
5:00 - 6:00 P						
6:00 - 7:00 P						
7:00 - 8:00 P						
8:00 - 9:00 P						
9:00 - 10:00 P						
10:00 - 11:00 P						
11:00 P - 12:00 A						

*1 cup = 8 oz or 240 ml

Column (3)
Degree of Leakage
D = Damp, few drops
W = Wet underwear or pad
S = Soaked pad or emptied bladder

Column (4)
Clothing Changed
P = Pad
C = Clothing

Figure 14-1.4. Day 1 of 3-day voiding diary, which we used at our clinic.

Clinical Applications of the Voiding Diary

Voiding Diaries and the Patient

Maintaining a voiding diary involves the patient in the management of their condition, and helps them become an active partner in the treatment process. Moreover, it provides them with an objective assessment of their condition during and after treatment.

Voiding Diaries and the Patient's History

The voiding diary represents a self-monitoring daily record of the patient's voiding behavior, whereas the history is a verbal explanation of the patient's complaint. It can be difficult for the patient to recall or judge their voiding frequency and pattern retrospectively, especially when their voiding behavior is irregular.[8] Patients are sometimes surprised at the actual number of voiding events that occur once they are asked to record them. The diary also provides an evaluation of treatment, which is more accurate than the history, because patients sometimes claim improvement, presumably to please their doctors, when their diaries show no change. However, the diary is only supplemental to the patient's history, and both should be used to identify the complaint and formulate an accurate diagnosis.

Several attempts have been made to use an analysis of the patient's history to help predict the type of incontinence; most have revealed that patient history is a poor predictor of the underlying cause of incontinence.[9,10] Sand et al.[11] reported that the history has a good sensitivity for genuine stress urinary incontinence but not for urge. McCormack et al.[12] revealed poor agreement between subjectively estimated urinary frequency and urinary frequency shown on a chart.

Using Voiding Diaries in Incontinent Patients

The primary use of voiding diaries in incontinent patients is documentation of incontinence episodes. Diaries may provide clues to the underlying cause of incontinence, particularly if the diary includes patient comments about the reason(s) or condition(s) associated with the incontinence episodes. Voiding diaries can be used in evaluation of the severity of urinary incontinence because the patient can report the number of pad used and the amount of leakage.

Voiding diaries have been tried in the evaluation of urge incontinence. In one study, a group of women with urge incontinence was compared with another group of normal individuals. There were significant differences between groups in frequency, mean voided volume, and largest single voided volume (Figure 14-1.5), although the overlap was large.[13] A similar study compared women with genuine stress urinary incontinence to a normal group, and surprisingly showed that total voided volume, frequency of micturition, and largest single voided volume were all significantly higher in the genuine stress urinary incontinence group than in the normal group.[14] Therefore, the diagnostic role of the diary is limited.

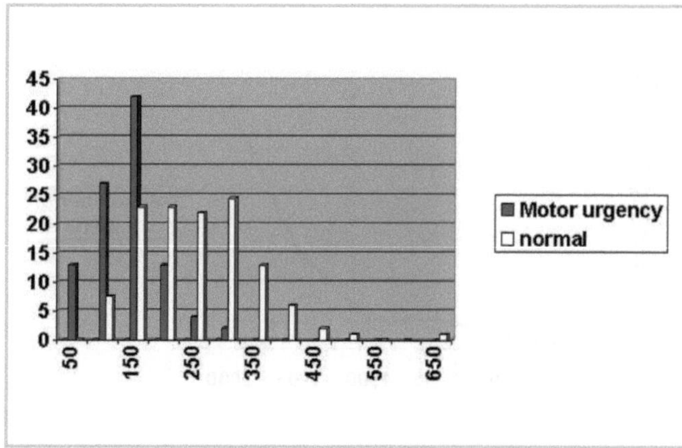

Figure 14-1.5. Average voided volumes in patients with motor urgency compared with normal group.

The voiding diary has been used to differentiate between urge and stress incontinence.[15] Most of these studies have shown not only significant differences between populations with urge and stress incontinence, but also considerable overlap. Total voided volumes, mean voided volumes, and largest single voided volumes were less in urge incontinent than stress urinary incontinent groups. Frequency of micturition during the day and at night was greater in the urge than the stress incontinent groups. Analysis of one of these studies showed that the frequency of micturition at night was the single parameter that best discriminates the two conditions. Combining daytime micturition frequency with the largest single voided volume or the mean voided volume increased the discrimination power.[15]

It is reasonable to evaluate incontinent patients with a voiding diary before other more invasive tests such as urodynamic investigation because it is a simple, noninvasive, and inexpensive tool. It evaluates the patient over a longer period of time, away from the laboratory. If urodynamic studies are indicated, a voiding diary can help the clinician choose which studies should be performed. A voiding diary is also reliable and valid when measuring the symptoms of overactive bladder, including urge and urge incontinence episodes, and nocturia.[16]

Application of Voiding Diaries in the Assessment of Voiding Dysfunction Treatment

In addition to providing baseline measurements before treatment, voiding diaries can be used to evaluate the progress and efficacy of treatment. Recently, new medica-

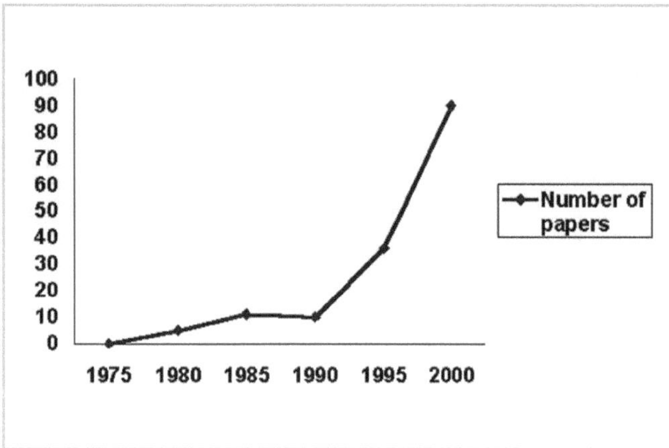

Figure 14-1.6. The published voiding diary papers.

tions for treatment of LUTS, including overactive bladder, have been developed in clinical trials. All of these medications are directed at symptom improvement; therefore, patient-completed voiding diaries are often used in evaluation of the effectiveness of these drugs.[17,18] This application also explains the recent increase in number of published studies making use of the voiding diary (Figure 14-1.6).

Voiding diaries can be used to evaluate treatment modalities other than pharmacotherapy, such as surgery and behavioral therapy. For example, they can provide a baseline assessment before anti-incontinence surgery, and evaluate outcome after the procedure. With behavioral therapy, changes and improvement occur gradually because of learning and the time it takes to strengthen muscles. These gradual changes are often unnoticed by the patient, but are measured and documented objectively in the diary. Objective feedback about such small changes can help the patient maintain the motivation and persistence needed for a successful outcome.

Application of Voiding Diaries in Research and Asymptomatic Populations

Voiding diaries have been used to define normal urinary measurements and physiologic differences between normal populations.

Children

Mattsson[19] studied 206 children, aged 7 to 15 years, considered to be asymptomatic. They were evaluated with a 24-hour frequency-volume chart. Ninety-five percent had a voiding frequency of 3 to 8. Approximately 10% voided once during the night. The morning voiding volume was the largest, and the last void before bedtime was the smallest. The total voided volume over 24 hours was between 325 and 2100 mL.

Females

Most of the published data on urinary diaries have been obtained in women. Several studies established normal values for healthy females.[20–23] Data obtained from frequency-volume charts of normal females showed that mean voided volume was 230 to 250 mL, mean frequency was 5.7 to 7.3, and total voided volume was 1272 to 1350 mL. Two of these studies analyzed diurnal and nocturnal data separately.[21,23] This may be important, because nighttime diuresis may exceed daytime diuresis and be responsible for nocturia, especially in the elderly. According to this study, the increase in urine volume during the night can be attributed to three physiologic reasons. First, the circadian rhythm of antidiuretic hormone, the rennin-angiotensin aldosterone mechanism, or atrial natriuretic hormone secretion may be abnormal. Second, the glomerular filtration rate or renal plasma flow may be altered to reduce the concentrating ability of the distal tubules. Third, an impaired cardiovascular system may not be able to supply sufficient amounts of blood to the kidneys during waking hours because of edema of lower extremities. The data obtained from the frequency-volume charts of normal females showed the mean voided volume was 230 to 250 mL, the mean frequency was from 5.7 to 7.3, and the total voided volume was 1272 to 1350 mL.[20–23]

Fitzgerald et al.[24] studied urinary habits with voiding diaries in asymptomatic women and reported that diary variables are affected by age, race, and fluid intake. Voids per liter intake varied with age and were higher among parous and Asian women. The number of nighttime voids depended only on patient age. The mean and maximum voided volumes were lower among black women.

Males

No data were found in the literature reporting reference values for urinary diaries in males. A study by Saito et al.[23] included males and females, but did not separate them into two subgroups.

References

1. Abrams P, Feneley R, Torrens M. Patient assessment. In: Abrams P, Feneley R, Torrens M, eds. Urodynamics. 1st ed. New York: Springer; 1983:6–27.
2. Abrams P, Cardozo L, Magnus F, et al. The standardization of terminology of lower urinary tract function: report from the standardization sub-committee of the International Continence Society. Neurourol Urodyn 2002;21:167–178.
3. Torrens MJ. The effect of selective sacral nerve blocks on vesical and urethral function. J Urol 1974;112:204–205.

4. Abrams P, Klevmark B. Frequency volume charts: an indispensable part of lower urinary tract assessment. Scand J Urol Nephrol Suppl 1996;179:47–53.
5. Baily R, Shepherd A, Tribe B. How much information can be obtained from frequency/volume charts? Neurourol Urodyn 1990;9:382–385.
6. Rabin JM, McNett J, Badlani GH. Computerized voiding diary. Neurourol Urodyn 1993;12:541–554.
7. Groutz A, Blaivas JG, Chaikin DC, et al. Voiding and incontinence frequencies: variability of voiding diary data and required daily length. Neurourol Urodyn 2002;21:205–209.
8. Planes Hansen C, Klarskov P. The accuracy of the frequency-volume chart: comparison of self-reported and measured volumes. Br J Urol 1998;81:709–711.
9. Cardozo LD, Stanton SL. Genuine stress incontinence and detrusor instability – a review of 200 patients. Br J Obstet Gynecol 1980;87: 184–190.
10. Webster GD, Sihelink SA, Stone AR. Female urinary incontinence: the incidence, identification, and characteristics of detrusor instability. Neurourol Urodyn 1984;3:235–242.
11. Sand PK, Hill R, Ostegard DR. Incontinence history as a predictor of detrusor stability. Obstet Gynecol 1988;71:257–260.
12. McCormack M, Infante-Rivard C, Schick E. Agreement between clinical methods of measurement of urinary frequency and functional bladder capacity. Br J Urol 1992;69:17–21.
13. Larsson G, Abrams P, Victor A. The frequency/volume chart in detrusor instability. Neurourol Urodyn 1991;10:533–543.
14. Larsson G, Victor A. The frequency/volume chart in genuine stress incontinent women. Neurourol Urodyn 1992;11:23–31.
15. Fink D, Perucchini D, Schaer G, Haller U. The role of the frequency-volume chart in the differential diagnostic of female urinary incontinence. Acta Obstet Gynecol 1999;78:254–257.
16. Brown JS, McNaughton KS, Wyman FJ, et al. Measurement characteristics of a voiding diary for use by men and women with overactive bladder. Urology 2003;61(4):802–809.
17. Norton P, Karram W, Wall LL, Rosenzweig B, Benson JT, Fantl JA. Randomized double-blind trial of terodiline in the treatment of urge incontinence in women. Obstet Gynecol 1994;84(3):386–391.
18. Zinner NR, Mattiasson A, Stanton SL. Efficacy, safety, and tolerability of extended-release once daily tolterodine treatment for overactive bladder in older versus younger patients. J Am Geriatr Soc 2002;50(5):799–807.
19. Mattsson SH. Voiding frequency, volumes and intervals in healthy schoolchildren. Scand J Urol Nephrol 1994;28:1–11.
20. Boedker A, Lendorf A, H-Nielsen A, et al. Micturition pattern assessed by the frequency/volume chart in a healthy population of men and women. Neurourol Urodyn 1989;8:421–422.
21. Kassis A, Schick E. Frequency-volume chart pattern in a healthy female population. Br J Urol 1993;72:708–710.
22. Larsson G, Victor A. Micturition patterns in a healthy female population, studied with a frequency/volume chart. Scand J Urol Nephrol Suppl 1988;114:53–57.
23. Saito M, Kondo A, Kato T, Yamada Y. Frequency-volume charts: comparison of frequency between elderly and adult patients. Br J Urol 1993;72:318–341.
24. Fitzgerald MP, Stablein U, Brubaker L. Urinary habits among asymptomatic women. Am J Obstet Gynecol 2002;187(5):1384–1388.

14-2

Scoring Systems

Wael Solh and Steven D. Wexner

There are several disease processes in colon and rectal surgery with highly subjective presentations and clinical histories. These processes present a continuum rather than a discrete set of symptoms. Disorders that lend themselves to these scoring systems include fecal incontinence, constipation, and inflammatory bowel disease. These conditions share the common denominator of the clinician needing to try to give an objective meaning to otherwise subjective symptoms. Moreover, a quality-of-life (QOL) tool may be valid and a useful tool for any patient with any colorectal disease.

These scoring systems provide an objective measure of disease severity and can be uniformly applied among various institutions, which is important in establishing the comparability of patients. Hence, these systems can measure the effectiveness of posttherapeutic intervention. They also allow interpretation and comparison of data among centers. Importantly, they permit prospective images of the clinical outcomes of individual patients. The success and therefore longevity of any scoring system is obviously contingent upon both its accuracy and it simplicity. Unfortunately, the two factors are often inversely related, as in an attempt to be easy to apply, a score may not provide enough meaningful or reproducible data. Conversely, to be very reproducible and accurate, an excessive level of detail may preclude an easy to use format. Ideally, any scoring system should not sacrifice either desired feature.

The Fecal Incontinence Severity Index

For many conditions, objective data may be used to assess severity. Although there are numerous physiologic measurements used for incontinence, none have been shown to accurately reflect either disease severity or response to therapy. These studies include anal manometry, electromyography, pudendal nerve terminal motor latency, and anal ultrasound.

A detailed patient history is the most accurate and reliable method of estimating the severity of fecal incontinence. However, to obtain objective data, numerous scoring or grading systems have been proposed in the literature. The majority of these systems only include the consistency of leakage but do not address the frequency of occurrence. Other scales mix historical data with data from physical examination or testing, as mentioned previously. These scores sacrifice discriminatory power for simplicity, but include numerical values that are arbitrarily assigned. These systems are summarized in Table 14-2.1.

We proposed and currently use the Cleveland Clinic Florida Scoring System (Table 14-2.2).[1] This scoring system permits objective comparison of levels of incontinence among groups of patients. It has gained such global popularity because it fulfills the criteria of both simplicity and accuracy. Moreover, it permits comparison of the pre- and postoperative values, and hence, treatment results. These evaluations can be made in single patients and among different centers. Furthermore, as evidenced by the table, the fecal incontinence score can be quickly calculated by the clinician office nursing staff. This or any other scoring system, however, should not substitute for a comprehensive history. It has been our practice to provide a detailed questionnaire to assess the degree and frequency of incontinence and its effect on the patient's overall QOL. Other fecal incontinence severity scales include that proposed by Rockwood et al.,[1] which is similar to the Cleveland Clinic Incontinence Scoring System (Table 14-2.3).

Fecal Incontinence Quality-of-Life Scale

Fecal incontinence can dramatically disrupt the lives of individuals who have this debilitating condition. The inability to control the passage of stool or flatus can produce embarrassment, and the fear of such episodes may severely limit daily activities. Therefore, one measure of the

Table 14-2.1. Incontinence summary scales

Author	Score Range	Assignment of Values
Rockwood et al.[1]	0–61	Weights assigned by patients, differ for type and frequency
Hull et al.[2]	0–31	Gas, staining, accidental bowel movements, and pad usage all receive equal weights
Jorge and Wexner[3]	0–20	Gas, liquid, solid incontinence, pad use, and lifestyle alteration all receive equal weight
Pescatori et al.[4]	0–6	Only frequency of the most severe type of incontinence is weighted
Vaizey et al.[5]	0–24	Gas, liquid, and solid incontinence and lifestyle receive equal weights. Other items are scored variably
O'Brien and Skinner[6]	0–120	
Miller et al.[7]	1–18	Incontinence for flatus scores 1–3, incontinence for liquid scores 4–6, incontinence for solids scores 7–8
Bai et al.[8]	0–13	Created for evaluation of adolescents after Hirschsprung's treatment. Additional items included in score. Reverse scoring
Rothenberger[9]	0–30	Incontinence for flatus scores 1–3, incontinence for liquid scores 4–6, incontinence for solids scores 7–8, lifestyle alteration scores 10–12
Lunniss et al.[10]	0–13	Gas, liquid, solid incontinence receive equal weight (1–3), urgency and difficulty cleaning score 1, soiling scores 1–2

Table 14-2.2. Incontinence scoring system

Type of Incontinence	Frequency: Never	Rarely	Sometimes	Usually	Always
Solid	0	1	2	3	4
Liquid	0	1	2	3	4
Gas	0	1	2	3	4
Wears pad	0	1	2	3	4
Lifestyle alteration	0	1	2	3	4

0 = perfect continence; 20 = complete incontinence; never = 0; rarely = <1/month; sometimes = <1/week, ≥1/month; usually = <1/day, ≥1/week; always = >1/day.

Table 14-2.3. Fecal incontinence severity index

	Two or More Times a Day	Once a Day	Two or More Times a Week	Once a Week	One to Three Times a Month
Gas					
Mucus					
Liquid					
Solid					

1 = most severe; 20 = least severe.

effectiveness of therapy to correct fecal incontinence is the degree to which a patient's QOL is enhanced. A QOL measure was specifically designed to assess the impact of treatment for fecal incontinence. This fecal incontinence QOL scale is composed of 29 items (Table 14-2.4) that form four scales: lifestyle (10 items); coping/behavior (9 items); depression/self-perception (7 items); and embarrassment (3 items).[11] Each of these four scales is capable of discriminating between patients with fecal incontinence and patients with other gastrointestinal diseases. Furthermore, the scales demonstrated a significant correlation with the Short Form 36 (SF-36) general QOL questionnaire.[12] These scales have also demonstrated stability over time with good reliability and validity.

The Constipation Scoring System

The clinical presentation of constipation includes a broad spectrum of symptoms, partially attributed to the myriad etiologies. Specifically, constipation may result from slow transit, pelvic outlet obstruction, or other mechanical, pharmacologic, metabolic, endocrine, and neurogenic reasons. Generally, physicians use the term "constipation" to define infrequent, incomplete, difficult, or prolonged evacuation or to describe stools that are too difficult to pass. However, many patients are more obsessed by the associated nonspecific symptoms of bloating, abdominal and pelvic pain, and nausea. Therefore, objective scoring systems have been developed to better describe this difficult problem, as well as to obtain a universally objective definition of "constipation." The Rome II criteria are the most widely accepted to define constipation (Chapter 7-4). However, the Rome II criteria do not qualify the severity of disease. Therefore, our constipation score is derived based on answers to questions in a symptom-based questionnaire (Table 14-2.5).[13]

A study of 232 patients with constipation confirmed the accuracy and validity of the applicability of this constipation scoring system.[13] As scores increased, a corresponding significant increase in severity of constipation was noted. This patient population included both patients with colonic inertia (diffuse marker delay on colonic transit studies without evidence of paradoxical contraction) and pelvic outlet obstruction syndromes (paradoxical puborectalis contraction, rectal prolapse, rectoanal intussusception, rectocele, or sigmoidocele).

Inflammatory Bowel Disease Index Scores

Disease activity indices have appeared in the literature for more than 20 years. The Crohn's Disease Activity Index (CDAI), for example, was derived on the basis of nine

Table 14-2.4. Fecal incontinence quality-of-life scale

1. In general, would you say your health is:
   1. Excellent
   2. Very good
   3. Good
   4. Fair
   5. Poor

2. For each of the items, please indicate how much of the time the issue is a concern for you due to accidental bowel leakage. [If it is a concern for you for reasons other than accidental bowel leakage, then check the box under N/A (not applicable).]

Due to Accidental Bowel Leakage	Most of the Time	Some of the Time	A Little of the Time	None of the Time	N/A
a. I am afraid to go out	1	2	3	4	
b. I avoid visiting friends	1	2	3	4	
c. I avoid staying overnight away from home	1	2	3	4	
d. It is difficult for me to get out and do things like going to a movie or to church	1	2	3	4	
e. I cut down on how much I eat before I go out	1	2	3	4	
f. Whenever I am away from home, I try to stay near a restroom as much as possible	1	2	3	4	
g. It is important to plan my schedule (daily activities) around my bowel pattern	1	2	3	4	
h. I avoid traveling	1	2	3	4	
i. I worry about being able to get to the toilet in time	1	2	3	4	
j. I feel I have no control over my bowels	1	2	3	4	
k. I can't hold my bowel movement long enough to get to the bathroom	1	2	3	4	
l. I leak stool without even knowing it	1	2	3	4	
m. I try to prevent bowel accidents by staying very near a bathroom	1	2	3	4	

3. Due to accidental bowel leakage, indicate the extent to which you AGREE or DISAGREE with each of the following items. [If it is a concern for you for reasons other than accidental bowel leakage, then check the box under N/A (not applicable).]

Due to Accidental Bowel Leakage	Strongly Agree	Somewhat Agree	Somewhat Disagree	Strongly Disagree	N/A
a. I feel ashamed	1	2	3	4	
b. I cannot do many of the things I want to do	1	2	3	4	
c. I worry about bowel accidents	1	2	3	4	
d. I feel depressed	1	2	3	4	
e. I worry about others smelling stool on me	1	2	3	4	
f. I feel like I am not a healthy person	1	2	3	4	
g. I enjoy life less	1	2	3	4	
h. I have sex less often than I would like to	1	2	3	4	
i. I feel different from other people	1	2	3	4	
j. The possibility of bowel accidents is always on my mind	1	2	3	4	
k. I am afraid to have sex	1	2	3	4	
l. I avoid traveling by plane or train	1	2	3	4	
m. I avoid going out to eat	1	2	3	4	
n. Whenever I go someplace new, I specifically locate where the bathrooms are	1	2	3	4	

4. During the past month, have you felt so sad, discouraged, hopeless, or had so many problems that you wondered if anything was worthwhile?
   1. Extremely so – to the point that I have just about given up
   2. Very much so
   3. Quite a bit
   4. Some – enough to bother me
   5. A little bit
   6. Not at all

Scales range from 1 to 5, with a 1 indicating a lower functional status of quality of life. Scale scores are the average (mean) response to all items in the scale (e.g., add the responses to all questions in a scale together and then divide by the number of items in the scale. Not applicable is coded as a missing value in the analysis of the questions).

Scale 1: Lifestyle, 10 items: Q2a, Q2b, Q2c, Q2d, Q2e, Q2g, Q2h, Q3b, Q3l, Q3m
Scale 2: Coping/behavior, 9 items: Q2f, Q2i, Q2j, Q2k, Q2m, Q3d, Q3h, Q3j, Q3n
Scale 3: Depression/self-perception, 7 items: Q1, Q3d, Q3f, Q3g, Q3i, Q3k, Q4 (question 1 is reverse coded)
Scale 4: Embarrassment, 3 items: Q2l, Q3a, Q3e

Table 14-2.5. Constipation scoring system (minimum score = 0; maximum score = 30)

	Score
Frequency of bowel movements	
1–2 times per 1–2 days	0
2 times per week	1
Once per week	2
Less than once per week	3
Less than once per month	4
Difficulty: painful evacuation effort	
Never	0
Rarely	1
Sometimes	2
Usually	3
Always	4
Completeness: feeling of incomplete evacuation	
Never	0
Rarely	1
Sometimes	2
Usually	3
Always	4
Pain: abdominal pain	
Never	0
Rarely	1
Sometimes	2
Usually	3
Always	4
Time: minutes in lavatory per attempt	
Less than 5	0
5–10	1
10–20	2
20–30	3
More than 30	4
Assistance: type of assistance	
Without assistance	0
Stimulant laxatives	1
Digital assistance or enema	2
Failure: unsuccessful attempts for evacuation	
Per 24 hours	
Never	0
1–3	1
3–6	2
6–9	3
More than 9	4
History: duration of constipation (years)	
0	0
1–5	1
5–10	2
10–20	3
More than 20	4

Table 14-2.6. Crohn's disease activity index

Item	Weight
Number of liquid or very soft stools	2
Abdominal pain score in 1 week	5
General well-being	7
Sum of findings per week:	20
Arthritis/arthralgia	
Mucocutaneous lesions	
Iritis/uveitis	
Anal disease (fissure, fistula, etc.)	
External fistula	
Fever >37.8°C	
Antidiarrheal use	30
Abdominal mass	10
47 minus hematocrit (males) or 42 minus hematocrit (females)	6
100 × (1 minus [body weight divided by standard weight])	1

Total score ranging from 0–600; score >450 – critically ill; score <150 – remission, inactive disease.

predictive variables (Table 14-2.6).[14] These include serum albumin, body weight, erythrocyte sedimentation rate, bowel resection, and extraintestinal symptoms related to Crohn's disease. Furthermore, the SF-36 is the most widely accepted general measure of QOL.[12] However, these scoring systems are cumbersome to use and require complicated analysis. Hence, they are rarely used in an office setting of gastroenterologists and colorectal surgeons.

The Cleveland Clinic Florida Global Quality of Life (CGQL) is a simpler tool, which has been validated in patients with ulcerative colitis as well as in those individuals with Crohn's disease (Table 14-2.7).[15] Furthermore, it correlates well with established scoring systems such as the SF-36 and the CDAI. However, conventional Crohn's disease activity indices do not reflect perianal disease activity, and thus do not allow prognostic implications from surgery. Therefore, we developed a standardized scoring questionnaire and applied it to a group of patients before surgical treatment of perianal Crohn's disease. The scoring system includes the presence of abscess, fistulae, ulcers and fissures, stenosis, incontinence, and concomitant intestinal disease (Table 14-2.8).[16] Weighted factors include acuity versus chronicity, de novo versus recurrent disease, and the presence of concomitant disease. This scoring system correlated well with the short-term outcome of surgical intervention in patients with perianal Crohn's disease and can be used as a predictor of surgical success; patients with a score of 10 or less had a good outcome, whereas all those with a score of 20 or greater had a poor outcome.

Table 14-2.7. Cleveland Clinic Florida global quality-of-life score

Please rate the following on a scale of 0 to 10 (where 10 is the best):
Current quality of life
Current quality of health
Current energy level

Each of the 3 items is scored on a scale of 0–10 and the total is divided by 30 to give a final score.

Table 14-2.8. Perianal Crohn's disease activity index

Feature	Score
Abscess	
None *or*	0
First occurrence, single abscess *or*	1
First occurrence, multiple abscesses *or*	3
First recurrence, single or multiple abscesses	4
Multiple recurrence, single or multiple abscesses	5
Maximum abscess score	8
Fistula	
None	0
Short-term (<30 d) fistula *or*	1
Long-term (>30 d) fistula *or*	2
Persistent postsurgery fistula *or*	3
Recurrent fistula	3
Multiple fistulas	3
Rectovaginal/rectourethral fistula *or*	4
Recurrent rectovaginal/rectourethral fistula	6
Maximum fistula score	14
Ulcer and fissure	
None	0
Short-term (<30 d) stenosis *or*	1
Long-term (>30 d) ulcer/fissure *or*	2
Single ulcer/fissure *or*	1
Multiple ulcers/fissures	2
Maximum ulcer/fissure score	4
Stenosis	
None	0
Short-term (<30 d) stenosis *or*	1
Long-term (>30 d) stenosis	2
Recurrent stenosis	4
Maximum stenosis score	6
Incontinence score	
No incontinence *or*	0
Incontinence score of 1–6 *or*	1
Incontinence score of 7–14 *or*	3
Incontinence score >14	5
Maximum incontinence score	5
Concomitant disease*	
None *or*	0,0,0
Moderate *or*	3,2,1
Severe	4,3,2
Active fistula	4,3,2
Maximum concomitant disease score	18

* Scores are for rectal, colonic, and small bowel disease, respectively.

Conclusion

Clearly, objective means of assessing otherwise subjective information is important. The recent progress made in design, validation, and use of these systems should elicit better assessment, therapeutic assignment, and efficacy evaluation than was previously possible. Patients being treated for these symptoms should have their subjective systems objectively scored at each encounter. Knowledge of these objective data will hopefully allow improved patient care.

References

1. Rockwood TH, Church J, Fleshman J, et al. Patient and surgeon ranking of symptoms associated with fecal incontinence. Dis Colon Rectum 1999;42(12):1525–1531.
2. Hull TL, Floruta C, Piedmonte M. Preliminary results of an outcome tool used for evaluation of surgical treatment for fecal incontinence. Dis Colon Rectum 2001;44:799–805.
3. Jorge JMN, Wexner SD. Etiology and management of fecal incontinence. Dis Colon Rectum 1993;36:77–97.
4. Pescatori M, Anastasio G, Bottini C, Mentasti A. New grading and scoring for anal incontinence: evaluation of 335 patients. Dis Colon Rectum 1992;35:482–487.
5. Vaizey CJ, Carapeti E, Cahill JA, Kamm MA. Prospective comparison of faecal incontinence grading systems. Gut 1999;44:77–80.
6. O'Brien PE, Skinner S. Restoring control: the Acticon Neosphincter artificial bowel sphincter in the treatment of anal incontinence. Dis Colon Rectum 2000;43:1213–1216.
7. Miller R, Bartolo DC, Locke-Edmunds JC, Mortensen NJ. Prospective study of conservative and operative treatment for faecal incontinence. Br J Surg 1988;75:101–105.
8. Bai Y, Chen H, Hao J, Huang Y, Wang W. Long-term outcome and quality of life after Swenson procedure for Hirschsprung's disease. J Pediatr Surg 2002;37:639–642.
9. Rothenberger DA. Anal incontinence: In: Cameron JL, ed. Current Surgical Therapy. 3rd ed. Philadelphia: BC Decker; 1989:186–194.
10. Lunniss PJ, Kamm MA, Phillips RK. Factors affecting continence after surgery for anal fistula. Br J Surg 1994;81:1382–1385.
11. Rockwood TH, Church J, Fleshman J, et al. Fecal incontinence quality of life scale. Quality of life instrument for patients with fecal incontinence. Dis Colon Rectum 2000;43(1):9–17.
12. Ware JE, Sherbourne CD. The MOS 36-item short-form health survey (SF-36). I. Conceptual framework and item selection. Med Care 1992;30(6):473–481.
13. Agachan F, Chen T, Reissman P, Wexner SD. A constipation scoring system to simplify evaluation and management of constipated patients. Dis Colon Rectum 1996;39(6):681–685.
14. Fazio VW, O'Riordan MG, Lavery IC, et al. Long term functional outcome of quality of life after stapled restorative proctocolectomy. Ann Surg 1999;230(5):575–586.
15. Pikarsky A, Gervaz P, Wexner SD. Perianal Crohn's disease: a new scoring system to evaluate and predict outcome of surgical intervention. Arch Surg 2002;137:774–778.
16. Yoshida EM. The Crohn's disease activity index, its derivatives and the inflammatory bowel disease questionnaire: a review of instruments to assess Crohn's disease [review]. Can J Gastroenterol 1999; 13(1):65–73.

14-3

Quality-of-Life Assessment Tools

G. Willy Davila

The nature of pelvic floor dysfunction as multifactorial with involvement of multiorgan systems makes severity and impact assessment an exceedingly challenging aspect of the evaluation of a symptomatic patient. Various factors such as organ system function, anatomic alterations, lifestyle impact, and psychological well-being may be impacted differently by the disease process, and more importantly, by the therapy received by a patient with pelvic floor dysfunction. Historically, outcome assessment was limited to continence and normal anatomy restoration. As recently as 10 years ago, outcome assessment was focused solely on objective parameters such as urodynamics. Aspects of day-to-day living of great importance to the patient, such as lifestyle alterations, work capability, and interpersonal intimacy, were not addressed by clinicians. It is inappropriate to assess outcomes of a multiorgan system dysfunction by assessing only one-dimensional factors. This is particularly true in the assessment of surgical outcomes. For example, a woman who undergoes reconstructive surgery for exteriorized vaginal prolapse, in whom the anatomic surgical result is very satisfactory, but who has resultant postoperative fecal incontinence caused by progressive denervation injury, will likely complain of a greater negative quality-of-life (QOL) impact after her surgical procedure. Fortunately, clinicians are beginning to question their patients regarding multiple aspects of pelvic floor symptomatology. Other examples of outcome variables that are appropriately evaluated using global or disease-specific instruments such as QOL questionnaires include the impact of surgical therapy on sexual function and the impact of pharmacologic therapy on other bodily functions. A clear example of the latter is the impact of anticholinergic therapy for overactive bladder on bowel function, particularly the well-recognized side effects of dry mouth and constipation.

Certain aspects of pelvic floor function that are best addressed in a questionnaire format include sexual and bowel function. It is not uncommon for a urogynecologic patient presenting with genital prolapse or urinary incontinence to not report fecal incontinence symptoms because of their perceived embarrassment during a face-to-face interview. Frequently, these patients do not report symptoms of fecal incontinence, although they are asked about them during their initial visit, until their evaluation is well underway and surgery has been scheduled. This can result in a delay in performance of the surgical procedure because of the need for evaluation of the patient's fecal incontinence. An even greater challenge is encountered in the evaluation of the impact of pelvic floor dysfunction on sexual function. It has been only recently that validated QOL questionnaires have been available to assess sexual function and the various aspects of sexual dysfunction. Because of the great variability in sexual function "normality," an individual perception of satisfactory, versus unsatisfactory, sexual function is best served by the use of a QOL questionnaire with subjective responses.

We routinely use QOL questionnaires as measures of disease severity, individual impact, and outcome assessment. We find these tools to be invaluable in the assessment of patients with pelvic floor dysfunction and recommend their use on a routine basis.

Available Quality-of-Life Assessment Tools

A global assessment tool for pelvic floor dysfunction is not yet available. Although a single tool would be highly desirable, its required complexity may make it impractical. Frequently used global QOL assessment tools, such as the SF-36, do not lend themselves particularly well to pelvic floor dysfunction, because the various aspects of pelvic floor dysfunction are not well covered. Disease-specific QOL questionnaires are required to assess the various parameters involved in pelvic floor dysfunction. Therefore, we currently use a battery of QOL questionnaires to comprehensively evaluate QOL impact. This allows us to evaluate individual aspects of QOL impact.

Patients are asked to complete the questionnaires after their initial consultation visit. The data are then entered into the clinical database for later analysis and comparison to questionnaire responses after therapy. We have found the responses to individual questions to be of equal, if not greater, value than the overall calculated impact score for a given questionnaire.

A particular challenge is the use of QOL tools for outcomes assessment. Managing to get a patient to complete a questionnaire can be difficult, especially if follow-up visit appointments are not kept. We have found that simply mailing the questionnaires to patients and requesting the return of completed questionnaires is associated with a less than 50% response rate. The best time to have a patient complete a QOL questionnaire may be during a follow-up visit, before the clinician sees the patient. If overall questionnaire scores are to be used for assessment, care must be taken to ascertain that all questions have been answered.

We have learned much about the use of QOL assessment tools over the last few years. Factors of significant importance are listed below:

1. Despite questionnaire validation, many questions are not clear to patients. Thus, the individual interpretation of a question will affect its answer. In clarifying the nature of a question, a clinician may add bias, especially if the questionnaire is completed in front of the clinician who is providing the care.
2. Many questions are not applicable to many patients. Work and sex/intimacy-related questions are particularly affected by differences in lifestyle. If overall questionnaire scores will be used, care must be taken to adjust the score to reflect questions that are not appropriate to that particular patient.
3. Many behavior-related voiding patterns are not corrected by medical treatment. Urinary frequency and nocturia may be affected by a patient's desired large-volume fluid intake or caffeine-containing beverage intake, which may continue after therapy.
4. Male factor is frequently the principal reason a patient is not sexually active. Thus, questions related to sexual activity should be assessed as a change from preoperative baseline rather than an overall score.
5. Sexual function has significant generational differences. For many elderly patients, requesting information regarding their sexual activity is considered offensive. Many patients do not consider sex pleasurable, and their practices – even at sexual peak – were very conservative. For example, many elderly women are not comfortable with genital self-contact.
6. Coexistence of functional bladder or bowel pathology will affect QOL impact. This is particularly true in patients with chronic constipation, with a high degree of fixation on bowel function. Because QOL questionnaires do not collect this information and are not modified based on their presence, the clinician must keep these issues in mind when evaluating QOL impact.

Recognized limitations of the utilization of QOL assessment tools include language barriers, cognitive dysfunction, and comparison at two remote times such that life situation may have changed (i.e., death of a spouse or new onset disease process or pelvic floor symptoms complex).

Over the last 10 years, a multiplicity of QOL questionnaires have been reported, validated, used in clinical trials, and correlated with disease severity and outcomes. We have been using the questionnaires listed below on a routine basis, but do realize others have become available and may be of equal, or greater, utility in urogynecologic practice.

Urinary Incontinence

Multiple tools have been evaluated and validated for the assessment of urinary incontinence impact on QOL. Among all of the aspects of pelvic floor dysfunction, this has been the most widely assessed, using QOL assessment tools. We have therefore chosen to utilize the most commonly accepted QOL questionnaires, which have been utilized in clinical practice as well as research. The short-form Incontinence Impact Questionnaire (IIQ-7) has been extensively evaluated, validated, and is well accepted.[1] Its questions include various aspects (domains) of QOL impact, including physical activities, social activities, travel, and emotional health. It is easy to complete, and the questions are readily understandable by most patients (Figure 14-3.1).

Because many patients' QOL impact from their incontinence is primarily related to one particular activity, we modified the IIQ by adding a question in which the patient identifies the particular activity that is most impacted by her urinary incontinence. This Individual IIQ (IIIQ) (Figure 14-3.2) has been particularly useful in younger women with stress incontinence who may be impacted during specific physical activities such as running, working out, or playing on the trampoline with their children, to name a few examples.[2] For overactive bladder patients, examples of frequently identified activities include going to the theater, airplane travel, and going to social events. It is very important to evaluate treatment outcome by including the same activity that the patient identified during her initial evaluation in the outcome assessments.

We also utilize the short form of a well-evaluated symptom inventory, the Urogenital Distress Inventory (UDI-6) in assessing women with urinary incontinence, as well as genital prolapse (Figure 14-3.3). The short form has also been widely studied and validated. Included subscales are irritative, obstructive/discomfort, and stress symptoms. Its questions are clear and well understood by most patients. It has been used clinically as well as in research, and complements the IIQ-7 well. We have thus adopted it in the evaluation of our patients.

44470

FORM B: INCONTINENCE IMPACT QUESTIONNAIRE

INSTRUCTIONS
You have certain activities you do in carrying on your life. We are interested in learning what effect, if any, urinary incontinence and/or prolapse has had on these activities. The questions below refer to ways in which your activities might have changed. For each question, mark (X) the response that indicates to what extent activities have been affected by urine leakage and/or prolapse.

MR#: 0 0 0 0 0

Patient Name: ______

Has urine leakage and/or prolapse affected your:					
1. Ability to physically do the following household chores: cooking, cleaning, laundry?	☐ not at all	☐ rarely	☐ frequently	☐ all of the time	☐ not applicable
2. Ability to participate in physical recreation: walking, swimming, other?	☐ not at all	☐ rarely	☐ frequently	☐ all of the time	☐ not applicable
3. Ability to travel to entertainment activities: movies, concerts, etc.?	☐ not at all	☐ rarely	☐ frequently	☐ all of the time	☐ not applicable
4. Ability to travel by car or bus more than 30 minutes from home?	☐ not at all	☐ rarely	☐ frequently	☐ all of the time	☐ not applicable
5. Participation in social/relationship activities outside of the home?	☐ not at all	☐ rarely	☐ frequently	☐ all of the time	☐ not applicable
6. Emotional health?	☐ not at all	☐ rarely	☐ frequently	☐ all of the time	☐ not applicable
7. Feeling frustrated?	☐ not at all	☐ rarely	☐ frequently	☐ all of the time	☐ not applicable

44470

CONFIDENTIAL

Figure 14-3.1. Short-form Incontinence Impact Questionnaire (IIQ-7).

44470

FORM C: INDIVIDUAL INCONTINENCE IMPACT QUESTIONNAIRE

MR#: 0 0 0 0 0

Patient Name: ____________________

INSTRUCTIONS
For each question, mark (**X**) the response that indicates the extent to which your bladder control problem affects your activities.

My bladder control problem affects my:					
1. Ability to do household chores (washing dishes, cleaning house, etc.)	☐ not at all	☐ rarely	☐ frequently	☐ all of the time	☐ not applicable
2. Ability to socialize and interact with friends and colleagues	☐ not at all	☐ rarely	☐ frequently	☐ all of the time	☐ not applicable
3. Quality and quantity of sleep	☐ not at all	☐ rarely	☐ frequently	☐ all of the time	☐ not applicable
4. Performance of routine exercise or participation in sports (walking, aerobics, tennis, swimming, jogging, etc.)	☐ not at all	☐ rarely	☐ frequently	☐ all of the time	☐ not applicable
5. Personal and intimate relationships (including hugging and sexual intercourse)	☐ not at all	☐ rarely	☐ frequently	☐ all of the time	☐ not applicable
6. Ability to participate in entertainment activities (sitting through the movies, playing cards, watching T.V. program)	☐ not at all	☐ rarely	☐ frequently	☐ all of the time	☐ not applicable
7. Ability to perform my job	☐ not at all	☐ rarely	☐ frequently	☐ all of the time	☐ not applicable
8. Ability to wear the clothes I want	☐ not at all	☐ rarely	☐ frequently	☐ all of the time	☐ not applicable
9. Ability to go places I want	☐ not at all	☐ rarely	☐ frequently	☐ all of the time	☐ not applicable
10. List an activity (not listed above) which is particularly affected by your urine loss:	☐ not at all	☐ rarely	☐ frequently	☐ all of the time	☐ not applicable

☐☐☐☐☐☐☐☐☐☐☐☐☐☐☐

** Please enter one letter per space*

44470

CONFIDENTIAL

Figure 14-3.2. Individual Incontinence Impact Questionnaire (IIIQ).

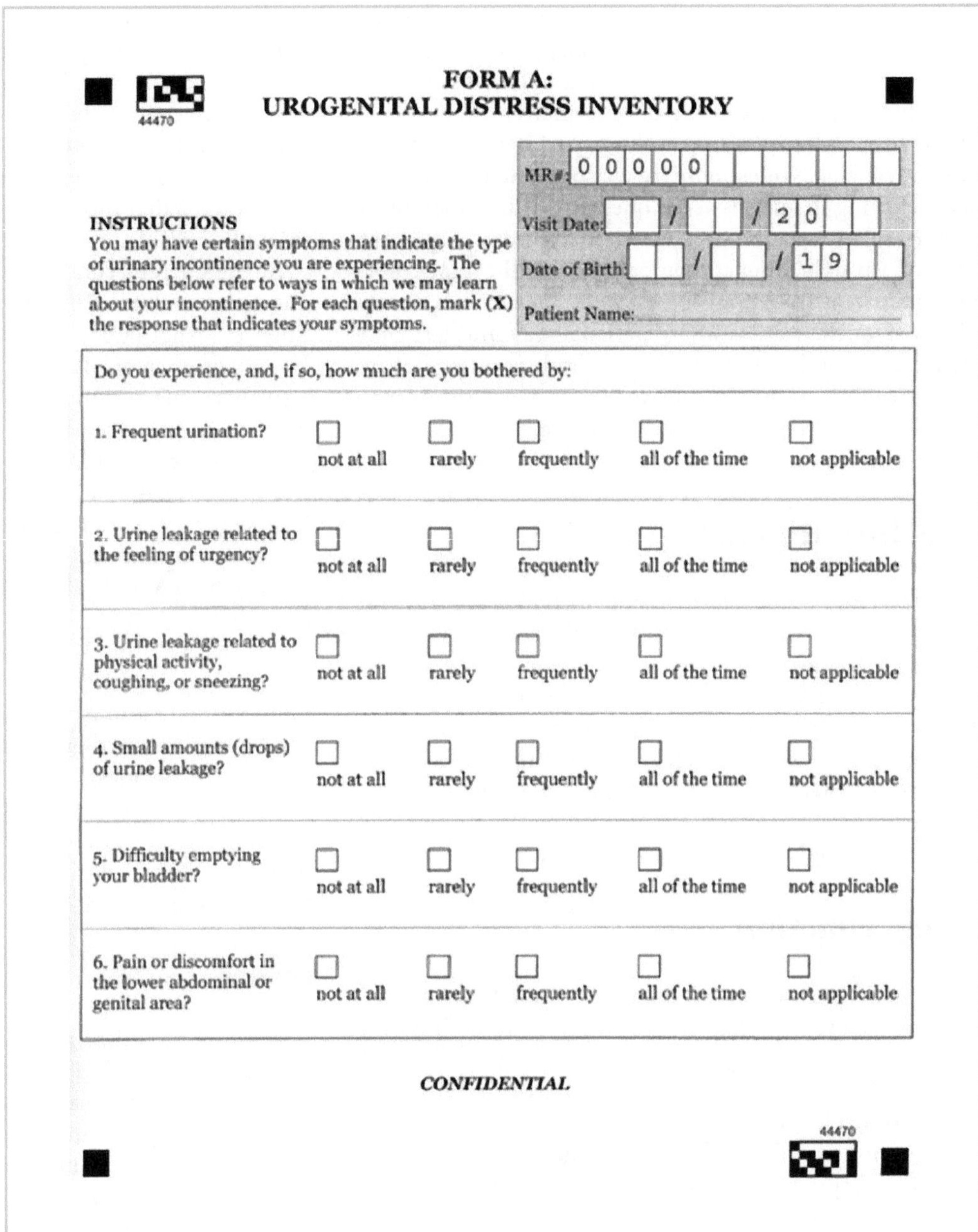

FORM A:
UROGENITAL DISTRESS INVENTORY

44470

MR#: 0 0 0 0 0

Visit Date: __ / __ / 20 __

Date of Birth: __ / __ / 19 __

Patient Name: ____________

INSTRUCTIONS
You may have certain symptoms that indicate the type of urinary incontinence you are experiencing. The questions below refer to ways in which we may learn about your incontinence. For each question, mark (X) the response that indicates your symptoms.

Do you experience, and, if so, how much are you bothered by:					
1. Frequent urination?	☐ not at all	☐ rarely	☐ frequently	☐ all of the time	☐ not applicable
2. Urine leakage related to the feeling of urgency?	☐ not at all	☐ rarely	☐ frequently	☐ all of the time	☐ not applicable
3. Urine leakage related to physical activity, coughing, or sneezing?	☐ not at all	☐ rarely	☐ frequently	☐ all of the time	☐ not applicable
4. Small amounts (drops) of urine leakage?	☐ not at all	☐ rarely	☐ frequently	☐ all of the time	☐ not applicable
5. Difficulty emptying your bladder?	☐ not at all	☐ rarely	☐ frequently	☐ all of the time	☐ not applicable
6. Pain or discomfort in the lower abdominal or genital area?	☐ not at all	☐ rarely	☐ frequently	☐ all of the time	☐ not applicable

CONFIDENTIAL

44470

Figure 14-3.3. Short-form Urogenital Distress Inventory (UDI-6).

Scoring

There has been a lack of uniformity in the literature in terms of how these questionnaires are scored. We score the questionnaires as described by the authors with assigned scores of 0 to 3 (not at all – 0, slightly – 1, moderately – 2, greatly – 3). The overall score for the IIQ and UDI is calculated by taking the average of all questions and multiplying by 33.3, for a total score of between 0 to 100.

We have found it very important to look at each individual question/domain addressed by the questionnaire and assess the impact score assigned by the patient for each domain. This is particularly useful in outcomes assessment.

We score the IIIQ differently, as: not at all – 0, rarely – 1, frequently – 2, all of the time – 3. Because this questionnaire includes questions regarding work (#7) and intimacy (#5), a global score will be dependent on a patient's own individual situation. A global score is calculated based on the following formula: Total points divided by total possible points (deleting "not applicable" questions) and multiplying by 100.

The score assigned to the individual activity that is impacted by the patient's incontinence is given particular importance. This will provide a greater value to that individual activity because other activities may not be as impacted and the overall assessment of outcome will be minimized. It also allows the patient to focus on that activity when assessing results of therapy.

We assess the UDI-6 with a global score as well as an assessment of its individual domains, as with the IIQ.

It is important to make sure that patients answer all of the included questions in the questionnaires, especially if global scores will be used for evaluation of impact/distress. When reviewing a questionnaire for completeness, we find that reviewing the unanswered questions with a patient to be particularly useful because they typically represent areas of variable impact, compared with the black and white nature of the "all of the time" or "never" responses. It is those unclear areas where patients may find less of a satisfactory result from therapy, as a result of other factors being involved, or greater clarity once a more significant factor has been corrected. Thus, those areas should not be ignored.

Fecal Incontinence

The Cleveland Clinic Florida fecal incontinence scoring system has been widely used in the colorectal surgery literature as a simple and accurate tool for assessment of severity. It has also been validated at other centers outside of Cleveland Clinic Florida. Although it is not entirely a QOL assessment tool, its value in assessing severity and outcomes of therapy is high. We therefore utilize it routinely in assessing impact and outcomes. This questionnaire is discussed in more detail in Chapter 14-2.

Sexual Function

Clearly, the assessment of sexual function "normality" is one of the greatest challenges to a pelvic floor clinician. The impact of sexual dysfunction on QOL has been the subject of many recent studies. This is particularly true as related to surgical therapy for prolapse and incontinence. There is not one tool or questionnaire that has gained foremost acceptance for assessment of this variable. Until such a tool is accepted, we have chosen to utilize two validated questionnaires for assessment of impact on sexual function. Although these tools contain more questions and/or include more complex answers than other QOL tools, we find them useful in assessing severity and outcomes.

Interestingly, we find that approximately 60% of our patients are not sexually active because of individual or partner factors. This adds a greater level of complexity to the usage of a sexual function assessment tool. Does every patient get asked to complete the questionnaire, or only those who are sexually active? Are the patients not sexually active because of a pelvic floor problem, or is it just a cofactor? Would they resume sexual activity once the pelvic floor problem is corrected? In addition, there are great generational differences regarding attitudes toward sex. This has resulted in many patients being offended by the questions included in a questionnaire asking about issues such as masturbation or climax achievement. Frequently, widows are offended that we are even asking them about sexual function. We thus routinely ask patients to complete the questions that are "applicable to you."

The age of erectile dysfunction drugs has brought post-reproductive age sexuality into the limelight. Unfortunately, the focus has been on male function. Assessing sexual function in this population of patients has demonstrated to us that there is a great variability in what is considered desirable.

At the current time, we are using two sexual function questionnaires, the Pelvic Organ Prolapse – Urinary Incontinence Sexual Function Questionnaire (PISQ-12) (Figure 14-3.4) in its short 12-item form, and the McCoy Female Sexuality Questionnaire (MFSQ) (Figure 14-3.5). These instruments were selected with the help of the sexual function member of our Pelvic Floor Center.

We selected the PISQ-12 because it has been validated in its long (31-item) as well as short forms, and has been used for postoperative outcome assessment.[3,4] It has a good balance between physical and emotional aspects of sexual function.

We selected the MFSQ in its 7-item short form as a validated and widely used questionnaire with usage in multiple populations.[5] It has a Likert scale design, which can be preferable in questions with a great degree of subjectivity. It was modified from its original composition of 19 items to a 9-item tool. Further modification led to the 7-item questionnaire we chose to use. It seems to complement the PISQ-12 fairly well, without significant overlap in the questions.

Scoring

There is not a uniformly accepted means of scoring these questionnaires. The PISQ-12 can be scored by assessing individual factors (behavioral/emotive, physical/partner-related). A global score is designated based on the sum of all answers or by multiplying the number of answered items by the mean, when answers were missing. The McCoy questionnaire is assessed based on individual items' scores.

As with other aspects of QOL assessment, looking at each individual answer can be very helpful in assessing individual needs and areas to focus therapy on.

Genital Prolapse

Until recently, there was not a validated genital prolapse QOL instrument. For the last 3 years, we have utilized the Prolapse Symptom Inventory and Quality of Life Scale (PSI-QOL) (Figure 14-3.6). This tool was used clinically after validation by Kobashi, Leach, and colleagues.[6] It contains 15 items: 11 regarding prolapse symptoms and 4 regarding QOL impact (relationships, sexual activity, physical activities, leisure activities).

We are aware of new questionnaires designed to assess QOL impact from genital prolapse, but have not implemented their use at our institution.

44470

FORM E: PELVIC ORGAN PROLAPSE/URINARY INCONTINENCE SEXUAL FUNCTION QUESTIONNAIRE

INSTRUCTIONS
The following is a list of questions about you and your partner's sex life. All information is strictly confidential. Your answers will be used only to help doctors understand what is important to patients about their sex lives.
Please check the box (X) that best answers the question for you. While answering the questions, consider your sexuality over the past six months.

MR#: 0 0 0 0 0

Patient Name: ____________

☐ **This questionnaire is Not Applicable to me**

1. How frequently do you feel sexual desire? This feeling may include wanting to have sex, planning to have sex, feeling frustrated due to lack of sex, etc.	☐ always	☐ usually	☐ sometimes	☐ seldom	☐ never
2. Do you climax (have an orgasm) when having sexual intercourse with your partner?	☐ always	☐ usually	☐ sometimes	☐ seldom	☐ never
3. Do you feel sexually excited (turned on) when having sexual activity with your partner?	☐ always	☐ usually	☐ sometimes	☐ seldom	☐ never
4. How satisfied are you with the variety of sexual activities in your current sex life?	☐ always	☐ usually	☐ sometimes	☐ seldom	☐ never
5. Do you feel pain during sexual intercourse?	☐ always	☐ usually	☐ sometimes	☐ seldom	☐ never
6. Are you incontinent of urine (leak urine) with sexual activity?	☐ always	☐ usually	☐ sometimes	☐ seldom	☐ never
7. Does fear of incontinence (either stool or urine) restrict your sexual activity?	☐ always	☐ usually	☐ sometimes	☐ seldom	☐ never
8. Do you avoid sexual intercourse because of bulging in the vagina (either the bladder, rectum, or vagina falling out)?	☐ always	☐ usually	☐ sometimes	☐ seldom	☐ never
9. When you have sex with your partner, do you have negative emotional reactions such as fear, disgust, shame, or guilt?	☐ always	☐ usually	☐ sometimes	☐ seldom	☐ never
10. Does your partner have a problem with erections that affects your sexual activity?	☐ always	☐ usually	☐ sometimes	☐ seldom	☐ never
11. Does your partner have a problem with premature ejaculation that affects your sexual activity?	☐ always	☐ usually	☐ sometimes	☐ seldom	☐ never

12. Compared to orgasms you have had in the past, how intense are the orgasms you have had in the past 6 months?

☐ much less intense ☐ less intense ☐ same intensity ☐ more intense ☐ much more intense

44470

CONFIDENTIAL

Figure 14-3.4. Pelvic Organ Prolapse/Urinary Incontinence Sexual Function Questionnaire.

FORM G: McCOY FEMALE SEXUALITY QUESTIONNAIRE

44470

INSTRUCTIONS
Please circle the number from 1 to 7 which most closely corresponds to your experience during the past 30 days. Your responses will be kept completely confidential.

MR#: 0 0 0 0 0

Patient Name:____________________

☐ **This questionnaire is Not Applicable to me**

1. Are you satisfied with your present frequency of sexual activity?

Extremely Unsatisfied						Extremely Satisfied
1	2	3	4	5	6	7

2. How many times a day have you had sexual thoughts or fantasies during the last month?

Never						Every Time
1	2	3	4	5	6	7

3. How enjoyable is sex for you?

Not at all Enjoyable						Very Enjoyable
1	2	3	4	5	6	7

4. How often during sex do you feel aroused or excited (for instance, increased heart beat/flushing/vaginal wetness/heavy breathing)?

Never						Every Time
1	2	3	4	5	6	7

5. How often do you have an orgasm during sex?

Never						Every Time
1	2	3	4	5	6	7

6. How often do you suffer from lack of vaginal lubrication (wetness) during sex?

Never						Every Time
1	2	3	4	5	6	7

7. How often do you suffer from pain during intercourse?

Never						Every Time
1	2	3	4	5	6	7

44470

CONFIDENTIAL

Figure 14-3.5. McCoy Female Sexuality Questionnaire.

44470

FORM F: PROLAPSE SYMPTOM INVENTORY (PSI) AND QUALITY OF LIFE SCALE (QOL)

MR#: 0 0 0 0 0

Patient Name: ____________

INSTRUCTIONS: Please indicate how frequently you have experienced the following symptoms by placing a check in the box which best describes your symptoms **over the past month.**

1. I feel that I have to push or strain to empty urine from my bladder	☐ All of the time	☐ Most of the time	☐ Some of the time	☐ Rarely	☐ Never
2. I assist my bladder to empty by placing my finger in the vagina or pressing on the skin outside the vagina	☐ All of the time	☐ Most of the time	☐ Some of the time	☐ Rarely	☐ Never
3. I experience leakage of urine (incontinence) when I engage in activities such as coughing, sneezing, laughing, lifting, walking, or changing position	☐ All of the time	☐ Most of the time	☐ Some of the time	☐ Rarely	☐ Never
4. I run to the bathroom so I will not wet myself	☐ All of the time	☐ Most of the time	☐ Some of the time	☐ Rarely	☐ Never
5. I wear protection (such as pads, diapers, or toilet paper) or change my undergarment to protect my clothes from urinary leakage	☐ All of the time	☐ Most of the time	☐ Some of the time	☐ Rarely	☐ Never
6. I experience constipation due to difficulty in moving the stool out of my rectum	☐ All of the time	☐ Most of the time	☐ Some of the time	☐ Rarely	☐ Never
7. I assist myself to empty my bowel by placing a finger in the vagina	☐ All of the time	☐ Most of the time	☐ Some of the time	☐ Rarely	☐ Never
8. I experience incontinence of stools	☐ All of the time	☐ Most of the time	☐ Some of the time	☐ Rarely	☐ Never
9. I feel as though there is a ball between my legs or that I am sitting on a ball	☐ All of the time	☐ Most of the time	☐ Some of the time	☐ Rarely	☐ Never
10. I feel a pressure in my vagina when standing which may lessen when lying down	☐ All of the time	☐ Most of the time	☐ Some of the time	☐ Rarely	☐ Never
11. Do you need to change position or posture to help you empty your bladder?	☐ All of the time	☐ Most of the time	☐ Some of the time	☐ Rarely	☐ Never
12. Some or all of the symptoms which are described above prevent me from pursuing new relationships with people	☐ All of the time	☐ Most of the time	☐ Some of the time	☐ Rarely	☐ Never
13. Some of all of the symptoms which are described above interfere with my sexual relationships	☐ All of the time	☐ Most of the time	☐ Some of the time	☐ Rarely	☐ Never
14. Some or all of the symptoms which are described above prevent me from engaging in physical recreation (dancing, sports, etc.)	☐ All of the time	☐ Most of the time	☐ Some of the time	☐ Rarely	☐ Never
15. Some or all of the symptoms which are described above prevent me from engaging in non-strenuous leisure activities (eating out, going to a show, etc.)	☐ All of the time	☐ Most of the time	☐ Some of the time	☐ Rarely	☐ Never

CONFIDENTIAL

44470

Figure 14-3.6. Prolapse Symptom Inventory and Quality of Life Scale.

In clinical use, we focus on the QOL impact questions (12–15) and look at their individual scores, as we obtain much of the symptom information during our routine clinical assessment. There is some overlap with data obtained in the IIQ, IIIQ, and UDI. Because the data collected are also a symptom inventory, we do not calculate a global score for this questionnaire.

Summary

Quality-of-life impact assessment should be a crucial part of the evaluation of a woman with pelvic floor dysfunction. There is great variability in the degree of impact a particular pelvic floor symptom may have on a given patient. In addition, the affected patient is the only person who can reliably determine how much a particular symptom affects her. Validated instruments such as QOL questionnaires are extremely useful for this purpose. These tools can be used to assess baseline severity/impact as well as response to therapy. As such, changes in an individual's responses compared with a baseline may be more important than the overall score at initial evaluation. For outcome assessment, we utilize these tools at least 3 months after surgical intervention. With nonsurgical therapy, such as pharmacologic intervention, improvements in scoring may be valuable after a shorter intervention period.

Global scores assigned to a questionnaire are very useful in clinical use. However, we have found responses to individual questions to be even more useful in optimizing an individual's care. We strongly encourage clinicians to incorporate QOL assessment tools into their evaluation and outcome assessment of patients with pelvic floor dysfunction.

References

1. Uebersax JS, Wyman JF, Shumaker SA, McClish DK, Fantl JA. Short forms to assess life quality and symptom distress for urinary incontinence in women: the Incontinence Impact Questionnaire and the Urogenital Distress Inventory. Continence Program for Women Research Group. Neurourol Urodyn 1995;14:131–139.
2. Davila GW, Primozich J. Quality of life assessment in urinary incontinence: evaluation of a disease specific, patient specific, short-form questionnaire. Drug Info J 1997;31:1416.
3. Rogers R, Coates KW, Kammerer-Doak D, Khalsa S, Qualls C. A short form of the Pelvic Organ Prolapse/Urinary Incontinence Sexual Questionnaire (PISQ-12). Int Urogynecol J Pelvic Floor Dysfunct 2003;14: 164–168.
4. Rogers R, Kammerer-Doak D, Villarreal A, Coates K, Qualls C. A new instrument to measure sexual function in women with urinary incontinence or pelvic organ prolapse. Am J Obstet Gynecol 2001;184: 552–558.
5. McCoy NL. The McCoy Female Sexuality Questionnaire. Qual Life Res 2000;9:739–745.
6. Kobashi KC, Gormley EA, Govier F, et al. Development of a quality of life assessment instrument for patients with pelvic prolapse [abstract]. Int Urogynecol J Pelvic Floor Dysfunct 2000;11(suppl 1):S39.

Index

T

MIX
Papier aus verantwortungsvollen Quellen
Paper from responsible sources
FSC® C105338

If you have any concerns about our products,
you can contact us on
ProductSafety@springernature.com

In case Publisher is established outside the EU,
the EU authorized representative is:
Springer Nature Customer Service Center GmbH
Europaplatz 3, 69115 Heidelberg, Germany

Printed by Libri Plureos GmbH
in Hamburg, Germany